Springer Series in
OPTICAL SCIENCES
90/2

Founded by H.K.V. Lotsch

Editor-in-Chief: W. T. Rhodes, Atlanta

Editorial Board: T. Asakura, Sapporo
K.-H. Brenner, Mannheim
T. W. Hänsch, Garching
T. Kamiya, Tokyo
F. Krausz, Vienna and Garching
B. Monemar, Linköping
H. Venghaus, Berlin
H. Weber, Berlin
H. Weinfurter, Munich

Springer
New York
Berlin
Heidelberg
Hong Kong
London
Milan
Paris
Tokyo

Physics and Astronomy

ONLINE LIBRARY

http://www.springer.de

Springer Series in
OPTICAL SCIENCES

The Springer Series in Optical Sciences, under the leadership of Editor-in-Chief *William T. Rhodes*, Georgia Institute of Technology, USA, and Georgia Tech Lorraine, France, provides an expanding selection of research monographs in all major areas of optics: lasers and quantum optics, ultrafast phenomena, optical spectroscopy techniques, optoelectronics, quantum information, information optics, applied laser technology, industrial applications, and other topics of contemporary interest.

With this broad coverage of topics, the series is of use to all research scientists and engineers who need up-to-date reference books.

The editors encourage prospective authors to correspond with them in advance of submitting a manuscript. Submission of manuscripts should be made to the Editor-in-Chief or one of the Editors. See also http://www.springer.de/phys/books/optical_science/

Mohammed N. Islam (Ed.)

Raman Amplifiers for Telecommunications 2

Sub-Systems and Systems

Foreword by Robert W. Lucky

With 286 Figures

 Springer

Mohammed N. Islam
Department of Electrical Engineering and Computer Science
University of Michigan at Ann Arbor
1110 EECS Building
1301 Beal Avenue
Ann Arbor, MI 48109-2122
mni@eecs.umich.edu
and
Xtera Communications, Inc.
500 West Bethany Drive, Suite 100
Allen, TX 75013
USA
mislam@xtera.com

Library of Congress Cataloging-in-Publication Data
Raman amplifiers for telecommunications 2: sub-systems and systems / editor, Mohammed N. Islam.
    p. cm. – (Springer series in optical sciences ; v. 90/2)
  Includes bibliographical references and index.

   1. Fiber optics.   2. Optical communications.   3. Raman effect.   4. Optical amplifiers.
I. Islam, Mohammed N.   II. Series.
TL5103.592.F52R35 2003
621.382′75–dc21   2003044945

ISBN 978-1-4419-2348-6   ISSN 0342-4111   Printed on acid-free paper.

9 8 7 6 5 4 3 2 1   SPIN 10948880

www.springer-ny.com

Springer-Verlag   New York Berlin Heidelberg
A member of BertelsmannSpringer Science+Business Media GmbH

To my loving wife,
Nasreen

Foreword

I remember vividly the first time that I heard about the fiber amplifier. At that time, of course, it was the erbium-doped fiber amplifier, the predecessor of the Raman amplifier that is the subject of this book.

It was an early morning in a forgotten year in Murray Hill, New Jersey at one of our Bell Labs monthly research staff meetings. About twenty directors and executive directors of research organizations clustered around a long table in the imposing executive conference room. Arno Penzias, the vice president of research, presided at the foot of the table.

Everyone who participated in those research staff meetings will long remember their culture and atmosphere. Arno would pick an arbitrary starting point somewhere around the table, and the designated person would head to the front of the table to give a short talk on "something new" in his or her research area. This first speaker would invariably fiddle helplessly with the controls embedded in the podium that controlled the viewgraph projector, but eventually we would hear machinery grinding in the back room as a large hidden mirror moved into place. We would all wait quietly, arranging and choosing our own viewgraphs from the piles that lay on the table in front of every participant.

The rules for the staff meeting were that each speaker was allowed seven minutes and three viewgraphs. However, in spite of Arno's best efforts to enforce this regimen, everyone took too long and used too many viewgraphs. Various attempts at using loud timers and other incentives all failed. No one could give a respectable talk on a research topic for which they had passionate feelings in seven minutes.

Another rule was that anyone could forfeit his talk by simply saying, "I pass." This forfeiture was always accepted without comment, but new directors always asked their friends about whether this would constitute a black mark against their performance. No one knew for sure, but rumor had it that it was unwise to pass unless you were truly destitute of material. After all, the implication would be that there was nothing new in your research organization for the last month—not a good indication of your management skills.

With no one passing, and everyone speaking too long, these staff meetings some-times seemed endless. Computer scientists would talk about new constructs in software, systems people would talk about new techniques for speech recognition, physicists would talk about some new laser, chemists would show diagrams of new organic materials, and so forth. It didn't take long for each talk to exceed the understanding of most listeners of whatever specialty was being discussed. I always left with a profound sense of the limitations of my own knowledge, but with an exhilarating inkling into the unfolding of science. It was, perhaps, the best of the old research model in Bell Labs, and in retrospect I can say that in this new competitive world I miss those old scientific-style management meetings.

It was in such a meeting that I first heard about the fiber amplifier. I don't know whether I had been paying attention, but I was immediately galvanized by the implications of this new discovery. One word came to me and blazed across my mind: that word was "transparency."

Surprisingly, in my experience I am not always immediately enthusiastic about a new technology upon initial exposure. One might think that the potentials of great breakthroughs are self-evident, but that does not seem to be the case. When I first heard about the invention of the laser, I had no premonition that lasers would become the primary instrument of the world's telecommunications traffic. When one of the inventors of public key cryptography told me his idea for having two keys, I scoffed at the naiveté of his concept. I remember thinking on first hearing about what is now the principal algorithm for data compression that I thought it was only a theoretical exercise. So my track record for such insights is not altogether good.

However, with the fiber amplifier I went to the other extreme. I foresaw a dramatic revolution in communications. I spoke up at the staff meeting that morning to say that this invention would transform the architecture of communications networks. This would lead to transparent networks, I said, and that this would not necessarily be good for AT&T. I got carried away with this vision, and went on to say that private networks could have their own wavelengths traveling transparently through the network, untouched by the common carrier in the middle. One private network might have "blue" light (figuratively speaking, of course, because we're not talking about visible wavelengths) whereas another would have "green." I foresaw a plug on the wall that passed only the chosen wavelength, which would be owned exclusively by that particular customer's network. AT&T would thus be deprived of the opportunity to process signals for value-added services. AT&T, in fact, wouldn't have any idea what was packed into those wavelengths.

Well, that hasn't exactly happened, but today's optical networks are moving towards increased transparency, and Raman amplifiers will accelerate this trend. The advantages of transparency are compelling. A great many constituent signals can be amplified cheaply in one fell swoop. More importantly, this amplification is independent of the bit rates, protocols, waveforms, multiplexing, or any other particulars of the transmission format. The design isn't "locked in" to any specific format, and as these details change, the amplification remains as effective as ever. In the case of the Raman amplifier, the bandwidth is so enormous that adjectives seem inadequate to describe its potential for bulk amplification.

Transparency in the network is so attractive that probably the only reason it isn't done is that it is so difficult to achieve. One reason is, of course, the necessity for periodically unbundling the signal to add or drop subcomponents. In the digital world this has usually meant a complete demultiplexing and remultiplexing of the overall signal, an expensive operation. The optical world opens up the possibility of selective transparency for certain wavelengths whereas others are unpacked to do add-drop multiplexing.

So network topology sets limits on transparency. But the other reason transparency is hard to achieve is the implicit accumulation of impairments as a signal incurs successive amplifications. It is ironic that the telephone network was essentially transparent for the first half-century of its existence. Until 1960 the long-haul transmission systems used analogue amplification to boost levels as the signal traversed the nation. The invention of the triode vacuum tube enabled the first transcontinental transmission system to be deployed in the 1920s. It was a marvelous feat to be able to send a band of signals 3000 miles across the country, passing through many amplifiers, accumulating noise and distortion along the way, but still providing intelligible speech at the other end. Some older readers will remember when long distance phone calls sounded crackly and "distant." Now, of course, it is impossible to tell how far away a connection is. They all sound local, because of digital transmission.

Digital transmission was the triumph of the 1960s. Though now it seems obvious, engineers found the philosophy of digitization hard to grasp for several decades after the invention of pulse code modulation by Reeves in 1939. There is a trade-off here: bandwidth against perfectibility. A 3 kHz voice signal, for example, is transformed by an analogue-to-digital converter into a 64,000 kbps stream of bits, greatly expanding the necessary transmission bandwidth. However, this digital signal can be regenerated perfectly, removing noise and distortion periodically as necessary. A miracle is achieved as the bits arrive across the country in the same pristine form as when they left.

So it was that all long distance transmission was converted to digital format. The introduction of the first lightwave transmission systems hurried this change, inasmuch as lightwave systems were deemed to be "intrinsically digital" because of their nonlinearities and the lack of amplifiers. No one cared much at the time—the early 1980s—but the entire design of the network was predicated on the transmission of 64 kbps voice channels. The multiplexing hierarchy, the electronic switching, the synchronization and timing, and the transmission format assumed that everything was packaged into neat little voice channels. That, of course, was before the rise of the Internet.

Now optical amplification has reversed this trend of the last half-century towards digitization based upon a hierarchy of voice signals. It isn't just that optical amplifiers have an enormous bandwidth. They do something those old triode vacuum tubes could never do: they amplify without substantially increasing the noise and distortion of the signal. Raman amplifiers are particularly good in this way. Moreover, because Raman amplification is distributed across the whole span of the fiber, the signal level never drops as low as it does when discrete amplifiers are employed. In a system using discrete amplifiers the signal level is at its lowest and most vulnerable right before the point of amplification.

Back at that research staff meeting I was concerned about the implications of transparency to the architecture of the network. A transparent network is, by definition, a "dumb" network. It doesn't do anything to the signal; it can't, because it doesn't know what the signal is. As an AT&T employee, that sounded threatening. As an Internet user, that sounded empowering. The Internet, after all, was designed around the so-called end-to-end principle. In the architecture of the Internet, intelligence is at the periphery of the network, and the network is as minimally intrusive as is necessary to achieve interconnection. It is an extremely important philosophical principle that was just beginning to be understood in the 1980s. Since then the argument has raged, and the concept of a "stupid" network has been put forward by a number of Internet designers as the ultimate desired objective. If that is so, then the optical amplifier has made possible the ultimate stupid network.

I can't leave this foreword without mentioning another observation on perhaps a more personal level. Raman amplifiers epitomize for me the transformation of communications from a world of electrical circuits to one of quantum mechanical phenomena. Of course you could argue that transistors themselves depend on quantum mechanical principles, and surely the laser does, and so forth. But for many practical and design purposes these devices could be modeled with traditional circuit equivalents. Since then, however, photonics has increasingly become a showpiece of modern physics. The erbium-doped fiber amplifier had to be understood as a quantum interaction of light with the erbium atom. Raman amplifiers, by contrast, involve the interaction of light with a material structure. We descend ever more into the realm of quantum phenomena, into a world of small and impressive miracles.

A number of my friends and associates at Bell Labs have contributed to this technology and even to this particular book. I'm very proud of the work that they and their peers in academia and other industries have done in the creation of photonics technology. I've seen it grow around me and have taken vicarious pride in their accomplishments. Sometimes I tell people that, yes, I know the inventors of this or that great technology, even though I may not have realized at the time the significance of the invention. In the case of Raman amplifiers I remember learning about Raman effects as one of the impairments to be overcome in optical transmission. Researchers in my organization were even then experimenting with Raman amplification, and although there was a glimmer of potential, I can't say that I was aware of what their future might bring. Perhaps now its day has come, and that's what this book is all about.

Robert W. Lucky
Fair Haven, New Jersey
March 2003

Preface

Technologies for fiber-optic telecommunications went through a major growth period—some might even say a revolution—roughly during the years 1994 to 2000. This growth came about due to the convergence of several market drivers and technologies. First were data traffic and the Internet, the key drivers of the demand for bandwidth. Prior to the explosion of data traffic and the Internet, voice traffic only grew at an average of 4% a year. The Internet, on the other hand, grew 100% a year or more starting in 1992 and sustained this phenomenal growth rate at least through about 2001. The second was the advent of the optical amplifier, which served the role in optical networks that the transistor had played in the electronics revolution. The optical amplifier was key because it allowed the simultaneous amplification of a number of channels, as opposed to electronic regenerators that operated channel by channel. The third technology was wavelength-division-multiplexing (WDM), which made a single strand of fiber act as many virtual fibers. WDM has allowed the capacity of fibers to be increased by more than two orders of magnitude over the past few years, providing plenty of bandwidth to fuel the growth of data traffic and the Internet. WDM served the role in optical networks that integrated circuits had played in the electronics revolution. Just as the transistor permitted the revolution associated with integrated circuits in electronics, the optical amplifier permitted the revolution associated with WDM in optical networks. Because a number of channels could be simultaneously amplified, the cost of deploying more wavelengths in WDM was gated by the terminal end costs rather than the regenerator costs. Hence far more cost-effective networks became available with the combination of optical amplifiers and WDM.

Raman amplification has been one of the optical amplifier technologies that had a slow start, but then experienced a wide deployment with increasing performance needs of optical networks. It would be reasonable to assume that almost every new or upgraded long-haul (~300 to 600 km between regenerators) and ultra-long-haul (>600 km between regenerators) will eventually deploy some form of Raman amplification technology. Any deployment concerns about discrete or distributed Raman amplification have been outweighed by the performance improvements permitted with Raman amplification. For example, distributed Raman amplification improves noise performance and decreases nonlinear penalties in WDM networks, thereby alleviating the two main constraints in dispersion-compensated, optically amplified systems. The improved noise performance can be used to travel longer distances between repeaters or to introduce lossy switching elements such as optical add/drop multiplexers or optical cross-connects. Discrete and distributed Raman amplifiers are wavelength agnostic, with the gainband being determined by the pump distribution. Also, discrete Raman amplification can efficiently be integrated with dispersion compensation. Hence, Raman amplification permits wide bandwidth and long reach simultaneously. For instance, commercial systems in 2002 provide 240 channels at 10 Gb/s over 100 nm bandwidth (capacity of 2.4 Tb/s) over 1500 km with static optical add/drop multiplexers at every inline amplifier site (roughly every 80 km). Of course, if less bandwidth is required, then the unrepeated distance can be even longer.

Although stimulated Raman scattering (SRS) dates back to 1928 [7], it was first studied in optical fibers by Roger Stolen and coworkers in 1972 [10, 9]. Much of the physics of Raman amplification was explored through the 1970s and early 1980s. Then, in 1984 Linn Mollenauer, Jim Gordon, and I suggested the use of Raman amplification in WDM soliton systems [5, 6]. We demonstrated the concept using color center lasers as the pump lasers, and there was a flurry of research on Raman amplification in fiber systems from about 1984 to 1988. However, by 1988 it became clear that erbium-doped fiber amplifiers (EDFAs) were closer to practical deployment, and the Raman work was mostly dropped in favor of rare earth-doped amplifiers. Admittedly, it was a bit difficult to imagine how to put large tabletop lasers such as color center lasers in a central office, in a hut, or under the sea.

Although it seemed that EDFAs would never be displaced as optical amplifiers in fiber-optic systems, by 1997 the scene began to change. Work particularly at Lucent Technologies by Steve Grubb, Per Hansen, Andy Stentz, and others began to show the promise of Raman oscillators pumped by cladding-pumped fiber lasers [1, 2, 4, 3, 8]. I realized the big payoff would be not in oscillators but in Raman amplifiers, and I spunoff from the University of Michigan a startup company called Xtera Communications. Xtera began by trying to develop S-band subsystems (Chapter 10), and later we redirected the business plan to a wideband, long-reach, all-Raman system (Chapter 14).

The key thing to understand is that stimulated Raman amplification had not changed. It was the technology required to make Raman amplifiers that had changed. The most fundamental change was the development and commercialization of practical, high-powered, laser diodes. Although we believed that commercial Raman amplifiers would require cladding-pumped fiber lasers, by 1999 it became clear that laser diodes with sufficient power would be available. This was an extremely important development because it would reduce the cost and size of Raman amplifiers while increasing their reliability. Second, dispersion-compensating fibers (DCF) were being commercialized for use with 10 Gb/s systems. It turns out that the DCF is an excellent gain medium for discrete Raman amplifiers, permitting the integration of dispersion compensation with optical amplification. Finally, all of the required passive components became available at least with fiber pigtails and with the ability to handle high pump powers.

By 2000, Raman amplified systems were becoming commercially available. Several startup companies (e.g., Corvis, Qtera, Xtera) were using Raman amplification as their differentiator. Even the stalwarts of the industry had to take notice of this important technological development. For example, Nortel Networks acquired Qtera, and Lucent Technologies began to develop their all-Raman system, which became commercially available in 2002. It finally looked as if Raman amplification had made inroads in long-haul and ultra-long-haul systems. The noise figure improvement of up to 7 dB was simply too large a system margin to ignore!

With heavy research and development of Raman amplified systems between 1997 and 2002, it would be fair to say that the physics and applications of Raman amplifiers were pretty well understood, at least in the arena of telecommunications. The Raman "buzz" was out there, and telecommunications engineers were constantly asking for a "good reference" that they could read to understand Raman amplification. Raman amplifiers were about to leave the eclectic world of research laboratories and PhDs and perhaps enter the commercial Main Street, and a book that summarized the key physical principles and applications was needed. After all, it would be difficult to deploy and maintain that which was unknown. Therefore, at OFC 2002 I finally agreed to put together this volume, *Raman Amplifiers for Telecommunications*.

For me the assembling of this book is another important step on a long journey. As a graduate student and when I first joined AT&T Bell Labs between 1983 and 1987, we were convinced that Raman amplification was going to be important. At MIT I worked with Linn Mollenauer and Jim Gordon on WDM soliton systems using Raman amplification, and then when I first joined Bell Labs I worked on Raman oscillators and amplifiers. Almost a decade later, I spent five years from 1997 to 2002 transferring to commercialization an all-optical, all-Raman amplified system through Xtera Communications. Now that Raman amplification is finally prime-time for systems, it is necessary to organize, articulate, and share the know-how so that telecommunications and systems engineers can deploy and exploit the technology.

Acknowledgments

This book was written and assembled while I was at Xtera Communications, on a leave of absence from the University of Michigan in Ann Arbor, as well as the first year after I returned to the University. Thanks are due to Professors Richard Brown and Duncan Steel at the University for permitting me to complete this book. Also, I am particularly indebted to Dr. Jon Bayless and Carl DeWilde for encouraging me to put this book together and having the foresight to understand the broader impact that a startup could have by allowing this endeavor.

Many at Xtera Communications worked on Raman amplification and helped in composing this volume. In particular, Amos Kuditcher helped significantly on Chapters 10 and 14. Special thanks are due to Monica Villalobos, without whose help this book could never have been written. Monica kept the process moving forward throughout the year with her usual methodical and professional style. She kept contact with all of the authors, collected all of the chapters, and helped in the hand-off to the publishers. I think that all of my coauthors would agree that Monica was a pleasure to work with throughout the process.

Finally, I am deeply appreciative of the love, support, and encouragement from my wife, Nasreen, and sons, Sabir and Shawn. The only regret I have in putting this book together is the time it took away from my family.

Mohammed N. Islam
Ann Arbor, Michigan
January 2003

References

[1] S. Grubb, T. Erdogan, V. Mizrahi, T. Strasser, W.Y. Cheung, W.A. Reed, P.J. Lemaire, A.E. Miller, S.G. Kosinski, G. Nykolak, and P.C. Becker, 1.3 μm cascaded Raman amplifier in germanosilicate fibers. In *Proceedings of Optical Amplifiers and Their Applications*, PD3-1, 187, 1994.

[2] S.G. Grubb, T. Strasser, W.Y. Cheung, W.A. Reed, V. Mizrachi, T. Erdogan, P.J. Lemaire, A.M. Vengsarkar, D.J. DiGiovanni, D.W. Peckham, and B.H. Rockney, High-power 1.48 μm cascaded Raman laser in germanosilicate fibers, In *Proceedings of Optical Amplifiers and Their Applications*, SA4, 197–199, 1995.

[3] P.B. Hansen, L. Eskildsen, S.G. Grubb, A.J. Stentz, T.A. Strasser, J. Judkins, J.J. DeMarco, R. Pedrazzani, D.J. DiGiovanni, Capacity upgrades of transmission systems by Raman amplification, *IEEE Photon. Technol. Lett.*, 9:2 (Feb.), 1997.

[4] P.B. Hansen, L. Eskildsen, S.G. Grubb, A.M. Vengsarkar, S.K. Korotky, T.A. Strasser, J.E.J. Alphonsus, J.J. Veselka, D.J. DiGiovanni, D.W. Peckham, D. Truxal, W.Y. Cheung, S.G. Kosinski, and P.F. Wysocki, 10 Gb/s, 411 km repeaterless transmission experiment employing dispersion compensation and remote post- and pre-amplifiers. In *Proceedings of the 21st European Conference on Optical Communications* (Gent, Belgium), 1995.

[5] L.F. Mollenauer, J.P. Gordon, and M.N. Islam, Soliton propagation in long fibers with periodically compensated loss, *IEEE J. Quantum Electron.* QE-22:157–173, 1986.

[6] L.F. Mollenauer, R.H. Stolen, and M.N. Islam, Experimental demonstration of soliton propagation in long fibers: Loss compensated by Raman gain, *Opt. Lett.* 10:229–231, 1985.

[7] C.V. Raman and K.S. Krishnan, A new type of secondary radiation, *Nature* 121:3048, 501, 1928.

[8] A.J. Stentz, S.G. Grubb, C.E. Headley III, J. R. Simpson, T. Strasser, and N. Park, Raman amplifier with improved system performance, *OFC '96 Technical Digest*, TuD3, 1996.

[9] R.H. Stolen and E.P. Ippen, Raman gain in glass optical waveguides, *Appl. Phys. Lett.*, 22:276, 1973.

[10] R.H. Stolen, E.P. Ippen, and A.R. Tynes, Raman oscillation in glass optical waveguide, *Appl. Phys. Lett.*, 20:62, 1972.

Quick Summary of Book

The book is organized into two volumes with three sections. Volume 1 begins with a chapter entitled "Overview of Raman Technologies for Telecommunications." The first major section (Volume 1, Section A), Raman Physics, contains eight chapters (Chapters 2–9). The second section (Volume 2, Section B) on Subsystems and Modules contains five chapters (Chapters 10–14). Finally, the third section (Volume 2, Section C), Systems Design and Experiments, contains an additional five chapters (Chapters 15–19). Almost half of the book is dedicated to Raman physics, because these are the principles that will remain time invariant. This is covered completely in Volume 1. The second section, Subsystems and Modules, describes applications of Raman technology that will be fairly time invariant, although the details and data of the applications will continuously evolve. Finally, the last section, Systems Design and Experiments, represents a snapshot of the state-of-the-art system demonstrations as of early 2003. This is the section that must necessarily change with time, but at least it can provide a good basis for comparison or updating from 2003. It is important to go all the way from basic physics to systems because they are intimately linked. The basic physics determines what can or cannot be done, and it points to the differential advantages that Raman amplification provides. On the other hand, the systems design and experiments ultimately define what is worth doing and where performance should be optimized. Fortunately, Raman amplification is very rich with physical principles as well as being one of the key enabling technologies for long-haul and ultra-long-haul submarine, terrestrial, soliton, and high-speed systems.

In selecting the topics to be covered in this book as well as the authors to invite, a broad, diverse, and insightful view was sought. As an example, the authors were chosen from industrial labs as well as universities. The industrial laboratories represented include Corning, Furukawa, Lucent Technologies, Nippon Telephone and Telegraph, Nortel Networks, OFS Fitel, Siemens, Tyco Telecommunications, and Xtera Communications. Also, the authors represent the international nature of Raman technology, with contributions from the United States, Europe, Japan, and Russia. Furthermore, young rising stars were invited to contribute chapters as well as the "giants in the field," starting with Roger Stolen, Linn Mollenauer, and Evgeny Dianov.

It is an honor that so many key researchers in Raman technology accepted the invitation to contribute to this book. The invitation was extended to researchers who have made significant contributions to the technology and whose work has consistently represented the highest quality and deepest insight. Obviously there are many other excellent researchers in the field, but the intent was to cover the main issues in Raman physics, subsystems and modules and systems design and experiments within the limited space of two volumes.

The book begins with "Overview of Raman Technologies for Telecommunications," which I authored (Chapter 1). Then, the physics section opens with a chapter by Stolen, "Fundamentals of Raman Amplification in Fibers" (Chapter 2), which is fitting since he did much of the original groundbreaking work on Raman amplification in fibers. Noise is a very important aspect of any optical amplifier, and Fludger contributes two chapters on the topic: "Linear Noise Characteristics" (Chapter 4) and "Noise due to Fast Gain Dynamics" (Chapter 8). The most significant technological development for commercial Raman amplifiers is the increase in laser diode power, and Namiki et al., in Chapter 5 describe "Pump Laser Diodes and WDM Pumping." The other major technological development is the availability of new fibers with efficient Raman gain, and two chapters are dedicated to this topic: in Chapter 6 Grüner-Nielsen and Qian describe dispersion compensating fibers for Raman applications, and in Chapter 7 Dianov describes more forward-looking work on new Raman fibers. The simplest Raman amplifier uses CW pumps and a counterpropagating geometry (i.e., where the pump and signal propagate in opposite directions). However, the performance of this basic Raman amplifier can be improved by a number of emerging techniques. In Chapter 3, Grant and Mollenauer describe the use of time-division multiplexing of pump wavelengths. Then, in Chapter 9 Radic discusses and compares forward, bidirectional, and higher-order Raman amplification.

In the second section, "Subsystems and Modules," four types of Raman devices are covered: discrete (or lumped) amplifiers, distributed amplifiers, lasers, and a combination of discrete and distributed amplifiers. In Chapter 10 I review work on discrete or lumped Raman amplifiers to open up new wavelength windows, particularly in the short wavelength S-band. Then, Headley et al. review in Chapter 11 work on Raman fiber lasers or oscillators. Next, in Chapter 12, Evans et al. discuss distributed Raman transmission, applications, and fiber issues. One of the most important applications of combined discrete and distributed amplifiers is to broadband transmission systems. One way to achieve the broadband amplifier is to combine erbium-doped fiber amplifiers with Raman amplifiers, and in Chapter 13 Masuda describes hybrid EDFA/Raman amplifiers. Another route to a broadband system is to use all-Raman discrete and distributed amplifiers, and in Chapter 14 on wideband Raman amplifiers I along with coworkers at Xtera illustrate this approach.

The third section of the book focuses on system design and experiments. Some of the challenges of the Raman effect are covered in the first two chapters, and system deployments of Raman amplifiers are discussed in the following three chapters. In Chapter 15 Bromage et al. detail multiple path interference and its impact on system design. Then, in Chapter 16 Krummrich discusses Raman impairments in WDM systems. As examples of areas where Raman amplifiers are a key enabling technology, three system experiments are included. First, in Chapter 17 Kidorf et al. describe the use of Raman amplifiers in ultra-long-haul submarine and terrestrial applications. Then, in Chapter 18 Mollenauer discusses ultra-long-haul, dense WDM using dispersion-managed dolitons in an all-Raman system. Finally, in Chapter 19 Nelson and Zhu illustrate 40 Gb/s Raman-amplified transmission.

Survey of Chapters

VOLUME 90/1

Overview of Raman Amplification in Telecommunications (Chapter 1)

As an overview for the book, this chapter surveys Raman amplification for telecommunications. It starts with a brief review of the physics of Raman amplification in optical fibers, along with the advantages and challenges of Raman amplifiers. It also discusses some of the recent technological advances that have caused a revived interest in Raman amplifiers. Then, distributed and discrete Raman amplifiers are described. Distributed Raman amplifiers improve the noise figure and reduce the nonlinear penalty of the amplifier, allowing for longer amplifier spans, higher bit rates, closer channel spacings, and operation near the zero dispersion wavelength. Discrete Raman amplifiers are primarily used to increase the capacity of fiber-optic networks. Examples of discrete amplifiers are provided in the 1310 nm band, the 1400 nm band, and the short-wavelength S-band.

Section A. Raman Physics

Fundamentals of Raman Amplification in Fibers (Chapter 2)

Raman scattering was first published by C.V. Raman in 1928, and he was awarded the 1930 Nobel Prize for the discovery. In 1972 stimulated Raman scattering was first observed in single-mode fibers, and the Raman gain coefficient was also measured that same year. The chapter focuses on various treatments of the Raman interaction, which can appear to be quite different. The quantum approach treats the problem as a transition rate involving photon number. In the classical approach, the Raman effect is a parametric amplifier with an interaction between signal, pump, and vibrational wave. Finally, the Raman interaction itself can be traced to a small time delay in the nonlinear refractive index. This chapter compares and contrasts these various treatments of the Raman effect in optical fibers. Also, a fundamental treatment of noise in fiber Raman amplifiers is included.

Time-Division Multiplexing of Pump Wavelengths (Chapter 3)

This chapter describes an approach to Raman pumping that uses time-division multiplexing of the pump wavelengths. TDM pumping has several advantages over CW pumping such as efficient gain leveling with a "smart" pump, the elimination of four-wave mixing between pumps, and the reduction of pump-to-pump Raman interactions. This technique only works with backward Raman pumping, where the pump and signal are counterpropagating. The rate of TDM pumping needs only to exceed 1 MHz, so electronic components for these speeds are widely available and very inexpensive. However, TDM Raman pumping does introduce a new set of problems. The higher gain for signal propagating in the backward direction leads to a larger backward spontaneous Raman noise level. Consequently, Rayleigh scattering of the backward propagating noise can significantly increase the forward noise level under high gain conditions.

Linear Noise Characteristics (Chapter 4)

Spontaneous emission is the inevitable consequence of gain in an optical amplifier. In this chapter, the definition of noise figure is shown to be useful only in characterizing shot noise and signal-spontaneous beat noise. The noise characteristics of both discrete and distributed Raman amplifiers are presented. Also, a general model that accurately predicts both signal propagation and the buildup of amplified spontaneous emission is discussed and compared to measurements. Further measurements and analysis of broadband Raman amplifiers show a clear dependence on temperature, which places a fundamental limit on their performance. Interactions between the pump wavelengths are also shown to play an important role, giving better system performance to longer signal wavelengths at the expense of the shorter wavelengths. Finally, an analysis of the relative linear noise performance of different transmission fibers is presented.

Pump Laser Diodes and WDM Pumping (Chapter 5)

This chapter discusses issues surrounding the pump laser diodes for broadband Raman amplifiers, which range from fundamentals to industry practices of Raman pump sources based on so-called 14XX nm pump laser diodes. The chapter also describes the design and issues of wavelength-division-multiplexed pumping for realizing a broad and flat Raman gain spectrum over the signal band. In addition, practical Raman pump units are illustrated, and the chapter also provides insights into ongoing issues on copumped Raman amplifiers and their pumping sources. The pump laser diodes discussed are InGaAsP/InP GRIN-SCH strained layer MQW structure with BH structure, which are the most widely used in the industry.

Dispersion-Compensating Fibers for Raman Applications (Chapter 6)

Dispersion-compensating fibers are the most widely used technology for dispersion compensation. Also, DCF is a good Raman gain medium, due to a relatively high germanium doping level and a small effective area. Dispersion-compensating Raman amplifiers integrate two key functions, dispersion compensation and discrete Raman amplification, into a single component. Use of DCF for broadband Raman amplifiers raises new requirements for the properties of the DCF including requirements for gain, double Rayleigh scattering, and broadband dispersion compensation. Dispersion slope compensation is now possible for all types of transmission fibers, and the next challenge for broadband dispersion compensation is dispersion curvature. Dispersion-compensating Raman amplifiers have been realized with high-gain, low-noise figure and low multipath interference arising from double Rayleigh back scattering.

New Raman Fibers (Chapter 7)

Standard transmission fibers with silica core doped with a small concentration of GeO_2 have a low value of the Raman gain and a peak Raman gain at a frequency shift of about 440 cm^{-1}. However, for a number of applications such as discrete Raman amplifiers and Raman fiber lasers, special fibers with much higher Raman gain and/or various Raman frequency shifts are often required. Early experiments show that low-loss, high GeO_2- and P_2O_5-doped silica fibers could be promising fiber gain Raman fibers. For example, phosphor-silicate glass has two Raman scattering bands shifted by 650 cm^{-1} and 1300 cm^{-1}, and the cross-section for these bands is 5.7 and 3.5 times higher compared to silica. However, these fibers have met with serious challenges during fabrication by well-developed techniques. Nonetheless, at present germano-silicate and phosphor-silicate Raman fibers are being widely used for constructing CW Raman fiber lasers, which can cover the whole spectral range of 1.2 to 1.75 microns. These CW medium power (1 to 10 W) lasers are a convenient laser source for pumping optical fiber amplifiers and some lasers.

Noise due to Fast Gain Dynamics (Chapter 8)

The time response of the Raman effect is associated with the vibrations of the molecules in the gain medium and is on the order of several hundred femtoseconds. Compared to current data rates, this energy transfer is practically instantaneous, resulting in very fast gain dynamics. In a copumped Raman amplifier, the gain dynamics are averaged due to chromatic dispersion between pump and signal wavelengths. This lessens the impact of the fast physical process and results in a more improved system performance than would otherwise be expected. In a counterpumped Raman amplifier, the different propagation directions of pump and signals average the gain over the cavity length. This much stronger averaging greatly reduces system penalties in counterpumped amplifiers. In this chapter models are developed for co- and counter-pumped Raman amplifiers that quantify both the transfer of relative intensity noise from the pump to the signal and also the signal-to-signal crosstalk, mediated by the pump (crossgain modulation). In addition, the system impact in terms of Q penalty is determined, as well as determining the actual energy transfer from pumps to signals and from crossgain modulation.

Forward, Bidirectional, and Higher-Order Raman Amplification (Chapter 9)

Distributed Raman amplification can be achieved by optical pumping at either end of the fiber. In a unidirectional transmission line, all signals travel in the same direction.

In contrast, bidirectional transmission can be used to realize two-way traffic within a single fiber line: counterpropagating signal traffic is launched and received at the opposite ends of the optical link. A bidirectionally pumped fiber span can support both uni- and bidirectional signal transmission. A unidirectionally pumped span, however, almost exclusively supports unidirectional signal traffic. This chapter explores and compares forward, bidirectional, and higher-order Raman amplification. Higher-order pumping refers to the introduction of shorter-wavelength pumps that are used to pump the pump; that is, the higher-order pump amplifies the first-order pump, which in turn pumps the signal band. Different pumping schemes provide different levels of performance, but each scheme has a trade-off of performance versus pump laser restrictions.

VOLUME 90/2

Section B. Subsystems and Modules

S-Band Raman Amplifiers (Chapter 10)

The design, implementation, and issues associated with S-band amplification are discussed in this chapter, with a special emphasis on lumped Raman amplifiers (LRAs). LRAs can be used in a split-band augmentation strategy with new or already deployed C- and/or L-band systems, which are usually amplified with EDFAs. To open up the S-band, the key enabling technology is the appropriate optical amplifier. Raman amplifiers appear to be a practical solution to the S-band amplifier, and they are a mature technology ready for deployment. Utilizing silica fiber as the gain medium, Raman amplifiers can be readily fusion spliced with the fiber used in the transport infrastructure. LRAs have also been demonstrated with performance on a par with commercial C-band EDFAs in terms of gain, noise figure, and bandwidth. In addition, LRAs can be implemented efficiently using DCF, which means that the lumped amplifier can be integrated with the dispersion compensation. The major challenge of Raman amplifiers has been their lower efficiency than EDFAs, but this discrepancy is narrowing through better gain fibers, higher laser diode pump powers, and the inherent better slope efficiency for Raman amplifiers at higher channel count. The bulk of this chapter focuses on the issues and experimental demonstration of S-band LRAs in fiber-optic transmission systems.

Raman Fiber Lasers (Chapter 11)

This chapter focuses on cascaded Raman fiber lasers (RFL), which use the stimulated Raman scattering in optical fibers to shift the wavelength of light from an input pump laser to another desired wavelength. Devices at almost any wavelength can be made by proper choice of a pump wavelength, and by cascading the pump through several Raman shifts. Although RFL had been demonstrated since the 1970s, the advent of fiber Bragg gratings made the devices practical. A broadband flat Raman gain profile can be obtained using multiple pump wavelengths, and it is advantageous to have all the required wavelengths emitted from one source. This motivated the development of a multiple wavelength RFL. Single cavities simultaneously lasing from two to six wavelengths have been demonstrated. Finally, distributed Raman amplification techniques have become more sophisticated with the proposed use of higher-order pumping schemes. The use of a RFL is especially suited to this application, inasmuch as large amounts of powers are required at the shortest pump wavelength.

Distributed Raman Transmission: Applications and Fiber Issues (Chapter 12)

The persistent demand for higher performance (capacity, system reach, data rate, etc.) has turned system designers to distributed Raman for its lower noise figure. Today's data-dominated traffic patterns require reach beyond 1000 km, and Raman amplification is one vital tool in pushing out the system reach. This chapter reviews the two most important properties of an optical amplifier—pump efficiency and noise figure—and compares Raman to erbium amplification. The concept of effective noise figure is covered, which leads to a generic system scaling relationship that aids in the prediction of Raman-assisted, system performance improvements. Raman transmission experiments at 10 Gb/s and 40 Gb/s are summarized, and design issues specific to these systems are covered. In addition, dispersion-managed fiber consisting of optical fiber spans that can be optimized for Raman transmission is introduced.

Hybrid EDFA/Raman Amplifiers (Chapter 13)

This chapter describes the technologies needed for cascading an EDFA and a fiber Raman amplifier to create a hybrid amplifier, the EDFA/Raman hybrid amplifier. Two kinds of hybrid amplifiers are defined in this chapter: the "narrowband hybrid amplifier," and the "seamless and wideband hybrid amplifier." The narrowband amplifier employs distributed Raman amplification in the transmission fiber together with an EDFA; this provides low-noise transmission in the C- or L-band. The noise figure of the transmission line is lower than it would be if only an EDFA were used. The wideband amplifier, on the other hand, employs distributed or discrete Raman amplification together with an EDFA. The wideband amplifier provides a low-noise and wideband transmission line or a low-noise and wideband discrete amplifier for the C- and L-bands. The typical gain bandwidth of the narrowband amplifier is ~30 to 40 nm, whereas that of the wideband amplifier is ~70 to 80 nm.

Wideband Raman Amplifiers (Chapter 14)

This chapter describes the design and implementation of wideband Raman amplifiers. All-Raman amplification enables the lowest cost and smallest footprint system, and Raman amplification provides a simple single platform for long-haul and ultra-long-haul fiber-optic transmission systems. Despite a significant list of advantages, a number of challenges exist for Raman amplification, including: pump–pump interactions, interband and intraband Raman gain tilt, noise arising from thermally induced phonons near the pump wavelengths, multipath interference from double Rayleigh scattering, coupling of pump fluctuations to the signal, and pump-mediated signal crosstalk. Fortunately, design techniques exist for overcoming all of these physical limitations, thus allowing for the relatively simple implementation of 100 nm Raman amplifiers. Although commercially available wideband Raman amplifiers have been limited to a bandwidth of 100 nm to date, laboratory experiments have shown amplifiers with much larger bandwidths. Bandwidths greater than 100 nm are usually achieved with such special techniques as new glass compositions or wavelength guard bands around the pump wavelengths. Finally, the application of wideband Raman amplification in high-performance transmission systems is reviewed. For example, an all-Raman amplifier structure with discrete and distributed amplification can give significant advantages of reach and capacity. Such a design has been implemented, and the transmission feasibility of 240 OC-192 channels over 1565 km standard single-mode fiber has been demonstrated

Section C. Systems Design and Experiments

Multiple Path Interference and Its Impact on System Design (Chapter 15)

Up to the end of the 1990s, the main causes of signal degradation in transmission were fiber nonlinearity and amplified spontaneous emission from optical amplifiers. With the advent of Raman amplification in fiber-optic communications systems, another source of signal degradation has become increasingly relevant: so-called multiple path interference or MPI. This chapter focuses on MPI and its impact on receiver and system design. Optical amplification can exacerbate MPI by providing gain for paths that would otherwise have too much attenuation to be significant. Sources of MPI include discrete reflections within or surrounding optical amplifiers, double Rayleigh scattering in optical amplifiers or in the transmission span, and unwanted transverse mode mixing in higher-order mode dispersion compensators. The properties of MPI and Rayleigh scattering are reviewed, and the techniques for measuring MPI level are then described. The impact of MPI on beat-noise limited receivers is discussed, along with techniques for system design optimization.

Raman Impairments in WDM Systems (Chapter 16)

In most chapters of this book, stimulated Raman scattering is invoked intentionally. Pump radiation is coupled into the fiber carrying the signal radiation to generate Raman gain. However, SRS also occurs unintentionally in WDM transmission systems. Due to the large number of channels inside the Raman gain bandwidth, total power can add up to levels where considerable amounts of SRS are generated, with the signal channels acting as pumps. In contrast to the beneficial effects of intentional Raman pumping, the unintended generation of SRS usually degrades system performance. This chapter addresses effects resulting from the unintended invocation of SRS and their impact on WDM signal transmission. A number of system impairments result from the interaction between signal channels due to SRS. Effects with time scales well below the bit period affect the mean values of the individual channel powers. On the other hand, fast interactions between individual bits change the variances of the respective channel powers and can be considered as noise. In addition, some selection criteria for transmission fibers with respect to Raman efficiency are provided.

Ultra-Long-Haul Submarine and Terrestrial Applications (Chapter 17)

Ultra-long-haul (ULH) optically amplified transmission systems (defined in this chapter as those spanning from 1500 to 12,000 km) are some of the most technically challenging systems designed today. Undersea cable systems require ULH, inasmuch as the distance across the Atlantic Ocean is approximately 6000 km and the distance across the Pacific Ocean is approximately 9000 km. For terrestrial networks, the ULH networks are needed because of the change in the nature of the traffic. Until a few years ago, voice traffic dominated the network, and a span distance of 600 km satisfied more than 60% of the connections for voice traffic. However, with the Internet dominating the traffic now, a span distance of 3000 km is required to satisfy 60% of the connections for Internet traffic. In terrestrial systems, the marriage of Raman amplification technology and EDFAs has demonstrated great benefit by expanding the bandwidth of amplifiers, extending the distance between amplifiers, and allowing longer distances to be spanned. For submarine systems where the systems are designed to achieve a desired capacity over often the longest transmission distances, shorter span length (than for terrestrial systems) often has to be chosen. For such shorter spans (~50 km), the benefits of Raman amplification are not nearly as substantial. Presently, the most promising candidate use of Raman amplification in submarine systems is the wideband hybrid Raman–EDFA. For systems that require a very wide bandwidth this seems like an attractive way to more than double the transmission bandwidth without doubling the component count.

Ultra-Long-Haul, Dense WDM Using Dispersion-Managed Solitons in an All-Raman System (Chapter 18)

In an all-Raman-amplified system, dispersion-managed solitons can provide for dense WDM, uniquely compatible with all-optical terrestrial networking, robust and error-free over many thousands of kilometers. This chapter discusses various aspects of system design, including optimal dispersion maps, nonlinear and noise penalties, and typical dense WDM system performance. For example, dispersion-managed solitons are described as well as their special, periodic pulse behavior, their advantages over other transmission modes, and the conditions required to create and to maintain them. Also studied is one serious nonlinear penalty they suffer, viz. the timing jitter from collisions with solitons of neighboring channels. Dispersion-managed solitons, in an all-Raman, dense WDM system at 10 Gb/s per channel, makes a natural and comfortable fit with existing terrestrial fiber spans and can provide for transmission that is robust and error-free out to distances of 7000 km or more. In addition, such transmission is uniquely suited to provide the backbone of an all-optical network.

40 Gb/s Raman-Amplified Transmission (Chapter 19)

High-capacity 40 Gb/s transmission systems offer scalable solutions for future traffic growth in the core network. The challenges of 40 Gb/s systems include optical signal-to-noise ratio, fine-tuned dispersion compensation, and polarization mode dispersion. Raman amplification is likely to be a key driver to ease the noise performance and increase the available bandwidth for 40 Gb/s DWDM systems. New fiber technologies provide high system performance and enable a simple and cost-effective dispersion-compensation scheme. More system margin can also be expected from high-coding-gain forward error correction. Optimized modulation formats and high-speed optoelectronics will make practical deployment of 40 Gb/s DWDM systems possible, facilitating multiple terabit transmission over Mm distance at low cost-per-bit-per-kilometer. The challenges of DWDM transmission at 40 Gb/s are addressed in this chapter, along with the technologies enabling 40 Gb/s terrestrial transmission. Also described are advanced experiments and demonstrations at 40 Gb/s using Raman amplification.

Contents

VOLUME 90/1

Contributors

Jean-Christophe Bouteiller
OFS Laboratories
25 Schoolhouse Road
Somerset, NJ 08873
USA
Email: jcbouteiller@ofsoptics.com

Jake Bromage
OFS
Crawford Hill Laboratories
791 Holmdel-Keyport Road
Holmdel, NJ 07733
USA
Email: bromage@ofsoptics.com

Carl DeWilde
Xtera Communications, Inc.
500 West Bethany Drive
Suite 100
Allen, TX 75013
USA
Email: cdewilde@xtera.com

Evgeny M. Dianov
A.M. Prokhorov General
   Physics Institute
Russian Academy of Sciences
38 Vavilov Str.
119991 Moscow
Russia
Email: dianov@fo.gpi.ru

Yoshihito Emori
Fitel Photonica Laboratory
Optical Subsystem Department
6 Yawata Kaigan Dori
Ichihara 290-8555
Japan
Email: yemori@ch.furukawa.co.jp

René-Jean Essiambre
Lucent Technologies
Bell Laboratories
Crawford Hill Laboratory
Room HOH L-129
791 Holmdel-Keyport Road
Holmdel, NJ 07733
USA
Email: rjessiam@lucent.com

Alan F. Evans
Optical Physics
Science and Technology
Corning Incorporated
SP-AR-02-4
Corning, NY 14831
USA
Email: evansaf@corning.com

Chris R.S. Fludger
Nortel Networks, UK
London Road
Harlow, Essex CM17 9NA
UK
Email : cfludger@nortelnetworks.com

Dmitri Foursa
Tyco Telecommunications
250 Industrial Way West
Eatontown, NJ 07724
USA
Email: dfoursa@tycotelecom.com

Andrew R. Grant
Lightwave Systems Research
Bell Laboratories
101 Crawfords Corner Road
Room 4C-316
Holmdel, NJ 07733-3030
USA
Email: ARGrant@lucent.com

Lars Grüner-Nielsen
Specialty Photonics Division
OFS Denmark
Priorparken 680
DK-2605 Brøndby
Denmark
Email: lgruner@ofsoptics.com

Clifford E. Headley
OFS Laboratories
25 Schoolhouse Road, Room A-16
Somerset, NJ 08873
USA
Email: cheadley@ofsoptics.com

Mohammed N. Islam
Department of Electrical Engineering
    and Computer Science
University of Michigan at Ann Arbor
1110 EECS Bldg., 1301 Beal Ave.
Ann Arbor, MI 48109-2122
USA
Email: mni@eecs.umich.edu
and
Xtera Communications, Inc.
500 West Bethany Drive
Suite 100
Allen, TX 75013
USA
Email: mislam@xtera.com

Howard Kidorf
82 Tower Hill Drive
Red Bank, NJ 07701
USA
Email: kidorf@kidorf.com

Andrey Kobyakov
Science and Technology
Corning Incorporated
SP-DV-02-08
Corning, NY 14831
USA
Email: KobyakovA@corning.com

Peter Krummrich
Siemens AG, Information and
    Communication Networks
Hofmannstrasse 51
81359 Munich
Germany
Email: peter.krummrich@siemens.com

Amos Kuditcher
Xtera Communications, Inc.
500 West Bethany Drive
Suite 100
Allen, TX 75013
USA
Email: akuditcher@xtera.com

Hiroji Masuda
NTT Network Innovation Laboratories
1-1 Hikarino-oka, Yokosuka
Kanagawa 239-0847
Japan
E-mail: masuda@exa.onlab.ntt.co.jp

Marc Mermelstein
OFS Laboratories
25 Schoolhouse Road, Room A-05
Somerset, NJ 08873
USA
Email: mermelstein@ofsoptics.com

Linn F. Mollenauer
Lucent Technologies
Bell Laboratories
101 Crawfords Corner Road
Room 4C-306
Holmdel, NJ 07733
USA

Shu Namiki
Manager
Furukawa Electric Company, Ltd.
Fitel Photonics Lab,Optical
    Subsystems Department
6 Yawata Kaigan Dori
Ichihara 290-8555
Japan
E-mail: snamiki@ch.furukawa.co.jp

Lynn E. Nelson
OFS
791 Holmdel-Keyport Road
Room L-137
Holmdel, NJ 07733
USA
Email: lenelson@ofsoptics.com

Morten Nissov
Transmission Research
Tyco Telecommunications
250 Industrial Way West
Room 1A225
Eatontown, NJ 07724
USA
Email: MNissov@tycotelecom.com

Yujun Qian
Specialty Photonics Division
OFS Fitel Denmark I/S
Priorparken 680
DK-2605 Brøndby
Denmark
Email: yujunqian@ofsoptics.com

Stojan Radic
Bell Laboratories
Lightwave Systems Research
Crawford Hill Laboratory R-231
791 Holmdel Keyport Road
Holmdel, NJ 07733-0400
USA
Email: radic@lucent.com

Roger H. Stolen
Department of Electrical
    and Computer Engineering
Virginia Polytechnic Institute
    and State University
Blacksburg, VA 24061
USA
Email: stolen@vt.edu

Naoki Tsukiji
Furukawa Electric Company, Ltd.
Optical Components Department
2-4-3 Okano, Nishi-ku
Yokohama 220-6374
Japan
Email: tsukiji@yokoken.furukawa.co.jp

Michael Vasilyev
Corning Incorporated
Photonic Research and Test Center
2200 Cottontail Lane
Somerset, NJ 08873
USA

Pete J. Winzer
Lucent Technologies
Thurn & Taxis Str. 10
Nuremberg 90411
Germany
Email: winzer@lucent.com

Benyuan Zhu
OFS
791 Holmdel-Keyport Road
Room L-111
Holmdel, NJ 07733
USA
Email: byzhu@ofsoptics.com

Chapter 10

S-Band Raman Amplifiers

Mohammed N. Islam

10.1. Introduction

In this chapter we focus on the use of discrete or lumped Raman amplifiers in the short-wavelength S-band. Recent advances in data communications have led to requirements for higher throughput of wavelength-division-multiplexed (WDM) transmission systems. Two approaches have been pursued to address the increasing throughput requirements, namely, increasing the spectral efficiency of WDM systems within existing transmission bands, and increasing optical bandwidth to utilize much more of the low-loss window in silica fiber than presently supported on WDM systems. The first approach has been pursued through coding and modulation techniques. The second approach to increasing throughput involves making available new transmission bands beyond those supported by conventional erbium-doped fiber amplifiers (EDFAs), which presently define the optical bandwidth of most WDM systems.

The low-loss window of single-mode silica fiber is shown in Fig. 10.1. Erbium-doped fiber amplifiers provide amplification in the C-band extending from 1530 to 1565 nm and the L-band extending from 1565 to as high as 1625 nm. On the short wavelength side of the C-band are the S-band stretching from 1480 to 1530 nm and the S+-band from 1450 to 1480 nm. The longer wavelength band stretching from 1625 to 1675 nm has also been suggested for WDM transmission.

Once the C- and L-bands are exhausted, the most likely band to be exploited is the S-band, with the potential for increasing the bandwidth of most transmission systems by 30 to 50 nm. As shown in Fig. 10.1, attenuation in the S-band is comparable to or better than attenuation in the L-band, and sensitivity to micro- and macrobending losses in the S-band is far less pronounced. In addition, dispersion of standard single-mode silica fiber in the S-band is lower than dispersion in the longer wavelength bands. Figure 10.2 shows dispersion in standard single-mode fiber. S-band dispersion is approximately 30% less than L-band dispersion in standard single-mode fiber. Although dispersion in the S-band is generally low, reducing dispersion compensation requirements, WDM transmission in that band may not be attractive on some fiber types for which the zero dispersion wavelength falls within the S-band. Because

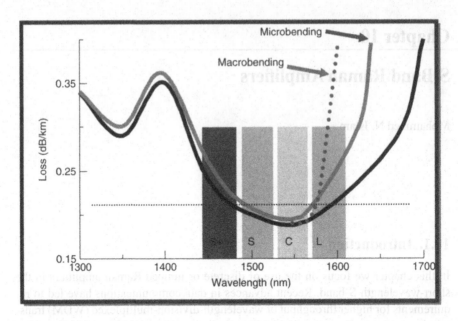

Fig. 10.1. Loss spectrum of standard single-mode fiber. The gray curves show micro- and macrobending loss.

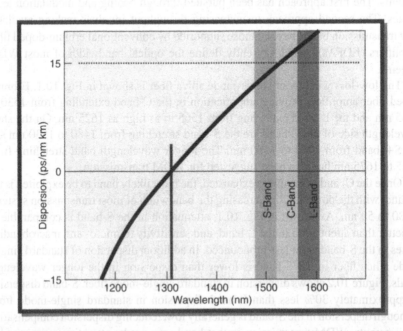

Fig. 10.2. Dispersion of standard single-mode fiber. Dispersion in the S-band is lower than in the C- and L-bands.

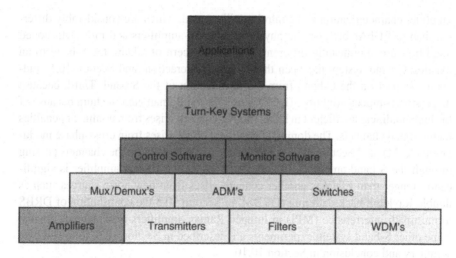

Fig. 10.3. Components of an optical network. The key technology for unlocking the S-band in fibers is the optical amplifier.

phase matching of parametric interactions between signals readily occurs near the zero dispersion wavelength, four-wave mixing may pose a challenge to WDM transmission on those fiber types. Figure 1.28 in Chapter 1 shows, however, that the zero dispersion wavelength falls within the S-band only for a few fiber types.

As illustrated in Fig. 10.3, the key technology for unlocking the S-band in fibers is the optical amplifier. All of the other building blocks are engineering changes on C- or L-band technologies. A number of S-band amplifiers are being studied, including rare earth-doped fiber amplifiers, semiconductor optical amplifiers, optical parametric amplifiers, and Raman amplifiers. Raman amplification has been demonstrated to be a viable approach to bandwidth extension. Because the spectral region where significant Raman gain occurs depends primarily on pump wavelength and power, amplification can be obtained practically anywhere within the low-loss window of silica fiber. In addition, transmission fiber and dispersion-compensating fiber are both good Raman gain media. Thus Raman amplification provides a means not only for compensating losses within the transmission fiber, but also for integrating amplification and dispersion compensation within a single module. The challenges that plagued deployment of Raman amplification and led to its abandonment in favor of EDFAs in the late 1980s have been resolved, and commercial products have already appeared.

In this chapter the S-band Raman amplified system is studied in detail. Characteristics of alternative amplifier technologies for the S-band are briefly reviewed in Section 10.2, and the advantages and disadvantages of Raman amplification are summarized in Section 10.3. The focus of this chapter is on the Raman amplification approach, and an experimental example of an S-band Raman amplified, long-haul system is reported in Section 10.4. For the system described, the next few sections study in greater detail some of the key issues in practical deployment of the S-band in augmentation of C-band and/or L-band amplifiers. First, in Section 10.5 the Raman amplifier characterization and control are illustrated, which are considerably different than in EDFAs because the physics of Raman amplifiers and rare earth-doped amplifiers are significantly different. A major concern of adding the S-band to an existing C-band system has been the interband interaction, and Section 10.6 studies the impact on the C-band from the introduction of the S-band. Third, because dispersion-compensating fiber (DCF) is used as the Raman gain medium because of the high nonlinear coefficient in DCF, another concern arises from nonlinear penalties on the signal channels. The dominant nonlinear effect arises from cross-phase modulation (XPM), and Section 10.7 quantifies the XPM penalty on the channels passing through the S-band amplifier. In addition, because the Raman amplifier is significantly longer than EDFA's, another concern arises from new noise sources such as double Rayleigh backscattering (DRBS). In Section 10.8 the contribution of DRBS to multipath interference (MPI) in lumped Raman amplifiers is explained. Finally, some other S-band system experiments are described in Section 10.9, followed by a summary and conclusion in Section 10.10.

10.2. Alternative S-Band Amplifier Technologies

Three technologies have been studied extensively in hopes of bringing to the lower wavelengths of the S-band the same cost-effective means of optical amplification that EDFAs provide for the C-band [1]. Semiconductor optical amplifiers (SOAs), thulium-doped fiber amplifiers (TDFAs) utilizing either fluoride or multicomponent silicates (MCS) as the host fiber, along with lumped, or discrete, Raman amplifiers (LRAs) have all been proposed as enablers for opening up transmission of wavelengths shorter than the traditional 1530 to 1560 nm range of EDFAs.

To ascertain which of these three optical amplification approaches holds the most promise for enabling S-band transmission, the principles of operation, implementation challenges, and possible applications for each approach are examined. All of these technologies have benefits to offer, however, only one provides cost-effective long-haul transmission of additional wavelength bands. In addition, recent research has shown the possibility of using EDFAs or optical parametric amplifiers to provide gain in the S-band. For example, optical parametric amplifiers have been considered for S-band amplification [2, 3]. EDFAs for use in the S-band are also briefly reviewed.

10.2.1. Semiconductor Optical Amplifiers (SOAs)

The principles of operation of the SOA differ from those of the other amplifier approaches in that an optical pumping source is not used. Optical amplification is achieved in an SOA through the application of an electric field to the input optical transmission signal traveling through an SOA waveguide structure. Within the SOA and electric field, an absorption or stimulated emission of a photon can occur. The absorption of a photon results in the generation of an electron-hole pair, and stimulated emission is brought about by a photon initiating the recombination of an electron-hole pair. To obtain amplification of an optical signal, the stimulated emission must be greater than the absorption of photons.

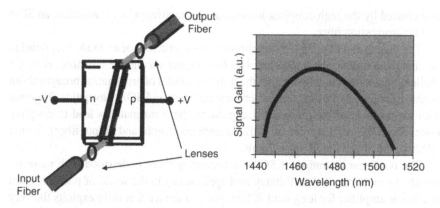

Fig. 10.4. Schematic structure and typical gain curve of a semiconductor optical amplifier.

The characteristics of the gain, the center wavelength, and the gain bandwidth offered by an SOA are dependent upon the type of semiconductor material used and the structure of the device. For optical transmission applications, current emphasis has focused on SOAs fabricated from indium phosphide (InP), indium gallium arsenide (InGaAs), and indium gallium arsenide phosphide (InGaAsP). The 3 dB gain bandwidth of a single device is typically 20 to 30 nm, although operation has been reported up to 40 to 60 nm away from the gain peak wavelength (Fig. 10.4). With current technology, maximum gains made available through the use of SOAs have been limited to 15 dB. The physical properties of the semiconductor material permit the construction of SOAs that can operate across a wide range of wavelengths from 1300 up to 1600 nm.

The very aspects that make SOAs so attractive as an optical amplifier also pose the greatest challenges to its eventual implementation for long-haul WDM applications. This is borne out by the nonlinear nature of the amplification mechanisms involved in an SOA being applied to a linear application of cascading optical amplifiers along a number of consecutive fiber spans. Challenges exist because SOAs suffer from low-gain, high-noise figures along with high bit rate and multiplex signal handling sensitivities. For brevity, only the noise and bit rate sensitivity issues are discussed.

The rapid response time of the SOA brings about a number of difficulties when attempting to achieve amplification of high bit rate WDM signals within a fiber. With recovery times on the same order as some of today's time-division-multiplexed bit rates, current SOAs only effectively handle signals below 10 Gbps before saturation, although 20 Gbps operation using lower powers to remain below saturation has been reported [4]. However, these lower power levels will not operate over the typical "standard" span length of 80 to 100 km.

With the fast gain response affecting the entire length of the signal within the waveguide of an SOA, changes in the magnitudes of the electric field or input signal(s) impart instantaneous changes (fluctuations) to the resulting signal(s). This leads to interchannel and intersymbol interference along with high noise figures, typically around 8 to 10 dB, prevalent in today's SOAs. This high noise figure (NF) is further exacerbated by the high coupling losses that are incurred when connecting an SOA to the transmission fiber.

The diagram in Fig. 10.4 shows the simplified structure of an SOA, highlighting the input and output fibers together with the internal waveguide structure. Here the challenge exists in bringing together the different modes of propagation brought about by the difference in the geometries between the fibers and SOA waveguide. Together with the problem of the mechanical alignment, these mismatches lead to coupling losses for SOAs on the order of 3 dB for each end (input and output fiber), further affecting the already high noise figure.

The nonlinear operation of SOAs is becoming more understood with more research. As such, the SOA will likely find applications in the world of photonics, not as a linear amplifier for long-haul WDM, but as a device that truly exploits the very nature of its operation. Current research has focused on the application of SOAs as modulators, optical switches, and wavelength converters. The size of the device together with the high level of integration makes it attractive for applications that are highly cost sensitive. Thus much of the recent work has also been for applications within the metropolitan optical network that require a lower level of amplification to overcome losses from multiple encounters with optical add/drop nodes and, in the future, optical switches.

10.2.2. Thulium-Doped Fiber Amplifiers (TDFAs)

The principle of operation of a TDFA is similar to that of the EDFA in that optical amplification of a signal is accomplished through the energy conversion from pump signals through a length of specially doped fiber. The differences lie in the type of gain fiber used, dopants employed, and pump configurations. An energy level schematic, shown in Fig. 10.5(a), depicts the optical amplification process in a TDFA, utilizing either a fluoride or MCS fiber gain medium. Two pumps, of either different [5] or the same wavelengths [6], are utilized to achieve optical amplification. An upconversion pump is used to overpopulate the quasistable level depicted as E_2, where the pump is then used to populate the other quasistable E_3 level, whereupon the stimulated release of photons then occurs.

A gain response such as that shown in Fig. 10.5 centered around 1460 nm is achieved by utilizing the pumping scheme discussed together with either a fluoride, fluoro-zirconate (ZBLAN), or a MCS such as antimony silicate fiber doped with high levels of thulium. Gain bandwidths of +20 dB have approached 35 nm in width, and maximum gains of 31 dB have been reported. Unlike the SOA, though, operation is limited over about a 65 nm range.

The fibers hosting the dopants pose the main challenge to the wide acceptance and commercial deployment of TDFAs. These fibers are more brittle than silica-based ones and, thus, greater care must be exercised throughout the manufacturing, deployment, and maintenance processes. Maintaining the same level of product reliability as expected with current silica-based technologies (e.g., EDFAs) will pose a significant challenge, as current acceptable practices must be changed.

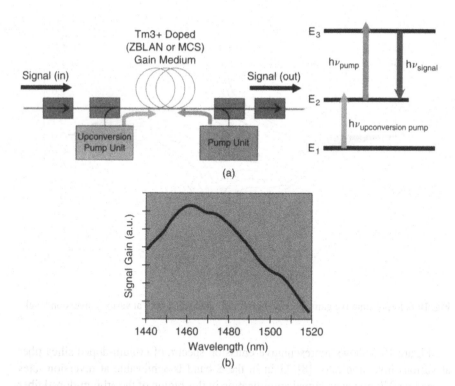

Fig. 10.5. (a) Schematic energy level diagram and pump configuration for a thulium-doped fiber amplifier; (b) typical gain spectrum of a thulium-doped fiber amplifier.

Being of different material composition than fiber deployed into the network's transport infrastructure, connecting TDFAs into the network poses another challenge. The less reliable and lossier mechanical splicing, or epoxy pigtailing must be employed, as fusion splicing is not an available option. This results in a difference of >0.3 dB per mechanical splice versus a <0.05 dB per fusion splice. The exotic compositions of these fibers will also reveal themselves in higher component costs. Thus there are challenges in reliability and cost that may prevent the widespread deployment of TDFAs.

10.2.3. Erbium-Doped Fiber Amplifiers

Although the peak gain of EDFAs occurs at wavelengths near 1530 nm, attempts have been made to extend operation into the S-band. Gain in EDFAs arises from inversion of the $^4I_{13/2}$ level of Er^{3+} with respect to the $^4I_{15/2}$ level [7]. The inversion is achieved by excitation of the $^4I_{11/2}$ upon pump absorption at 980 nm followed by relaxation to the $^4I_{13/2}$ level, or by excitation of the $^4I_{13/2}$ level via pump absorption at 1480 nm. The subsequent emission from the $^4I_{13/2}$ level extends into the S-band at sufficiently high inversion rates.

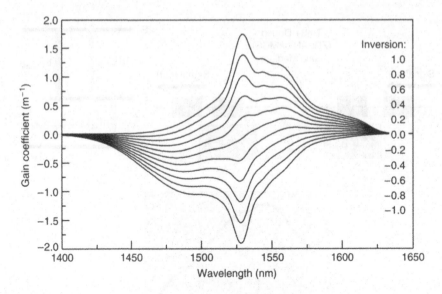

Fig. 10.6. Representative gain spectra of erbium-doped silica fiber at various inversion levels.

Figure 10.6 shows representative emission spectra of erbium-doped silica fiber at various inversion rates [8]. Gain in the S-band is achievable at inversion rates exceeding 0.7; however, signal amplification in this region of the erbium-doped fiber emission spectrum competes with amplification of spontaneous emission in the longer wavelength bands where gain is higher. Amplified spontaneous emission (ASE) at the peak emission wavelength, 1530 nm, is particularly strong and tends to saturate the amplifier, severely limiting achievable S-band gain.

Adequate gain in the S-band may be achieved by suppressing the ASE. Because the ASE is generated over the length of the amplifier, ASE suppression would only be effective if applied over a substantial fraction of the erbium-doped fiber. One method for increasing S-band gain by ASE suppression utilizes short-pass filters inserted between sections of gain of gain fiber. This configuration has been demonstrated using five sections of erbium-doped fiber and bidirectional pumping at 980 nm [9]. An average gain of 21 dB was achieved for 40 channels (−20 dBm per channel) in the wavelength range 1489 to 1519 nm.

A second method for suppressing ASE in the long-wavelength bands utilizes a host fiber with a long-wavelength cutoff located near the peak wavelength of the ASE [10]. With a cutoff wavelength near 1525 nm, the core of the host fiber does not guide ASE at 1530 nm and longer wavelengths. Saturation of the amplifier by ASE at peak emission wavelengths is effectively eliminated, shifting the gain peak to shorter wavelengths. This approach effectively distributes ASE filtering along the entire length of the gain medium and high-gain S-band amplifiers.

Except for a few components, S-band EDFAs are based on standard EDFA technology developed for the C-band and share the same pump characteristics and gain dynamics. Compared to conventional EDFAs, these amplifiers require longer lengths of gain fiber and have a higher noise figure. Dividing the gain fiber into sections and inserting filters between them introduces lumped loss along the length of the amplifier and increases the amplifier noise figure. For example, in the previous example utilizing five fiber sections and short-pass filters, a noise figure of 7 dB was reported for an aggregate output power of 17 dBm.

10.2.4. Lumped or Discrete Raman Amplifiers

Because it utilizes silica fiber as the gain medium, Raman amplification does not suffer from the same implementation or deployment challenges as SOAs and TDFAs. Amplifiers can be fusion spliced with the same type of fibers utilized within the fiber-optic transport infrastructure. Discrete Raman amplifiers have also been demonstrated to have performance on a par with EDFAs, in terms of gain, NF, and bandwidth.

Raman amplification arises from the transfer of power from a pump beam to a signal beam of different frequency. The frequency difference between the pump and signal beams corresponds to the energy of a vibrational mode of the medium in which the pump–signal interaction occurs. Gain characteristics of Raman amplifiers depend on the medium and the pump wavelength and power. In silica fiber, the gain bandwidth is approximately 40 THz with maximum gain occurring for signals with frequency offset from the pump by 13.2 THz. Gain shape depends on the number of pump wavelengths utilized and their relative power.

Raman amplification has emerged as a viable technology for enabling WDM transmission in the S-band. The following section examines the advantages and challenges of Raman amplification. Then an S-band system using lumped Raman amplification is illustrated in Section 10.4. Thereafter, detailed analysis is provided of various issues that require resolution before practical deployment of the S-band in an augmentation strategy with the C- and/or L-band systems.

10.3. Advantages and Challenges of S-Band Raman LRA

Because silica fiber is a good gain medium, Raman amplification benefits from an established technology base with an extensive research and deployment history. Dispersion-compensating fiber is a particularly convenient gain medium. Implementing the LRA in the DCF permits the deployment of a module that simultaneously provides gain to overcome the fiber and component losses in addition to having integrated dispersion compensation.

In addition, gain is achievable over a broad wavelength range simply by pump wavelength and power selection. The gain flexibility of Raman amplification has led to discrete amplifiers for many different wavelength ranges and demonstrations of flat gain over wide gain bandwidths. Recent progress in diode laser technology has made available high-power pump modules at various wavelengths, enabling construction of Raman amplifiers with gain at practically any wavelength within the S as well as S+ bands with performance comparable to C-band EDFAs.

10.3.1. Challenges of Raman Approach

Perhaps the most significant challenge that Raman amplifiers have had to overcome is the relatively poor efficiency compared to EDFAs. However, recent advances in the understanding of the Raman gain efficiency of optical fibers and the development of DCF has led to a natural solution to the gain fiber issue. Increases of over tenfold in Raman gain efficiencies have been reported for commercially available DCFs. In addition, semiconductor laser diode powers have risen steadily over the past few years. Newer generations of high-power laser diodes are today being demonstrated to have greater than 1 W of output power [11]. New diode-array-cladding-pumped lasers are reaching output power levels on the order of 10 W and above. The challenge imposed by the poor efficiency of Raman amplification is quickly fading with these increases in fiber Raman gain efficiency and laser pump output powers. Moreover, as discussed in Chapter 1, at high channel count or high signal/pump powers, the slope and overall efficiency of LRAs can actually exceed even 1480 nm pumped EDFAs.

Noise issues within Raman amplifiers brought about by double Rayleigh backscattering (DRS) and the extremely fast response of the Raman effect itself are also effectively being addressed. DRS can be controlled through the use of improved isolation between the multiple stages of a discrete Raman amplifier. The extremely fast response of the amplification process can be averaged out through the utilization of counterpropagating pumping. Thus noise figures are decreasing towards 5 dB, on a par with commercial EDFAs.

Packaging challenges for Raman amplifiers still exist as the lengths required to achieve appreciable gain can easily reach tens of kilometers. However, as the fiber's Raman gain efficiencies increase, these lengths decrease. In addition, for systems with bit rates of 10 Gb/s or higher, dispersion compensation is required, and DCF is the most commonly deployed form of compensation. Because the LRA gain can be achieved in the same DCF, the long fiber length limitation is muted by using a LRA with integrated dispersion compensation.

10.4. S-Band Long-Haul Transmission Using LRAs

As an example of the application of an S-band lumped Raman amplifier (SLRA), consider the following experiment of the first S-band long-haul WDM transmission using a cascade of dispersion-compensating LRAs. Twenty non-return-to-zero (NRZ) channels, spanning the entire S-band, were transmitted over 10 spans of standard single-mode fiber (SSMF), each achieving bit error rate (BER) $<10^{-12}$ without forward error correction.

In particular, Puc and co-workers [12] demonstrate the cascade of 11 rack-mounted S-band dispersion-compensating LRAs (SLRAs) to transmit 20 S-band channels modulated at 10.67 Gb/s over 867 km of standard single-mode fiber. The margins accumulated in this demonstration show the capability for such a system to achieve 80-channel transmission over 10×25 dB SSMF with standard out-of-band forward error correction (OOB-FEC) and presently available SLRAs.

Fig. 10.7. Photograph of six rack-mounted S-band lumped Raman amplifiers (SLRAs). In addition to the SLRAs, the rack also holds band couplers and optical supervisory units. OSA: optical spectrum analyzer; APM: administrative processor module; BCM: band coupling module.

Figure 10.7 shows a photograph of a standard seven-foot rack containing six SLRAs. Each SLRA is a two-stage amplifier containing a gain-flattening filter (GFF) with a midstage access (Fig. 10.8(a)). The pump module corresponds to four laser diodes wavelength and polarization multiplexed together (Fig. 10.8(b)). The pump wavelengths are selected to achieve sufficiently flat gain across the S-band.

Figure 10.9 shows the functional block diagram of the amplifier and its typical gain and noise performance at a total input power of −14 dBm. It can be seen that high values of gain (up to 28 dB) can be achieved with a very small gain ripple (<1 dB), in addition to the noise figure values on the order of 5.5 dB across the S-band (1493 to 1523 nm). The SLRA gain fiber has a high negative dispersion in the S-band, as well as a negative dispersion slope, providing coarse dispersion and dispersion slope compensation throughout the band. Each SLRA compensates for the dispersion of about 75 km of SSMF.

A block diagram of the system experimental setup is shown in Fig. 10.10. Eleven dispersion-compensating SLRAs are used to transmit 20 channels in the S-band (be-

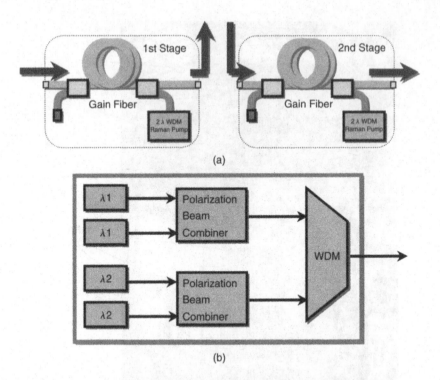

(a)

(b)

Fig. 10.8. (a) Configuration of a two-stage S-band lumped Raman amplifier. Provision is made for midstage access; (b) schematic diagram of pump module. Two pairs of laser diodes each polarization multiplexed are combined with a WDM coupler.

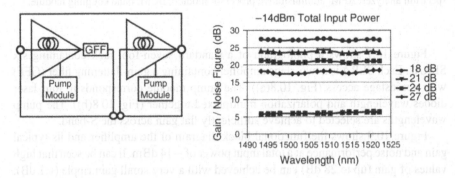

Fig. 10.9. Functional block diagram and typical gain and noise figure spectra of the S-band Raman amplifier. Total input power is −14 dBm.

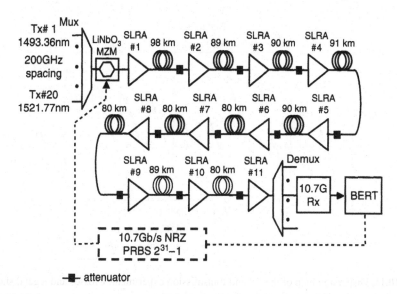

Fig. 10.10. Block diagram of S-band transmission experiment. Output power is 14 dBm for all amplifiers (1 dBm/channel) and average span loss is 21 dB. SLRA: S-band Raman amplifier.

tween 1493.36 and 1521.77 nm) nominally spaced by 200 GHz and modulated at 10.67 Gb/s over 10 spans of SSMF, for a total length of 867 km. Each span contains on average six connectorized joints and additional loss elements to bring the average span loss to 21 dB.

The average amplifier output power is only +14 dBm, resulting in a launched power per channel of +1 dBm, and each SLRA is capable of at least +19 dBm output power. All the channels are launched in parallel polarization and modulated with a $2^{31} - 1$ pseudorandom bit sequence NRZ pattern at 10.67 Gb/s through an external LiNbO$_3$ modulator. The receiver has an adjustable decision threshold and a clock extractor, allowing all the BER readings to be made at a maximum likelihood setting.

The dispersion map is sketched in Fig. 10.11 for the system configuration of Fig. 10.10. The line consists of sections of SSMF, with a dispersion of about 15ps/nm.km at 1510 nm. A section of dispersion-compensating fiber is inserted between the two stages of the fifth SLRA. The zero dispersion wavelength of the dispersion map is at 1508 nm. The crosses and the legend in Fig. 10.11 show the measured residual dispersion at the center and two extreme wavelengths.

The received spectrum is shown in Fig. 10.12. The average received optical signal-to-noise ratio OSNR (0.1 nm bandwidth) is about 20.7 dB. At this nominal level, all the channels operated at BER $< 10^{-12}$ without any error correction at 10.67 Gb/s line rate. In addition, Puc et al. measure BER versus OSNR curves by raising the noise floor of the spectrum while maintaining the peak signal power constant. The results are summarized in Fig. 10.13. The most dispersive channel is #20, and its received

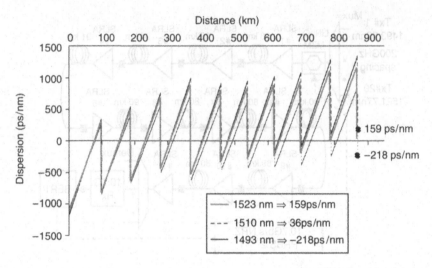

Fig. 10.11. Dispersion map of the S-band transmission experiment. Crosses and legend show residual dispersion at the center and extreme wavelengths.

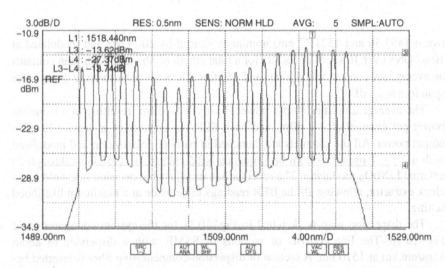

Fig. 10.12. Spectrum after 867 km in the S-band transmission experiment. Resolution bandwidth is 0.5 nm.

optical eye diagram is shown as an inset in Fig. 10.13. Moreover, almost no dispersion penalty is observed.

Similarly, the channels did not suffer from the four-wave mixing (FWM) effect. Even when five channels are spaced by 50 GHz, no FWM intermodulation product is observed down to 28 dB below the channel level (Fig. 10.14). At the nominal launched channel power level (+1 dBm), the self-phase modulation (SPM) appears to

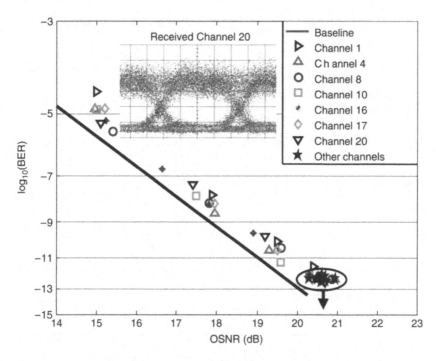

Fig. 10.13. Variation of BER with OSNR after 10 spans for 20 channels. Inset shows the eye diagram of the most dispersive channel.

be the largest impairment aside from amplified spontaneous emission. The estimated SPM penalty is less than 1 dB for any of the transmitted channels. No cross-phase modulation (XPM) penalty is apparently observed.

For an 80-channel system using the full power level of the SLRAs (+19 dBm), the power per channel would be 0 dBm. Furthermore, if a standard OOB-FEC such as Reed–Solomon 255/239 is employed, one can expect that a reduction of the OSNR by 5 dB would result in a similar BER performance as if no FEC were used. Therefore the present demonstration translates into a capability of transmitting 80 channels at 10.67 Gb/s with OOB-FEC (OC-192 line rate) over 10 spans of 25 dB.

To summarize this section, a novel, S-band long-haul WDM transmission scheme using dispersion-compensating lumped Raman amplifiers has been successfully demonstrated. This 10-span transmission demonstration confirms the viability of SLRAs as a key enabling technology for a cost-effective and reliable expansion of optical networks into the S/S$^+$-band regions and potentially to other wavelength windows.

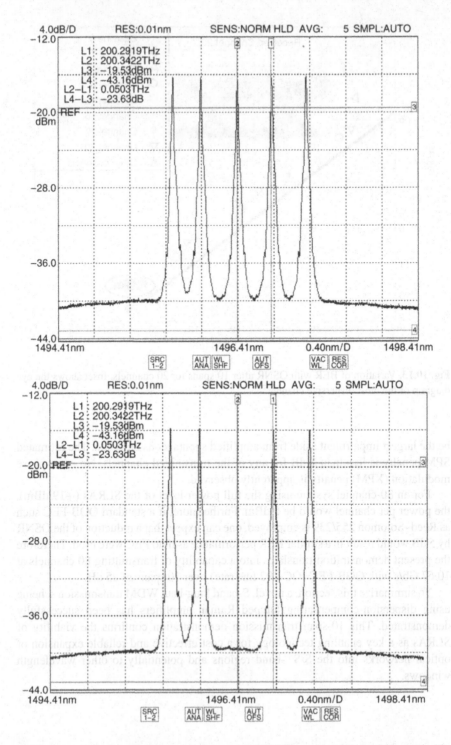

Fig. 10.14. Four-wave mixing test over 10 spans at 50 GHz spacing. Channel bit rate is 10 Gb/s.

10.5. Amplifier Characterization and Control

Having demonstrated the S-band system experiment in the last section, it is worth looking now in more detail at the performance and control of the Raman amplifiers [13]. The gain fiber, which is a dispersion-compensating fiber, is first characterized. The gain fiber has high negative dispersion in the S-band as well as a negative dispersion slope, which provides for coarse dispersion and slope compensation throughout the band. Figure 10.15 shows the unsaturated gain coefficient in the DCF as compared with the gain in SiO_2 glass. The peak gain coefficient is more than an order of magnitude higher than in fused silica, and the full-width at half maximum of the gain curve is about 37 nm wide at 1500 nm. Figure 10.16 examines the attenuation in the gain fiber. The figure of merit for the Raman amplifier fiber is $g_R/\alpha_p A_{eff}$, so that low loss is as important as high gain coefficient. The loss as a function of wavelength is shown in Fig. 10.16. It is important to note that a low OH^- attenuation at ~1380 nm is critical for S-band efficiency, because this is the loss seen by some of the pump wavelengths.

Moreover, Fig. 10.17 shows the splice loss spectral variation between the gain fiber and the transmission fiber for two different splices. Knowledge of the splice loss variation is important for accurate prediction of amplifier performance. As an example, Fig. 10.18 shows the calculated (lines) and measured (squares) net gain versus wavelength. The top dotted curve is the modeling results without including the spectrally dependent loss of the splice, and the lower solid curve includes the spectral dependence. As can be seen, reliable prediction of experimental performance requires accurate knowledge of the fiber parameters as well as their spectral dependence.

Beyond the loss and gain, another important aspect of the amplifier modeling is the gain saturation performance. For instance, Fig. 10.19 shows the simulated (lines) and experimental data (points) for signal input powers ranging from −18 dBm to 0 dBm.

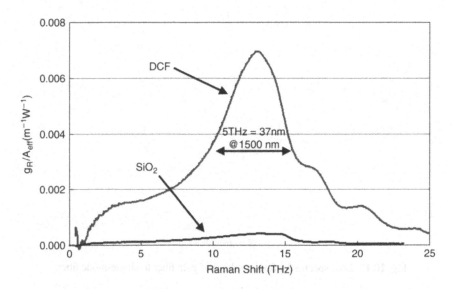

Fig. 10.15. Raman gain spectra in dispersion-compensating fiber (DCF) and silica.

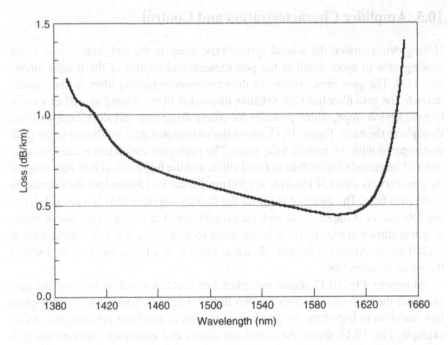

Fig. 10.16. Attenuation in gain fiber used in S-band Raman amplifier. Low absorption at 1380 nm (hydroxyl absorption peak in silica) is critical for high-power conversion efficiency.

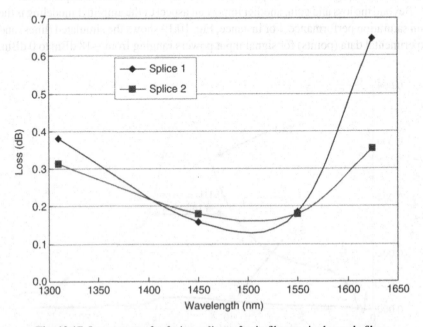

Fig. 10.17. Loss spectra for fusion splices of gain fiber to single-mode fiber.

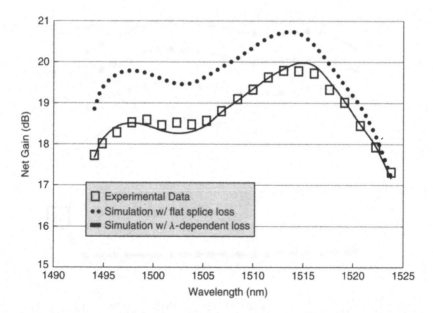

Fig. 10.18. Calculated (lines) and measured (squares) gain spectra. The dotted line is for flat splice loss spectrum and the solid line is for wavelength-dependent loss. Knowledge of splice loss spectral variation is important for performance prediction.

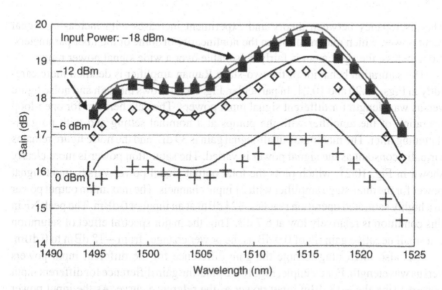

Fig. 10.19. Measured and simulated gain spectra in SLRA for input powers of −18, −12, −6, and 0 dBm. Solid lines: simulation results; points: experimental data.

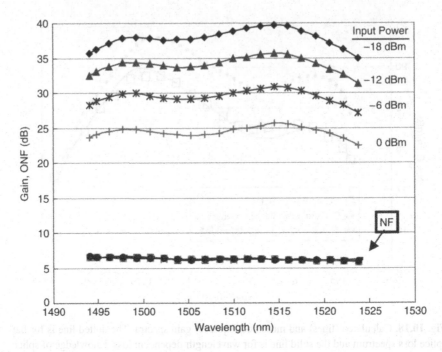

Fig. 10.20. Gain spectra without gain-flattening filter at various input powers. The peak gain is as high as +39 dB, and the NF remains virtually constant as the gain is varied.

The discrepancy between theory and experiment increases with increasing signal input power, which may result from the nonlinearity of some of the fiber parameters. Nonetheless, the agreement is fairly reasonable over a wide signal power range.

The saturation behavior of the two-stage Raman amplifier is detailed more carefully in Figs. 10.20 and 10.21. In particular, Fig. 10.20 shows the gain and noise figure versus wavelength for different signal input powers. The data here are for open-loop operation of the amplifier with the pumps at a nominal setting and with no gain-flattening filter. The maximum small-signal gain is 39 dB, and the noise figure remains virtually constant as the signal power is varied. The saturation power is more clearly shown in Fig. 10.21, which plots the total output signal power versus input signal power for the dual-stage amplifier with 21 input channels. The maximum output power in a highly saturated operation reaches +24 dBm at an input of 0 dBm. The peak NF in this condition is relatively low at 6.7 dB. Thus the major spectral effect of saturation is a small negative gain tilt of 0.9 dB as the power changes from −18 dBm to 0 dBm.

It is also interesting to note the gain difference for the different input powers versus wavelength. For example, Fig. 10.22 plots the gain difference for different input powers using the −18 dBm input power as the reference curve. As the input power increases, some inhomogeneous gain saturation behavior is observed. In general, Raman amplifiers act homogeneous-like due to pump depletion of the common pump. However, the situation becomes more complicated with multiple-wavelength pumps, which themselves interact through the Raman effect. Hence, some inhomogeneity of the gain curve is observed.

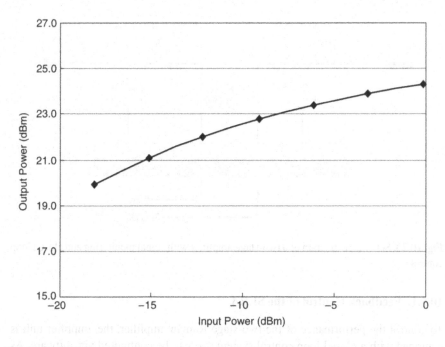

Fig. 10.21. Variation of total output signal power with total input signal power for 21 input channels.

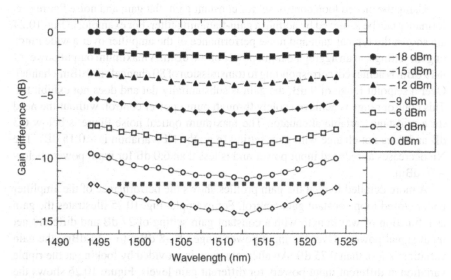

Fig. 10.22. Gain difference spectra for different input powers. Reference is gain at −18 dBm input power.

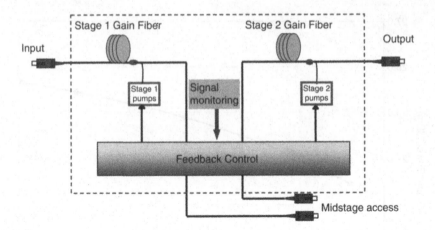

Fig. 10.23. Schematic diagram of a two-stage amplifier with signal monitoring and closed loop control.

10.5.1. Feedback Control of the SLRA

To control the performance of the two-stage Raman amplifier, the amplifier unit is equipped with a closed loop control system that can be configured via software. As shown in Fig. 10.23, the feedback control circuit has as an input signal monitoring at the midstage, and the output controls the pump laser powers in the two stages. The closed loop control acting on the pumps can be set for various conditions such as constant gain, constant output power, and provisional gain or power tilt.

Using the closed loop control set for constant gain, flat gain and noise figure performance can be achieved by adding a gain-flattening filter. For example, in Fig. 10.24 are shown the typical gain and noise performance of the amplifier over a wide range of gain settings, at an aggregate input power of −8 dBm. A maximum output power of +19 dBm is achieved, corresponding to transmission of 80 channels at 0 dBm/channel. Over a setpoint range of 9 dB, the gain is substantially flat and does not exhibit tilt. This control range is achieved solely through pump power control, without the need for a midstage variable attenuator. The maximum optical noise figure is below 6.3 dB and does not change with gain setting (e.g., the NF variation is <0.15 dB). The NF decreases at reduced input power and is less than 6.0 dB for input powers below −10 dBm.

A more detailed look at the gain profiles shows the performance of the amplifier under closed loop constant gain control. For example, Fig. 10.25 illustrates the gain as a function of wavelength with a constant gain setting of 27 dB and different net input signal powers. Over the input power range of −8 dBm to −14 dBm, the gain variation is less than 0.25 dB. An alternate view is provided by looking at the ripple variation at different input powers for different gain levels. Figure 10.26 shows the ripple at constant signal input power for different gain levels. Again, the closed loop gain control is found to be effective.

Another aspect of the closed loop operation is to use the control circuit to provide provisional gain tilt. As an illustration, Fig. 10.27 shows that controlling the relative

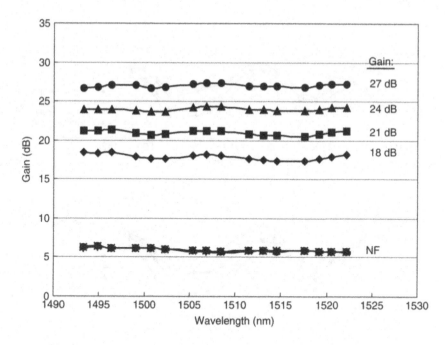

Fig. 10.24. Gain spectra under constant gain closed loop control for different gain levels.

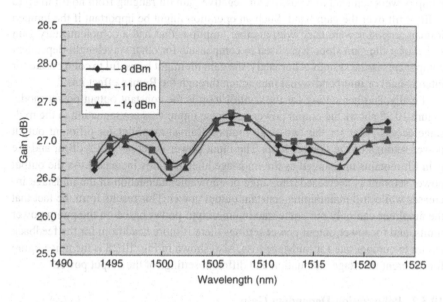

Fig. 10.25. Gain spectra under constant gain closed loop control for different input signal powers. Gain setting of the loop is 27 dB.

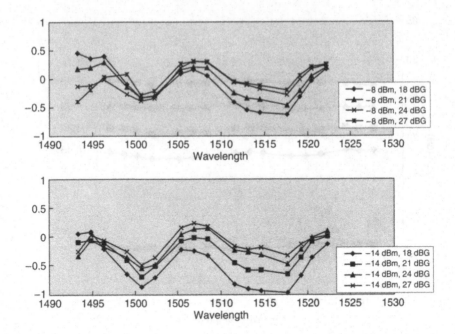

Fig. 10.26. Gain ripple variation under constant gain closed loop control for −8 dBm input power (upper graph) and −14 dBm input power (lower graph).

pump powers can lead to a positive or negative gain tilt ranging from no tilt to up to 3 dB of tilt over the gain band. Such an operation might be important if the lumped Raman amplifier were used with another amplifier that had a complementary gain slope, of if the gain slope were used to compensate for other wavelength-dependent components in the system. Alternately, the gain tilt might be useful to compensate for interchannel or interband signal interaction through the Raman effect itself.

Finally, another setting for the control loop is the constant output power mode. Figure 10.28 shows the output power and noise figure versus attenuation in the midstage access point for the two-stage lumped Raman amplifier for different output power settings of the control loop. The input power is set at −12.5 dBm, and the gain tilt remains unchanged as the midstage attenuation is increased. As the output power setpoint is decreased, the range of allowable attenuation in the midstage increases while still maintaining constant output power. This results from the fact that the amplifier can only put some maximum output power based on the pump power limits, and for lower output power settings there is more headroom for the feedback circuit to compensate for midstage loss. Also shown in Fig. 10.28 is the noise figure for different midstage attenuation and different settings of the output power.

10.5.2. Polarization-Dependent Gain

Polarization-dependent gain (PDG) in Raman amplifiers is primarily dependent on the polarization state of the pump. In the lumped Raman amplifier, the pump modules employ polarization multiplexing to realize low PDG. Typical values of PDG are

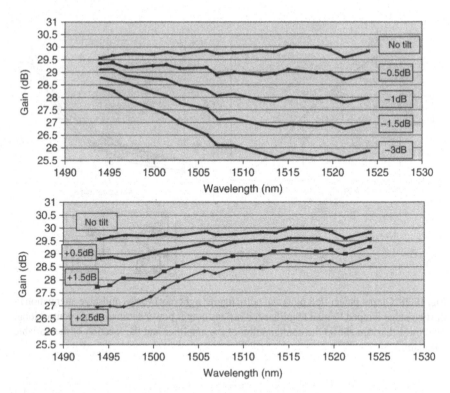

Fig. 10.27. Gain spectra at various negative (upper plot) and positive (lower plot) gain slopes. The different slopes are obtained through pump power control.

<0.05 dB. Figure 10.29 shows the increase in PDG when the ratio X/Y of x-polarized to y-polarized components of the pump is deliberately varied from 0 to 100%. It is seen that a low value of PDG of <0.05 dB is obtained for pump polarizations <20%, and the PDG remains below 0.1 dB for the ratio $X/Y > 50\%$. In particular, the data represent the PDG for a single-stage, backward pumped Raman amplifier with a signal gain of 15 dB.

10.5.3. Stimulated Brillouin Scattering in Amplifier

One of the concerns raised about Raman amplification is the penalty from stimulated Brillouin scattering. Stimulated Brillouin scattering is related to interaction with acoustic phonons (whereas Raman amplification is related to the optical phonons), and the phase matching is in the backward direction. Therefore, to study the impact of Brillouin scattering, the backscattered light is measured as a function of input power. Figure 10.30 shows the reflected power as a function of input pump power for a single-stage, unpumped gain fiber. The threshold for Brillouin scattering is often given as $P^{TH} = 21(A_{eff}/L_{eff}g_B)$. For this gain fiber, the stimulated Brillouin scattering threshold is ~4 mW for an unmodulated distributed-feedback laser source with a linewidth of 3 MHz. In addition, Fig. 10.31 shows the back-reflected power

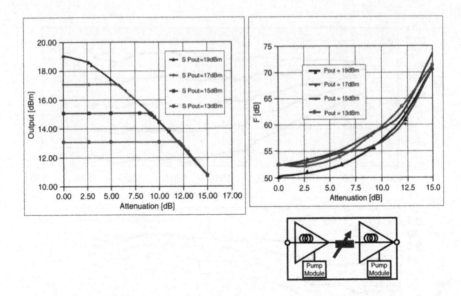

Fig. 10.28. Output power (left plot) and noise figure (right plot) versus midstage attenuation in the two-stage lumped Raman amplifier for various output power settings of the control loop. The lower diagram shows the configuration of the amplifier and the location of the variable attenuator.

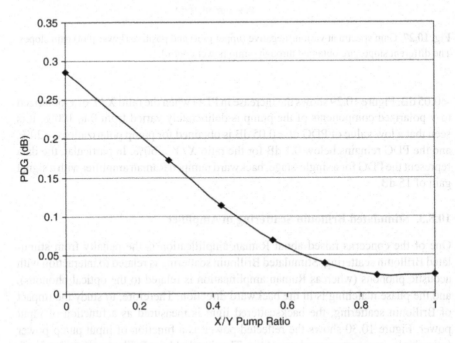

Fig. 10.29. Polarization-dependent gain (PDG) versus ratio X/Y of x-polarized to y-polarized components of the pump.

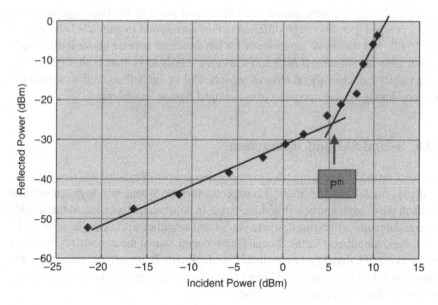

Fig. 10.30. Reflected power versus incident power for a single-stage unpumped gain fiber.

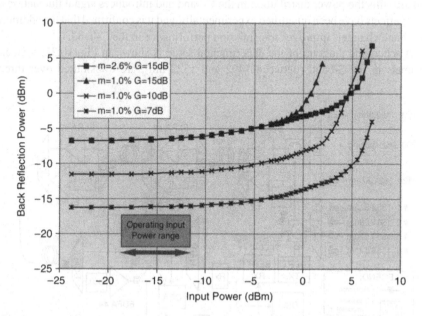

Fig. 10.31. Back reflection power in the output stage versus input signal power for the two-stage amplifier at various gain levels.

versus input signal power measured in the output stage of the two-stage amplifier with a pumped gain fiber. The different curves correspond to gain levels between 7 and 15 dB, and the signal input power for the threshold appears above 0 dBm input power. Thus the stimulated Brillouin scattering threshold is more than 10 dB above the typically launched signal channel powers, and so the Brillouin effect should not be a problem in typical operation of the lumped Raman amplifiers.

10.6. S- and C-Band Interaction

The work on the S-band is intended to extend the capacity of existing and new systems. To deploy the S-band amplifiers, it is expected that the S-band will augment existing bands in the C-band and possibly the L-band. In particular, it is expected that a split-band architecture will be used, where the S-band amplifier sits in parallel with the C- and L-band amplifiers. As the S-band is introduced, one of the concerns is the effect on the existing C-band that may already be deployed. To understand the effects, the impact of S-band channels on C-band channels is experimentally investigated in a three-span dual-band transmission [14].

The existing C-band is affected by the addition of the S-band through the interband stimulated Raman scattering. The S-band composite signal acts on the C-band as a broadband copropagating Raman pump. This interaction manifests itself in several different ways: it increases the optical signal-to-noise ratio throughout the C-band, and also tilts the power distribution in the C-band and introduces signal fluctuations. These effects have been quantified experimentally, and it is confirmed that the addition of S-band channels improves transmission performance in the C-band.

A schematic diagram of the experimental setup is shown in Fig. 10.32. Sixteen channels in the S-band (between 1492 and 1522 nm), are transmitted over three

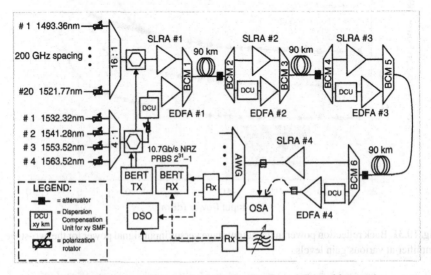

Fig. 10.32. Experimental setup for investigating S- and C-band interaction.

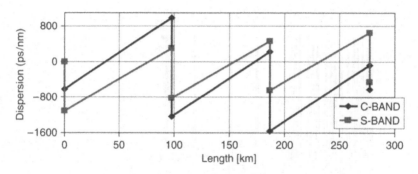

Fig. 10.33. Dispersion map for C- and S-bands.

spans of SSMF, for a total length of 280 km, using four dispersion-compensating S-band lumped Raman amplifiers. The S-band WDM signals are multiplexed and demultiplexed with the C-band WDM signals at each repeater. The C-band erbium-doped fiber amplifiers are operated in constant gain mode to minimize the tilt induced by the EDFAs. The launch power in the C-band is ~3 dBm/channel, and the average span loss is ~21 dB. The S-band amplifiers are operated in constant output power mode and provide a negative gain tilt (vs. wavelength) to compensate for the SRS-induced positive tilt. For test purposes, the C-band WDM signal consists of only four equally spaced, dispersion-compensated channels between 1532 and 1564 nm. Both S- and C-band signals are modulated at 10.67 Gb/s with a pseudorandom bit sequence (PRBS) of length $2^{31} - 1$. In a separate experiment, the S-band is modulated with a $2^{15} - 1$ PRBS and the ripple induced in the C-band is monitored on a sampling oscilloscope.

The inline amplifiers have the following characteristics. For the S-band lumped Raman amplifier, the output power is up to 19 dBm, the noise figure is <6 dB, the control circuit is set to adjust the gain tilt, and there is both dispersion and dispersion slope compensation. For the C-band amplifier, a commercial EDFA is used that maintains the gain at 23 dB. The EDFA has an output power up to 17 dBm and a noise figure <6 dB. With the EDFA an external dispersion and dispersion slope compensation is required. The corresponding dispersion maps for the C- and S-bands are illustrated in Fig. 10.33.

The spectrum at the receiver for the S-band and the C-band is shown in Fig. 10.34 as measured by an optical spectrum analyzer. The BER is measured by $<10^{-12}$ for all of the channels at a bit rate of 10.67 Gb/s. Also shown by the arrows are the C-band channel levels at different S-band launch powers. As the S-band channel power increases, the C-band power level increases due to the stimulated Raman scattering effect.

Figure 10.35 depicts the measured C-band power tilt versus S-band launch power with an 11 dBm C-band launch power level. The dashed curves in the same figure present numerically simulated data with C-band launch power of 11, 9, and 7 dBm. When prorated, the results show about same amount of tilt as in [15], but somewhat less tilt than in [16]. The difference is explained by a complete depolarization between

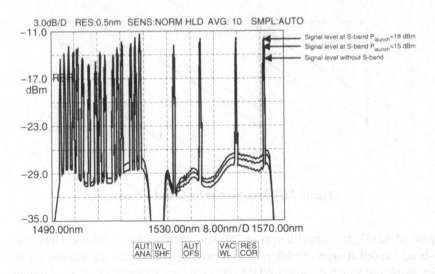

3.0dB/D RES:0.5nm SENS:NORM HLD AVG: 10 SMPL:AUTO

Signal level at S-band P$_{launch}$=18 dBm
Signal level at S-band P$_{launch}$=15 dBm
Signal level without S-band

Fig. 10.34. Spectra at the receiver for the S-band and C-band for S-band launch powers of 18 and 15 dBm and for no S-band signals.

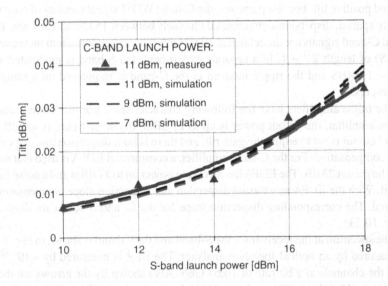

C-BAND LAUNCH POWER:

▲   11 dBm, measured
– – 11 dBm, simulation
– – 9 dBm, simulation
– – 7 dBm, simulation

Fig. 10.35. Measured and simulated C-band power tilt versus S-band launch power at various composite C-band launch powers. Composite C-band power for the measurement is 11 dBm (solid curve). The dashed curves show simulation results for C-band launch powers of 11, 9, and 7 dBm.

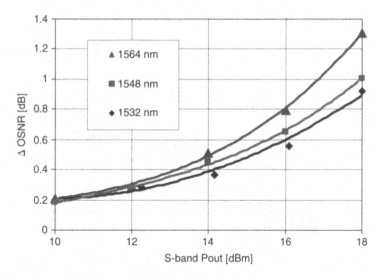

Fig. 10.36. C-band OSNR improvement versus S-band output power at various wavelengths for C-band launch power of 11 dBm.

S- and C-band signals. Also, a number of consecutive measurements disclosed no difference in the tilt magnitude between modulated and unmodulated data, which is in agreement with [17]. Moreover, the results show a good agreement with the numerical results. The compensation of this tilt could be achieved by either or any combination of a band-flattening filter in the band-combiner module, preemphasis of the signals, or negative gain tilt in the inline amplifiers.

OSNR improvement due to the presence of the C-band signals is shown in Fig. 10.36. The measurements are taken for various S-band launch power levels with a C-band launch power of 11 dBm. The C-band OSNR improves proportionally with wavelength and S-band signal power at the expense of S-band signal power. For a fully loaded S-band (launch power >18 dBm), the C-band OSNR improvement would be more than 1 dB. Also, the improvement is roughly proportional (in dB) to the number of spans.

Due to SRS dynamic crosstalk, ripples appear on the C-band channels induced by the S-band channels. Measurement results of peak-to-peak ripple that appeared on C-band signals when all S-band channels are modulated with the same PRBS (worst-case scenario) are summarized in Fig. 10.37. For example, Fig. 10.38 illustrates a typical ripple pattern. The baseband ripple, which is proportional to the bandwidth times the modulation depth, remains approximately constant. Also, the 3 dB ripple bandwidth is less than 5 MHz.

The BER penalty is determined by comparing BER versus OSNR measurements in C-band channels with and without S-band transmission. Figure 10.39 summarizes the results in terms of S-band-induced receiver sensitivity penalty at BER $= 10^{-10}$ for an S-band launch power of 18 dBm. For comparison, the corresponding OSNR improvement and the resulting BER improvement are plotted in the same diagram. The

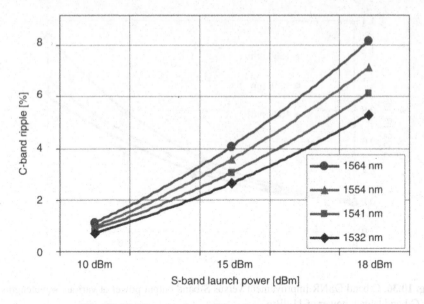

Fig. 10.37. S-band induced C-band peak-to-peak ripple.

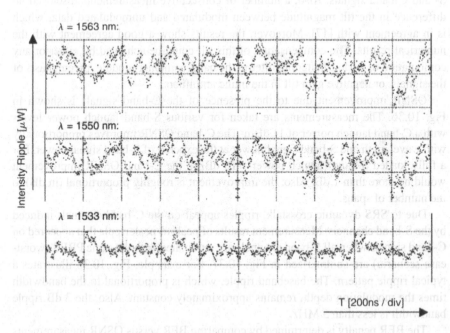

Fig. 10.38. Typical C-band ripple pattern.

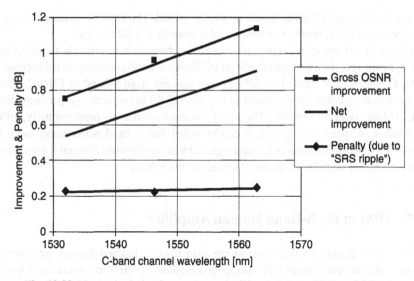

Fig. 10.39. Net transmission improvement in C-band due to addition of S-band.

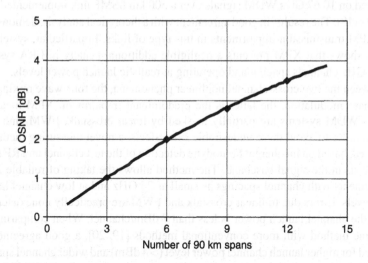

Fig. 10.40. OSNR improvement versus number of spans. S-band launch power is 18 dBm.

curve labeled net OSNR improvement corresponds to the gross OSNR improvement minus the penalty due to SRS dynamic crosstalk. It is interesting to observe that the BER penalty is relatively flat across the C-band. A roughly constant product of the ripple magnitude and spectral width can explain this result.

Finally, simulations are used to study the net improvement of transmission performance for an increasing number of 90 km spans. For instance, Fig. 10.40 plots the OSNR improvement in the C-band versus the number of 90 km spans when

S-band signals with 18 dBm launch power are added. The net OSNR improvement is proportional to the number of spans, and is roughly 0.3 dB per span.

Because of a large signal walk-off between the channels in both bands, the effect of SRS dynamic crosstalk between the S- and C-bands is small compared to the increase in the C-band OSNR caused by SRS energy transfer from S-band to C-band. The net result is an average improvement of roughly 0.7 dB in transmission performance in the C-band over three spans. The improvement should grow proportionally with the number of spans. Conversely, it can be noted that C-band performance would deteriorate with the addition of L-band signals by about the same amount it would be improved by the addition of identical signals in the S-band.

10.7. XPM in the S-Band Raman Amplifier

Because a small-core DCF is used as the gain fiber in lumped Raman amplifiers, another concern arises about XPM in the gain medium. Therefore, experiments were performed to assess the impact of XPM on the performance of WDM transmission using a cascade of lumped Raman amplifiers. In particular, XPM measurements were conducted on 10.67 Gb/s WDM signals over a 500 km SSMF link implemented with six LRAs [18]. The results, in good agreement with a theoretical analysis, demonstrate small XPM transmission impairments in this type of links. In particular, systematic analysis shows that XPM presents a negligible additional penalty in LRA systems with 50 GHz channels spacing and operating at realistic launch power levels.

Between the two cross-channel nonlinear phenomena, the four-wave mixing and cross-phase modulation, the latter is the predominant impairment. XPM measurements in WDM systems are partially masked by linear crosstalk, FWM, and ASE noise. In order to avoid these undesirable side effects, a novel measuring method is introduced, based on incoherent homodyne detection of the test channel and RF spectral analysis in the signal baseband. The method allows the taking of reliable XPM measurements with channel spacings as small as 25 GHz and at low channel launch power levels. Errors due to linear crosstalk and FWM are practically nonexistent, as long as the channel launch power is less than 7 dBm/channel. When comparing the homodyne method with more conventional methods [19, 20], a good agreement is measured for higher launch channel power level (>4 dBm) and wider channel spacing (≥50 GHz).

The set-up block diagram is shown in Fig. 10.41. Fourteen DFB lasers with wavelengths spanning the entire S-band (between 1493.36 and 1521.77 nm) and nominally spaced by 100 GHz are combined using a 4:1 coupler and a 100 GHz arrayed waveguide grating (AWG) multiplexer. The nominal test channel (probe) is surrounded with four "interfering" channels with adjustable wavelength capability in order to adjust for different channel spacing. Six dispersion-compensating LRAs are used to transmit WDM channels in the S-band over 5 × 100 km spans of SSMF, for a total length of 500 km. The probe channel launch power is maintained at 0 dBm, and the launch power of the rest of channels varies from 0 dBm to 6 dBm, depending on test scenarios.

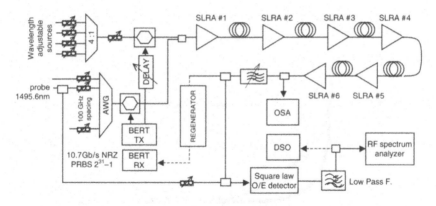

Fig. 10.41. Block diagram of experimental setup for XPM measurements.

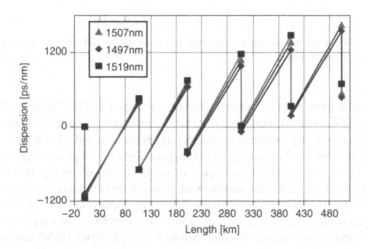

Fig. 10.42. Testbed dispersion map. The transmission line consists of five spans of standard single-mode fiber.

All of the channels are launched in parallel polarization and are modulated with a $2^{31} - 1$ PRBS NRZ pattern at 10.67 Gb/s through external LiNbO$_3$ modulators. Although the probe channel would not be normally modulated, the setup allows the modulation of the probe, as well as the direct measurement of the bit error rate penalties. The probe and the interfering channels are modulated by two different modulators, driven by the same pattern, but separated by an adjustable delay to test for the worst-case scenario [19]. All laser sources are dithered by a small internal modulation at 10 kHz in order to prevent a possible stimulated Brillouin scattering contamination at higher signal power levels.

The dispersion map is sketched in Fig. 10.42 for all three test wavelengths, which are at 1498, 1507, and 1519 nm. The line consists of five spans of SSMF, with a dispersion of about 15 ps/nm.km at 1510 nm.

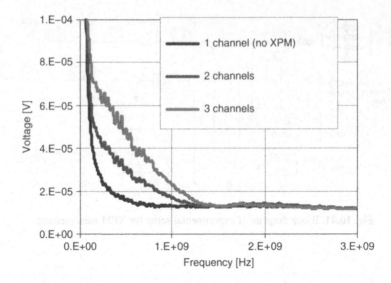

Fig. 10.43. Baseband RF spectrum of unmodulated probe for 50 GHz channel spacing and 4 dBm/channel launch power.

The homodyne receiver consists of the probe signal and the "local oscillator" combiner, O/E converter, a low-pass filter (8 GHz bandwidth), and an AC-coupled RF spectrum analyzer. By switching the interfering channels on and off, it is determined that more than 99.5% of the induced XPM appears within the frequency range from 30 MHz to 3 GHz. The baseband spectrogram of unmodulated probe in Fig. 10.43 shows three spectral density lines: the baseline with no interfering channels, a spectrum with one interfering channel, and a spectrum with two interfering channels. The numerical integration of the spectral density lines within the entire baseband frequency range yields the rms "noise" power with and without XPM. The signal OSNR and average power are recorded for each measuring point in order to determine the normalized XPM "noise" variance.

For analysis of the experimental results, the approach explained in [20] is followed. In the absence of FWM and for a CW probe channel, the following simple relations hold with a good accuracy.

$$\sigma^2 = \sigma_1^2 + \sigma_{XPM}^2$$

$$\sigma_{XN}^2 = \left(\frac{\sigma_{XPM}}{S}\right)^2 \cong \left(\frac{\sigma_{XPM}}{\sigma_1}\right)^2 \times \frac{1}{OSNR} \times \frac{B_e}{B_o}$$

$$\Delta Q[dB] = 20 \log\left(\frac{Q}{Q_{XPM}}\right) = 10 \log\left(\sigma_{XN}^2 \times Q^2 + 1\right).$$

Here the total noise variance σ^2 is a linear sum of noise variances σ_1^2 (signal–spontaneous noise is the predominant one) and signal fluctuations due to XPM, denoted as σ_{XPM}. S corresponds to the signal baseband power. The XPM "noise" variance σ_{XN}, normalized with respect to the baseband signal power, can be calculated either numerically or analytically [21]. B_o and B_e denote the receiver's optical and electrical bandwidths, respectively. ΔQ is the system penalty, defined as the ratio of Q-values without and with XPM; Q and Q_{XPM} are system Q-values without and with XPM.

Commercial system design software is used to simulate a number of test scenarios, and to numerically compute the XPM-induced signal fluctuations. The simulation assumes linearly polarized and polarization-aligned channels.

The measurements are taken at the beginning, at the middle, and at the end of the 30 nm wide band. The amount of XPM crosstalk remains unchanged, to within the accuracy of measurements, across the band. Hence, for simplicity, only the results measured at the beginning of the band are presented.

Figure 10.44 depicts the results for 50 GHz channel spacing and for different signal power values. The square marks in the diagram present the measurements taken by the direct BER measurement method. The solid line curves are obtained by numerical simulation. The good agreement between the numerical and experimental results indicates that the channels suffered little relative polarization rotation throughout the 500 km long testbed, an indication of a true worst-case scenario test.

Figure 10.45 summarizes the experimental data for different channel spacings. There is good agreement between numerical analysis of the test scenarios and the experiment. Based on this agreement, the XPM simulations are extended to a 15 span LRA link. The results show that for the nominal channel launch power of 0 dBm, the XPM Q penalty should be less than 0.5 dB.

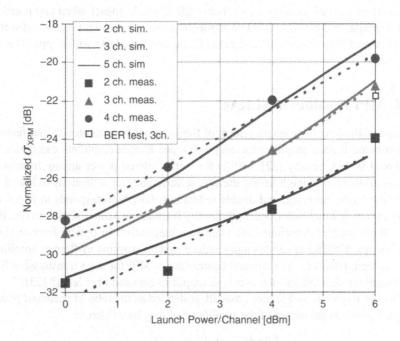

Fig. 10.44. Measured and simulated XPM noise variance over 500 km for 50 GHz channel spacing.

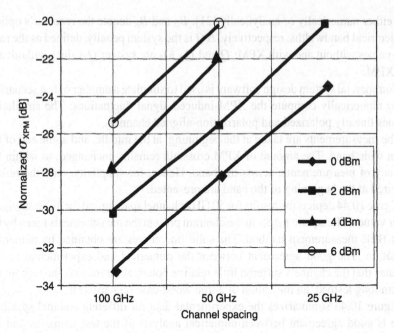

Fig. 10.45. XPM total system penalty versus channel spacing for various channel powers.

From Figs. 10.44 and 10.45, it can be seen that for channel spacings down to 50 GHz and realistic system conditions with launch channel power levels of <2 dBm, the worst-case XPM penalty is less than 1 dB (a small impact when compared to SNR degradation due to amplified spontaneous noise). Therefore it is concluded that XPM is not a limiting factor in the design of long terrestrial links with lumped Raman amplifiers.

10.8. MPI Penalties in SLRAs

Because of the long fiber length typical of Raman amplifiers, multipath interference in distributed Raman amplification and in lumped Raman amplifiers can create an additional system penalty [22]. MPI is forward scattered power arising from such sources as double Rayleigh backscattering, splice/connector reflections followed by single Rayleigh scattering, and double reflections. DRBS corresponds to two scattering events in which subwavelength density fluctuations scatter signal waves. Because Rayleigh loss constitutes the main loss mechanism in optical fiber at short wavelengths, DRBS can significantly reduce system margins in Raman amplified links. Indeed, DRBS is the dominant contributor to MPI for most transmission link configurations, and DRBS increases with amplifier gain and fiber length [23].

Double Rayleigh backscatter crosstalk is defined as the ratio of scattered power to signal power at the output of the amplifier and may be written as

$$\text{DRBS crosstalk} = \frac{P_{DRBS}}{P_s} = k^2 \int_0^L dx\, G^{-2}(x) \int_x^L dy\, G^2(y). \qquad (10.1)$$

The backscatter coefficient k is the product of the capture coefficient S and the Rayleigh loss coefficient α_r at the signal wavelength. In Eq. (10.1), P_{DRBS} is the average DRBS output power, P_s the average signal output power, and $G(z)$ the net gain of the amplifier as a function of position. For the case of a single backward propagating undepleted Raman pump, the net gain $G(z)$ is

$$G(z) = \exp\left[-\alpha_s z + \frac{g_0}{\alpha_p}\left(e^{\alpha_p(z-L)} - e^{-\alpha_p L}\right)\right]$$

$$g_0 = \frac{\ln(G(L)e^{\alpha_s L})}{L_{eff}}, \qquad L_{eff} = \frac{1-e^{-\alpha_p L}}{\alpha_p}, \qquad (10.2)$$

where $G(L)$ is the net gain at the signal output end of the fiber. Using Eq. (10.2), the double integral in Eq. (10.1) can be evaluated after making an additional approximation of constant gain to obtain

$$\text{DRBS crosstalk} = \left(\frac{kL}{2\ln(G(L))}\right)^2 \left[G(L)^2 - 1 - 2\ln(G(L))\right]. \qquad (10.3)$$

Measurements of DRBS crosstalk have been made and the results are compared to predictions from Eq. (10.3) [23]. A calibrated electrical spectrum analyzer is used to make MPI spectral measurements, as described by Fludger and Mears [24]. Background noise due to thermal noise, amplified spontaneous emission, and shot noise is subtracted from the measured spectra to obtain the MPI spectra, as shown in Fig. 10.46. The spikes evident in the spectra at low MPI levels arise from electrical noise on the current driver for the distributed feedback (DFB) laser diode used in the measurement. Calibration of the electrical spectrum analyzer (i.e., conversion of the results from the electrical to optical domain) is accomplished by using a filtered ASE source with known relative intensity noise (RIN) spectrum. After conversion to the optical domain, the result is integrated over the frequency range from 0.5 to 100 MHz to obtain the MPI noise variance. A continuous-wave DFB laser source at 1510.4 nm with a RIN less than -150 dB/Hz at frequencies greater than 0.5 MHz is used. A delayed self-homodyne interferometer is used to calculate a transfer function relating MPI crosstalk values to MPI integrated noise values shown in Fig. 10.46.

Figure 10.47 compares measured and calculated DRBS crosstalk for a range of amplifier net gain values. Experimental MPI data are measured for a single-stage S-band Raman amplifier, a two-stage S-band Raman amplifier, and an optical link of five 80 km spans of standard single-mode fiber (25 dB loss in each span) with six cascaded two-stage S-band Raman amplifiers. Figure 10.47 also shows calculated MPI data and demonstrates good agreement between calculated and measured results. Data in Fig. 10.47 show that for a stage-gain of up to 14 dB, MPI crosstalk in all configurations is less than -30 dB.

System performance is represented by the quality factor $Q = (I_1 - I_0)/(\sigma_1 + \sigma_0)$, where I_1 and I_0 are the mean detector currents for marks and spaces in the data stream; σ_1 and σ_0 represent standard deviations of the mark and space current distributions,

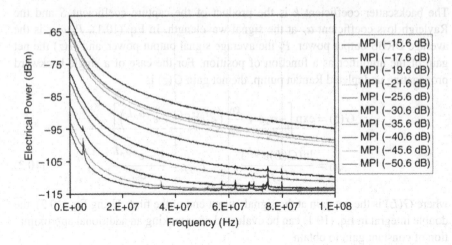

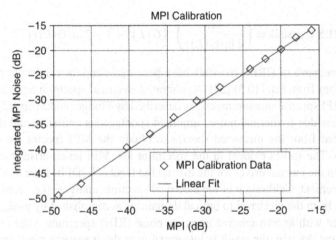

Fig. 10.46. Characteristic MPI power spectra (left plot) for different values of MPI crosstalk and MPI crosstalk versus integrated MPI noise (right plot).

respectively. MPI noise is added to other noise sources such as signal–spontaneous beat noise and thermal noise so that

$$\sigma_1^2 = \sigma_{s-sp}^2 + \sigma_{th}^2 + \sigma_{mpi}^2,$$

where σ_{mpi}^2 is proportional to the total MPI crosstalk:

$$\sigma_{mpi}^2 = 0.5 * d * Xtalk * I_1^2.$$

Here, $Xtalk$ stands for DRBS crosstalk and the factor of 0.5 accounts for the fact that for unpolarized MPI noise, only half of the MPI power beats with the signal in the receiver. The duty cycle d is equal to 1 (0.5) for the NRZ (RZ) data format. If, for

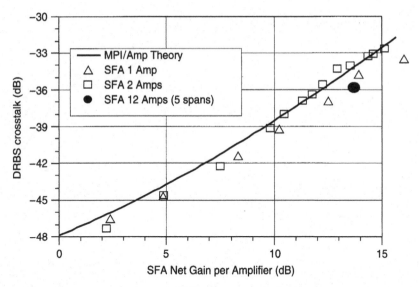

Fig. 10.47. DRBS crosstalk as a function of net gain in an S-band amplifier. Solid line: calculated DRBS crosstalk for a single-stage amplifier. Points: measured DRBS crosstalk for a single-stage amplifier (triangles), a two-stage amplifier (squares), and a five-span (25 dB) link with six double-stage amplifiers with about 12.5 dB gain/amplifier. The Rayleigh scatter parameter for dispersion-compensation fiber is chosen to be 7.1×10^{-4} km^{-1} and fiber loss at the pump wavelength is 0.65 dB/km. For the two-stage and link cases, the ordinate represents the MPI crosstalk per stage.

simplicity, I_0 and σ_0 are neglected, the MPI penalty for an NRZ system becomes:

$$\text{MPI Penalty} = 20 \log_{10} \left[\frac{Q(Xtalk)}{Q_{base}} \right]$$

$$Q(Xtalk) = \frac{I_1}{\sqrt{\sigma_1^2 + 0.5 * Xtalk * I_1^2}}, \qquad Q_{base} = \frac{I_1}{\sigma_1},$$

where Q_{base} denotes the "baseline" quality factor, that is, the quality factor without MPI.

Figure 10.48 shows variation of the system penalty with MPI crosstalk for three different values of Q_{base}. The MPI penalty for a 10-span system with lumped Raman amplifiers and average span loss of 23 dB is about -34.6 dB per stage. MPI crosstalk for multispan systems accumulates linearly with the number of amplifiers. Thus for a 10-span system with 11 two-stage amplifiers the total MPI crosstalk is about -21.2 dB. Accounting for the finite bandwidth of MPI noise due to signal modulation (10 Gb/s) and the electrical bandwidth of the system (7 GHz) decreases the MPI crosstalk values by about 0.5 dB. Thus MPI penalty for a 10-span system is only 0.5 dB$_{20}$ according to the curve of $Q_{base} = 15.5$ dB$_{20}$ (corresponding to a Q of 6 in linear units and a bit-error rate of 10^{-9}) in Fig. 10.48. This result is consistent with the findings of Puc et al. [12], where the worst-case total system penalty resulting from

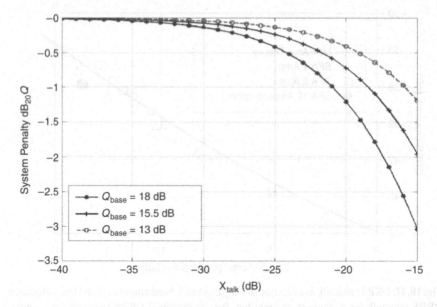

Fig. 10.48. Variation of system penalty with MPI crosstalk for three different values of baseline quality factor Q_{base}.

WDM transmission over 867 km standard single-mode fiber using 11 cascaded S-band lumped Raman amplifiers was measured and found to be lower than 1 dB.

10.9. Other S-Band Experiments

Raman amplification has been used to extend the bandwidth of TDFAs. Kani and Jinno [26] demonstrated a hybrid TDFA/Raman amplifier with a 2 dB gain bandwidth of 50 nm. The amplifier comprised a TDFA followed by a two-stage Raman amplifier pumped at 1415 nm (Fig. 10.49(a)). Gain and noise figure spectra are shown in Fig. 10.49(b). It can be seen from the figure that gain greater than 25 dB and a noise figure less than 6 dB could be achieved from 1460 to 1510 nm. These results were obtained with a single-channel measurement using −30 dBm input power.

In another experiment on hybrid TDFA/Raman amplifiers, Masum-Thomas et al. [25] demonstrated a bandwidth of 70 nm. The amplifier comprised a TDFA in series with a Raman amplifier pumped at 1413 nm, the Raman amplifier providing gain at longer wavelengths than the TDFA gain band. Gain and noise figure spectra are shown in Fig. 10.50 for two configurations, one in which the TDFA precedes the Raman amplifier and another in which the Raman amplifier precedes the TDFA. In both cases, gain greater than 15 dB is achieved over the wavelength range spanning 1445 to 1520 nm. As indicated by the figure, the noise figure in this hybrid configuration using a single Raman pump wavelength depends on whether the Raman amplifier precedes or is preceded by the doped-fiber amplifier. Where the TDFA precedes the Raman amplifier, the high noise figure of the TDFA dominates the overall noise figure, which is higher than the noise figure in the configuration where the Raman amplifier

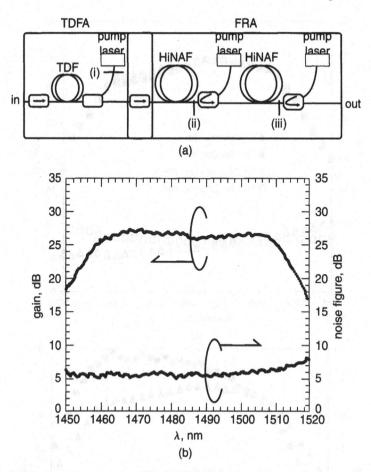

Fig. 10.49. Configuration (a) and gain and noise figure spectra (b) of a hybrid TDFA/Raman amplifier reported by [26]. Source: J. Kani and M. Jinno "Wideband and Flat-Gain Optical Amplification from 1460 to 1510 nm by Serial Combination of a Thulium-Doped Fluoride Fiber Amplifier and Fiber Raman Amplifier" IEEE Electronics Letters, Vol 35, Issue 12 (© 1999 IEEE)

is the first stage. When the Raman amplifier is in the first stage, net noise figures as low as 5.5 dB could be obtained at long wavelengths. Also shown in the figure are results for co- and counterpumping of the Raman stage in the case where the Raman amplifier is the first stage. Copumping yields channel gains ranging from 14 to 39 dB over the 1445 to 1520 nm wavelength band, and counterpumping yields a lower gain variation over the band, from 18 to 34 dB.

An experiment employing only Raman amplification in the S band was reported by Kani et al. [27]. Raman amplification in the wavelength range spanning 1490 and 1530 nm is demonstrated using an amplifier configuration comprising two stages of germania-doped fiber of total length 10 km, a pump laser operating at 1420 nm, and an interstage isolator to minimize DRBS (Fig. 10.51). Figure 10.52 shows the gain and noise figure spectra for −30 dBm input signal power and 750 mW pump power.

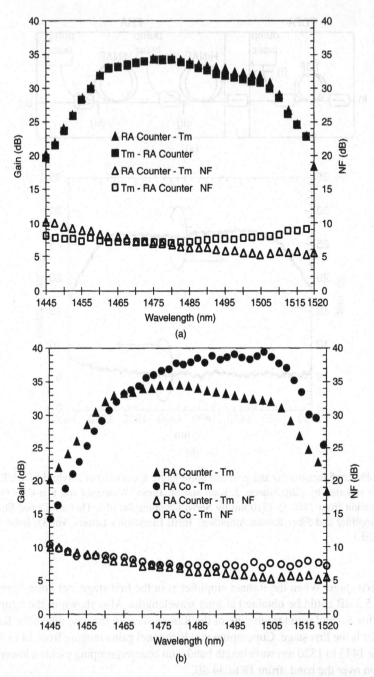

Fig. 10.50. Gain (upper curves) and noise figure (lower curves) spectra for hybrid TDFA/Raman amplifiers reported in [25]: (a) spectra for counterpumped Raman amplifier preceding TDFA (triangles) and preceded by TDFA (squares); (b) spectra for counterpumped (triangles) and copumped (circles) Raman amplifier preceding the TDFA. Source: J. Masum-Thomas, D. Cappa, A. Moroney: "A 70 nm Wide S-Band Amplifier by Cascading TDFA and Raman Fiber Amplifier" OFC Technical Digest, Postconference Edition, pg WDD9 (© 2001 OSA)

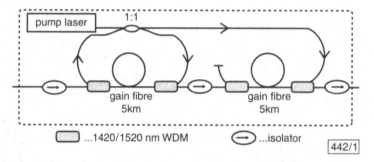

Fig. 10.51. Configuration of all-Raman S-band amplifier reported in [27]. Source: J. Kani, M. Jinno, K. Oguchi: "Fiber Raman Amplifier for 1520 nm Band WDM Transmission" IEEE Electronics Letters, Vol. 34, Issue 17m 3 September 1998 (© 1998 IEEE)

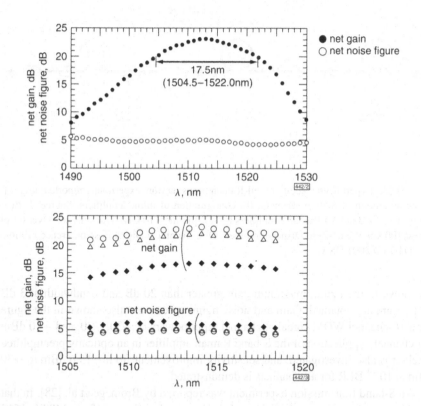

Fig. 10.52. Gain and noise figure spectra for −30 dBm input signal power (upper plot) and for 16-channel WDM signal (lower plot) [27]. Source: J. Kani, M. Jinno, K. Oguchi: "Fiber Raman Amplifier for 1520 nm Band WDM Transmission" IEEE Electronics Letters, Vol. 34, Issue 17m 3 September 1998 (© 1998 IEEE)

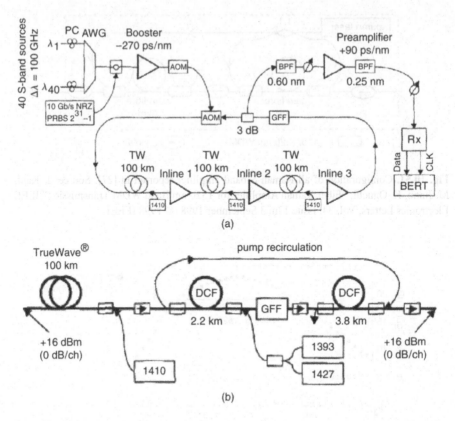

Fig. 10.53. Experimental setup of all-Raman transmission experiment reported in [28]: (a) configuration of loop experiment; (b) configuration of inline amplifiers. Source: J. Bromage et al. "S-Band All-Raman Amplifiers for 40/spl times/ 10 Gb/s Transmission Over 6 /spl times/ 100 km of Non-Zero Dispersion Fiber" OFC Technical Digest Postconference Edition, pg PD4-1 (© 2001 OSA)

As shown in the figure, maximum gain greater than 20 dB and bandwidth at 3 dB of 17.5 nm are obtained. Gain and noise figure spectra are also shown in the figure for a 16-channel WDM signal at channel input powers of −30, −20, and −10 dBm per channel. Application of the S-band Raman amplifier in an optically preamplified receiver is also investigated, and sensitivity improvement from −15.6 dBm to −30 dBm at 10^{-9} BER for all channels is demonstrated.

An S-band transmission experiment was reported by Bromage et al. [28]. In that experiment, transmission of 40 channels in the wavelength range from 1488 to 1518 nm was carried out over two laps of a 300 km nondispersion-shifted fiber loop. The experimental setup, shown in Fig. 10.53(a), includes three Raman amplified spans of 100 km long nondispersion-shifted fiber with a zero dispersion wavelength lower than that of conventional dispersion-shifted fibers. In addition, there are three inline amplifiers comprising dispersion-compensating Raman amplifiers, as well as a dispersion-compensating booster amplifier at the transmitter and a dispersion-compensating preamplifier at the receiver. The line fiber is pumped close to transparency, and the

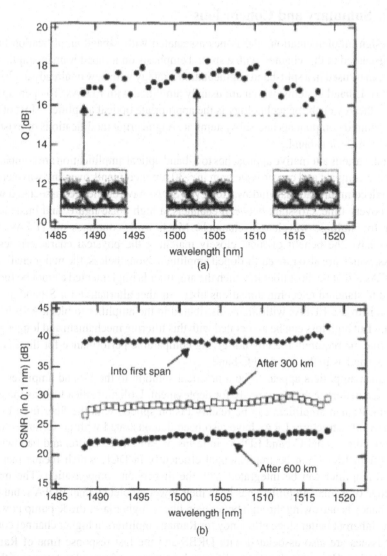

Fig. 10.54. (a) Q and (b) OSNR spectra after 600 km transmission [28]. Source: J. Bromage et al. "S-Band All-Raman Amplifiers for 40/spl times/ 10 Gb/s Transmission Over 6 /spl times/ 100 km of Non-Zero Dispersion Fiber" OFC Technical Digest Postconference Edition, pg PD4-1 (© 2001 OSA)

dispersion-compensating amplifiers, configured as two-stage amplifiers with mid-stage isolators to minimize DRBS (Fig. 10.53(b)), compensate for residual span loss. The Q-spectra and optical signal-to-noise ratio at the end of the 600 km transmission are illustrated in Fig. 10.54. The inline amplifiers and preamplifiers provide 95% span compensation with a residual 220 ps/nm dispersion across the signal band after 600 km. Optical signal-to-noise ratios greater than 20 dB are obtained for all channels, and the worst-case Q value is 16 dB. Because line fiber with a low zero dispersion wavelength is used, no significant four-wave mixing sidebands are observed.

10.10. Summary and Conclusions

The design, implementation, and issues associated with S-band amplification have been discussed in this chapter, with a special emphasis on lumped Raman amplifiers. LRAs can be used in a split-band augmentation strategy with new or already deployed C- and/or L-band systems, which are usually amplified with EDFAs. To open up the S-band, the key enabling technology is the appropriate optical amplifier. Most of the other required components and subsystems are engineering modifications of existing parts in the C- or L-band.

First, various alternative approaches to S-band optical amplification are compared. SOAs are compact and can provide gain in ~30 nm increments over any wavelength in the telecommunications window. However, SOAs have challenges associated with fast recovery time, crosstalk between channels, a high noise figure, and inadequate power for multiple wavelength channels. Much work continues on TDFAs, and TDFAs have the benefit of most closely matching the physical characteristics of EDFAs, which are also rare earth-doped amplifiers. Nonetheless, the major challenge of TDFAs is that the host fiber in which the amplifier is implemented cannot be fusion spliced to standard telecommunications fiber. Another alternative for S-band amplification is to use EDFAs with filters distributed in the amplifier to block ASE in the C-band. But high loss can be associated with this filtering mechanism and longer gain fibers may be required, thereby not giving comparable performance for the EDFAs in the S-band as in the standard C-band.

Raman amplifiers appear to be a practical solution to the S-band amplifier, and they are a mature technology ready for deployment. Utilizing silica fiber as the gain medium, Raman amplifiers can be readily fusion spliced with the fiber used in the transport infrastructure. LRAs have also been demonstrated with performance on a par with commercial C-band EDFAs in terms of gain, noise figure, and bandwidth. In addition, LRAs can be implemented efficiently in DCF, which means that the lumped amplifier can be integrated with the dispersion compensation. The major challenge of Raman amplifiers has been their lower efficiency than EDFA's, but this discrepancy is narrowing through better gain fibers, higher laser diode pump powers, and the inherent better slope efficiency for Raman amplifiers at higher channel count. Other issues are also associated with DRBS and the fast response time of Raman amplification. However, simple amplifier architectures can be used to overcome these limitations. By using counterpropagating pumping, a longer effective upper-state lifetime corresponding to the transit time can be implemented. In addition, by splitting the amplifier into a multiple-stage amplifier and using isolation between stages, the DRBS penalty can be controlled.

The bulk of this chapter focuses on the issues and experimental demonstration of SLRAs in fiber-optic transmission systems. To start with, in Section 10.4 a detailed system experiment is described that uses cascaded SLRAs. Then, Sections 10.5 through 10.8 study various key issues associated with practical deployment of SLRAs in an augmentation strategy with C-band systems.

The system experiment of Section 10.4 uses SLRAs with integrated dispersion compensation to transmit 20 NRZ channels over the entire S-band over 10 spans of SSMF. The S-band channels are modulated at 10.67 Gb/s, and they are transmitted over 867 km of SSMF. Each channel achieves a BER $<10^{-12}$ without FEC, and there is negligible dispersion penalty and negligible four-wave mixing penalty measured. The margins accumulated in this demonstration show the capability for such as system to achieve 80-channel transmission over 10×25 dB SSMF with standard OOB-FEC.

Section 10.5 dissects in more detail the amplifier characteristics and control of the SLRAs used in the above system experiment. The gain fiber is chosen to provide both dispersion and dispersion slope compensation, and accurate modeling of the amplifier performance requires detailed knowledge of the loss, gain, and saturation as a function of wavelength. Also, to control the performance of the two-stage SLRA, the amplifier unit is equipped with a closed loop control system. The closed loop control acting on the pump laser powers can be set for various conditions such as constant gain, constant output power, and provisional gain or power tilt. Moreover, since a polarization-independent amplifier is desired, a PDG <0.05 dB can be obtained for pump polarizations $<20\%$. Finally, the stimulated Brillouin scattering threshold is found to be more than 10 dB above the typically launched signal channel powers.

When used in an augmentation strategy, the existing C-band will be affected by the addition of the S-band through the interband SRS, which is studied in Section 10.6. The interband SRS is quantified experimentally, and it is confirmed that the addition of the S-band channels improves transmission performance in the C-band. Also, because of a large walk-off between the channels in both bands, the effect of SRS dynamic crosstalk between the S- and C-bands is small compared to the increase in the C-band OSNR caused by the SRS energy transfer. For example, in a three-span link an overall improvement of ~ 0.7 dB in transmission performance is measured for the C-band signals, and the improvement grows proportionally with the number of spans.

Section 10.7 addresses the concern that there may be undesirable nonlinear transmission penalties arising from the gain fiber in a SLRA, which is a small-core size DCF. XPM is the largest nonlinear effect, and XPM measurements were conducted on 10.67 Gb/s WDM signals over a 500 km SSMF link implemented with six SLRAs. Systematic analysis shows that XPM presents a negligible additional penalty in LRA systems with 50 GHz channel spacing and operating at realistic launch power levels.

The last concern addressed is the MPI penalties in LRAs. DRBS is the dominant contributor to MPI for most transmission link configurations, and DRBS increases with amplifier gain and fiber length. It is found that the MPI penalty for a 10-span system with LRAs and an average span loss of 23 dB is about -34.6 dB per stage. Because the MPI crosstalk for multispan systems accumulates linearly with the number of amplifiers, for the 10-span system with 11 LRAs the total MPI crosstalk is about -21.2 dB. This corresponds to a penalty of only about 0.5 dB$_{20}$ for a Q_{base} of 15.5 dB$_{20}$.

Finally, in Section 10.9 other S-band Raman amplifier and system experiments are briefly reviewed. For example, hybrid TDFA/Raman amplifiers have been demonstrated with a bandwidth between 50 and 70 nm. Another experiment uses an all-Raman, two-stage, lumped amplifier over the wavelength range between 1490 and 1530 nm. In addition, an S-band transmission experiment with a SLRA is demonstrated using 40 channels in the wavelength range between 1488 and 1518 nm that is carried over 600 km of nondispersion-shifted fiber. Optical signal-to-noise ratios greater than 20 dB are obtained for all channels, and the worst-case Q value is 16 dB.

As the experiments on S-band amplifiers demonstrate, the SLRA appears ready for commercial deployment, and most of the fundamental systems issues have been addressed. There appear not to be any fundamental limitations that block the opening up of the S-band. Rather, the issue is probably more of a practical and economic one. Is it better to augment a C- and/or L-band system with the S-band, or is it better to use a wideband, all-Raman amplified system (c.f. Chapter 14)?

Although the answer to this question is not straightforward, there may be at least three scenarios where the S-band LRAs prove out for system deployment. First, when the goal is to upgrade an already deployed C- and/or L-band system, the split-band configuration with SLRAs can be attractive, particularly if the triband couplers are in place. Also, in systems that have particular difficulty in using the C-band for dense WDM, it may be better to shift transmission to the S-band. For example, it would be much easier to use the S-band for dense WDM systems when the fiber is dispersion-shifted fiber, which has the zero dispersion wavelength right in the middle of the C-band. Finally, for systems using older vintage fiber with large variation and mean bend-induced loss in the L-band, using a combination of the S- and C-bands should be advantageous.

There may also be completely new uses of the S-band beyond simple fiber-optic transport for increasing capacity. A different application of the S-band might be if a traffic-bearing fiber were to be used as a backup path for another fiber [29]. For example, consider a system with traffic only in the C-band. Rather than using an idle fiber as the backup or redundant path to the fiber, another traffic-bearing fiber that supports both the C- and S-bands can serve as the redundant path. If the original C-band traffic-bearing fiber is disabled, the data can be translated to the S-band and augment the C-band traffic on the backup fiber. This type of application would require identical performance of the S-band as in the C-band, which has been demonstrated for the SLRAs.

Acknowledgements

Special thanks to the Xtera staff for their excellent work on the S-band Raman amplifier, including A. B. Puc, M. W. Chbat, J. D. Henrie, N. A. Weaver, H. Kim, A. Kaminski, A. Rahman, F. Barthelemy, and S. Burtsev.

References

[1] M. Islam and M. Nietubyc, *WDM Solutions* 3: 53, 2001.
[2] M.-C. Ho, M.E. Marhic, Y. Akasaka, and L.G. Kazovsky. In *Proceedings of CLEO2000*, CThC6 (San Francisco, May), 2000.
[3] Y. Akasaka, K. K. Y. Wong, M.-C. Ho, M. E. Marhic, and L. G. Kazovsky. In *OSA Trends in Optics and Photonics (TOPS)* Vol. 54, *Optical Fiber Communication Conference*, Technical Digest, Postconference Edition, Washington DC: Optical Society of America, WDD31, 2001.
[4] L. H. Spiekman, G.N. van den Hoven, T. van Dongen, M. J. H. Sander-Jochem, J. H. H. M. Kemperman, and J. J. M. Binsma. In *Proceedings of the 26th European Conference on Optical Communication 2000*, Berlin: VDE Verlag, vol.1, 35, 2000.

[5] B. N. Samson, D. T. Walton, A. J. G. Ellison, J. D. Minelly, J. P. Trentleman, J. E. Dickinson, and N. J. Traynor. In *Optical Amplifiers and their Applications*, OSA Technical Digest, Washington DC: Optical Society of America, PDP-6, 2000.

[6] S. Aozasa, K. Hoshino, T. Kanamori, K. Kobayashi, T. Sakamoto, and M. Shimizu, *IEEE Photon. Technol. Lett.*, **12**: 1331, 2000.

[7] E. Desurvire, *Erbium-Doped Fiber Amplifiers Principles and Applications*, New York: Wiley, 1994.

[8] Y. Sun, A. K. Srivastava, J. Zhou, and J. W. Sulhoff, *Bell Labs Tech. J.* **4**: 187, 1999.

[9] E. Ishikawa, M. Nishihara, Y. Sato, C. Ohshima, Y. Sugaya, and J. Kumasako. In *Proceedings of ECOC2001* (Amsterdam), PD.A.1.2, 2001; M. Nishihara, Y. Sugaya, and E. Ishikawa. In *OSA Trends in Optics and Photonics (TOPS)* Vol. 77, *Optical Amplifiers and Their Applications*, OSA Technical Digest, Postconference Edition, Washington DC: Optical Society of America, OWB4-1, 2002.

[10] M. A. Abore, Y. Zhou, G. Keaton, and T. Kane. In *OSA Trends in Optics and Photonics (TOPS)* Vol. 77, *Optical Amplifiers and Their Applications*, OSA Technical Digest, Postconference Edition, Washington DC: Optical Society of America, PD4-1, 2002.

[11] A. Mathur, M. Ziari, and V. Dominic. In *Optical Fiber Communication Conference*, Technical Digest, Postconference Edition, Washington DC: Optical Society of America, PD-15, 2000.

[12] A. B. Puc, M. W. Chbat, J. D. Henrie, N. A. Weaver, H. Kim, A. Kaminski, A. Rahman, and H. Fevrier. In *OSA Trends in Optics and Photonics (TOPS)* Vol. 54, *Optical Fiber Communication Conference*, Technical Digest, Postconference Edition, Washington DC: Optical Society of America, PD39-1, 2001.

[13] P. Gavrilovic. In *Proceedings of the 14th Annual Meeting of the IEEE Lasers and Electro-Optics Society* (Cat. No. 01CH37242) (IEEE, Piscataway, NJ), 471, 2001.

[14] A. Puc, F. Barthelemy, M. Chbat, and H. Kim. In *Proceedings of the 14th Annual Meeting of the IEEE Lasers and Electro-Optics Society* (Cat. No. 01CH37242) (IEEE, Piscataway, NJ), 413, 2001.

[15] V. J. Mazurczyk, G. Shaulov, and E. A. Golovchenko, *Photon. Technol. Lett.*, **12**: 1573, 2000.

[16] S. Bigo, S. Gauchard, A. Bertaina, and J.-P. Harmaide, *Photon. Technol. Lett.*, **11**: 671, 1999.

[17] A. G. Grandpierre, D. N. Christodoulides, W. E. Schiesser, C. M. McIntosh, and J. Toulouse. In *OSA Trends in Optics and Photonics (TOPS), Nonlinear Guided Waves and Their Applications Topical Meeting*, Technical Digest, Postconference Edition, Washington DC: Optical Society of America, MC 91, 2001.

[18] A. Puc, F. Barthelemy, H. Kim, M. W. Chbat, S. Burtsev, and Ned A. Weaver, unpublished.

[19] S. Bigo, G. Bellotti, and M. W. Chbat, *IEEE Photon. Technol. Lett.*, **11**: 605, 1999.

[20] S. Burtsev, R. Abramov, J. Hurley, S. Kumar, D. Lambert, and G. Luther. In *OSA Trends in Optics and Photonics (TOPS)* Vol. 54, *Optical Fiber Communication Conference*, Technical Digest, Postconference Edition, Washington DC: Optical Society of America, MF2-1, 2001.

[21] R. Hui, K.R. Demarest, and C.T. Allen, *J. Lightwave Technol.* **17**: 1018, 1990.

[22] S.A.E. Lewis, S.V. Chernikov, and J.R. Taylor, *IEEE Photon. Technol. Lett.*, **12**: 528, 2000.

[23] S. Burtsev, W. Pelouch, and P. Gavrilovic. In *OSA Trends in Optics and Photonics (TOPS)* Vol. 70, *Optical Fiber Communication Conference*, Technical Digest, Postconference Edition, Washington DC: Optical Society of America, 120, 2002.

[24] C. R.S. Fludger and R.J. Mears, *J. Lightwave Technol.* **19**: 536, 2001.

[25] J. Masum-Thomas, D. Crippa, A. Maroney. In *OSA Trends in Optics and Photonics (TOPS)* Vol. 54, *Optical Fiber Communication Conference*, Technical Digest, Postconference Edition, Washington DC: Optical Society of America, WDD9-1, 2001.

[26] J. Kani and M. Jinno, *Electron. Lett.*, **35**: 1004, 1999.

[27] J. Kani, M. Jinno, and K. Oguchi, *Electron. Lett.*, **34**: 1745, 1998.

[28] J. Bromage, J.-C. Bouteiller, H. J. Thiel, K. Brar, J. H. Park, C. Headley, L. E. Nelson, Y. Qian, J. DeMarco, S. Stulz, L. Leng, B. Zhu, and B. J. Eggleton. In *OSA Trends in Optics and Photonics (TOPS)* Vol. 54, *Optical Fiber Communication Conference*, Technical Digest, Postconference Edition, Washington DC: Optical Society of America, PD4-1, 2001.

[29] M. N. Islam and O. Boyraz, *IEEE J. Select. Topics Quantum Electron.*, **8**: 527, 2002.

Chapter 11

Raman Fiber Lasers

C. Headley, M. Mermelstein, and J.-C. Bouteiller

11.1. Introduction

The use of stimulated Raman scattering (SRS) as a means of amplifying signals in telecommunication systems has been demonstrated since 1976 [1]. Yet despite its advantages over erbium-doped fiber, Raman amplification was not used in the first generation of deployed optically amplified systems. One of the principal reasons for this was the lack of reliable high-power pump sources needed for Raman amplification. It was in this environment that the cascaded Raman fiber laser (RFL) was invented.

A cascaded RFL uses stimulated Raman scattering (SRS) in optical fibers to shift the wavelength of light from an input pump laser to another desired wavelength. Devices at almost any wavelength can be made by proper choice of a pump wavelength, and by cascading the pump through several Raman shifts. Although RFLs had been demonstrated since the 1970s [2–9], the advent of fiber Bragg gratings (FBG) [10, 11] made the devices practical [12, 13]. Following the initial demonstration, there have been numerous experimental [13–18] and theoretical results [19–23].

A broadband flat Raman gain profile can be obtained using multiple pump wavelengths [24, 25]. It is advantageous to have all the required wavelengths emitted from one source. This motivated the development of multiple-wavelength RFLs [26–34]. Single cavities simultaneously lasing from two to six wavelengths have been demonstrated. These cavities have the ability to distribute the total output power among the different wavelengths to desired values.

Finally, distributed Raman amplification techniques have become more sophisticated with the proposed use of higher-order pumping schemes [35–41]. These entail launching two or more pump wavelengths separated by approximately one Stokes shift from each other. This improves the system performance by more uniformly distributing the gain along the fiber. The use of a RFL is especially suited to this application because large amounts of power are required at the shortest wavelength. In response to this application multiple-order RFLs have been proposed [9, 39, 41].

This chapter describes a cascaded RFL in detail. In the next section a single-wavelength RFL is discussed both theoretically and experimentally. Section 11.3

examines multiple-wavelength RFL, and finally Section 11.4 looks at multiple-order pump sources.

11.2. Single-Wavelength Raman Fiber Lasers

11.2.1. Overview

A schematic of a complete RFL pump module is shown in Fig. 11.1. It has three parts. The first is a set of multimode 9XX nm diodes that are the optical pumps [42]. The light from the diodes is coupled into a single multimode fiber with a tapered fiber bundle. This and alternative coupling approaches are discussed shortly [43–46]. The next section of the RFL pump module is a rare earth-doped cladding-pumped fiber laser (CPFL) [47–49], which converts the multimode diode light into single-mode light at another wavelength. In the final section the single-mode light is converted to the desired wavelength by a cascaded RFL. In this chapter, the whole device, including the diodes and CPFL, is referred to as a RFL pump module, with cascaded RFL (boxed section in Fig. 11.1) used exclusively for the wavelength-converting part of the RFL. Early on in the literature the cascaded RFL was sometimes referred to as a cascaded Raman resonator [12, 13]. In this section, a single-wavelength RFL module is described.

Diodes that are coupled to multimode fibers typically emit light over a larger facet area compared to those designed to couple into single-mode fibers. Significantly more power can therefore be extracted from them. Multimode diodes capable of emitting 1 to 3 W in the 9XX nm region, from a 100 μm diameter fiber core with 0.22 NA (numerical aperture) are typically used as the optical pumps in a RFL module [42]. The lower facet intensity increases the yield in the manufacturing of these diodes thereby lowering their cost. The reduced thermal stress on the diodes due to the lower intensity also means that these diodes have the potential to be run without thermal electric cooling. This reduces electrical power consumption and provides a saving in operational costs.

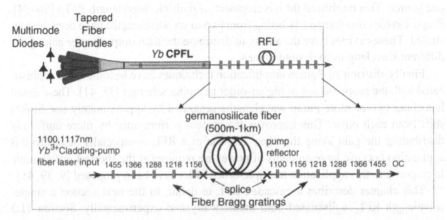

Fig. 11.1. Exemplary schematic of a RFL pump module with the cascaded RFL highlighted.

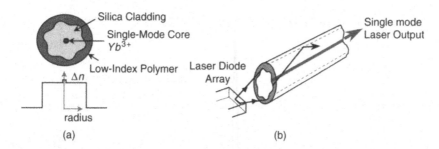

Fig. 11.2. (a) Shape and index profile of a CPF; and (b) schematic of light propagating down the fiber.

The light from the multimode diode is next coupled into a CPFL. Several different approaches have been presented in the literature, including V-groove side pumping [43, 44], multimode couplers [45], and single-clad coiled fibers [46]. The approach shown in Fig. 11.1 is a tapered fiber bundle. In this approach, several fibers are adiabatically fused into one multimode core. Efficient coupling into the single multimode fiber is governed by the Brightness Theorem, and is obtained if $NA_o/D_o > NA_i/D_i$, where $D_{i/o}$ and $NA_{i/o}$ are the input/output fiber diameters and numerical apertures respectively.

A schematic of the index profile of a cladding-pumped fiber (CPF) is shown in Fig. 11.2(a) [50, 51]. It consists of a rare earth-doped core surrounded by a silica glass cladding. What differentiates the fiber from typical fiber is that surrounding the glass is a polymer whose index is lower than that of silica. This allows light to be guided by the silica cladding as well as the core. Light from a multimode diode is transmitted along the cladding of the fiber as indicated in Fig. 11.2(b). As the light propagates through the core it is absorbed by the rare-earth dopant. The light emitted by the rare earth-doped ions can be trapped and guided in the single-mode core. By placing a FBG with the appropriate reflectivity at either end of the single-mode core of the CPF, a CPFL is formed. It should be noted that if the silica cladding were circular, some modes could propagate without crossing the single-mode core, reducing the efficiency of the device. The noncircular shape forces mode mixing, so all the modes eventually cross the core. The CPFL acts as a brightness and wavelength converter, coupling the high NA multimode light from the diodes into a small area low NA fiber. Typically, the wavelengths that are chosen for the CPFL range from 1064 to 1117 nm in Yb-doped fibers.

The choice of diode pump wavelengths for a Yb-doped CPFL is based on a trade-off, which can be seen from examining the Yb absorption spectrum shown in Fig. 11.3. The use of 975 nm pumps allows for a much higher absorption coefficient and hence a more efficient device. However, due to the narrow width of the absorption spectrum around this wavelength, the diodes may have to be thermally stabilized. Alternatively, pumping at 915 nm reduces the efficiency of the device, but alleviates concerns about the thermal stability of the multimode diodes.

Rare-earth dopants for the single-mode core of the CPF that allow efficient tunable lasing in the 14XX region are not available. Therefore a cascaded RFL is used to shift

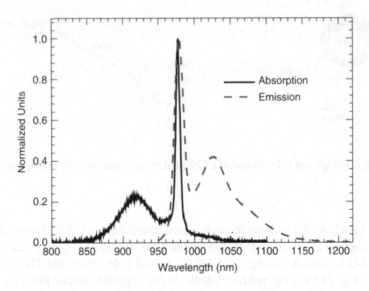

Fig. 11.3. The emission and absorption spectra of Yb-doped fiber.

the light at 11XX nm to 14XX nm [12–19]. Light is shifted through multiple Stokes shifts to the desired wavelength by stimulated Raman scattering. The RFL consists of input and output grating sets separated by a fiber with an enhanced Raman gain coefficient [53]. In reference to Fig. 11.1, which is an exemplary 1455 nm RFL, light from the 1100 nm laser enters the cavity. As it propagates down the fiber it is converted into light at the next Stokes shift, 1156 nm. Any 1100 nm light that is not converted will be reflected by a high reflector (HR) (~100%) on the output grating set. The light at 1156 nm is confined in the cavity by two HR gratings on either end of the cavity fiber. The 1156 nm light is converted to 1218 nm light. The process continues in a similar manner with nested pairs of HR at all intermediate Stokes shifts forming the intermediate cavities. When the desired output wavelength is reached, the output grating set contains a grating whose reflectivity is less than 100% so that light is coupled out of the cavity. This grating is called the output coupler (OC).

Experimental measurements of the performance of a Ge-doped RFL pump module are shown in Fig. 11.4. This device is pumped with an 1100 nm Yb-doped CPFL, which is in turn pumped by 915 nm multimode diodes. Figure 11.4(a) is a plot of the output power at 1455 nm for the RFL pump module versus input power at 915 nm. The slope efficiency and threshold of this device are 40% and 740 mW respectively. Figure 11.4(b) shows the slope efficiency for only the cascaded RFL device is 52%, with $P_{th} = 425$ mW. For P-doped RFL the best slope efficiency in the literature is 48% [17], though it should be pointed out that this type of laser is in an early stage of development.

The output spectrum is shown for the same device in Figs. 11.4(c) and (d). The intermediate Raman orders can be seen in Fig. 11.4(c). The ratio between the peak power of the desired output wavelength and the intermediate Stokes order with the highest output power is called the suppression ratio. This is approximately 20 dB

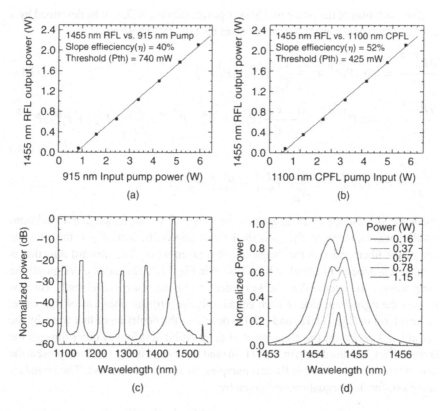

Fig. 11.4. Plot of the RFL ouput power versus (a) 915 nm multimode diode power and (b) CPFL power. Output spectra of a RFL showing (c) the spectrum of the intermediate Stokes shifts at an ouput power of 1 W, and (d) the spectral distribution around 1455 nm.

in Fig. 11.4(c), and changes with pump power. The spectral linewidth of the RFL significantly broadens as the power is increased, as seen in Fig. 11.4(d).

Raman fiber lasers have been used in several of the pioneering experiments in distributed Raman amplification. For example, the first demonstrations of (1) capacity upgrades using Raman amplification by Hansen et al. [52], (2) multiwavelength pumping for large bandwidth by Rottwitt and Kidorf [24], and (3) higher-order pumping by Rottwitt et al. [35] all used single-wavelength Raman fiber lasers. Many other systems results have also established RFL as a viable Raman pump source.

11.2.2. Cavity Design

The behavior of a RFL depends on parameters such as fiber length, OC reflectivity, and splice losses. Numerical simulations are used in order to understand the trade-offs in making these choices. The mathematical basis for the numerical model is given below, and the results of the numerical simulations are used to elucidate the effect of some design parameters on a RFL. Finally, experimental results are presented.

The evolution of the pump and Stokes power inside a RFL can be described by a set of nonlinear ordinary differential equations [19–23]. These are:

$$\frac{dP_p^{F/B}}{dz} = \mp\alpha_p P_p^{F/B} \mp \frac{\nu_p}{\nu_1}\frac{g_R^1}{A_{\text{eff}}^1}(P_1^F + P_1^B)P_p^{F/B} \tag{11.1a}$$

$$\frac{dP_i^{F/B}}{dz} = \mp\alpha_i P_i^{F/B} \mp \frac{\nu_i}{\nu_{i+1}}\frac{g_R^i}{A_{\text{eff}}^i}(P_{i+1}^F + P_{i+1}^B)P_i^{F/B} \pm \frac{g_R^{i-1}}{A_{\text{eff}}^i}(P_{i-1}^F + P_{i-1}^B)P_i^{F/B} \tag{11.1b}$$

$$\frac{dP_n^{F/B}}{dz} = \mp\alpha_n P_n^{F/B} \pm \frac{g_R^n}{A_{\text{eff}}^n}(P_{n-1}^{F/B} + P_{n-1}^{F/B})P_n^{F/B}, \tag{11.1c}$$

where the superscript F/B designates the power P in the forward and backward traveling waves, respectively, g_R is the Raman gain coefficient, A_{eff} is the effective area of the fiber, and ν is the frequency of a given wave. The forward direction is from the Yb-doped CPFL to the OC as shown in Fig. 11.1. The index p designates the pump wave, i the intermediate Stokes orders, and n the lasing wavelength. The first term on the right-hand side of each equation is the intrinsic fiber loss of that wave; the next term in Eqs. (11.1a) and (11.1b) describes the depletion of that wave by the forward and backward traveling wave of the next highest Stokes order through the Raman effect. The third term in Eq. (11.1b) and the second in Eq. (11.1c) represent the gain of the ith wave through Raman pumping by the $(i-1)$th wave. The boundary conditions for these equations are given by

$$P_p^F(0) = P_{in} \qquad\qquad P_p^B(L) = R_p^b \cdot P_p^F(L)$$

$$P_i^F(0) = R_i^f \cdot P_i^B(0) \qquad P_i^B(L) = R_i^b \cdot P_i^F(L)$$

$$P_n^F(0) = R_n^f \cdot P_n^B(0) \qquad P_n^B(L) = R_{oc} \cdot P_n^F(L). \tag{11.2}$$

where R is the reflectivity of the front/back (f/b) Bragg grating, L is the length of the RFL cavity, and R_{oc} is the reflectivity of the OC. For notational simplicity $P_{\text{out}} = (1 - R_{oc})P_n^F(L)$ is used.

Several important effects are omitted in Eqs. (11.1) for simplicity: (1) no terms are included for spontaneous emission, (2) interaction between nonsequential Stokes lines is neglected (e.g., the first Stokes line interacting with the third), as well as (3) the possibility of generating the next Stokes line beyond the nth wave. Nonetheless these models provide an excellent qualitative description of the effect of various parameters on a RFL. The model used to obtain the results presented here includes all of these effects.

In the simulations presented here a 1117 to 1480 nm RFL such as that shown in Fig. 11.1 was modeled. Experimentally measured values of the Raman gain coefficient and the fiber loss for a Ge-doped fiber were used. The splices between the fiber containing the gratings and the Raman enhanced fiber, and the gratings themselves were initially assumed to have no loss. The three measures of the RFL performance used were the slope efficiency η_s, pump threshold power P_{th}, and the overall (total)

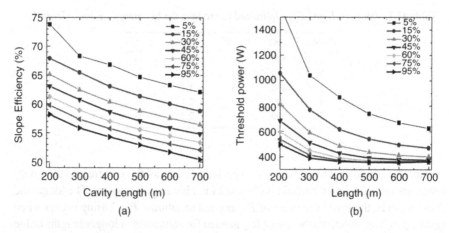

Fig. 11.5. Simulation results showing the effect of fiber length on (a) slope efficiency, and (b) threshold power for different output coupler reflectivities.

efficiency η_T. These quantities are defined from the linear fit of a graph of the launched pump power P_{in} versus output power P_{out} as

$$P_{out} = \eta_s (P_{in} - P_{th}),$$

$$\eta_T = \frac{P_{out}}{P_{in}}. \tag{11.3}$$

Because the focus is on improving the performance of the cascaded RFL, P_{in} is the power from the CPFL entering the cascaded RFL.

Figure 11.5(a) is a plot of η_s as a function of L for different R_{oc}. As L is increased, η_s decreases almost linearly for the length range under consideration. This decrease is due to the increased intrinsic fiber loss as its length is increased. The effect of changing R_{oc} can also be extracted from Fig. 11.5(a). By moving down a vertical line at a constant length it is seen that η_s decreases fairly linearly as R_{oc} is increased. The slightly closer spacing of the lines at higher reflectivities suggests a slight curvature to the fall-off. This decrease in η_s is expected because a higher R_{oc} means less power generated in the cavity is extracted from it.

The trade-offs in designing a RFL become apparent in comparing Figs. 11.5(a) and 11.5(b). Figure 11.5(b) is a plot of the cascaded RFL's threshold as a function of length. All of the changes that had a negative effect on η_s now have a positive effect on P_{th}. Increasing the length of fiber reduces P_{th}. This is simply because the length-integrated Raman gain increases with increasing fiber length, lowering the threshold. Eventually the benefit on P_{th} of increasing fiber length saturates as the length-integrated gain cannot increase any more. Simulations show, for long enough fiber lengths, P_{th} can eventually increase as the increased intrinsic fiber loss exceeds any increased benefit from the length-integrated Raman gain. As with η_s, the effect of R_{oc} on P_{th} can also be obtained from Fig. 11.5(b). By moving down a vertical line at a constant value of length it is seen that P_{th} decreases nonlinearly as R_{oc} is increased,

Table 11.1. Values of L and R_{oc} Optimized to Produce the Maximum η_T for a Given P_p

P_{in}(W)	L(m)	R_{oc}(%)	P_{out}(W)	η_T(%)
1	500	70	0.4	40
2	400	45	0.9	46
3	400	25	1.5	51
4	400	15	2.1	53
5	300	15	2.7	55

eventually saturating. This behavior is expected because as R_{oc} is increased the cavity losses are reduced, and threshold is obtained for a lower pump power. The longer the fiber length is, the lower the value of R_{oc} needed to saturate P_{th}. Lasing occurs when gain equals loss, therefore the lower R_{oc} needed for saturation at longer lengths is due to the increased integrated Raman gain at these lengths. Saturation occurs because beyond the point at which gain equals loss little value is obtained by retaining more power in the cavity (by increasing R_{oc}).

The behavior described in the previous paragraphs points to the need to optimize between the slope efficiency and threshold power with L and R_{oc}. The overall efficiency is a measure of this trade-off. The results of simulations to maximize η_T by varying L and R_{oc}, for different values of P_{in} are summarized in Table 11.1. It shows that the optimized parameters depend on the desired operating point. For low P_{out}, it is more important to design the cavity to minimize P_{th} inasmuch as it represents a large percentage of P_{in}. Increasing L and R_{oc} does this. For higher powers, η_s is more important because the device is operating far above threshold; hence smaller values of L and R_{oc} are required. It is also seen that the maximum η_T increases with P_p. This is again a reflection of the decrease in the percentage of P_{in} light used to reach P_{th}.

Further insight is obtained by examining the behavior of η_T as a function of L and R_{oc} for a fixed P_{in}. Plots of η_T versus L and R_{oc} with $P_{in} = 5$ W and 2 W are shown in Figs. 11.6(a) and 11.6(b). Unless otherwise stated, the parameters used are those in Table 11.1. The plots indicate the sensitivity of the optimum design points listed in Table 11.1. Qualitatively, it is seen that η_T is fairly insensitive to L and R_{oc} around the optimum parameters. It is noteworthy, however, that a laser designed to operate at 2 W(5 W) will not perform optimally at 5 W(2 W). Therefore a laser required to operate over a wide power range will have to be a compromise between the two designs for optimum performance.

11.2.3. Raman Fiber

The fiber choice for a RFL is based on five considerations, increasing the gain coefficient, reducing the effective area, decreasing the fiber loss, reducing splice losses, and the ability to write Bragg gratings in the fiber.

The largest influence on the Raman gain coefficient is the dopant used and the quantity of the dopant [54, 55]. To date, those RFLs built have used either Ge or

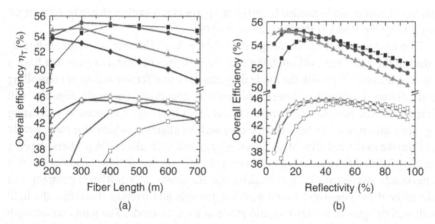

Fig. 11.6. Overall efficiency as a function of (a) fiber length and (b) output coupler reflectivity for an input pump power of 5 W (solid symbols) and 2 W(open symbols). Plot (a) is for $R_{oc} = 5\%$ (squares), 15% (circles), 45% (triangles) and 70% (diamonds). Plot (b) is for 5 W $L = 200$ m (squares), 300 m (diamonds), 400 m (circles), and for 2 W 300 m (squares), 400 m (circles), and 500 m (triangles).

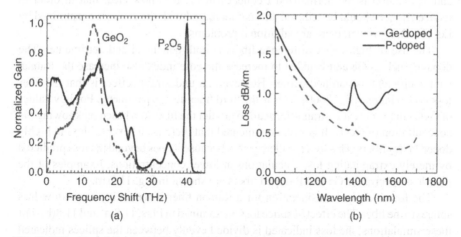

Fig. 11.7. For P- and Ge-doped fibers (a) normalized gain, and (b) loss spectra. (P-doped data courtesy of Professor E. M. Dianov, General Physics Institute Russia; Ge-doped fiber is RF courtesy of OFS Fitel.)

P-doped fibers. Figure 11.7 (a) is a diagram of the normalized gain spectrum of P-doped and Ge-doped fiber. Germanium has been the traditional dopant used in optical fibers, and the techniques for incorporating Ge into fibers are well understood. On the other hand, P-doped fiber with a larger Stokes shift (39.9 THz) could reduce the number of Raman shifts needed (e.g., two shifts from 1060 to 1480 nm) as compared to Ge-doped fiber (six shifts from 1060 to 1480 nm). Further flexibility is gained in P-doped fiber because lasing can be achieved from the silica peak at 13.2 THz. The number of Stokes shifts needed to arrive at a particular wavelength can be minimized by utilizing the P and Si Raman gain peaks [19].

The next goal of the fiber design is to reduce A_{eff}. This can be accomplished by reducing the core area. Although the reduction in the core radius r does not have a one-to-one correlation with the mode diameter, it is an effective means of controlling mode diameter. There is a lower limit on r as eventually the mode diameter will increase and the bend loss become excessive. This loss can be reduced by increasing the index difference, Δn between the core and the cladding, which also more tightly confines the mode, reducing A_{eff}. This is why in order to make low A_{eff} fibers, the core cladding index difference is high. However, there is another design constraint and that is the cutoff wavelength λ_c. It is intuitive that the most efficient Raman pumping will take place if the pump and signal wavelengths spatially overlap. Therefore the light at all wavelengths in the fiber should guide in a single-mode. The pump wavelength for a RFL then determines the design λ_c. An expression for λ_c for step index fibers is given as [56]

$$\lambda_c = \frac{2\pi r}{V_c}[(n_1 + n_2)\Delta n]^{1/2}, \tag{11.4}$$

where n_1 and n_2 are the refractive indices of the core and cladding, respectively, and $V_c = 2.405$ is the normalized frequency at λ_c. It is now clear that in order to reduce A_{eff}, r should be reduced and Δn increased with the changes constrained by Eq. (11.4). However, there are additional problems.

The use of high-index difference fibers is both beneficial and detrimental. The dopants such as Ge and P added to increase the core index also increase the Raman gain coefficient as outlined earlier. However, an undesirable effect is that the fiber loss is also increased [57–60]. This is the third fiber design parameter. For low values of GeO_2 in the core it is straightforward to predict the fiber loss based on knowing the concentration of GeO_2. It is well documented that there is an "excess" loss in highly doped GeO_2 fibers. This loss is called excess because it goes beyond the loss predicted by merely extrapolating loss calculations at lower concentrations. Examples of the fiber loss spectra for Ge and P-doped fibers are shown in Fig. 11.7(b).

The fourth design consideration for a Raman fiber is the ability to get low-loss splices to the fiber. The effect of splice loss is examined in Figs. 11.8(a) and 11.8(b). For these simulations, the loss indicated is divided evenly between the splices indicated on the cascaded RFL in Fig. 11.1. As was indicated earlier a splice loss of 0 dB was used in the previous simulations. Just adding 0.05 dB splice loss (considered an excellent splice loss) to each end of the cavity produces a predicted decrease in η_T of 4% (5%) for a RFL optimized for 5 W(2 W). The drop in overall efficiency is worse at lower powers because the increased cavity loss raises P_{th}, which represents a larger percentage of the pump power. Because in most applications a constant output power is what is required, Fig. 11.8(b) shows the percentage of pump power increase required to maintain the same output power compared to a 0 dB splice loss. For a 0.1 dB splice loss in order for a 5 W (2 W) device to maintain the same operating power a 7% (7.5%) increase in pump power is needed. The reduction in A_{eff} results in a significant mode field mismatch between the Raman gain fiber and other fibers. For splices

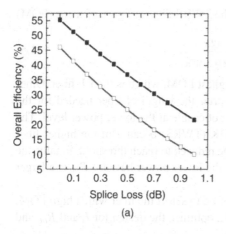
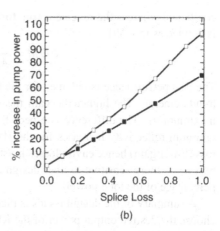

(a)   (b)

Fig. 11.8. (a) Change in η_T as a function of splice loss for a 5 W (closed squares) and a 2 W (open squares) optimized RFL as described in Table 11.1; (b) the percentage increase in pump power required to maintain the same output power as with 0 dB splice loss for a given splice loss, under the same conditions described in (a).

outside the cavity, the different melting points and diffusion rates between high index-difference fibers and standard fibers present splicing challenges. In telecommunication applications, the RFL will have to be spliced to a standard fiber and designing a fiber that minimizes this loss is important.

The need to minimize the splice losses in the cavity leads to the final design parameter for a Raman fiber laser cavity, and that is the ability to write low-loss Bragg gratings in the same fiber that is used to provide gain in the cavity [16]. Without this ability the losses such as those shown in Fig. 11.8 would quickly accumulate leading to significantly reduced cavity efficiency.

The practical implications of design trade-offs are now considered. Table 11.2 shows the optimized L and R_{oc} for three different fiber types at 5 and 2 W. The fibers are all manufactured by OFS Fitel and are high slope dispersion-compensating fiber (HSDK), TrueWave® RS (TWRS), and a Raman enhanced fiber (RF). Fiber type RF was used for the prior simulations. Note that the HSDK fiber has a higher gain and loss

Table 11.2. Values of L and R_{oc} Optimized to Produce the Maximum η_T for a Given P_p and for Fibers with Various Raman Gain and Loss Coefficients

P_{in} (W)	Fiber Type	g_R/A_{eff} (km^{-1}W^{-1})	$\alpha_{1550\,nm}$ (dB/km)	FOM (W^{-1}dB^{-1})	L (m)	R_{oc} (%)	P_{out} (W)	η_T (%)
5	HSDK	3.3	0.64	5.1	200	25	2.6	51
	RF	2.4	0.30	8.0	300	15	2.7	55
	TWRS	0.7	0.21	3.4	600	55	2.4	48
2	HSDK	3.3	0.64	5.1	250	60	0.8	40
	RF	2.4	0.30	8.0	400	45	0.9	46
	TWRS	0.7	0.21	3.4	800	95	0.7	35

coefficient whereas the reverse is true for the TWRS fiber. A figure of merit (FOM) is defined as [53, 59]

$$FOM = \frac{g_R}{A_{eff}\alpha_{1550\ nm}}.$$ (11.5)

The best performance is obtained for the highest FOM, which is the RF fiber. Other trends emerge. The higher the gain the shorter the length of fiber needed for the maximum η_T. The difference is especially noticeable at the lower power level. The optimum reflectivity increases for the HSDK (TWRS) because of the higher fiber loss (lower gain) hence cavity losses must be reduced to reach threshold. Finally, the best η_T is more sensitive to fiber design at the lower power because P_{th} is a larger percentage of the total power.

A summary of the design issues is therefore to select the fiber with a high FOM, choose the desired output power of the RFL, optimize the design for L and R_{oc} and then work hard to reduce splice losses.

The focus of this section has been on the cascaded RFL. However, in optimizing the whole RFL pump module consideration should be given to the CPFL, in particular to the choice of the CPFL wavelength. The efficiency of this device will depend on the output wavelength selected. As an example, in considering whether to use P-doped or Ge-doped fiber the starting wavelengths for a 1480 nm device are 1060 and 1117 nm, respectively, which may tip the choice of the most efficient RFL pump module in the favor of Ge-doped fibers [61].

11.2.4. Temporal Behavior

Raman scattering is a very fast (<1 ps) process, thus any pump fluctuations occurring on a time scale slower than 1 ps can cause fluctuations in the signal gain. This imposes stringent requirements on the noise of a pump laser [62, 63]. In a counter-pumped configuration, because the pump and signal travel in opposite directions there is a strong averaging of pump power fluctuations. In the copumped configuration, the signal and pump propagate through the fiber together, and the only averaging effect is through the dispersive delay caused by walk-off between the pump and signal. One measure of the noise in a pump is the relative intensity noise (RIN) of the pump. The RIN values at a frequency ν are equal to the mean square optical power fluctuations divided by the square of the mean signal power in a 1.0 Hz bandwidth. It has been shown that for a 100 km span of standard single-mode fiber the RIN required of a pump source is -119 dB/Hz and -81 dB/Hz for the co- and counter-pumped cases, respectively, across a broad frequency range that extends up to 100 s MHz [63]. Although the exact numbers will vary depending on fiber type and length these numbers are good estimates of the required pump RIN.

Figure 11.9 is a plot of the RIN of a 1366 nm RFL similar to the one shown in Fig. 11.1. The mode spacing of the CPFL (4.2 MHz) and the RFL (200 kHz) are evident. The RIN levels measured allow counter-pumping, but are too high for copumping. Based on work with semiconductor diodes it can be speculated that it is due to the large number of modes beating with each other in the fiber [65, 66]. This being the case, novel approaches are needed in order to significantly reduce the RIN of RFL.

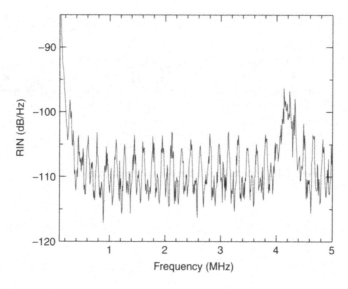

Fig. 11.9. Noise spectrum of a 1366 nm RFL from 0.1 to 5 MHz.

11.3. Multiple Wavelength Raman Fiber Lasers

11.3.1. Overview

An important property of stimulated Raman scattering is that the wavelength at which gain occurs, for a given fiber, depends only on the pump wavelength. Therefore, a broad flat Raman gain profile can be obtained using multiple pump wavelengths [24, 25]. Multiple-wavelength RFL (MWRFL) distribute the large amount of power at one wavelength among multiple wavelengths. In this section MWRFL are described [26–34]. Following a broad overview, the stability of the device at an operating point and the ability to achieve a given operating point are examined. Finally the noise properties are reviewed.

Multiple-wavelength lasing in RFL was first demonstrated in a dual wavelength device. This device used a ring cavity configuration with carefully selected WDM couplers and fiber Bragg gratings [26]. Linear cavities showing two and three wavelengths were demonstrated shortly thereafter [27–31]. A three-wavelength RFL (3λRFL) similar to that used in [29] is shown schematically in Fig. 11.10. It is pumped by a Yb-doped CPFL at 1100 nm. The RFL consists of a spool of enhanced Raman gain single-mode fiber. Light is shifted from 1100 nm to 1347 nm in the Raman fiber with four nested pairs of high reflectors with fixed reflectivities of approximately 100%. The output grating set also contains a HR for the 1100 nm pump. Resonant cavities at 1428, 1445, and 1466 nm (λ_1, λ_2, λ_3) are created by a set of adjustable-reflectivity OCs [28] and a matching set of broadband HR gratings. Examples of the emission spectra obtained from this device are shown in Fig. 11.11. The ability to partition the power between the different wavelengths is evident.

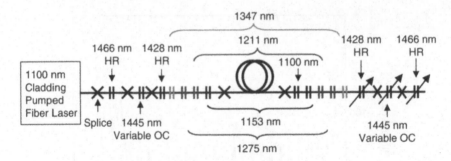

Fig. 11.10. Schematic of a three-wavelength cascaded RFL.

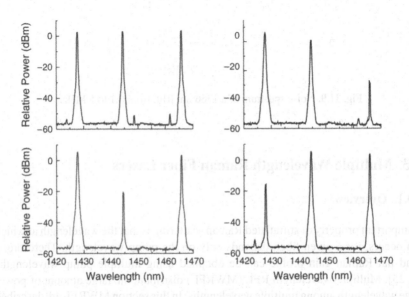

Fig. 11.11. Output spectra of a three-wavelength RFL for various settings of the OC reflectivity.

The output power at each of the three wavelengths and the total power as a function of the incident 1100 nm pump power is shown in Fig. 11.12. It is noteworthy that the total output power is nonlinear. This is because of the differences in quantum energy of the different wavelengths, and the fact that the power is redistributed among the different wavelengths as the CPFL power is increased. In the published literature the slope efficiency was estimated by drawing an asymptote to the curved line [30]. For Ge-doped fibers a threshold power of 400 mW and the total slope efficiency of 38% from 1100 to 14XX nm light has been reported [28, 29]. In work published using P-doped fibers a threshold of 2.8 W and a slope efficiency of 50% were reported [31]. Both works held out the promise of further improvements in the device.

The ability to control the distribution of power in a MWRFL is critical for a practical device. Two techniques for accomplishing this have been proposed. In the

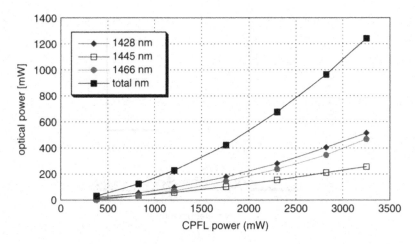

Fig. 11.12. Total output power and power at each wavelength in a three-wavelength RFL, as a function of input CPFL power.

first, the HR and OC at a specific wavelength are deliberately misaligned, reducing the efficiency of the cavity at that wavelength and hence its power [26, 31]. In the second approach, the reflectivities of the OC are varied with the same net effect [28–30]. Spectra for the OC at various reflectivities are shown in Fig. 11.13(a), with the corresponding output laser spectra at one of the wavelengths shown in Fig. 11.13(b). Note that in this approach there are only small changes in the center wavelength as the OC (power) at that wavelength is adjusted.

11.3.2. Operating Point Stability and Power Partioning

In the exemplary 3λRFL shown in Fig. 11.10, the Stokes radiation at 1347 nm provides Raman gain to all three lasing wavelengths. The radiation at 1428 nm provides gain to the radiations at 1445 and 1466 nm and the radiation at 1445 nm pumps the radiation at 1466 nm. Therefore the output powers at all three lasing wavelengths are interdependent. This is seen in Fig. 11.14, which shows the power at the three wavelengths as a function of the controlling voltage applied to one OC. In Fig. 11.14(a), as the voltage on the 1445 nm OC is increased, thereby decreasing the OC reflectivity, the power emitted at 1445 nm decreases. Simultaneously there is a rise in power at 1428 nm because the depletion of its power by the 1445 nm light is decreasing. The power at 1466 nm remains fairly constant. Similar results are seen when the OC reflectivity of the 1466 nm line is reduced. Now the powers in both at 1428 and 1445 nm increase as they are no longer being depleted by the 1466 nm line. Note that in both cases the total power remains constant.

The interdependence of the optical powers at the different wavelengths raises two questions. If the powers emitted from the laser are set to some predetermined values, will the MWRFL remain at that operating point? If the answer is yes the next question is: Can a wide range of such operating points be obtained? The next

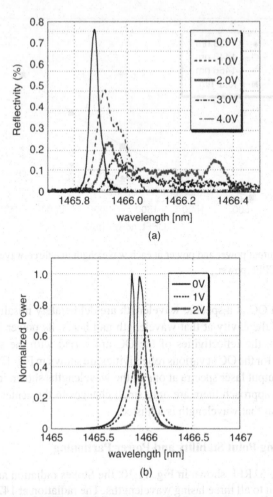

Fig. 11.13. (a) Spectra of 1466 nm OC for various reflectivity settings; and (b) output spectra at 1466 nm for various OC settings.

two subsections answer these questions by examining the stability of the source, and the ability to properly partition the powers among the different wavelengths. These results are not general in the sense that the stability of the source will depend on the physical mechanism used to redistribute the wavelengths (e.g., varying reflectivity or misaligning gratings), as well as how that mechanism is exploited (e.g., strain or temperature tuning of gratings).

11.3.2.1. Operating Point Stability

The stability of a 3λRFL was examined in Ref. [30] to see if small changes in the operating point voltages applied to the OC do not cause large abrupt changes in the power distribution. A 3λRFL was placed in the experimental setup shown in Figure 11.15.

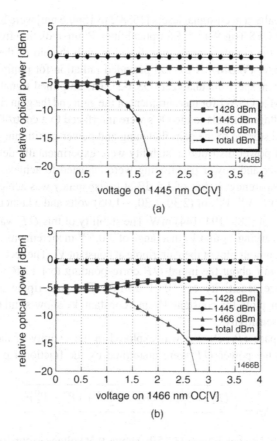

Fig. 11.14. Output at each of the wavelengths as the reflectivity of (a) 1445 nm OC and (b) 1466 nm OC are adjusted. Also shown is the total power as a function of OC voltage.

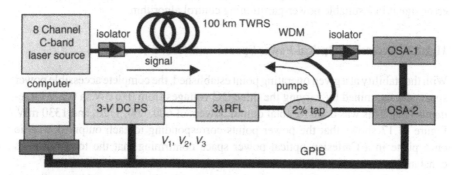

Fig. 11.15. Experimental setup for evaluating the operating point stability and accessible power space for a MWRFL.

Eight C-band equally spaced signal lasers [1538.2 to 1560.6 nm] were launched down a 100 km span of OFS TrueWave® RS optical fiber. Pump radiation from the 3λRFL was launched in the counterpropagating direction with the aid of the wavelength-division-multiplexer (WDM). A 2% tap provided radiation for pump power monitoring by the optical spectrum analyzer (OSA-2). The amplified signal lasers passed through the WDM to OSA-1 for measurement of the gain and the gain ripple ΔG. The OSAs and all voltage controls to the OCs were interfaced to a controlling computer for collection and storage of the gain-flattening and power-partitioning data.

The gain and power distribution stability were experimentally determined. The total optical power and power partitioning were adjusted to achieve a gain ripple minimum at transparency. A ΔG of 1.4 dB at transparency was achieved at an initial voltage $OP(V_1^0, V_2^0, V_3^0)$ of (2.30, 1.30, −1.05) volts and a launch pump power $OP(P_1^0, P_2^0, P_3^0)$ of (283, 194, 144) mW. The stability of this OP was interrogated by fixing V_1 and varying V_2 and V_3 in a range of ±0.5 V in increments of 25 mV. Figure 11.16(a) is a contour plot of the gain ripple at constant V_1. The rectangle indicates a region of ±50 mV about the initial OP corresponding to 1.1 dB< ΔG < 1.8 dB yielding a voltage sensitivity of ∼0.01 dB/mV. The gain ripple minimum was 1.13 dB. Contour plots for constant V_2 and constant V_3 show similar results with reduced voltage sensitivity.

The power space (P_1, P_2, P_3) was explored in the vicinity of the power OP. Deviations from the power OP were quantified by the fractional power partition deviation:

$$\Delta\rho = \frac{\sqrt{(P_1 - P_1^0)^2 + (P_2 - P_2^0)^2 + (P_3 - P_3^0)^2}}{P_0}, \tag{11.6}$$

where P_0 is the total power. Figure 11.15(b) shows that voltage excursions of ±50 mV generate <4% variations in the 3λRFL spectral power distribution indicating a $\Delta\rho$ voltage sensitivity of ∼0.04 %/mV. Data at constant V_2 and constant V_3 show comparable results with lower voltage sensitivity. Hence, ΔG and/or $\Delta\rho$ can serve as an error signal in a suitable power-partitioning control algorithm.

11.3.2.2. Accessible Optical Power Space

With the stability at a given operating point established, the complete accessible power space was examined by varying the three OC voltages from 0 to 3 V in 0.2 V increments [30]. This was done for total output powers of 600, 860, 1120, and 1330 mW. Figure 11.17 shows that the power points corresponding to each output power lie on a plane in a Cartesian optical power space confirming that the total power is constant.

The three-dimensional power state can be conveniently represented in two dimensions by the simplex diagram shown in Figure 11.18. An equilateral triangle is constructed with vertices (x, y) equal to $(0, \sqrt{3}/2)$, $(-1/2, 0)$, and $(1/2, 0)$, representing the power states $(P_1, 0, 0)$, $(0, P_2, 0)$ and $(0, 0, P_3)$, respectively. This construction

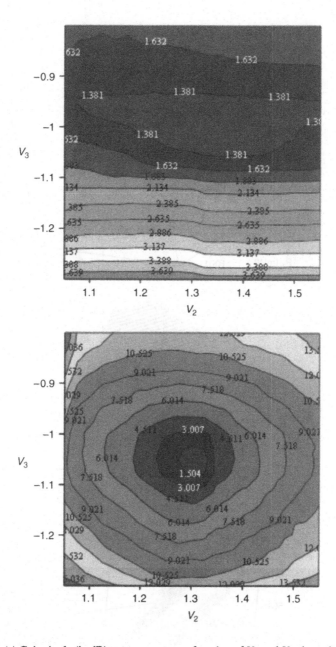

Fig. 11.16. (a) Gain ripple (in dB) contour map as a function of V_2 and V_3 about the operating point with V_1 constant. Minimum gain ripple is 1.1 dB. Box interior indicates ±50 mV; (b) fractional power partition deviation $\Delta\rho$ (%) as a function of V_2 and V_3 about the OP with V_1 constant. Box interior indicates ±5 mV. The dot shows the operating point.

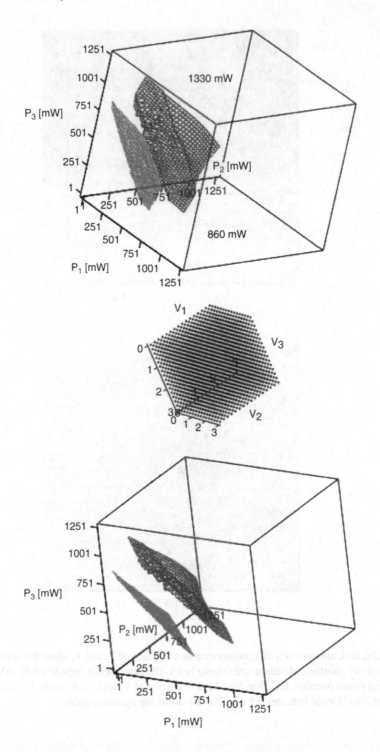

Fig. 11.17. Two views of 3λRFL optical power planes in 3-D optical power space for 860 and 1330 mW total power, with the voltage space covered shown in the center.

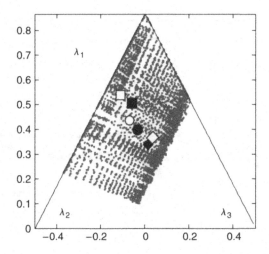

Fig. 11.18. Power simplex plot showing accessible power states and experimental and simulated operating points. (1) Small dots represent all accessible power states; (2) open symbols are experimental power states corresponding to C-band gain ripple minima of 0.7, 1.1, and 1.6 dB in 60, 100, and 140 km span of TWRS optical fiber; (3) solid symbols are simulated power states for 60, 100, and 140 km spans of TWRS and SMF28 optical fiber.

can be generalized to an arbitrary power state (P_1, P_2, P_3) according to the following equations.

$$x = 0 \cdot \left(\frac{P_1}{P_0}\right) - \frac{1}{2} \cdot \left(\frac{P_2}{P_0}\right) + \frac{1}{2} \cdot \left(\frac{P_3}{P_0}\right). \tag{11.7a}$$

$$y = \frac{\sqrt{3}}{2} \cdot \left(\frac{P_1}{P_0}\right) + 0 \cdot \left(\frac{P_2}{P_0}\right) \cdot \left(\frac{P_3}{P_0}\right). \tag{11.7b}$$

The small dots in Fig. 11.18 show the power states spanned by the 3λRFL at a total launch power of 620 mW. The simplex coverage is not a sensitive function of total power. Data were taken with a voltage resolution of (0.20, 0.20, 0.05) volts for (V_1, V_2, V_3). The open symbols correspond to experimentally determined power distributions corresponding to minimum gain ripple, for 60, 100, and 140 km span lengths of OFS TrueWave® RS fiber. The adjacent solid symbols are simulation results, for the same fiber, and for Corning SMF-28™ at similar lengths. The experimental datapoints and simulation datapoints fall within the laser reach. This leads to the conclusion that the laser can access an adequate pump combination space.

Finally, an interesting point is how quickly a given operating point can be obtained. A complete answer to this depends on the electronics used, the physical mechanism for partitioning the powers, and the starting and finishing operating points. One aspect of the problem that can be answered is the number of iterations in the OC reflectivity needed in tuning the cavity at a given wavelength. Using the experimental setup shown in Fig. 11.15, a given operating point was selected. Initially, the total power was determined by adjusting the injection current to the multimode laser diodes. The reflectivity of each OC was then adjusted until the desired power at the wavelength

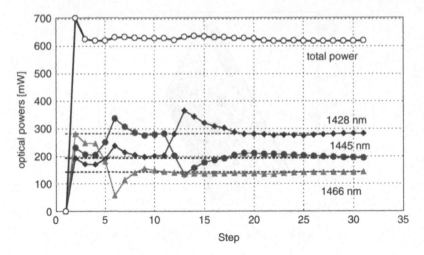

Fig. 11.19. Evolution of the three wavelengths and the total pump power to a programmed operating point.

corresponding to that OC is reached. The next OC was then adjusted until the power at its wavelength was at the required power. Because changing the OC at each wavelength changes the power in the other wavelengths, this process was repeated until all the radiation at each wavelengths were at the desired power level. Figure 11.19 is a plot of the number of steps required to reach the operating point. Each step corresponds to one adjustment of an OC. In this example as little as 20 steps were required to reach the desired operating point. The exact number of steps will depend on the prior and final operating points, and it may be possible to implement faster search algorithms, however, it is reasonable to assume the number of steps needed is on the order of tens.

11.3.3. Temporal Behavior of a MWRFL

As was discussed in Section 11.2.4 the temporal fluctuations of a RFL are an important consideration for Raman amplification. Of particular concern for a MWRFL is the possible presence of additional noise due to the simultaneous lasing of the different wavelengths. This question was examined in Ref. [29] by looking at the RIN spectra of a 3λRFL, lasing at 1427, 1455, and 1480 nm. The three wavelengths were separated, and individual RIN spectra measured. The results in a 100 kHz bandwidth are shown in Fig. 11.20. This bandwidth was chosen to include the low and intermediate frequency regime where averaging of the pump power fluctuations does not completely remove this noise source in a counter-pumped configuration. The solid lines in Figure 11.20 exhibit the RIN in units of dB/Hz as a function of frequency for each spectral component of the multiwavelength Raman fiber laser and for all three spectral components simultaneously. It is seen that the RIN values lie below −90 dB/Hz and that spectral components with lower mean powers exhibit higher noise levels. The solid trace exhibiting the lowest noise level is that corresponding to the RIN of the

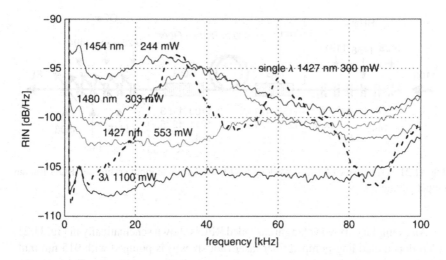

Fig. 11.20. RIN spectra for the three wavelengths of a 3λRFL, as well as all three wavelengths measured simultaneously. The RIN spectra for a single-wavelength RFL is shown for comparison.

total optical power. A careful examination of the corresponding time records shows that the photovoltage standard deviation for all three spectral components is equal to the incoherent summation of the standard deviations for the individual spectral components, indicating that the fluctuations in each spectral component are statistically independent. The dotted line corresponds to the RIN measured for a single wavelength 1427 nm Raman fiber laser identical to the multi-wavelength laser but without the 1466 and 1480 nm OCs. This laser was operated at an optical power of 300 mW corresponding to the typical power in a single component of the multiwavelength fiber laser. The single-wavelength noise level is comparable to that measured for the three spectral components of the three-wavelength laser. Therefore, MWRFL exhibit noise levels comparable to single-wavelength devices, and are therefore adequate for counter-pumped Raman amplification.

11.3.4. Six Wavelength Raman Fiber Lasers

The discussions of the previous sections focused primarily on three-wavelength RFL. If more pump wavelengths can be extracted from a single cavity, then a larger signal bandwidth can be amplified with the required gain flatness. This was demonstrated using six-wavelength RFL (6λRFL) [32–34]. These devices were able to simultaneous amplify both the C- and L-band signals. In the MWRFL shown in Fig. 11.10, this would require generating additional pump laser components removed in frequency from the Raman gain peak, of the 1347 nm light, by nearly twice the gain curve linewidth where the gain has fallen to approximately 10% of its peak value. Therefore, significant gain must be provided to these additional longer wavelength pumps from the shorter output wavelength pumps, suggesting that a controlled power distribution may have been difficult to achieve.

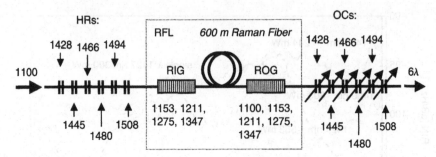

Fig. 11.21. A schematic of a six-wavelength RFL. RIG-Raman input grating set, ROG-Raman output grating set.

An exemplary six-wavelength cascaded RFL is shown schematically in Fig. 11.21. A Yb-doped cladding-pumped fiber laser (not shown) is pumped with 915 nm multimode semiconductor laser diodes to generate radiation at 1100 nm. This radiation is subsequently injected into the 6λRFL where successive Raman shifts generate radiation at 1347 nm. The radiation at 1347 nm is coupled to resonant cavities at the six lasing wavelengths: 1428, 1445, 1466, 1480, 1494, and 1508 nm. The resonant cavities occupy a single fiber and are constructed with six broadband HR Bragg gratings and six adjustable OCs. Stokes radiation at 1347 nm provides Raman gain for lasing at all six wavelengths. Laser radiation at the shorter wavelengths provides additional gain to the longer wavelength spectral components. Hence the spectral power distribution is dependent upon the power in each individual spectral component.

The 6λRFL was used to counter-pump a 100 km span of TrueWave® RS fiber [33]. Figure 11.22 shows the six-wavelength pump power distribution required to achieve

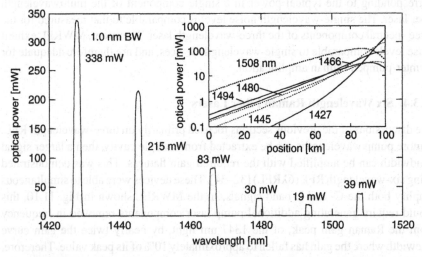

Fig. 11.22. 6λRFL pump power distribution used to achieve a C+L-band gain ripple of 1.7 dB in a 100 km span of TrueWave® RS optical fiber. Numbers adjacent to peaks indicate optical power in mW. Inset shows simulated pump powers as a function of span position.

an optimal gain flatness of 1.7 dB pk–pk in the C+L-bands. This power distribution was achieved by judiciously varying the voltages of the six OCs. Note that the long-wavelength pumps have significantly less power than the short-wavelength pumps. This is because the short-wavelength pumps amplify the long-wavelength pumps as the radiation propagates along the fiber length. This is illustrated by the inset in Fig. 11.22, which shows a simulation for the pump power as a function of position in the fiber span corresponding to the launched power distribution.

11.4. Dual-Order Raman Fiber Lasers

Second-order Raman pumping has been proposed as a way to improve optical signal-to-noise ratio, and hence system performance, by more uniformly distributing the Raman gain along the transmission span [30–35]. Such a pumping scheme requires two pump wavelengths, the first separated by approximately one Stokes shift from the signal, and the other two Stokes shifts from the signal. The former is referred to as the first-order pump (FOP), and the latter as the second-order pump (SOP). Schemes in which the FOP and the SOP are co- and counterpropagating with respect to each other have been demonstrated. The arrangement that is most relevant to RFL is when both pumps are counterpropagating relative to the signal. The SOP amplifies the FOP, which in turn amplifies the signal. There is also some direct pumping of the signal by the SOP. A drawback of second-order pumping schemes is that significantly more power is needed as compared to using first-order pumping. For example, to pump an 80 km span of standard single-mode fiber to transparency, power levels into the fiber from 0.8 to 1 W are required compared to just 580 mW for three FOP [39]. The need for large amounts of power and multiple wavelengths makes the RFL an ideal device for this application.

A single RFL cavity capable of providing FOP and SOP light was presented in Ref. [39]. Conceptually, the pump source is based on the multiwavelength Raman fiber laser technology described in Section 11.3 and an embodiment is shown in Fig. 11.23. The multimode diodes and CPFL are the same as those shown in Fig. 11.1. As in the previously described RFL, light from the CPFL at 1100 nm will be converted to 1365 nm. The OC at 1365 nm allows a portion of the light at this wavelength to exit the cavity. For a fixed-reflectivity OC at 1365 nm, the reflectivity must be chosen carefully. There is enough gain at 1455 nm so that the lasing threshold is low. However, only a small amount of power is required at 1455 nm. The choice of this reflectivity ultimately limits the tuning of the ratio of the power between the two pump wavelengths. The OC at 1455 nm is adjustable, enabling the device to control

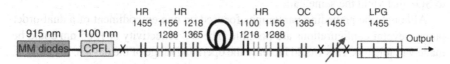

Fig. 11.23. Schematic of a dual-order RFL.

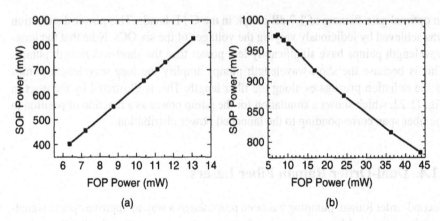

Fig. 11.24. Power in the second-order pump versus first-order pump when (a) the power ratio between the two pumps is constant, and (b) when the gain in an 80 km span of fiber is held constant.

the ratio of second-order to first-order powers. In order to have stable operation of the 1455 nm laser line at only 10 mW of output power, it was operated at 70 mW and then attenuated by a narrow long period grating. The ability to manipulate the ratio between the two pumps is shown in Fig. 11.24(a), which is a plot of the power in the SOP versus FOP. As the total power is increased, the ratio between the power in the two pumps is held constant.

It was shown that by careful selection of the pump wavelengths the same bandwidth covered by three first-order pumps could be covered by one second-order pump wavelength and one first-order seed wavelength [39]. In order to maximize the energy transfer from the second-order to the first-order pump, a frequency spacing equal to the peak of the Raman coefficient at 13.2 THz would be chosen. However, the gain bandwidth can be maximized if the first-order wavelength is slightly offset from the peak. This is because there is a small peak in the Raman gain spectrum at 24 THz, which for the second-order pump falls in the signal bandwidth. The wavelength of the first-order pump should be chosen so that the location of its peak gain at 13.2 THz complements the direct pumping by the second-order pump.

Quantifying the performance of the dual order source in order to compare its performance to that of a single or multiwavelength device is difficult. This is because for a fixed 1455 nm OC reflectivity, as the power is increased more power is emitted at 1455 nm. With the long-period grating in place this would mean that the efficiency of the device would decrease as a larger percentage of the power is being lost. However, from a telecommunications standpoint, these power configurations would be irrelevant. As Figure 11.24(b) shows, there are several different ratios of power in the FOP to SOP that yield the same gain.

Although the preceding paragraphs focused on one embodiment of a dual-order pump, several configurations are possible. The OC reflectivity at 1365 nm could be made variable too. This would increase the operating range of the device. In addition, other techniques to attenuate the light at 1455 nm are possible. Finally, this scheme can be extended to even higher-order pumping, using, for example, three- or four-pump wavelengths.

References

[1] C. Lin and R. H. Stolen, Backward Raman amplification and pulse steepening in silica fibers, *Appl. Phys. Lett.*, 29:428, 1976.

[2] E. P. Ippen, Low-power quasi-cw Raman oscillator, *Appl. Phys. Lett.*, 16:303, 1970.

[3] R. H. Stolen, E. P. Ippen, and A. R. Tynes, Raman oscillation in glass optical waveguide, *Appl. Phys. Lett.*, 20:62, 1972.

[4] R. H. Stolen and E. P. Ippen, Raman gain in glass optical waveguides, *Appl. Phys. Lett.*, 22:276–278, 1973.

[5] K. O. Hill, B. S. Kawasaki, and D. C. Johnson, Low-threshold cw Raman laser, *Appl. Phys. Lett.*, 29:181, 1976.

[6] C. Lin, R. H. Stolen, W. Pleibel, and P. Kaiser, "A high efficiency tunable cw Raman oscillator," *Appl. Phys. Lett.*, 30:162, 1977.

[7] C. Lin, R. H. Stolen, W. G. French, and T. G. Malone, A tunable 1.1 μm fiber Raman oscillator, *Appl. Phys. Lett.*, 31:97, 1977.

[8] C. Lin, L. G. Cohen, R. H. Stolen, G. W. Tasker, and W. G. French, Near-infrared sources in the 1–1.3 μm region by efficient stimulated Raman emission in glass fibers, *Opt. Commun.*, 20:426–428, 1977.

[9] R. K. Jain, C. Lin, R. H. Stolen, and A. Ashkin, A tunable multiple Stokes cw fiber Raman oscillator, *Appl. Phys. Lett.*, 31:89, 1977.

[10] K. O. Hill, Y. Fujii, D. C. Johnson, and B. S. Kawasaki, Photosensitivity in optical waveguides: Application to reflection filter fabrication, *Appl. Phys. Lett.*, 32:10, 647, 1978.

[11] R. Kashyap, *Fiber Bragg Gratings*, San Diego: Academic, 1999.

[12] S. Grubb, T. Erdogan, V. Mizrahi, T. Strasser, W. Y. Cheung, W. A. Reed, P. J. Lemaire, A. E. Miller, S. G. Kosinski, G. Nykolak, and P. C. Becker, 1.3 μm cascaded Raman amplifier in germanosilicate fibers. In *Proceedings of Optical Amplifiers and Their Applications*, PD3-1, 187, 1994.

[13] S.G. Grubb, T. Strasser, W.Y. Cheung, W.A. Reed, V. Mizrachi, T. Erdogan, P.J. Lemaire, A.M. Vengsarkar, D. J. DiGiovanni, D. W. Peckham, and B. H. Rockney, High-power 1.48 μm cascaded Raman laser in germanosilicate fibers. In *Proceedings of Optical Amplifiers and Their Applications*, SA4, 197–199, 1995.

[14] D. Innis, D. J. DiGiovanni, T. A. Strasser, A. Hale, C. Headley, A. J. Stentz, R. Pedrazzani, D. Tipton, S. G. Kosinski, D. L. Brownlow, K. W. Quoi, K. S. Kranz, R. G. Huff, R. Espindola, J. D. Le Grange, and G. Jacobovitz-Veselka, Ultrahigh-power single-mode fiber lasers from 1.065 to 1.472 μm using Yb-doped cladding-pumped and cascaded Raman lasers. In *Proceedings of the Conference on Lasers and Electro-Optics*, CPD-31, 1997.

[15] V. I. Karpov, E. M. Dianov, A. S. Kurkov, V. M. Paramonov, V. N. Protopopov, M. P. Bachynski, and W. R. L. Clements, LD-pumped 1.48 μm laser based on Yb-doped double-clad fiber and phosphorosilicate-fiber Raman converter. In *Proceedings of the Optical Fiber Communication Conference*, WM3, 202–204, 1999.

[16] E. M. Dianov, I. A. Bufetov, M. M. Bubnov, A. V. Shubin, S. A. Vasiliev, O. I. Medvedkov, S. I. Semjonov, M. V. Grekov, V. M. Paramonov, A. N. Gur'yanov, V. F. Khopin, D. Varelas, A. Iocco, D. Costantini, H. G. Limberger and R-P. Salathé, CW highly efficient 1.24 μm Raman laser based on low-loss phosphosilicate fiber. In *Proceedings of the Optical Fiber Communication Conference*, PD25, 1999.

[17] V. I. Karpov, E. M. Dianov, V.M. Paramonov, O. I. Medvedkov, M. M. Bubnov, S. L. Semyonov, S. A. Vasiliev, V. N. Protopopov, O. N. Egorova, V. F. Hopin, A. N. Guryanov, M. P. Bachynski, and W. R. L. Clements, Laser-diode-pumped phosphosilicate-fiber Raman laser with an output power of 1 W at 1.48 μm, *Opt. Lett.*, 24:13, 887–889, 1999.

[18] E. M. Dianov, I. A. Bufetov, M. M. Bubnov, M. V. Grekov, S. A. Vasiliev, and O. I. Medvedkov, Three cascaded 1407-nm Raman laser based on phosphorus-doped silica fiber, *Opt. Lett.*, 25:6, 402–404, 2000.

[19] W. A. Reed, W. C. Coughran and S. G. Grubb, Numerical modeling of cascaded cw Raman fiber amplifiers and lasers. In *Optical Fiber Communication Conference*, WD1, 107–109, 1995.

[20] M. Rini, I. Cristiani, and V. Degiorgio, Numerical modeling and optimization of cascaded cw Raman fiber lasers, *IEEE J. Quantum Electron.*, QE 36:10, 1117–1122, 2000.

[21] A. Bertoni and G. C. Reali, 1.24-μm cascaded Raman laser for 1.31-μm Raman fiber amplifiers, *Appl. Phys. B* 67:5–10, 1998.

[22] G. Vareille, O. Audouin, and E. Desurvire, Numerical optimization of power conversion efficiency in 1480 nm multi-Stokes Raman fibre lasers, *Electron. Lett.* 34:7, 675–676, 1998.

[23] S. D. Jackson and P. H. Muir, Theory and numerical simulation of nth-order cascaded Raman fiber lasers, *J. Opt. Soc. Am. B* 18:9, 1297–1306, 2001.

[24] K. Rottwitt and H. D. Kidorf, A 92 nm bandwidth Raman amplifier. In *Proceedings of the Optical Fiber Communication Conference*, PD6, 1998.

[25] Y. Emori and S. Namiki, 100 nm flat Raman amplifiers pumped and gain equalized by 12-wavelength-channel WDM high power laser diodes. In *Proceedings of the Optical Fiber Conference*, PD19-1, 1999.

[26] D. I. Chang, D. S. Lim, M. Y. Jeon, H. K. Lee, K. H. Kim, and T. Park, Dual wavelength cascaded Raman fiber laser, *Electron. Lett.*, 36:1365–1368, 2001.

[27] S. B. Paperny, V. I. Karpov, and W. R. L. Clements, Efficient dual-wavelength Raman fiber laser, In *Proceedings of the Optical Fiber Communication Conference*, WDD15-1, 2001.

[28] M. D. Mermelstein, C. Headley, J.-C. Bouteiller, P. Steinvurzel, C. Horn, K. Fedder and B. J. Eggleton, A high-efficiency power-stable three-wavelength configurable Raman fiber laser. In *Proceedings of the Optical Fiber Communication Conference*, PD3-1, 2001.

[29] M. D. Mermelstein, C. Headley, J.-C. Bouteiller, P. Steinvurzel, K. Feder, and B.J. Eggleton, Configurable three-wavelength Raman fiber laser for Raman amplification and dynamic gain flattening, *IEEE Photon. Technol. Lett.*, 13:12, 1286–1288, 2001.

[30] M. D. Mermelstein, C. Horn, Z. Huang, P. Steinvurzel, K. Feder, M. Luvalle, J.-C. Bouteiller, C. Headley, and B. J. Eggleton, Configurability of a three-wavelength Raman fiber laser for gain ripple minimization and power partitoning. In *Proceedings of the Optical Fiber Communication Conference*, TuJ2-1, 2002.

[31] X. Normandin, F. Leplingard, E. Bourova, C. Leclère, T. Lopez, Jean-Jacques Guérin, D. Bayart, Experimental assessment of phospho-silicate fibers for three wavelength (1427 nm, 1455 nm, 1480 nm) reconfigurable Raman lasers. In *Proceedings of the Optical Fiber Communication Conference*, TuB2-1, 2002.

[32] M. D. Mermelstein, C. Horn, J.-C. Bouteiller, P. Steinvurzel, K. Feder, C. Headley, and B. J. Eggleton, Six wavelength Raman fiber laser for C+L-band Raman amplification. In *Proceedings of the Conference on Lasers and Electro-Optics*, CThJ1, 2002.

[33] M. D. Mermelstein, C. Horn, S. Radic, and C. Headley, Six-wavelength Raman fibre laser for C- and L-band Raman amplification and dynamic gain flattening, *Electron. Lett.*, 38:636–638, 2002.

[34] F. Leplingard, S. Borne, L. Lorcy, T. Lopez, J.-J. Guérin, C. Moreau, C. Martinelli, D. Bayart, Six output wavelength Raman fibre laser for Raman amplification, *Electron. Lett.*, 38:886–887, 2002.

[35] K. Rottwitt, A. Stentz, T. Nielsen, P. Hansen, K. Feder, and K. Walker, Transparent 80 km bidirectionally pumped distributed Raman amplifier with second order pumping. In *Proceedings of the European Conference on Optical Communication*, II-144-145, 1999.

[36] V. Dominic, A. Mathur, and M. Ziari, Second-order distributed Raman amplification with a high-power 1370 nm laser diode, In *Proceedings of Optical Amplifiers and their Applications*, OMC6, 2001.

[37] L. Labrunie, F. Boubal, E. Brandon, L. Buet, N. Darbois, D. Dufournet, V. Havard, P. Le Roux, M. Mesic, L. Piriou, A. Tran, and J.-P. Blondel, 1.6 Terabits (160 × 10.66 Gbit/s) unrepeated transmission over 321 km using second order pumping distributed Raman amplification. In *Proceedings of the Optical Amplifiers and their Applications*, PD3-1, 2001.

[38] P. Le Roux, F. Boubal, E. Brandon, L. Buet, N. Darbois, V. Havard, L. Labrunie, L. Piriou, A. Tran and J.-P. Blondel, 25 GHz spaced DWDM 160 × 10.66 Gbit/s (1.6 Tbit/s) unrepeated transmission over 380 km. In *Proceedings of the European Conference on Optical Communication*, PD. M. 1.5, 2001.

[39] J.-C. Bouteiller, K. Brar, S. Radic, J. Bromage, Z. Wang, and C. Headley, Dual-order Raman pump providing improved noise figure and large gain bandwidth. In *Proceedings of the Optical Fiber Communication Conference*, Postdeadline Paper FB3, 2002.

[40] S. B. Paperny, V. I. Karpov, and W. R. L. Clements, Third-order cascaded Raman amplification. In *Proceedings of the Optical Fiber Communication Conference*, FB4-1, 2002.

[41] M. Prabhu, N. S. Kim, L. Jianren, and K. Ueda, Simultaneous two-color CW Raman fiber laser with maximum output power of 1.05 W/1239 nm and 0.95W/1484 nm using phosphosilicate fiber, *Opt. Commun.* 182:305–309, 2000.

[42] Alfalight, Boston Laser Inc., IRE-Polus Group, JDSU, LaserTel, SLI, Spectra Physics Products List.

[43] L. Goldberg, B. Cole, and E. Snitzer, V-groove side-pumped 1.5 μm fibre amplifer, *Electron. Lett.*, 33:2127–2129, 1997.

[44] L. Goldberg and J. Koplow, Compact, side-pumped 25 dBm Er/Yb co-doped double cladding fibre amplifier, *Electron. Lett.*, 34:2027–2028, 1998.

[45] IRE-Polus Group Products List.

[46] A. B. Grudin, J. Nilsson, P. W. Turner, C. C. Renaud, W. A. Clarkson, and D. N. Payne, Single clad coiled optical fibre for high power lasers and amplifiers. In *Proceedings of the Conference on Lasers and Electro-Optics*, CPD26-1, 1999.

[47] V. P. Gapontsev, P. I. Sadovsky, and I. E. Samartsev, 1.5 μm erbium glass lasers, *Proceedings of the Conference on Lasers and Electro-Optics*, CPDP-38, 1990.

[48] J. D. Minelly, E. R. Taylor, K. P. Jedrzejewski, J. Wang and D. N. Payne, "Laser-diode pumped neodymium-doped fibre laser with output power >1W," *Proceedings of the Conference on Lasers and Electro-Optics*, CWE-6, 1992.

[49] H. M. Pask, J. L. Archambault, D. C. Hanna, L Reekie, P.St.J. Russell, J. E. Townsend, and A. C. Tropper, Operation of cladding-pumped Yb^{3+}-doped silica fibre lasers in 1μm region, *Electron. Lett.*, 30:11, 863–865, 1994.

[50] E. Snitzer, H. Po, F. Hakimi, R. Tumminelli, and B. C. McCollum, Double-clad, offset core Nd fiber laser. In *Proceedings of the Optical Fiber Communication Conference*, PD5, 1988.

[51] H. Po, E. Snitzer, L. Tumminelli, F. Hakimi, N. M. Chu, and T. Haw, Doubly clad high brightness Nd fiber laser pumped by GaAlAs phased array. In *Proceedings of the Optical Fiber Communication Conference*, PD7, 1989.

[52] P. B. Hansen, L. Eskildsen, S. G. Grubb, A. J. Stentz, T. A. Strasser, J. Judkins, J. J. DeMarco, R. Pedrazzani, and D. J. DiGiovanni, Capacity upgrades of transmission systems by Raman amplification, *IEEE Photon. Technol. Lett.*, 9:2, 262–264, 1997.

[53] Y. Qian, J. H. Povlsen, S. N. Knudsen, and L. Grüner-Nielsen, On Rayleigh backscattering and nonlinear effects evaluations and Raman amplification characterizations of single-mode fibers. In *Proceedings of the Optical Amplifiers and their Applications Conference*, OMD18, 2000.

[54] F. L. Galeener, J. C. Mikkelsen, Jr., R.H. Geils, and W.J. Mosby, The relative Raman cross sections of vitreous SiO_2, GeO_2, B_2O_3, and P_2O_5, *Appl. Phys. Lett.*, 32:34–36, 1978.

[55] N. Shibata, M. Horigudhi, and T. Edahiro, Raman spectra of binary high-silica glasses and fibers containing GeO_2, P_2O_5 and B_2O_3, *J. Non-Crystalline Solids*, 45:115–126, 1981.

[56] D. Marcuse, *Light Transmission Optics*, New York: Van Nostrand Reinhold, Chap. 8, 1982.

[57] V. M. Mashinsky, E. M. Dianov, V. B. Neustruev, S. V. Lavrishchev, A. N. Guryanov, V. F. Khopin, N. N. Vechkanov, and O. D. Sazhin, UV absorption and excess optical loss in preforms and fibers with high germanium content. In *Fiber Optic Materials and Components*, H. H. Yuce, D. K. Paul, R. A. Greenwell, ed. *Proc. SPIE*, 2290:105–112, 1994.

[58] E. M. Dianov, V. M. Mashinsky, V.B. Neustruev, O. D. Sazhin, A. N. Guryanov, V. F. Khopin, N. N. Vechkanov, and S. V. Lavrishchev, Origin of excess loss in single-mode optical fibers with high GeO_2-doped silica core, *Optic. Fiber Technol.* 3:77–86, 1997.

[59] L. Grüner-Nielsen, High index fibers, thesis for Industrial Ph.D., EF 546/Ph.D. No. 94-0146-ATV, Danish Academy of Technical Sciences, May, 1998.

[60] M.E. Lines, W.A. Reed, D.J. DiGiovanni, and J.R. Hamlins, Explanation of anomalous loss in high delta singlemode fibres, *Electron. Lett.*, 1009–1010, 1999.

[61] A. S. Kurkov, V. M. Paramonov, O. I. Medvedkov, S. A. Vasiliev, and E. M. Dianov, Raman fiber laser at 1.45 μm: Comparison of different schemes. In *Proceedings of Optical Amplifiers and Their Applications*, OMB5, 16–18, 2000.

[62] C. R. S. Fludger, B. Handerek, and R. J. Mears, Pump to signal RIN transfer in Raman fibre amplifiers, *J. Lightwave Tech.*, 19:8, 1140–1148, 2001.

[63] M. D. Mermelstein, C. Headley, and J.-C. Bouteiller, RIN transfer analysis in the pump depletion regime for Raman fiber amplifiers, *Electron. Lett.*, 38:403–405, 2002.

[64] N. Tsukiji, N. Hayamizu, H. Shimizu, Y. Ohki, T. Kimura, S. Irino, J. Yoshida, T. Fukushima, and S. Namiki, Advantage of inner-grating-muti-mode laser (iGM-laser) for SBS reduction in copropagating Raman amplifier. In *Proceedings of Optical Amplifiers and their Applications*, OMB4, 2002.

[65] L. L. Wang, R. E. Tench, L. M. Yang, and Z. Jiang, Linewidth limitations of low noise, wavelength stabilized Raman pumps. In *Proceedings of Optical Amplifiers and their Applications*, OMB5, 2002.

[66] B. J. Eggleton, J. A. Rogers, P. B. Westbrook, and T. A. Strasser, Electrically tunable power efficient dispersion compensating fiber Bragg grating, *IEEE Photon. Technol. Lett.*, 11:854–856, 1999.

Chapter 12

Distributed Raman Transmission: Applications and Fiber Issues

Alan Evans, Andrey Kobyakov, and Michael Vasilyev

12.1. Historical Introduction

After discovery of a new physical effect, the speed to widespread commercialization is dictated by complementary technologies and market need. In the case of the Raman effect, the spectral shift of scattered light from "molecules in dust-free liquids or gases" observed in 1928 [1, 2] has long been used for the spectroscopic characterization of materials. Yet application of it to amplification in optical fiber has had a long wait: both for the development of the key enabling technologies, namely, high-power pump lasers and low-loss optical fiber, and for the commercial need for significantly reduced cascaded optical noise buildup in fiber transmission systems.

Even before the availability of high-power sources become a critical issue, the field had to wait for low-loss optical fiber. Raman gain followed soon after 20 dB/km loss fiber was demonstrated and in fact was measured on that same fiber [3]. Another early achievement was the spectral characterization of gain in different waveguide materials [4]. In the 1980s researchers turned to distributed Raman amplification for lossless soliton propagation: first for compensating the loss-induced pulse broadening of 10 ps solitons through 10 km of fiber [5], followed by periodic amplification over 4000 km [6] and then 20 Gb/s single-channel transmission over 70 km [7]. During that same time, the rich history of Raman-preamplified repeaterless transmission for improved receiver sensitivity began [8]. In 1987 the first semiconductor laser diodes (at 60 mW) [9] were used as pumps and soon after greater than 100 mW semiconductor laser diodes were combined with a then-novel pump/signal wavelength-division-multiplexer [10]. However, the commercial development of erbium-doped fiber amplifiers (EDFAs) in the early 1990s quickly supplanted Raman effect as the fiber-optic amplifier technology of choice primarily due to erbium's high signal gain and high pump efficiency.

Although the telecommunications industry rapidly began deployment of EDFAs by the mid1990s, the work on Raman technology still continued, forgetting particularly the amplification in the 1310 nm wavelength band for cable television and as a high-power fiber laser via cascaded Raman gain in fiber Bragg grating resonators. During that same period, the length of repeaterless submarine links was being extended by 1480 nm wavelength pumping of remotely located erbium coils. Pumping through the transmission fiber from either the transmitter or receiver end or both provided additional gain from Raman amplification translating to an increase in loss budget and repeaterless distance. The earliest transmission of significant length and bit rate was in 1991 demonstrating an increase in 620 Mb/s transmission reach from 301 to 375 km using a remotely pumped erbium amplifier [11]. Another milestone was the first 10 Gb/s repeaterless transmission in 1995 using both remotely located post- and preamplifiers and high-power laser diodes over 411 km of optical fiber [12].

These early transmission experiments improved understanding of the system benefits of distributed amplification. Soon Raman-assisted repeatered transmission combining distributed Raman with discrete inline EDFAs were used to control the balance between fiber nonlinearities and noise. At the Optical Fiber Communications Conference of 1999, two system papers demonstrated distributed Raman amplification could keep four-wave-mixing in check even for 50 GHz spaced channels across the zero dispersion wavelength of dispersion-shifted fiber [13, 14]. Optical signal-to-noise ratio (OSNR) was maintained through distributed Raman amplification and launch power was lowered to -12.8 dBm [13]. The flood gate was opened with lower noise-distributed Raman amplification pushing all transmission performance dimensions: total capacity [15–17], spectral efficiency [16, 18, 19], data rate [20, 21], number of spans (i.e., total system reach) [22–24], and span loss [25–27]. Of these, ultra-long (>1500 km) 10 Gb/s and long-haul (>600 km) 40 Gb/s have become leading contenders for commercial deployment due to the overall system cost savings they can provide. Raman for 10 Gb/s transmission holds the promise of longer optical transmission before the need for expensive electrical regeneration. Longer distances are required as the Internet rather than voice dominates transmission traffic. Raman for 40 Gb/s transmission holds the promise of maintaining optical links at current values ($\sim$80 to 100 km) while pushing system reach well beyond 600 km.

This chapter begins by focusing on the two most important properties of an optical amplifier—pump efficiency and noise figure—and compares Raman to erbium amplification. The concept of effective noise figure is reviewed and leads to a generic system scaling relationship that aids in the prediction of Raman-assisted, system performance improvements. The next sections review Raman transmission experiments at 10 and 40 Gb/s and cover design issues specific to these systems. The last section of the chapter introduces dispersion-managed fiber consisting of optical fiber spans that can be optimized for Raman transmission.

12.2. Raman Amplifier Issues

12.2.1. Pump Efficiency

Gain is the fundamental property of interest for an amplifier and high pump efficiency is what makes a particular technology practical. For point-to-point, preamplified transmission, gain from inline amplifiers compensates the loss from the preceding span coming from the combination of cabled fiber, splices, and connectors. Span lengths in US terrestrial long-haul fiber links are between 50 and 120 km, requiring gain between 10 and 27 dB with 80 to 100 km being more typical with gain between 20 and 25 dB.

EDFAs have always been able to provide this level of gain at low pump power from readily available semiconductor laser diodes. The relevant figure of merit is pump efficiency, measured in terms of the percentage of pump photons converted to signal photons. The range for a typical coil of erbium-doped fiber is 65 to 80%. Including all discrete loss elements of an EDFA, the efficiency drops to 20 to 30%. In comparison, the transmission fiber efficiency for distributed Raman is less than 1%: it is a weak, nonlinear scattering process with a small gain per unit length (known as the Raman gain coefficient g_R). In addition, the need to depolarize the Raman pump light to avoid polarization-dependent gain means only half the pump photons are available for stimulated scattering at any given location in the fiber. Finally, there is a several more dBs drop in efficiency to account for the insertion loss of the Raman module at both the signal and pump wavelengths and connector/patch panel loss between the module and the transmission fiber. As an example, to provide all the gain of an average length of standard single-mode fiber (90 km), the total pump power within the fiber would have to be over 1 W. Add to this the well-known issues of efficient pump-to-signal noise transfer, pump-mediated signal crosstalk, and double-Rayleigh backscatter (DRBS) crosstalk and it is not surprising that EDFAs rapidly displaced Raman in the early 1990s.

Several things have changed to the benefit of Raman in the intervening years. First, counterpropagating the Raman pump light with respect to the signals greatly mitigates pump-to-signal noise transfer and pump-mediated signal crosstalk. Second, better understanding of the gain, fiber effective area, and receiver bandwidth dependence of DRBS crosstalk allows for system designs that avoid its onset. Third, low degrees of polarization via polarization-multiplexing of equal intensity pump diodes are easily implemented to control polarization-dependent gain. Finally, better packaging techniques allow for much higher optical power density on optical components in the high-power pump path.

Most important, great progress has been made on increasing the output power Raman sources. As the available pump power has increased, so has interest in deployment of Raman amplifiers. Cascaded Raman fiber lasers were the early front runner as Raman pumps and now provide power well over 1 W and up to six independently power-tunable wavelengths [28–30]. Most early repeaterless experiments used these sources. Semiconductor laser diodes offer power in reliable, compact, and standard packages. Recent reports of 350 mW per diode [31] mean 1 W into the transmission fiber is achievable for a typical four-diode, two-wavelength figuration and an illustrative 1.5 dB pump path loss to the fiber [32].

12.2.2. Effective Noise Figure

Progress in pump power, crosstalk, and PDG alone is not enough to reignite Raman as a viable, economic amplification technology for fiber-optic communications. Revived interest is due to the fact that Raman amplification in transmission fiber has lower noise buildup than conventional EDFAs. One reason is that the distributed nature of Raman

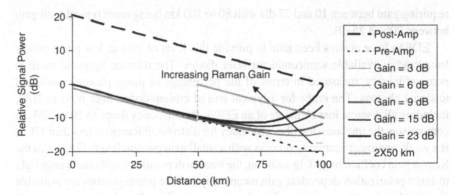

Fig. 12.1. Signal power evolution through 100 km of transmission fiber for various amplification configurations. The solid curves of increasing signal power near the output end of the fiber are for counterpropagating Raman pumping at several power/gain values; 2 × 50 km equals two fiber spans of 50 km each.

amplification allows it to act as a low-noise preamplifier to the subsequent inline discrete amplifier. Another is that twice the number of amplifiers in a hybrid distributed Raman/discrete EDFA system means much less gain and noise per amplifier.

All optical amplifiers have the inseparable processes of stimulated and spontaneous emission. The former provides the desired signal gain through coherent addition of pump photons to signal photons; amplification of the later (i.e., amplified spontaneous emission, ASE) gives the unavoidable ASE–ASE and ASE–signal beat noise at the receiver. It is well known that distributed amplification gives a value of ASE density and fiber nonlinear impairments between that of post- and preamplification [33]. The signal power evolution through a span of transmission fiber in Fig. 12.1 helps to explain why. In postamplification, the signal undergoes gain and the added noise before launch into the fiber. The high signal launch (and high path-average) power yields high fiber nonlinearities but excellent OSNR because both the signal and a fixed amount of ASE power propagate together through the same fiber loss. Postamplification OSNR at the end of the span is $P_s/(\rho_{ASE}\Delta v)$ where ρ_{ASE} is the ASE power density of the amplifier before the fiber span, P_s is the signal launch power, and Δv is the bandwidth of the optical filter before the receiver. In preamplification, the signal undergoes fiber loss before entering a discrete inline amplifier. The low signal launch power yields low fiber nonlinearities but poor OSNR because the signal is attenuated before the addition of the same amount of ASE power. Preamplification OSNR is reduced by the factor $10^{T_f/10}$ where $T_f(dB) < 0$ is the fiber transmittance. Based on signal power evolution, it should not be surprising that distributed amplification is intermediate between these extremes; however, nonideal, counterpropagating Raman pumps allow for a continuous knob to retune the noise/fiber nonlinearity balance. This is especially relevant inasmuch as today's dense wavelength-division-multiplexed systems are effectively preamplified because fiber nonlinearities strongly limit launch power.

Understanding the distributed nature of Raman comes from comparing it with erbium. In EDFAs, the strong absorption and emission cross-section in the three-level erbium ion transition means strongly attenuated pump power and efficient gain conversion within a few tens of meters of erbium-doped gain fiber. In Raman, there is no real energy transition and Raman scattering weakly attenuates the pump. Most of the pump is lost to Rayleigh scattering which sets an effective penetration length of the Raman gain into transmission fibers of ~20 km effectively distributing the gain over a large fraction of the transmission span. So, it's the low pump efficiency that leads to distributed gain over a substantial length of the fiber. Higher Raman pump power raises the minimum signal power and pushes it farther into the fiber as shown in Fig. 12.1. Although some amount of Raman gain exists at every distance within the span, the location of this signal minimum can be considered as the input of an effective amplifier. It has been shown that the generation of ASE in the distributed, backward-pumped Raman amplifier can be approximated by injection of one noise photon per mode at this effective input [34]. Given this, we expect the OSNR to improve by approximately the ratio of minimum signal powers in Raman and preamplified EDFA cases, i.e., higher Raman gain yields lower noise impact, as shown in Fig. 12.2.

Another simple but intuitive interpretation of the effect of Raman is that the addition of *any* amplifier within a transmission span will reduce noise. Cumulative ASE density is proportional to the product of the number of spans and the amplifier gain and noise figure *in linear units*. Consider the example of a single 100 km span of 20 dB loss compared to two 50 km spans of 10 dB each. If the loss-compensating amplifiers have the same noise figure, the total ASE density drops by a factor of $1 \times 100/(2 \times 10) = 5$. The equal amplifier spacing of this example is optimal. For Raman, the farther the gain penetrates into the fiber, the more the hybrid system looks like two equally spaced amplifiers.

It is convenient to replace the complicated distributed, distance-dependent gain and ASE density with an equivalent lumped amplifier at a discrete location in the fiber not only for the simple interpretation described previously, but also for the bookkeeping of cascaded gain and noise buildup throughout a multispan optical link. Lumping the Raman gain and noise at the end of the transmission span allows the use of the usual cascaded noise figure equation provided an effective Raman noise figure is used: $F_T = F_{eff} + (F_{EDFA}/T_L - 1)/G_R$, where F_T is the total noise figure, F_{eff} is the Raman effective noise figure, G_R is the ratio of signal power at the span output with and without the Raman pumps on (i.e., on/off gain),

$$F_{EDFA} = \frac{1}{G_{EDFA}} \left(\frac{\rho_{ASE,EDFA}}{h\upsilon} + 1 \right)$$

is the EDFA noise figure, and $T_L < 1$ is the fractional discrete loss between the Raman gain at the end of the fiber and the EDFA amplifier module (in linear units). A more descriptive name for F_{eff} is differential noise figure because a differential gain (i.e., G_R rather than $G_R \times$ span loss) is used in its definition: $\rho_{ASE,Raman} \equiv h\upsilon(G_R F_{eff} - 1)$, where $\rho_{ASE,Raman}$ is the Raman ASE power density, h is Plank's constant, and υ is the optical frequency. $\rho_{ASE,Raman}$ is the physical quantity of interest and F_{eff} loses its

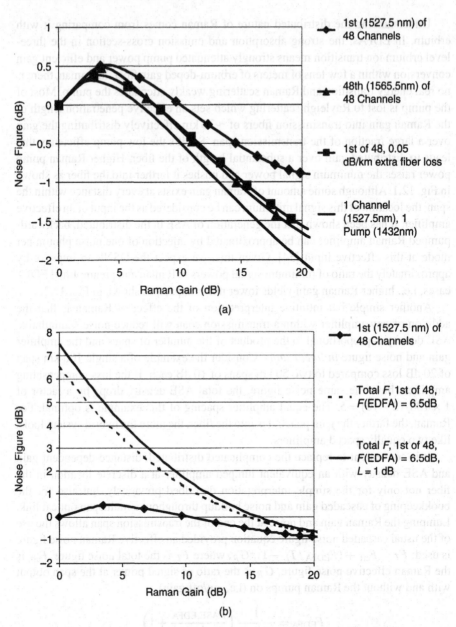

Fig. 12.2. (a). Raman effective noise figure as a function of Raman gain for different pump and signal configurations. Lines with diamonds, squares, and triangles have pumps at 1432 nm and 1461 nm wavelength. The solid line has a single pump at 1432 nm and a single signal at 1527.5nm. (b). Raman-only (line with diamonds) and total noise figure as a function of Raman gain with added 0 dB (solid line) and 1 dB (dashed line) extra insertion loss between the Raman and EDFA modules and assuming an EDFA noise figure of 6.5 dB.

physical meaning as the ratio of output to input electrical signal-to-noise ratio. Hence the effective noise figure can be less than three, or even zero, dB but the total (and physically accurate) noise figure of a distributed Raman amplifier needs to add the front end loss of the amplifying fiber and is much greater than the quantum noise limit.

The use of effective noise figure can sometimes lead to counterinitiative results. Consider the numerical example shown in Fig. 12.2(a) where the effective noise for 90 km of nonzero dispersion-shifted fiber (NZDSF) pumped at 1432 and 1461 nm is plotted versus Raman on-off gain. The solid curve is for a single 1527.5 nm channel at the gain peak of a single 1432 nm pump. This represents the best-case, lowest noise figure. Adding more pumps and signals decreases the pump penetration depth. The highest noise figure occurs for the shortest wavelength, in this case the first of 48 C-band channels from 1527.5 to 1565.5 nm, for a number of reasons: (1) pump-to-pump Raman gain allows the higher wavelength pump to penetrate deeper into the fiber so the higher wavelength signals have more distributed gain than the lower wavelength signals; (2) the phonon energy distribution means that the ground state is more heavily populated for lower wavelength signals; (3) the spectral attenuation of the pump and C-band signals is greater for shorter wavelengths; (4) the signal-to-signal Raman gain means that for spectrally flat gain, the gain of shorter wavelengths occurs closer to the output end of the fiber. One effect, gain saturation, increases the noise figure for all wavelengths but changes the tilt in noise figure in the opposite direction: to higher values at longer signal wavelength (see Chapter 4 for more details on the spectral dependence of the noise figure). Another interesting point about the effective noise figure is that it *decreases* with increasing fiber attenuation. In the figure, the fiber loss at both the pump and signal wavelengths has increased by 0.05 dB/km and the resulting first channel noise figure, shown in squares, decreases by over 0.5 dB at 20 dB on-off gain.

The final issue addressed in this section is the total noise figure of hybrid distributed Raman/discrete EDFAs. Assuming a constant 6.5 dB noise figure for the subsequent EDFA independent of input power, the total noise figure for the first channel of our 48 channel example is shown as the solid line in Fig. 12.2(b). Adding a 1 dB loss between the Raman and EDFA module to account for the 14XX/1550 nm wavelength-division-multiplexer or patch panel connector loss gives the dashed line in the same figure. This example illustrates that minimizing the "midstage" loss of this hybrid amplifier is particular important at low Raman gain.

12.3. Raman Q Scaling

12.3.1. Rule of Thumb

This section describes a simple Q scaling relationship [35] that is useful for predicting the relative system performance improvement when the total amplifier noise figure decreases due to distributed Raman amplification. As a relative scaling, it is robust against most system parameters. However, it assumes that systems are limited by the noise from signal–spontaneous ASE beat noise and interchannel crossphase modulation (XPM) and four-wave-mixing (FWM). Predictions from this simple scaling will not be accurate for systems limited by dispersion, gain nonuniformity, polarization mode dispersion, or single-channel nonlinearities. However, where the scaling is known for these effects, the relationship can be generalized. Additional loss due to network elements is particularly easy to add.

System Q, a measure of the electrical signal-to-noise ratio, is directly related to the bit error rate (BER). Many of the contributions to Q can be ignored in a well-designed system: detector thermal noise, signal and ASE shot noise, amplifier spontaneous–spontaneous beat noise, and laser relative intensity noise. In typical 10 Gb/s long-haul dense wavelength-division-multiplexing (WDM) systems, signal–spontaneous emission beat noise is the dominant noise mechanism and XPM and FWM are often the dominant nonlinear effects. Considering only ASE-generated signal–spontaneous beat noise and treating XPM and FWM as Gaussian noise, system Q has the following functional dependence on channel power P, referenced to the output power of the inline optical amplifier.

$$Q = \frac{P}{\sqrt{aP + bP^4}}. \tag{12.1}$$

The coefficient a governs the signal–spontaneous beat noise and the coefficient b governs the combined XPM and FWM noise terms. As four-photon parametric processes, the XPM and FWM can be shown to depend on P^4. A typical Q dependence on launch power is shown in Fig. 12.3, using Eq. (12.1). The a and b coefficients have been set equal to 6.54×10^{-3} so that at -1 dBm launch power, the maximum Q, in 20 log dB units, is 18 corresponding to a BER of 10^{-15}. Equation (12.1) can be re-written as

$$Q = \frac{\sqrt{3}Q_0(P/P_0)}{\sqrt{2(P/P_0) + (P/P_0)^4}}. \tag{12.2}$$

The significance of this form is that the peak value of $Q = Q_0$ occurs at $P = P_0$. The relationship between a and b, and P_0 and Q_0 is

$$a = \frac{2P_0}{3Q_0^2} \qquad P_0 = \left(\frac{a}{2b}\right)^{1/3}$$

$$b = \frac{1}{3P_0^2 Q_0^2} \qquad Q_0 = \left(\frac{4}{27a^2 b}\right)^{1/6} \tag{12.3}$$

with the coefficient a scaling as

$$a \propto NF_T G_T,$$

where F_T is the total amplifier noise figure (linear units), N is the number of spans, and G is the total amplifier gain (linear units); the coefficient b scales as

$$b \propto \frac{N}{A_{\text{eff}}^2}. \tag{12.4}$$

The interesting aspect is that the maximum Q scales as

$$Q_0 \propto \frac{1}{F^{1/3}}, \frac{1}{G^{1/3}}, \frac{1}{N^{1/2}}, \quad \text{and } A_{\text{eff}}^{1/3}.$$

One nonobvious conclusion is that a 1 dB improvement in the amplifier noise figure does not give a 1 dB improvement in Q^2 as one would expect from considering only

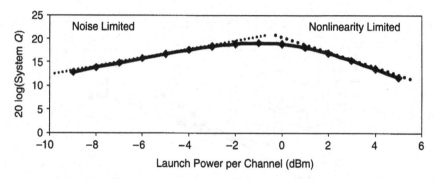

Fig. 12.3. A graph of Q versus launch power per channel from Eq. (12.1) with the maximum $Q = 18$ for -1 dBm launch power.

the signal–spontaneous beat noise. In fact, inclusion of nonlinear impairments implies that the improvement is only 0.66 dB. In general, inclusion of the nonlinear Q power scaling is important in scaling to a different number of spans, gain per span, noise figure, channel spacing and effective area.

12.3.2. Scaling Verification

To verify the scaling for Raman applications, a number of numerical transmission simulations were performed. The base case contained no Raman amplifiers with other parameters as follows: 6 spans of nonzero dispersion-shifted fiber of length 100 km, 32 channels at 10 Gb/s, 100 GHz channel spacing from 1536 nm to 1562 nm, fiber loss of 0.25 dB/km, 100% slope-compensating fiber of 0.4 dB/km loss, ignoring polarization mode dispersion (PMD) and gain ripple. The base case $20 \log Q$ was 20.4 dB. For this Q, coefficients a and b in Eq. (12.3) were assumed to be equal to one. From this base case, Raman was added with 10 dB gain and -1.6 dB effective noise figure, neglecting gain ripple and additional fiber nonlinearities that could be caused by a change in the channel power evolution; that is, only the improvement in cascaded amplifier noise figure was included. The system Q, with and without Raman, was calculated changing a number of system parameters: the number of spans was varied from 5 to 8 without Raman and 5 to 30 with Raman; the channel spacing was changed to 50 GHz for 5 to 8 spans; the data rate was increased to 40 Gb/s with 100 and 200 GHz spacing; and span length was changed to 5×120 and 4×150 km. For each system, the launch power per channel was reoptimized.

The success of scaling is illustrated in Fig. 12.4 which plots 17 system transmission simulations of $10 \log(Q)$ versus $10 \log(4/(27a^2b))$ on the right y-axis and upper x-axis and optimum launch power per channel versus $10 \log(a/(2b))$ on the left y-axis and lower x-axis. As expected, all points fall on a line with slope 1/6 for the Q dependence and 1/3 for the power dependence. As previously mentioned, this should be valid scaling for any NRZ or RZ modulated 10 Gb/s system, limited by crosschannel nonlinearities. It may also work for other modulation formats or data rates.

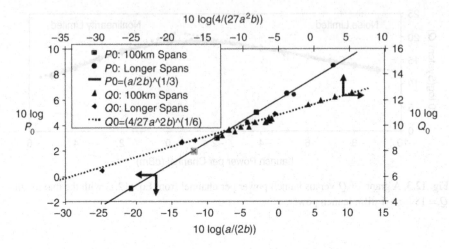

Fig. 12.4. Plot of 17 system simulations with respect to scaling parameters. The 10 logs of Eq. (12.4) are shown as solid and dotted lines for power and Q, respectively.

12.4. 10 Gb/s

Hybrid Raman/EDFA systems combine the lower noise of distributed Raman amplifiers with the higher gain and pump efficiency of discrete EDFAs. The resulting Q improvement can be used to improve various system parameters: span length, as in repeaterless submarine experiments of the early to mid1990s; increased channel count as in the repeated dispersion-shifted fiber demonstrations; spectral density as in the more recent results greater than 0.4 bit/s Hz; increased number of spans described in the following section; and higher data rate as described in Section 12.5. The scaling rule of thumb can be used to explain why these systems work.

12.4.1. 10 Gb/s System Design

The predominant system application of Raman amplification is increasing the number of spans to meet the needs of today's changing long-haul traffic patterns. Internet-initiated connections are much less distance-dependent than are voice connections, and IP packages are rapidly dominating total transmission bandwidth. The 10 Gb/s NRZ system simulations described in the last section are shown in Fig. 12.5. In Fig. 12.5(a), Q versus launch power per signal shows that the addition of Raman with an effective noise figure of -1.6 dB shifts the maximum Q launch power lower by 2 dB. The optimum launch power has a broad maximum with Q decreasing at lower power due to OSNR degradation and at higher power due to fiber nonlinearities. In Fig. 12.5(b), Q versus number of spans shows that the addition of Raman increases Q for a fixed number of spans by ~4 dB or increases system reach for a fixed Q by 2.5 times.

A key design consideration is the appropriate gain partitioning between the distributed and discrete gain of Raman-assisted transmission. Factors include the optical signal-to-noise ratio, fiber nonlinearities, DRBS-induced multipath interference

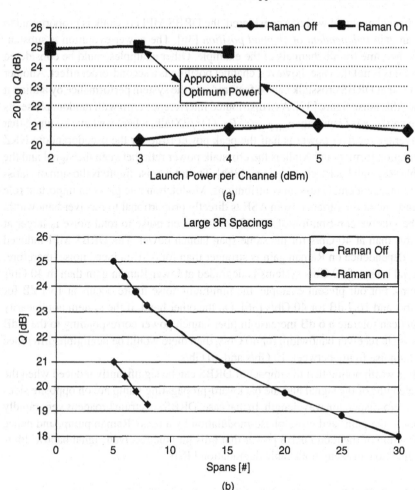

Fig. 12.5. (a). 10 Gb/s NRZ simulations showing the dependence of Q on launch power per channel for five span NRZ transmission with (squares) and without (diamonds) Raman amplification for effective noise figure of -1.6 dB. (b). 10 Gb/s NRZ simulations showing the dependence of Q on the number of 100 km spans of NZDSF with (circles) and without (diamonds) Raman amplification for effective noise figure of -1.6 dB.

(MPI) crosstalk, and even the cost of amplification. Although EDFA costs are rather insensitive to design gain, distributed Raman gain (in dB) costs increase in proportion to the pump power (in mW) because they operate in the linear, undepleted pump regime. The lower pump efficiency of Raman means that the cost of gain is higher than EDFAs. Therefore, total amplifier costs increase as the fractional gain in distributed Raman increases. Cost-effective gain partitioning is set by increasing the distributed Raman gain to the necessary level to achieve acceptable system Q, but no higher.

Although double Rayleigh backscattering is covered in another chapter, it is important enough to warrant some comments here. For the 48 channel, 0 dBm launch per channel example we have considered, the DRBS MPI crosstalk is proportional to Raman gain *independent of channel position* [36]. The power evolution is substantially the same for all channels of the example. Other examples could be considered where this is not the case, however, channel position is a second-order effect. Another point is that DRBS crosstalk can be factored into the system performance by adding it to the ASE noise figure [37]. As a distributed interference term with random phase, its noise variance is Gaussian: $\sigma_{MPI^2} = r^2 P_{Rx}(1)^2/2\text{Xtalk}$, where we have assumed that the average receiver power is half the peak power (true for the pseudorandom NRZ modulation format) and Xtalk is the crosstalk power ratio between the signal and the double-Rayeigh backscattered signal [36b]. The total noise figure is the sum of statistically independent, Gaussian contributions. Modulation rate plays an important role because the noise variance from ASE is directly proportional to receiver bandwidth. So the relative contribution of DRBS MPI-induced noise to total noise is larger at 10 Gb/s than at 40 Gb/s for the same span launch power. The DRBS MPI-induced noise dependence on Raman gain is stronger than for ASE-induced noise, therefore, the total noise of 10 Gb/s systems is degraded at lower Raman gain than in 40 Gb/s systems. For our present example, the minimum noise figure occurs at 13.1 dB for 10 Gb/s and 16.7 dB for 40 Gb/s [36]. On the other hand, if the system nonlinearity budget can tolerate a 6 dB increase in fiber launch power corresponding to the 6 dB increase in receiver bandwidth for 40 Gb/s, then there would be negligible difference in total noise figure between 10 Gb/s and 40 Gb/s.

It is worth noting that the impact of DRBS can be significantly reduced when the wavelengths of the signal and the backward-propagating pump are on opposite sides of the zero-dispersion wavelength. In that case, DRBS spectrum broadens very rapidly owing to the gain- and cross-phase-modulation by a noisy Raman pump, and hence the fraction of the total DRBS power that falls into the received signal bandwidth is too small to cause any noticeable degradation [36a].

12.4.2. 10 Gb/s Raman Transmission Experiments

Recent improvements in transmission performance enabled by Raman amplification are best seen in a compilation of the recent transmission experiments as shown in Table 12.1 [13, 14, 22, 24, 26, 38–51]. Although there has been much work at lower data rates and single-span transmission dating back into the 1980s, the first single-span 10 Gb/s Raman-assisted transmission was not until 1995 [12] and multispan with 7200 km reach followed two years later [38]. As previously mentioned, several early papers applied distributed Raman to the problem of four-wave-mixing in dispersion-shifted fiber where the zero dispersion wavelength is in the middle of the C-band. Fujitsu researchers instead pushed on the reach performance dimension, showing 1 Tb/s (100 × 10 Gb/s) over 10,000 km [22], and doubling to 2 Tb/s over the somewhat shorter 7200 km reach the following year [23]. These have been followed by a couple of S-band transmission experiments at modest total capacity and single wideband transmission [45, 46]. All-Raman amplification, where the transmission fiber is pumped close to transparency by a combination of counter- and copropagating pumps and, optionally, the dispersion-compensating fiber having separate Raman pumps,

Table 12.1. Summary of Repeatered 10 Gb/s Raman Transmission Experiments

Total Capacity Gb/s	Reach (km)	Raman Gain	Claim to Fame	Group	Year	Ref.
100	7200	2 dB higher OSNR in 514 km chain	First high capacity multispan WDM at 10 Gb/s	Lucent	1997	[38]
400	5 × 120	12 dB	4 dB power margin in multispan WDM system	Lucent	1999	[39]
490	4 × 84	440 mW	50 GHz spacing in DSF	Lucent	1999	[13]
250	8 × 84	440 mW	100 GHz but farther than 50 GHz	Lucent	1999	[13]
320	80 × 80	15.3 dB backward, 4.3 dB forward	50 GHz spacing in DSF	NTT	1999	[14]
1000	200 × 50	1.1 dB	1 Tb/s over 10,000km, C+L	Fujitsu	1999	[22]
1000	4 × 80	18 dB backward, 4 dB forward	25 GHz spacing in DSF	NTT	1999	[40]
320	6 × 125	10 dB	6 passes through a 640 × 640 WS × C	Corning	2000	[41]
1000	3 × ~133	10 dB	25 GHz spacing in L-band	Lucent	2000	[42]
1280	6 × 140	5.8 dB in C, 5.4 dB in L	Best loss × span and Distance × Length	Fujitsu	2000	[26]
320	3 × 250	1 W	Submarine medium haul	Alcatel	2000	[43]
2110	90 × 80	7 dB	Distance × span length product using asym DMF	Fujitsu	2000	[23]
400	6 × 100	All-Raman 20 − 24 dB +650/250 mW	Highest S-band distance × bit rate product	Lucent	2001	[45]
200	10 × 87	21 dB	Discrete S-band slope–DCF	Xtera	2001	[46]
400	40 × 100	23.6 dB	All-distributed Raman >80 km spans using sym DMF	Lucent	2001	[47]
2400	185 × 40	1.5 dB OSNR improvement	Tb/s greater than 2500 km, single 74 nm amplification	Fujitsu	2001	[48]
1280	40 × 100	20 dB co + counter in NZDSF + 12 dB in DCF	53 nm all-Raman over 100 km spans	Lucent	2001	[49]
800	52 × 80	20.5 dB	Reconfigurable OADMS with DMF and all-distributed Raman	Corning	2002	[50]
2560	~270 × 40	not mentioned	Widest transoceanic bandwidth, single Raman pump	Tyco	2002	[24]
800	52 × 80	20.5 dB	Capacity × distance record (no FEC) for all-distributed Raman using sym DMF	Corning	2002	[51]

offers complete bandwidth flexibility in terms of placement and width [45, 47, 49, 51]. An intriguing hybrid Raman/EDFA combination offered simple yet very wide band amplifiers and a very long reach system [24].

A few conclusions are possible from Table 12.1. First, most major system houses are participating in using Raman amplifiers to demonstrate proof-of-concept, record-breaking transmission. Second, there is no standard Raman gain value: it ranges from 1 to 22 dB depending on the trade-off with other system parameters. Third, tremendous progress has been made over the last few years in system performance: terabits per second over much greater than 1000 km. It should be noted though, that Raman is only one of the recently added tools; forward error correction (FEC), lower gain ripple amplifiers, and better dispersion slope compensation and dispersion-managed fiber spans are others.

12.5. 40 Gb/s

For 40 Gb/s bit rate systems, noise accumulation is a challenging design issue, as it seriously limits amplifier spacing and system reach. Cost-effective terrestrial transmission systems require amplifier spacing of at least 80 km and at least five to six spans between electrical regeneration. Distributed Raman-assisted transmission offers the necessary low-noise amplification for economical, field-deployed 40 Gb/s systems. In addition, discrete Raman, possibly in the form of a pumped dispersion-compensating fiber, overcomes the bandwidth limitations created by the discrete energy levels of erbium in EDFAs. Lastly, carefully designed all-Raman systems combining distributed and discrete amplifiers could reduce the number of discrete components compared to EDFA-based systems. These components are a source of PMD, polarization-dependent loss (PDL), and group delay ripple that impairments need careful consideration for 40 Gb/s.

12.5.1. Noise

For a given receiver sensitivity, the 25% reduction in bit period in going from 10 to 40 Gb/s implies a 6 dB increase in launch power for an equivalent amplifier Q (i.e., the system Q of Eq. (12.1) for $b = 0$). However, this increase is usually not possible due to system degradation from increased fiber nonlinearities. In fact, as previously mentioned, this nonlinearity-limiting launch power causes today's long-haul systems to operate in a preamplified mode rather than the postamplified mode. Without any change in technology, in particular the amplification noise figure, the scaling of amplifier Q implies that the number of spans decreases by a factor of four when moving from 10 to 40 Gb/s. Or for a fixed number of spans, the amplifier gain needs to decrease by a factor of four. For example, with a fiber attenuation of 0.25 dB/km, a 6 dB gain reduction is 24 km of span length. Yet with distributed Raman amplification the increased receiver bandwidth can be counteracted by an overall noise figure reduction. In fact, for very high Raman gain, noise improvement greater than 6 dB can be used to increase the number of spans. Looking at the Raman noise figure, the solid curve of Figure 12.2(b), a total noise figure improvement greater than 6 dB occurs for Raman gain greater than ~10 dB, meaning a 40 Gb/s hybrid Raman/EDFA system will have less noise buildup than a 10 Gb/s, EDFA-only system. That noise improvement can be converted to a fractional reach increase as shown with the Q scaling equation.

12.5.2. Other Amplifier Design Issues

Although optical amplification is inherently bit rate-independent, PMD, PDL, and group delay ripple within the optical components, as well as differences in gain and loss requirements, typically prevent the transparent bit rate upgrade of systems. Differences in the gain/loss map arise from gain partitioning reoptimization among the distributed and discrete amplifiers, different dispersion-compensation maps, and different required optical networking functionality and their associated insertion losses. Whereas the PMD requirement scales inversely with bit rate and its interaction with PDL sets specifications at the limit of current component manufacturing capability. Some components, particularly those based on grating and dielectric thin films, have high-frequency wavelength variations in their waveguide propagation constants. Such group delay ripple becomes a problem for the wider signal bandwidth of 40 Gb/s.

The exact requirements for PMD, PDL, and group delay ripple is system specific and dependent on contributions from the other elements of the optical link (transmission and dispersion-compensating fiber, dynamic spectral equalizers, wavelength add-drop multiplexers and optical switches, etc.). The contribution from Raman amplifiers is expected to be less than from EDFAs due to lower component counts making them an attractive option for 40 Gb/s systems. Typical distributed Raman amplifiers only have a single WDM to couple pump light into the transmission fiber. Given that roughly half the gain comes from this amplifier, the discrete amplifier, whether EDFA or Raman-based, should have fewer stages of amplification. In addition, discrete Raman-based pumping of dispersion compensating fiber eliminates the erbium-doped fiber coils that can be a source of PDL.

12.5.3. 40 Gb/s Raman Transmission Experiments

Raman amplifier technology was well timed to be introduced into 40 Gb/s transmission experiments as higher speed transmitters, electrical amplifiers, and receivers were becoming available. A complete list (to the best of the author's knowledge) of repeatered, 40 Gb/s Raman transmission experiments is given in Table 12.2 [15–17, 27, 52–68]. Experiments by Lucent and NEC in 1999 proved that Raman allowed for the same amplifier spacing as lower bit rates: multiple spans of 80 to 100 km [52, 53]. The following year, a record capacity 3 Tb/s transmission over 3 × 100 km of NZDSF was enabled by combining Raman and a well-matched dispersion slope compensating fiber with C+L -band transmission [15]. Pushing a new performance dimension, Nortel showed that 160 to 200 km amplifier spacing and a record 1000 km reach for 1.28 Tb/s was possible [27]. In that same European Conference of Optical Communications 2000 session, Alcatel pushed total capacity to 5.12 Tb/s using

Table 12.2. Summary of Repeatered 40 Gb/s Raman Transmission Experiments

Total Capacity (Gb/s)	Reach (km)	# of λs	Raman Gain	Claim to Fame	Group	Year	Ref.
1600	4 × 100	40	23 dB, 520 mW	no OTDM, Pmux, or FEC, 0.4 bit/Hz	Lucent	1999	[69]
320	5 × 80	8	9 dB, 110 mW	RDF, 80 km amp + 40 Gb/s	NEC	1999	[53]
3280	3 × 100	82	25 dB in C, 24 dB in L	Record capacity	Lucent	2000	[15]
1280	3 × 100	32	5 to 8 dB	Full ETDM, 0.4 bit/s/Hz spectral efficiency	Alcatel	2000	[54]
5120	3 × 100	128	12 dB in C, 10 dB in L	0.64 bit/s/Hz spectral efficiency	Alcatel	2000	[16]
1280	6 × 160/200	32	5 to 10 dB	Record span length and loss	Nortel	2000	[27]
640	6 × 120	16	7 to 10 dB	CS-RZ, 0.4 bit/s/Hz over 120 km spans	Nortel	2001	[55]
3080	12 × 100	77	23.5/22.5 dB	3 Tb/s using dual C and L Raman	Lucent	2001	[56]
10920	2 × 58	273	5.5 dB	First 10 Tb/s—Raman in S-band	NEC	2001	[17]
3200	3 × 82	80	8 dB	3 Tb/s in field fiber	WorldCom, Siemens	2001	[57]
640	25 × 80	16	8 dB	Gain partitioning optimized (1/2 span loss)	KDD	2001	[58]
1280	60 × 40	32	12.2 dB	Tb/s greater than 2000 km, all-Raman + DMF	Alcatel	2001	[59]
5000	12 × 100	125	15 dB in NZDSF + 8 dB in DCF	2 Tb/s higher capacity at same distance than line 9	Alcatel	2001	[60]
1600	20 × 100	40	not mentioned	Longest reach over terrestrial span length using dispersion-managed spans	OFS	2001	[61]
3200	3 × 100	80	14 dB in NZDSF + 6 dB in DCF	3.2 Tb/s in C-band only with 0.8 bit/s-Hz, no PDM	Alcatel	2001	[62]
1600	24 × 100	40	15.5	Combination of capacity, span length, and distance over commercial available fiber	OFS	2002	[63]
2560	40 × 100	64	not mentioned	Capacity × distance record of 10 Petabit-km/s, Raman used but not main enhancement	Lucent	2002	[64]
10240	3 × 100	256	not mentioned	10 Tb/s over 100 km spans, counterpropagating second-order Raman + pumped DCF	Alcatel	2002	[65]
1280	1160 × 52	32	11.5 dB all-Raman	First 40 Gb/s transoceanic distance, quadruple-hybrid span	NEC	2002	[66]
1600	36 × 100	40	7.5 dB forward + 18.5 dB backward all-Raman	Record number of 100 km spans at 40 Gb/s	Mintera	2002	[67]
3200	20 × 100	80	All-Raman: ~6dB co-, ~15 dB counter, ~6 dB in DCF	Higher capacity/lower reach than previous	OFS	2002	[68]

vestigial sidebandlike filtering to reach a spectral efficiency of 0.64 bits/s/Hz [16]. The large signal-to-signal Raman interaction across the entire C+L 72 nm band was partially compensated using distributed Raman amplification. Then in 2001, 3 Tb/s was pushed to 1200 km with FEC, tight residual dispersion control, and Raman [56]; 10 Tb/s in one fiber was reached using S+C+L and 273 wavelengths [17]; and multiple Tb/s (80 × 40 Gb/s) field trials brought the technology closer to commercial reality [57]. Six transmission demonstrations in the first half of 2002 continued to add new complementary technologies to enhance performance. Raman itself keeps amplifier spacing at the commercially viable 80 to 100 km range. In combination with other technologies such as FEC, dispersion slope compensation, dispersion-managed fiber spans, gain equalization, and alternate modulation formats it enables up to 10 Tb/s capacity, greater than 0.4 bit/s/Hz spectral efficiency, and reach over 3000 km. Raman has become an indispensable tool in the system designer's toolbox.

Another trend for Raman is toward more complex pumping schemes: forward pumping of the transmission fiber, pumping the dispersion-compensating fiber, or even second-order co- or counterpropagating pumping. For one thing, these tricks allow for all-Raman transmission without any EDFAs. For another, the signal power evolution is flattened, leading to a more optimal balance between the constraining effects of ASE beat noise for low launch power and fiber nonlinearities for high launch power. In fact, flat signal power along the fiber is ideal. The dispersion-managed fiber described in the following section is another approach to attain this ideal.

12.6. Raman Amplification in Dispersion-Managed Fibers

As we discussed above, the distributed optical amplification improves OSNR compared to lumped preamplification and reduces fiber nonlinearities compared to lumped postamplification. (We used the same convention for definition of pre- and post-amplification as in Section 12.2.2.) The optimum trade-off between the nonlinearities and noise is achieved by the distributed amplifier that balances gain and loss at every point in the fiber, making the signal power constant versus distance. One of the key trends in the evolution of Raman amplifiers is the progress toward approaching the performance limit of this ideal distributed amplifier. Combining backward (i.e., counterpropagating with the signal) and forward (copropagating) Raman pumping, as previously described, has recently been complemented by other methods that greatly facilitate achieving optimum signal profile in the span. The first is the use of effective-area management in the span (e.g., by employing the dispersion-managed fiber, or DMF) to distribute the Raman gain more evenly along the fiber [47, 70–74]. The second is the demonstration of feasibility of second- and higher-order Raman pumping in a distributed amplifier, which is another way of reaching the same goal [30, 75, 76a].

This section deals with Raman amplification in DMFs that have recently gained a significant amount of interest [23, 47, 50, 51, 66, 77, 78]. The potential of DMFs as the fiber of choice in next-generation high-capacity transmission systems has been demonstrated lately by several research groups (see Tables 12.1 and 12.2). Below we discuss peculiarities of noise performance caused by inhomogeneous fibers with non-axially uniform optical properties and show that an optimally pumped DMF closely (within a few dB) approaches the performance of an ideal Raman amplifier based on a single fiber.

DMFs, which are hybrid fiber spans consisting of sections having positive $(+D)$ and negative $(-D)$ dispersion, were originally designed for the dispersion and dispersion slope compensation. It turns out, however, that they also offer significant benefits for Raman amplification. High pumping efficiency in the small effective area $-D$ fiber, combined with improved noise performance discussed below are the main factors that contribute to the DMF advantage over a single fiber type. The reduction of the midstage loss requirement of an EDFA with the removal of the dispersion-compensation module allows for a further reduction in the effective noise figure of a Raman-assisted transmission system. In addition, DMFs were shown to be more robust with respect to intrachannel nonlinearities [79].

12.6.1. Improved Noise Performance of DMF-Based Raman Amplifiers

Apart from more complicated DMF configurations [66], two types of DMF design are most common: $+D/-D$ (asymmetric) and $+D/-D/+D$ (symmetric). As we have described previously, a more uniformly distributed Raman gain entails less ASE penalty. This also holds true for the DMF where the desirable deeper penetration of the pump into the span can be achieved by proper management of the effective area.

The idea of effective area management is to extend the region of useful Raman gain deeper into the fiber span (Fig. 12.6(a)) by using large effective area fiber (e.g., $+D$ section of DMF) near the pump launch point and small effective area fiber (e.g., $-D$ section of DMF) at some distance from it. This results in improved OSNR (Fig. 12.6(b)). The symmetric DMF design with $-D$ fiber in the middle of the span is particularly attractive. As can be seen from Fig. 12.6(b), the minimum signal power in the symmetric configuration is higher than that in the asymmetric configuration, leading to superior noise performance in the $+D/-D/+D$ design.

The asymmetric design has other disadvantages. Although placing the $-D$ fiber at the end of the DMF span maximizes the Raman gain efficiency, the smaller effective area in the $-D$ fiber enhances intensity-dependent nonlinear impairments. In addition, because most of the Raman gain is generated in the small effective area $-D$ fiber at the exit end of the span, this type of DMF is highly DRBS-prone (DRBS is inversely proportional to the square of effective area). In contrast, for symmetric DMF only a fraction of the total Raman gain is generated in the $-D$ fiber, so the DRBS crosstalk is reduced, which, in turn, reduces the minimum effective noise figure and shifts the minimum to a higher Raman gain. These trends are illustrated in Fig. 12.7 [70] where the measured effective lumped noise figure of a distributed Raman amplifier is shown for single fiber, symmetric DMF, and asymmetric DMF. A dramatic noise figure improvement ($\sim$2.5 dB) of symmetric DMF over asymmetric DMF and single fiber can be seen.

As discussed in Section 12.2, a more homogeneously distributed Raman gain provides better noise performance. On the other hand, it increases the path-average power (Fig. 12.6(b)) of the signal and therefore induces more nonlinear penalties. The question arises as to whether the noise figure advantage of the DMF is large

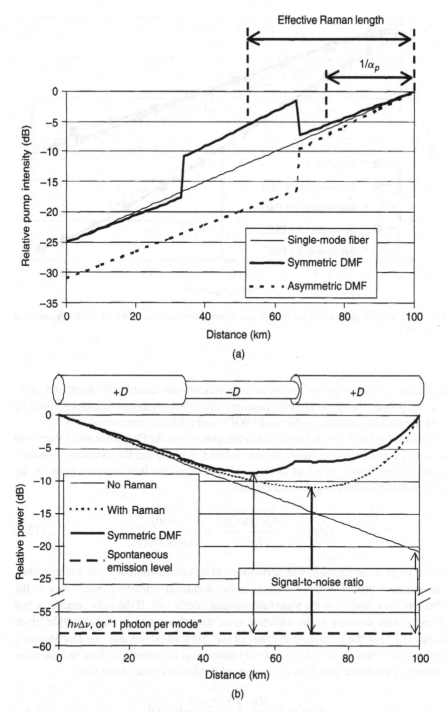

Fig. 12.6. Pump (a) and signal (b) evolution in a single fiber (with and without backward Raman pumping) and symmetric DMF with backward Raman pumping. A schematic drawing of symmetric DMF is shown above the graph.

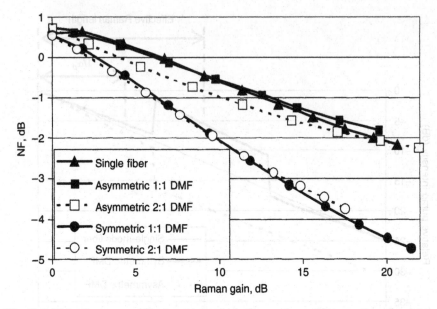

Fig. 12.7. Noise figure versus Raman gain in single fiber and two different x:1 configurations of the DMF.

enough to compensate for nonlinear impairments associated with increased path-average power. The answer requires a more elaborate analysis that includes accounting for bidirectional pumping, effects of DRBS, and postspan component loss.

To resolve this issue, we note that in the span where the fiber parameters vary with distance, constant signal power versus distance is no longer the optimum evolution profile. If we ignore the effect of Rayleigh scattering, the best trade-off between the ASE and nonlinearity is achieved for [80]

$$G(z) = \sqrt{\frac{A_{\text{eff}}(z)\alpha_s(z)}{n_2(z)} \times \frac{n_2(0)}{A_{\text{eff}}(0)\alpha_s(0)}}, \qquad (12.5)$$

where $G(z)$ is the signal power normalized to its launch value P_{in} (for a single fiber, the optimum profile is $G(z) = 1$), n_2 is the nonlinear refractive index, A_{eff} is the effective area, and α_s is the signal attenuation coefficient. If the only parameter that changes with distance is the effective area, then the optimum signal profile given in Eq. (12.5) is the geometric average of the constant-power and constant-intensity profiles. For transparent (i.e., $G(L) = 1$) spans with the same span loss and the same amount of nonlinear penalties characterized by the nonlinear phase shift

$$\phi_{NL} = P_{\text{in}}\frac{2\pi}{\lambda}\int_0^L \frac{n_2(z)}{A_{\text{eff}}(z)}G(z)\,dz,$$

the OSNRs of the ideal distributed single-fiber amplifier and DMF with the ideal signal profile (12.5) are related by

$$\frac{\text{OSNR}_{\text{idealDMF}}}{\text{OSNR}_{\text{ideal}}} = \frac{n_2 A_{\text{eff}}^+}{n_2^+ A_{\text{eff}}} \frac{(x + \alpha_s^-/\alpha_s^+)(x + 1)}{[x + \sqrt{\alpha_s^- n_2^- A_{\text{eff}}^+/(\alpha_s^+ n_2^+ A_{\text{eff}}^-)}]^2}, \tag{12.6}$$

where x is the length ratio of $+D$ and $-D$ span portions, and "+", "−", and no superscripts denote $+D$, $-D$, and single-fiber parameters, respectively. For $x = 2$ and parameters listed in Table 12.3 below, the ideal DMF case offers 0.8 dB OSNR advantage over the ideal single-fiber case.

When Rayleigh scattering is taken into account, a more elaborate analysis is required. The next section describes optimization of a realistic bidirectionally pumped DMF, including the effects of DRBS, Rayleigh scattered ASE, and postspan component loss.

12.6.2. Optimization of Raman Amplifiers Based on DMFs

The on-off Raman gain found from the undepleted pump approximation in an inhomogeneous fiber can be written as

$$G_R(z) = \exp\left[\int_0^z \gamma(z')\alpha_p(z')p_p(z')\,dz'\right], \tag{12.7}$$

where piecewise functions of distance

$$\gamma(z) = \frac{g_R(z)P_R}{A_{\text{eff}}(z)\alpha_p(z)},$$

α_p, and g_R are, respectively, the dimensionless Raman interaction strength in a nonpolarization-maintaining fiber, the loss at the pump wavelength, and the Raman gain coefficient. The evolution of the total (forward and backward) normalized pump has the form

$$p_p(z) = (1 - k)\exp\left(-\int_z^L \alpha_p(z')\,dz'\right) + k\exp\left(-\int_0^z \alpha_p(z')\,dz'\right), \tag{12.8}$$

where $k = P_f(0)/P_R$ is the ratio of forward pump power to the total launched pump power P_R, and L is the span length. The total gain $G(z)$ is related to the on-off gain by $G(z) = G_R(z)p_s(z)$. It is also assumed that linear signal attenuation

$$p_s(z) = \exp\left(-\int_0^z \alpha_s(z')\,dz'\right)$$

Table 12.3. Fiber Parameters Used in Calculations

Fiber Type	α_s [dB/km]	α_p [dB/km]	$\alpha_{\text{RBS}} \times 10^{-4}$ [1/km]	A_{eff} [μm^2]	$g_R/A_{\text{eff}} \times 10^{-3}$ [$\text{m}^{-1}\text{w}^{-1}$]	$n_2 \times 10^{-20}$ [m^2/W]
+D	0.19	0.22	0.4	110	0.2	1.8
−D	0.25	0.29	1.6	30	1.0	2.3
SF	0.21	0.26	1.0	55	0.64	2.2

and pump attenuation $p_p(z)$ include lumped loss at the point of fiber splice $z = z_{\text{splice}}$. Similarly to single fiber [81], the effective lumped noise figure of an inhomogeneous fiber can be calculated as

$$F_{\text{DMF}}^{\text{ASE}} = \frac{1}{G_R(L)} + 2n_{sp}T_F \left[\int_0^L \frac{\gamma(z)\alpha_p(z)p_p(z)}{G(z)}\, dz \right.$$
$$\left. + \int_0^L \frac{\alpha_{\text{RBS}}(z)}{G^2(z)} \int_z^L \gamma(z')\alpha_p(z')p_p(z')G(z')\, dz'\, dz \right],$$

(12.9)

where $n_{sp} = \{1 - \exp\lfloor -h\Delta v/(k_B T_{\text{fib}})\rfloor\}^{-1}$ is the spontaneous emission factor that depends on the fiber temperature T_{fib} and the pump–signal frequency difference Δv (h and k_B are Planck and Boltzmann constants, respectively); $T_F < 1$ is transmittance of a passive fiber, and α_{RBS} is the Rayleigh backscatter coefficient. For completeness, we have included the effect of backscattered ASE (second term in brackets in Eq. (12.9) that contributes at high Raman gains [82].

As a reference point for the DMF, we consider a single fiber with the same total span loss as the DMF (equal to fiber loss plus splice loss) and noise only from forward-propagating ASE. This means that our reference system is an ideal distributed amplifier with optimum trade-off between ASE and nonlinearity in the absence of DRBS (see above), that is, with a flat gain profile $G(z) = 1$. To ensure equal nonlinear penalties in both systems, we scale signal power to the DMF in such a way that the nonlinear phase shifts ϕ_{NL} in both spans are the same. Thus, if P_{ref} is the input signal power to the single fiber span, the signal power to the DMF span is $P_{\text{DMF}} = P_{ref}/R_{NL}$, where for equal transmittances T_F of the DMF and a single fiber

$$R_{NL} = \frac{\alpha_{s,ref}}{\ln(1/T_F)} \int_0^L \frac{\bar{n}_2(z)}{\bar{A}_{\text{eff}}(z)} G_{\text{DMF}}(z)\, dz \qquad (12.10)$$

In Eq. (12.10), $\alpha_{s,ref}$ is the absorption coefficient at the signal wavelength in a single fiber, $\bar{n}_2(z) = n_2^{\text{DMF}}(z)/n_{2,ref}$, and $\bar{A}_{\text{eff}}(z) = A_{\text{eff}}^{\text{DMF}}(z)/A_{\text{eff},ref}$. Under these assumptions, the ratio of electric signal-to-noise-ratios (SNR) is given by [81] as

$$R_{\text{SNR}} = \frac{\text{SNR}_{\text{DMF}}}{\text{SNR}_{ref}} = \frac{1}{R_{NL}} \frac{F_{ref} - 1}{F_{\text{DMF}}^{\text{total}} - 1},$$

where we have neglected the effect of electrical shot noise. For a single fiber, we have $F_{ref} = 1 + 2n_{sp}\ln(1/T_F)$. In the DMF-based system, we also account for the postspan loss T_L of system components such as a gain-flattening filter or an add-drop multiplexer, so that $T_F G_R(L)T_L = 1$ and the total noise figure after one span $F_{\text{DMF}}^{\text{total}} = F_{\text{DMF}}/T_F + 1 - T_L$. After substitution of these relations into the expression for R_{SNR} we obtain

$$R_{\text{SNR}} = \frac{1}{R_{NL}} \frac{2n_{sp}\ln(1/T_F)}{(F_{\text{DMF}}/T_F) - T_L}, \qquad (12.11)$$

where the noise figure of the DMF,

$$F_{\text{DMF}} = F_{\text{DMF}}^{\text{ASE}} + \frac{5}{9}\frac{P_{ref}}{R_{NL}}\frac{T_F X}{hv B_e^{\text{eff}}}$$

includes the contribution from the DRBS crosstalk,

$$X = \int_0^L \frac{\alpha_{RBS}(z)}{G^2(z)} \int_z^L \alpha_{RBS}(z') G^2(z') \, dz' \, dz,$$

and B_e^{eff} is the effective electrical filter bandwidth that accounts for optical signal bandwidth [81].

12.6.3. Single Fiber Versus DMF: A Performance Comparison

Parameters of a generic DMF [83] and single fiber used in calculations are listed in Table 12.3. We assume the ratio between the lengths of $+D$ and $-D$ sections equals two, the splice loss at signal and pump wavelengths equals 0.3 dB, $B_e^{\text{eff}} = 8$ GHz, $n_{sp} = 1.13$, and $P_{ref} = 0.5$ mW.

First, we compare the performance of the DMF and single fiber without postspan loss ($T_L = 1$). As can be seen from Fig. 12.8, the optimum amount of forward pumping weakly depends on span length and amounts to $k_{opt} \approx 0.3 - 0.4$ for the DMF case and $k_{opt} \approx 0.5$ for the single fiber case. For short spans, the performance advantage due to forward pumping is small (<0.3 dB). It increases with the span length and reaches ~ 2 dB for $L = 100$ km (Fig. 12.8(a)). The corresponding R_{SNR} for the single fiber with realistic gain profile and impaired by DRBS is shown in Fig. 12.8(b). Comparison of Fig. 12.8(a) and 12.8(b) shows 0.5 dB SNR advantage for an 80 km span DMF and 1.1 dB for a 100 km span DMF. In Fig. 12.9 we plot R_{SNR} for 100 km long DMF (Fig. 12.9(a)) and single fiber (Fig. 12.9(b)) spans, pumped to transparency, as a function of forward pumping ratio k and the postspan loss T_L. One can see that for low postspan loss, the SNR advantage of the DMF is about 1.2 dB (for $T_L = 0$ dB) but drops to about 0.7 dB for $T_L = 6$ dB. Figure 12.9 shows that use of forward pumping is especially important in the systems with high postspan loss.

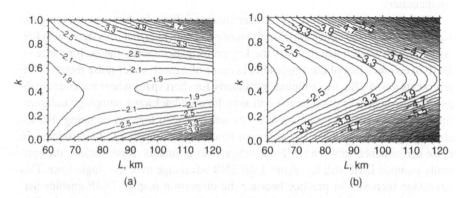

Fig. 12.8. Relative electrical signal-to-noise ratio, R_{SNR}, calculated from Eq. (12.11) for (a) DMF and (b) single fiber without postspan loss as a function of span length L and amount of forward pumping κ; 0 dB corresponds to ideal single-fiber amplifier with $G(z) \equiv 1$.

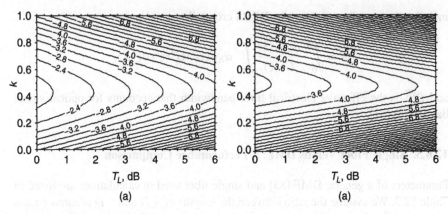

Fig. 12.9. Relative electrical signal-to-noise ratio, R_{SNR}, for (a) DMF and (b) single fiber for constant span length $L = 100$ km as a function of postspan loss, T_L and amount of forward pumping κ.

12.7. Conclusions

This chapter has tried to put Raman amplification in a historical context. In the early days of high gain WDM transmission, it was not a clear winner for amplification. The persistent demand for higher performance (capacity, system reach, data rate, etc.) has turned system designers back to distributed Raman for its lower noise figure. The weak nonlinear interaction of Raman that results in poor pump efficiency also gives the distributed gain benefit of lower noise. Today's data-dominated traffic patterns require reach beyond 1000 km. Raman is one vital tool in pushing out system reach together with more exotic modulation formats, improved cascaded forward error-correction algorithms, lower PMD components and fiber, tighter amplifier gain ripple specifications and gain equalizers, and better fixed and dynamic dispersion compensaters.

Dispersion-managed fiber spans offer another vehicle for better distributing the signal gain. From the above DMF analysis one can conclude that in systems with low signal power when noise figure is the key performance factor, the performance of DMF overcomes that of the single fiber because of a much better noise figure. DMF represents a cost-effective solution particularly for short spans where backward-only pumping is nearly as good as the optimum forward/backward pumping configuration. It also avoids the pump noise issues inherent with forward pumping. For increased signal power when the nonlinear penalties are important, the difference in noise performance between DMF and single fiber is smaller. Nevertheless, the optimally pumped DMF still has about 1 dB SNR advantage over the single fiber. This advantage increases in practice because the dispersion map of DMF enables better tolerance to nonlinearities than with single fibers, and because the single fiber also requires dispersion compensation at the amplifier site that introduces additional penalties.

References

[1] C. V. Raman and K. S. Krishnan, A new type of secondary radiation, *Nature*, 121:3048, 501, 1928.

[2] G. Landsberg and L. Mandelstam, *Naturwissen*, 16:28, 557, 1928.

[3] R. H. Stolen and E. P. Ippen, Raman gain in glass optical waveguides, *Appl. Phys. Lett.*, 22:6, 276–281, 1973.

[4] F. L. Galeener, J. C. Mikkelsen, R. H. Geils, and W. J. Mosby, The relative Raman cross sections of vitreous SiO2, GeO2, B2O3 and P2O5, *Appl. Phys. Lett.*, 32:1, 34–36, 1978.

[5] L. F. Mollenauer, R. H. Stolen, and M. N. Islam, Experimental demonstration of soliton propagation in long fibers: Loss compensated by Raman gain, *Optics Lett.*, 10:5, 229–231, 1985.

[6] L. F. Mollenauer and K. Smith, Demonstration of soliton transmission over more than 4000 km in fiber with loss periodically compensated by Raman gain, *Optics Lett.*, 13:8, 675–677, 1988.

[7] K. Iwatsuki, K. Suzuki, S. Nishi, M. Saruwatari, and K. Nakagawa, 20 Gb/s optical soliton data transmission over 70 km using distributed fiber amplifiers, *IEEE Photon. Technol. Lett.*, 2:12, 905–907, 1990.

[8] J. Hegarty, N. A. Olsson, and L. Goldner, CW pumped Raman preamplifier in a 45 km-long fibre transmission system operating at 1.5 μm and 1 Gbit/s, *Electron. Lett.*, 21:7, 290–292, 1985.

[9] N. Edagawa, K. Mochizuki, and Y. Iwamoto, Simultaneous amplification of wavelength-division-multiplexed signals by a highly efficient fibre Raman amplifier pumped by high-power semiconductor lasers, *Electron. Lett.*, 23:5, 196–197, 1987.

[10] Y. Tamura, S. Shikii, H. Maeda, E. Nishimura, and M. Kawahara, Fiber Raman amplifier module with semiconductor laser pump source. In *Proceedings of the ECOC 87: Thirteenth European Conference on Optical Communication*. Technical Digest, Helsinki, Finland, 1987.

[11] K. Aida, J. Nakajima, and K. Nakagawa, Long-span repeaterless IM/DD optical transmission experiment over 300 km using optical amplifiers. In *Proceedings of the ICC 91 International Conference on Communications Conference Record*, New York, 1991.

[12] P. B. Hansen, L. Eskildsen, S. G. Grubb, A. M. Vengsarkar, S. K. Korotky, T. A. Strasser, J. E. J. Alphonsus, J. J. Veselka, D. J. DiGiovanni, D. W. Peckham, D. Truxal, W. Y. Cheung, S. G. Kosinski, and P. F. Wysocki, 10 Gb/s, 411 km repeaterless transmission experiment employing dispersion compensation and remote post- and pre-amplifiers. In *Proceedings of the 21st European Conference on Optical Communications*, Gent, Belgium, 1995.

[13] P. B. Hansen, A. Stentz, T. N. Nielsen, R. Espindola, L. E. Nelson, and A. A. Abramov, Dense wavelength-division multiplexed transmission in "zero-dispersion" DSF by means of hybrid Raman/erbium-doped fiber amplifiers. In *Proceedings of the OFC/IOOC'99, Optical Fiber Communication Conference* and the *International Conference on Integrated Optics and Optical Fiber Communications*, Piscataway, NJ, 1999.

[14] N. Takachio, H. Suzuki, H. Masuda, and M. Koga, 32*10 Gb/s distributed Raman amplification transmission with 50-GHz channel spacing in the zero-dispersion region over 640 km of 1.55- mu m dispersion-shifted fiber. In *Proceedings of the OFC/IOOC'99, Optical Fiber Communication Conference* and the *International Conference on Integrated Optics and Optical Fiber Communications*, Piscataway, NJ, 1999.

[15] T. N. Nielsen, A. J. Stentz, K. Rottwitt, D. S. Vengsarkar, Z. J. Chen, P. B. Hansen, J. H. Park, K. S. Feder, T. A. Strasser, S. Cabot, S. Stulz, D. W. Peckham, L. Hsu, C. K. Kan,

A. F. Judy, J. Sulhoff, S. Y. Park, L. E. Nelson, and L. Gruner-Nielsen, 3.28 Tb/s (82*40 Gb/s) transmission over 3*100 km nonzero-dispersion fiber using dual C- and L-band hybrid Raman/erbium doped inline amplifiers. In *Proceedings of the Optical Fiber Communication Conference*, Technical Digest Postconference Edition, Trends in Optics and Photonics, Vol. 37, Washington, DC, 2000.

[16] S. Bigo et al., 5.12 Tbit/s (128 × 40 Gbit/s WDM) transmission over 3 × 100 km Ter-alight fiber. In *Proceedings of the European Conference on Optical Communications*, Munich, Germany, Sept. 3–7, 2000.

[17] K. Fukuchi et al., 10.92-Tb/s (273 × 40-Gb/s) triple-band/ultra-dense WDM optical-repeatered transmission experiment. In *Proceedings of the Optical Fiber Communications*, Anaheim, CA, March 22, 2001.

[18] W. S. Lee, 2.56 Tb/s capacity, 0.8 b/Hz-s DWDM transmission over 120 km NDSF using polarization-bit-interlevaed 80 Gb/s OTDM signal. In *Proceedings of the Optical Fiber Communications*, 2001.

[19] B. Mikkelsen, G. Raybon, B. Zhu, R. J. Essiambre, P. G. Bernasconi, K. Dreyer, L. W. Stulz, and S. N. Knudsen, High spectral efficiency (0.53 bit/s/Hz) WDM transmission of 160 Gb/s per wavelength over 400 km of fiber. In *Proceedings of the OFC 2001, Optical Fiber Communication Conference and Exhibition*, Technical Digest, Washington, DC, 2001.

[20] B. Mikkelsen, G. Raybon, R.-J. Essiambre, A. J. Stentz, T. N. Nielsen, D. W. Peckham, L. Hsu, L. Gruner-Nielsen, K. Dreyer, and J. E. Johnson, 320-Gb/s single-channel pseu-dolinear transmission over 200 km of nonzero-dispersion fiber, *IEEE Photon. Technol. Lett.*, 12:10, 1400–1402, 2000.

[21] U. Feiste et al., 160 Gb/s transmission over 116 km field-installed fiber using 160 Gbit/s OTDM and 40 Gbit/s ETDM. In *Proceedings of the Optical Fiber Communications*, Anaheim, CA, 2001.

[22] T. Haito et al., 1 Terabit/s WDM transmission over 10,000 km. In *Proceedings of the European Conference on Optical Communications*, 1999.

[23] T. Tanaka et al., 2.1-Tbit/s WDM transmission over 7,221 km with 80-km repeater spacing, In *Proceedings of the European Conference on Optical Communications*, 2000.

[24] D. G. Foursa et al., 2.56 Tb/s (256 × 10 Gb/s) transmission over 11,000 km using hybrid Raman/EDFAs with 80 nm of continuous bandwidth. In *Proceedings of the Optical Fiber Communications*, Anaheim, CA, 2002.

[25] M. X. Ma, H. D. Kidorf, K. Rottwitt, F. W. Kerfoot, III, and C. R. Davidson, 240-km repeater spacing in a 5280-km WDM system experiment using 8*2.5 Gb/s NRZ transmission, *IEEE Photon. Technol. Lett.*, 10:6, 893–895, 1998.

[26] T. Terahara, T. Hoshida, J. Kumasako, and H. Onaka, 128*10.66 Gbit/s transmission over 840 km standard SMF with 140 km optical repeater spacing (30.4 dB loss) em-ploying dual-band distributed Raman amplification. In *Proceedings of the Optical Fiber Communication Conference*. Technical Digest Postconference Edition. Trends in Optics and Photonics, Vol. 37, Washington, DC, 2000.

[27] Y. Zhu et al., 1.28 Tbit/s (32 × 40 Gbits/s) transmission over 1000 km with only 6 spans. In *Proceedings of the European Conference on Optical Communications*, 2000.

[28] M. D. Mermelstein, C. Headley, J.-C. Bouteiller, P. Steinvurzel, C. Horn, K. Feder, and B. J. Eggleton, A high-efficiency power-stable three-wavelength configurable Raman fiber laser. In *Proceedings of the OFC 2001, Optical Fiber Communication Conference and Exhibition*. Technical Digest, Washington, DC, 2001.

[29] M. D. Mermelstein, C. Horn, S. Radic, and C. Headley, Six-wavelength Raman fibre laser for C- and L-band Raman amplification and dynamic gain flattening, *Electron. Lett.*, 38:13, 636–638, 2002.

[30] S. B. Papernyi, V. I. Karpov, and W. R. L. Clements, Third-order cascaded Raman amplification. In *Proceedings of the Optical Fiber Communications*, Anaheim, CA, 2002.

[31] Y. Qian et al., High-performance AlGaInAs/InP 14xx-nm semiconductor pump lasers for optical amplifications. In *Proceedings of the Asia-Pacific Optical Communication (APOC) Conference*, October, 2002.

[32] P. Kim, J. Park, and N. Park, Performance optimization of distributed Raman amplifier using optical pump time domain reflectometry. In *Proceedings of the Optical Fiber Conference*, 2002.

[33] A. Yariv, Signal-to-noise considerations in fiber links with periodic or distributed optical ampification, *Optics Lett.*, 15:9, 1064–1066, 1990.

[34] R. G. Smith, Optical power handling capacity of low loss optical fibers as determined by stimulated Raman and Brillouin scattering, *Appl. Optics*, 11:2489–2494, 1972.

[35] J. Grochocinski, personal communication, 1999.

[36] A. Belahlou et al., Fiber Design Considerations for 40 Gb/s Systems, Fiber design considerations for 40 Gb/s systems, *J. Lightwave Technol.*, 20(12):2290–2305, 2002.

[36a] S. Gray, M. Vasilyev, and K. Jepsen, "Spectral broadening of double Rayleigh backscattering in a distributed Raman amplifier," presented at Optical Fiber Communication conference, 2001.

[36b] A. Kobyakov, S. Gray, and M. Vasilyev, Quantitative analysis of Rayleigh cross-talk in Raman amplifiers, *Electron. Letters*, 39:9, 732-733, 2003.

[37] S. Burtsev, W. Pelouch, and P. Gavrilovic, Multi-path interference nose in multi-span transmission links using lumped Raman amplifiers. In *Proceedings of the Optical Fiber Communications*, 2002.

[38] M. Nissov, C. R. Davidson, K. Rottwitt, R. Menges, P. C. Corbett, D. Innis, and N. S. Bergano, 100 Gb/s (10*10 Gb/s) WDM transmission over 7200 km using distributed Raman amplification. In *Proceedings of the Eleventh International Conference on Integrated Optics and Optical Fibre Communications 23 European Conference on Optical Communications IOOC-ECOC97* (Conf. Publ. No.448), London, 1997.

[39] A. K. Srivastava, L. Zhang, Y. Sun, J. W. Sulhoff, and C. Wolf, System margin enhancement with Raman gain in multi-span WDM transmission. In *Proceedings of the OFC/IOOC'99, Optical Fiber Communication Conference* and *the International Conference on Integrated Optics and Optical Fiber Communications*, Piscataway, NJ, 1999.

[40] H. Suzuki, J. Kani, H. Masuda, N. Takachio, K. Iwatsuki, Y. Tada, and M. Sumida, 25 GHz-spaced, 1 Tb/s (100*10 Gb/s) super dense-WDM transmission in the C-band over a dispersion-shifted fibre cable employing distributed Raman amplification. In *Proceedings of ECOC'99, 25th European Conference on Optical Communication*, Paris, 1999.

[41] K. S. Jepsen, U. Gliese, B. R. Hemenway, S. Yuan, K. S. Chen, J. E. Hurley, L. Guiziou, J. W. McCamy, N. Boos, D. J. Tebben, B. Dingel, M. J. Li, S. Gray, G. E. Kohnke, L. Jiang, V. Srikant, A. F. Evans, and J. M. Jouanno, Network demonstration of 32 lambda *10 Gb/s across 6 nodes of 640*640 WSXCs with 750 km Raman-amplified fiber. In *Proceedings of the Optical Fiber Communication Conference*. Technical Digest postconference edition. Trends in Optics and Photonics Vol. 37, Washington, DC, 2000.

[42] A. K. Srivastava, S. Radic, C. Wolf, J. C. Centannil, J. W. Sulhoff, K. Kantor, and Y. Sun, Ultra-dense terabit capacity WDM transmission in L-band. In *Proceedings of the Optical Fiber Communication Conference*. Technical Digest postconference edition. Trends in Optics and Photonics Vol. 37, Washington, DC, 2000.

[43] J.-P. Blondel, F. Boubal, E. Brandon, L. Buet, L. Labrunie, P. Le Roux, and D. Toullier, Network application and system demonstration of WDM systems with very large spans

(error-free 32*10 Gbit/s 750 km transmission over 3 amplified spans of 250 km). In *Proceedings of the Optical Fiber Communication Conference*. Technical Digest post-conference edition. Trends in Optics and Photonics Vol. 37, Washington, DC, 2000.

[44] T. Tanaka and et al., 2.1-Tbit/s WDM transmission over 7,221 km with 80-km repeater spacing. In *Proceedings of the European Conference on Optical Communications*, 2000.

[45] J. Bromage, J.-C. Bouteiller, H. J. Thiele, K. Brar, J. H. Park, C. Headley, L. E. Nelson, Y. Qian, J. DeMarco, S. Stulz, L. Leng, B. Zhu, and B. J. Eggleton, S-band all-Raman amplifiers for 40 * 10 Gb/s transmission over 6 * 100 km of non-zero dispersion fiber. In *Proceedings of the OFC 2001, Optical Fiber Communication Conference and Exhibition*. Technical Digest, Washington, DC, 2001.

[46] A. B. Puc, M. W. Chbat, J. D. Henrie, N. A. Weaver, H. Kim, A. Kaminski, A. Rahman, and H. Fevrier, Long-haul WDM NRZ transmission at 10.7 Gb/s in S-band using cascade of lumped Raman amplifiers. In *Proceedings of the OFC 2001, Optical Fiber Communication Conference and Exhibition*. Technical Digest, Washington, DC, 2001.

[47] S. N. Knudsen, B. Zhu, L. E. Nelson, M. O. Pedersen, D. W. Peckham, and S. Stulz, 420 Gbit/s (42*10 Gbit/s) WDM transmission over 4000 km of UltraWave fibre with 100 km dispersion-managed spans and distributed Raman amplification, *Electron. Lett.*, 37:15, 965–967, 2001.

[48] N. Shimojoh et al., 2.4-Tbit/s WDM transmission over 7400 km using all Raman amplifier repeaters with 74-nm continuous single band. In *Proceedings of the European Conference on Optical Communications*, 2001.

[49] D. F. Grosz et al., Demonstration of all-Raman ultra-wide-band transmission of 1.28 Tb/s (128 × 10Gb/s) over 4000 km of NZ-DSF with large BER margins. In *Proceedings of the European Conference on Optical Communications*, 2001.

[50] I. Thomkos et al., 80 × 10.7 Gb/s ultra-long-haul (+4200 km) DWDM network with reconfigurable "broadcast & select" OADMs. In *Proceedings of the OFC 2002, Optical Fiber Communication Conference and Exhibition*. Technical Digest, Washington, DC, 2002.

[51] M. Mehendale, M. Vasilyev, A. Kobyakov, M. Williams, and S. Tsuda, All-Raman transmission of 80 * 10 Gbit/s WDM signals with 50 GHz spacing over 4160 km of dispersion-managed fibre, *Electron. Lett.*, 38:13, 648–649, 2002.

[52] T. N. Nielsen, A. J. Stentz, P. B. Hansen, Z. J. Chen, D. S. Vengsarkar, T. A. Strasser, K. Rottwitt, J. H. Park, S. Stulz, S. Cabot, K. S. Feder, P. S. Westbrook, and K. G. Kosinski, 1.6 Tb/s (40*40 Gb/s) transmission over 4*100 km nonzero-dispersion fiber using hybrid Raman/Er-doped inline amplifiers. In *Proceedings of ECOC'99. 25th European Conference on Optical Communication*, Paris, 1999.

[53] R. Ohhira, Y. Yano, A. Noda, Y. Suzuki, C. Kurioka, M. Tachigori, S. Moribayashi, K. Fukuchi, T. Ono, and T. Suzaki, 40 Gbit/s × 8-ch NRZ WDM transmission experiment over 80 km*5-span using distributed Raman amplification in RDF. In *Proceedings of ECOC'99. 25th European Conference on Optical Communication*, Paris, 1999.

[54] S. Bigo, E. Lach, Y. Frignac, D. Hamoir, P. Sillard, W. Idler, S. Gauchard, A. Bertaina, S. Borne, L. Lorcy, N. Torabi, B. Franz, P. Nouchi, P. Guenot, L. Fleury, G. Wien, G. Le Ber, R. Fritschi, B. Junginger, M. Kaiser, D. Bayart, G. Veith, J.-P. Hamaide, and J.-L. Beylat, 1.28 Tbit/s WDM transmission of 32 ETDM channels at 40 Gbit/s over 3*100 km distance. In *Proceedings of the 26th European Conference on Optical Communication*, Berlin, 2000.

[55] Y. Zhu et al., 16-channel 40 Gb/s carrier-suppressed RZ ETDM/DWDM transmission over 720 km NDSF without polarization channel interleaving. In *Proceedings of the Optical Fiber Communications*, Anaheim, CA, 2001.

[56] B. Zhu, L. Leng, L. E. Nelson, Y. Qian, S. Stulz, C. Doerr, L. Stulz, S. Chandrasekar, S. Radic, D. Vengsarkar, Z. Chen, J. Park, K. Feder, H. Thiele, J. Bromage, L. Gruner-Nielsen, and S. Knudsen, 3.08 Tb/s (77*42.7 Gb/s) transmission over 1200 km of non-zero dispersion-shifted fiber with 100-km spans using C and L-band distributed Raman amplification. In *Proceedings of the OFC 2001. Optical Fiber Communication Conference and Exhibition*. Technical Digest, Washington, DC, 2001.

[57] D. Chen, S. Wheeler, D. Nguyen, A. Farbert, A. Schopflin, A. Richter, C.-J. Weiske, K. Kotten, P. M. Krummrich, A. Schex, and A. S. C. Glingener, 3.2 Tb/s field trial (80*40 Gb/s) over 3*82 km SSMF using FEC, Raman and tunable dispersion compensation. In *Proceedings of the OFC 2001. Optical Fiber Communication Conference and Exhibition*. Technical Digest, Washington, DC, 2001.

[58] I. Morita, K. Tanaka, and N. Edagawa,Benefit of Raman amplification in ultra-long-distance 40 Gbit/s-based WDM transmission. In *Proceedings of the OFC 2001. Optical Fiber Communication Conference and Exhibition*. Technical Digest, Washington, DC, 2001.

[59] L. du Mouza, G. Le Meur, H. Mardoyan, E. Seve, S. Cussat-Blanc, D. Hamoir, C. Martinelli, D. Bayart, F. Raineri, L. Pierre, B. Dany, O. Leclerc, J. P. Hamaide, L. A. de Montmorillon, F. Beaumont, P. Sillard, P. Nouchi, A. Hugbart, R. Uhel, and G. Granpierre, 1.28 Tbit/s (32*40 Gbit/s) WDM transmission over 2400 km of TeraLight, Reverse TeraLight(C) fibres using distributed all-Raman amplification, *Electron. Lett.*, 37:21, 1300–1302, 2001.

[60] S. Bigo et al., Transmission of 125 WDM channels at 42.7 Gbit/s (5 Tbit/s capacity) over 12 × 100 km of Teralight Ultra fibre. In *Proceedings of the European Conference on Optical Communications*, 2001.

[61] B. Zhu et al., 1.6 Tb/s (40 × 42.7 Gb/s) transmission over 2000 km of fiber with 100-km dispersion-managed spans. In *Proceedings of the European Conference on Optical Communications*, 2001.

[62] H. Bissessur et al., 3.2 Tb/s (80 × 40 Gb/s) C-band transmission over 3 × 100 km with 0.8 bit/s/Hz efficiency. In *Proceedings of the European Conference on Optical Communications*, 2001.

[63] B. Zhu, L. E. Nelson, L. Leng, S. Stulz, S. Knudsen, and D. Peckham, 1.6 Tbit/s (40 × 42.7 Gbit/s) WDM transmission over 2400 km of fibre with 100 km dispersion-managed spans, *Electron. Lett.*, 38:13, 647–648, 2002.

[64] A. H. Gnauck et al., 2.5 Tb/s (64 × 42.7 Gb/s) transmission over 40 × 100 km NZDSF using RZ-DPSK format and all-Raman-amplified spans. In *Proceedings of the Optical Fiber Communications*, 2002.

[65] Y. Frignac et al., Transmission of 256 wavelength-division and polarization-division-multiplexed channels at 42.7 Gb/s (10.2 Tb/s capacity) over 3 × 100 km of Teralight fiber. In *Proceedings of the Optical Fiber Communications*, 2002.

[66] H. Sugahara, K. Fukuchi, A. Tanaka, Y. Inada, and T. Ono, 6,050 km transmission of 32 × 42.7 Gb/s DWDM signals using Raman-amplified quadruple-hybrid span configuration. In *Proceedings of the Optical Fiber Communications*, Anaheim, CA, 2002.

[67] F. Liu et al., 1.6 Tb/s (40 × 42.7 Gbit/s) transmission over 3600 km UltraWave fiber with all-Raman amplified 100 km terrestrial spans using ETDM transmitter and receiver. In *Proceedings of the Optical Fiber Communications*, 2002.

[68] B. Zhu et al., 3.2 Tb/s (80 × 42.7 Gb/s) transmission over 20 × 100 km of non-zero dispersion fiber with simultaneous C+ L-band dispersion compensation. In *Proceedings of the Optical Fiber Communications*, 2002.

[69] T. N. Nielsen et al., 1.6 Tb/s (40 × 40Gb/s) transmission over 4 × 100 km nonzero-dispersion fiber using hybrid Raman/erbium-doped inline amplifiers. In *Proceedings of the European Conference on Optical Communications*, 1999.

[70] M. Vasilyev, B. Szalabofka, S. Tsuda, J. M. Grochocinski, and A. F. Evans, Reduction of Raman MPI and noise figure in dispersion-managed fibre, *Electron. Lett.*, 38:6, 271–272, 2002.

[71] R. Hainberger, J. Kumasako, K. Nakamura, T. Terahara, H. Osaka, and T. Hoshida, Comparison of span configurations of Raman-amplified dispersion-managed fibers, *Photon. Technol. Lett.*, 14:471, 2002.

[72] K. Shimizu, K. Kinjo, N. Suzuki, K. Ishida, S. Kajiya, K. Motoshima, and Y. Kobayashi, Fiber-effective-area managed fiber lines with distributed Raman amplification in 1.28-Tb/s (32 × 40 Gb/s), 202-km unrepeatered transmission. In *Proceedings of the Optical Fiber Communications*, 2001.

[73] T. Miyakawa, I. Morita, K. Tanaka, H. Sakata, and N. Edagawa, 2.56 Tbit/s (40 Gbit/s*64 WDM) unrepeatered 230 km transmission with 0.8 bit/s/Hz spectral efficiency using low-noise fiber Raman amplifier and 170 mu m/sup 2/-Aeff fiber. In *Proceedings of the OFC 2001. Optical Fiber Communication Conference and Exhibition.* Technical Digest, Washington, DC, 2001.

[74] Y. Zhu, I. Hardcastle, W. S. Lee, C. R. S. Fludger, A. Hadjifotiou, C. Li, D. Qiao, H. Sun, K. T. Wu, and J. McNicol, Experimental comparison of Raman-amplified dispersion-managed fibre types using 16 × 40 Gbit/s transmission over 500 km, *Electron. Lett.*, 38:895, 2002.

[75] Y. Hadjar and N. J. Traynor, Quantitative analysis of second order distributed Raman amplification. In *Proceedings of the Optical Fiber Communications*, 2002.

[76] J.-C. Bouteiller, K. Brar, S. Radic, J. Bromage, Z. Wang, and C. Headley, Dual-order Raman pump providing improved noise figure and large gain bandwidth. In *Proceedings of the Optical Fiber Communications*, 2002.

[76a] J.-C Bouteiller, K. Brar, and C. Headley, "Quasi-constant signal power transmission," presented at European Conference on Optical Communication, 2002.

[77] I. Tomkos, M. Vasilyev, J.-K. Rhee, A. Kobyakov, M. Ajgaonkar, and M. Sharma, Dispersion map design for 10 Gb/s ultra-long-haul DWDM transparent optical networks. In *Proceedings of the Seventh Optoelectronics and Communications Conference*, Yokohama, 2002.

[78] I. Tomkos, M. Vasilyev, J.-K. Rhee, M. Mehendale, B. Hallock, B. Szalabofka, M. Williams, S. Tsuda, and M. Sharma, Ultra-long-haul DWDM network with 320 × 320 wavelength-port "broadcast & select" OXCs. In *Proceedings of the European Conference on Optical Communication*, Copenhagen, 2002.

[79] S. Kumar, J. C. Mauro, S. Raghavan, and D. Q. Chowdhury, Intrachannel nonlinear penalties in dispersion-managed transmission systems, *IEEE J. of Select. Topics Quantum Electron.* 8:3, 626–631, 2002.

[80] M. Vasilyev, Raman-assisted transmission: Toward ideal distributed amplification. In *Proceedings of the Optical Fiber Communications*, Atlanta, 2003.

[81] R.-J. Essiambre, P. Winzer, J. Bromage, and C. H. Kim, Design of bidirectionally pumped fiber amplifiers generating double Rayleigh backscattering, *IEEE Photon. Technol. Lett.*, 14:7, 914–916, 2002.

[82] A. Kobyakov, M. Vasilyev, S. Tsuda, G. Giudice, and S. Ten, Raman noise figure in dispersion-managed fibers. In *Proceedings of the ECOC*, 2002; Analytical model for Raman noise figure in dispersion-managed fibers, *IEEE Photon Technol. Lett.*, 15:1, 30–32, 2003.

[83] S. N. Knudsen, Design and manufacture of dispersion compensating fibers and their performance in systems. In *Proceedings of the Optical Fiber Communications*, 2002.

Chapter 13

Hybrid EDFA/Raman Amplifiers

Hiroji Masuda

13.1. Introduction

This chapter describes the technologies needed for cascading an erbium-doped fiber amplifier (EDFA) and a fiber Raman amplifier (FRA or RA) to create a hybrid amplifier (HA), the EDFA/Raman HA. Two kinds of HA are defined in this chapter: the narrowband HA (NB-HA) and the seamless and wideband HA (SWB-HA). The NB-HA employs distributed Raman amplification in the transmission fiber together with an EDFA and provides low noise transmission in the C- or L-band. The noise figure of the transmission line is lower than it would be if only an EDFA were used. The SWB-HA, on the other hand, employs distributed or discrete Raman amplification together with an EDFA, and provides a low-noise and wideband transmission line or a low-noise and wideband discrete amplifier for the C- and L-bands. The typical gain bandwidth ($\Delta\lambda$) of the NB-HA is $\sim$30 to 40 nm, whereas that of the SWB-HA is $\sim$70 to 80 nm.

The basic configurations of these HAs are introduced in the next section (Section 13.2) and are compared with those of other amplifiers. Section 13.3 shows the performance limitations of EDFAs with regard to gain bandwidth and noise figure (NF). Properties of the NB-HA are described in Section 13.4. An analysis of noise properties and an experimental example are also shown. Section 13.5 describes the properties of the SWB-HA. An analysis of gain bandwidth, a classification of amplifier type, four amplifier configurations, and gain and noise characteristics are shown based on both calculations and experiments. Section 13.6 introduces two other types of hybrid amplifiers, the TDFA/Raman amplifier and the tellurite/silica Raman amplifier. Finally, conclusions are drawn in Section 13.7.

13.2. Basic Amplifier Configurations

Figure 13.1 shows some basic configurations of a transmission line with an inline amplifier. An EDFA is used as the repeater between two installed transmission fibers, and amplifies the input signal light (Fig. 13.1(a)) [1]. The signal light usually consists

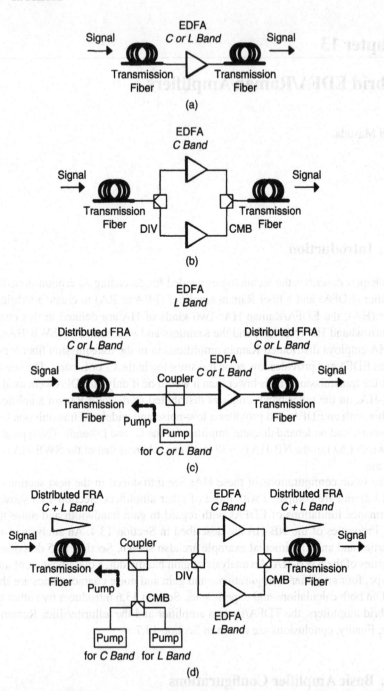

Fig. 13.1. Basic configurations of a transmission line with an inline amplifier: (a) an EDFA; (b) a two-gain band amplifier (EDFA) with C- and L-band EDFAs in parallel; (c) a hybrid EDFA/distributed Raman amplifier with C- or L-band; and (d) a hybrid EDFA/distributed Raman amplifier with C- and L-bands in parallel (CMB: combiner, DIV: divider).

of wavelength-division-multiplexed (WDM) multichannels, and the EDFA offers C- or L-gain band coverage. The typical gain bands of C- and L-gain band EDFAs are the wavelength ranges of about 1530 to 1560 nm and 1570 to 1600 nm, respectively. Figure 13.1(b) shows a two-gain band amplifier (EDFA) with C- and L-gain band EDFAs in parallel with each other [2, 3]. The combiner and divider connected to the EDFAs multiplex and demultiplex the WDM signal channels according to their wavelengths. The two-gain band EDFA has a gain bandwidth that is about twice that of the C- or L-band EDFA (Fig. 13.1(a)). However, its cost and the number of optical components are about twice those of the C- or L-band EDFA. The NB-HA that offers C- or L-band coverage is shown in Fig. 13.1(c) [4–7]. The NB-HA consists of a C- or L-band distributed RA (DRA), which is a transmission fiber itself, and a C- or L-band EDFA set after the transmission fiber as a repeater. Finally, Fig. 13.1(d) shows a C- and L-two-gain band HA [8, 9]. The two-gain band HA consists of a two-wavelength pumped DRA (C- and L-band) and a two-gain band EDFA. The pump lights for the C- and L-bands are multiplexed by a combiner and launched into the transmission fiber via a coupler. The transmission line with a discrete (or lumped) Raman amplifier (LRA) instead of the EDFA of Fig. 13.1(a), which is not shown for simplicity, can be also constructed as shown in Chapter 10. Moreover, using just a DRA is also possible as described in Chapters 12 and 14.

We can evaluate the performance of a cascaded transmission line with multiple transmission fibers and inline amplifiers by considering a single span of transmission line [1]. The single transmission line with an EDFA (Figs. 13.1(a) and 13.1(b)), an LRA, or a DRA has noise and output power characteristics determined by each of the amplifiers. On the other hand, the noise characteristics of a single transmission line with an EDFA/Raman HA are determined by both the DRA and EDFA as described in detail in Section 13.4. The optical SNR (OSNR) of a transmission line with an HA is higher than would be the case if the EDFA of Fig. 13.1(a) or 13.1(b) were used. In other words, the HA has a lower effective NF (F_{eff}) than the EDFA. Moreover, the output power characteristics of the HA are determined by the EDFA used. EDFAs offer available output powers of up to $\sim$30 dBm; the main determining factor is the pump power [1]. However, the available output power of the Raman amplifier (DRA and LRA) can range up to $\sim$20 dBm, and the main determining factor is the nonlinearity of the long silica fiber used as the gain medium [10].

The basic configurations of a transmission line with an SWB-HA are shown in Fig. 13.2. Each of the EDFAs and RAs covers a different portion of the C- and L-bands. The gain-slope of the EDFA is opposite to that of the RA as shown in Fig. 13.14 of Section 13.5. The EDFA and RA are connected in a series. Figure 13.2(a) shows the HA with an LRA followed by an EDFA, whereas Figure 13.2(b) shows the HA with an EDFA followed by an LRA. An HA with pre- and post-EDFAs and an intermediate LRA is shown in Fig. 13.2(c). Finally, Figure 13.2(d) shows an HA with a DRA followed by an EDFA. The HAs of Figs. 13.2(a), 13.2(c), and 13.2(d) have high output power because they use an EDFA at the output side. The HA of Fig. 13.2(b), however, has moderate output power because it uses an LRA at the output side. The NF of the HA is mainly determined by the NF of the first-stage amplifier if the gain of the first-stage amplifier is large enough (i.e., > 10 dB). Otherwise (i.e., gain < 10 dB),

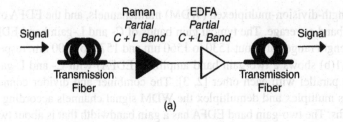

(a)

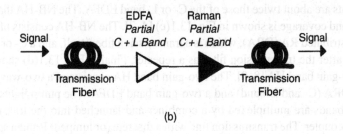

(b)

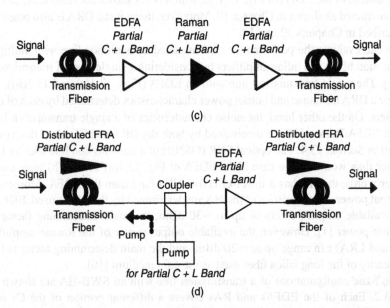

(c)

(d)

Fig. 13.2. Basic configurations of a transmission line with a seamless wideband hybrid EDFA/Raman amplifier. Each of the EDFA and Raman amplifiers covers a partial C- and L-band with different gain slopes. The EDFA and Raman amplifier are serially connected. The hybrid amplifiers are: (a) a discrete Raman amplifier followed by an EDFA; (b) an EDFA followed by a discrete Raman amplifier; (c) pre- and post-EDFAs and an intermediate discrete Raman amplifier; and (d) a distributed Raman amplifier followed by an EDFA.

the NF is determined by the NFs of both the first- and second-stage amplifiers. The configurations and performances of the HAs are described in detail in Section 13.5.

13.3. Performance Limitations of EDFAs

This section describes the limitations of the gain bandwidths and NFs of EDFAs. Figure 13.3 shows the three basic EDFA configurations. An EDFA with a single-stage amplification configuration and an external gain equalizer (GEQ) is shown in Fig. 13.3(a). An EDFA with a two-stage (three-stage) amplification configuration and one (two) internal GEQ is shown in Figs. 13.3(b) and 13.3(c) [11–14]. The relevant optical components such as isolators and signal–pump couplers have been omitted for simplicity. The gains of the EDFs and the losses of the GEQs are identified by G, G_1, G_2, G_3, and L, L_1, L_2, respectively. We can expand the gain bandwidth of an EDFA with the two- or three-stage amplification configuration as shown below.

The gain (in units of dB) of an EDFA at a signal wavelength of λ_s ($G_{dB}(\lambda_s)$) is expressed as

$$G_{dB}(\lambda_s) = \frac{10}{\ln(10)} \eta \rho L\{(\sigma_{abs}(\lambda_s) + \sigma_{emi}(\lambda_s) - \sigma_{ESA}(\lambda_s))N_{2ave} - \sigma_{abs}(\lambda_s)\},$$

$$(13.1a)$$

$$N_{2ave} = \frac{1}{L} \int_0^L N_2(z) \, dz,$$

$$(13.1b)$$

where η is the overlap factor, ρ is the Er concentration, L is EDF length, $\sigma_{abs}, \sigma_{emi}, \sigma_{ESA}$ are the absorption, emission, and excited-state-absorption (ESA) cross-sections,

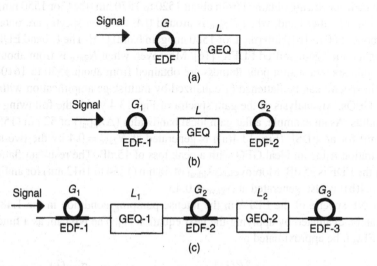

Fig. 13.3. Basic EDFA configurations: (a) single stage configuration with an external gain equalizer (GEQ); (b) two-stage configuration with an intermediate GEQ; (c) three-stage configuration with two intermediate GEQs (G: gain, L: loss).

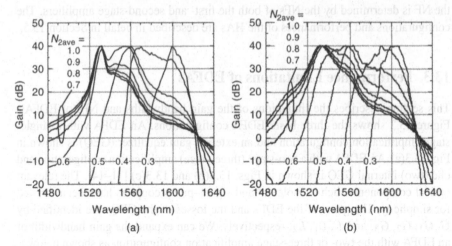

Fig. 13.4. Gain spectra at a peak gain of 40 dB for (a) silica EDF (EDSF); (b) flouride EDF (EDFF) (N_{2ave}: average fractional population of the upper state).

respectively, and N_{2ave} is the path-averaged fractional population of the upper state ($0 < N_{2ave} < 1$) [1]. Equation (13.1) indicates that the gain is proportional to the concentration-length product ρL, and that the gain spectrum depends on N_{2ave}.

Figure 13.4 shows the gain spectra of (a) a silica EDF (erbium-doped silica fiber: EDSF) and (b) a fluoride EDF (erbium-doped fluoride fiber: EDFF) at N_{2ave} values ranging from 0.3 to 1 in steps of 0.1. The peak gain of 40 dB can be achieved by setting ρL at an appropriate value. A variety of N_{2ave} values can be generated experimentally by altering the pumping condition (wavelength and power). When N_{2ave} is around 0.7, flat gain spectra are obtained from about 1520 to 1570 nm (the C or 1550 nm gain band). On the other hand, when N_{2ave} is around 0.4, flat gain spectra are obtained from about 1560 to 1610 nm (the L or 1580 nm gain band) [15]. The L-band EDFA is often called the "gain-shifted EDFA" [16]. Moreover, when N_{2ave} is from about 0.5 to 0.6, gain spectra with a poor flatness are obtained from about 1530 to 1610 nm. The gain spectra can be flattened or equalized by multistage amplification with one or two GEQs. An analysis of the gain spectra of Fig. 13.4 yields the following gain bandwidths. As an example, a flat gain ideal bandwidth ($\Delta\lambda_{ideal}$) of 52 nm (1559 to 1611 nm) for an EDSF (Fig. 13.4(a)) is generated at $N_{2ave} = 0.4$ by the two-stage configuration using an ideal GEQ with a peak loss of 15 dB. The resultant flattened gain of the EDF is 25 dB. Moreover, $\Delta\lambda_{ideal}$ of 58 nm (1554 to 1612 nm) for an EDFF (Fig. 13.4(b)) is also generated at $N_{2ave} = 0.45$.

The NF spectra of the EDFA in the intense-pumping condition in the 1.48 μm band can be evaluated by applying the next equation [1]. The NF can, as a function of λ_s, $F(\lambda_s)$, be approximated as

$$F(\lambda_s) = \frac{2R(\lambda_s)}{R(\lambda_s) - R(\lambda_p) - \sigma_{ESA}(\lambda_s)/\sigma_{abs}(\lambda_s)}, \quad R(\lambda) \equiv \frac{\sigma_{emi}(\lambda)}{\sigma_{abs}(\lambda)}, \quad (13.2)$$

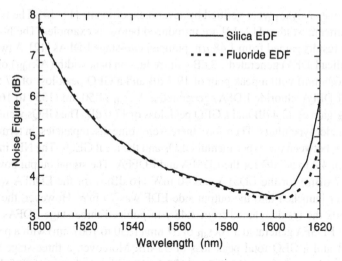

Fig. 13.5. Noise figure spectra of silica and flouride EDFs (EDSF and EDFF) pumped at 1.48 μm.

where λ_p is the pump wavelength, and $R(\lambda)$, $\lambda = \lambda_s$ or λ_p, is the ratio of cross-sections. On the other hand, $F(\lambda_s)$ is about 2 (or 3 dB) in the intense-pumping condition in the 0.98 μm band. Figure 13.5 shows the NF spectra for an EDSF and EDFF as calculated by Eq. (13.2). $F(\lambda_s)$ under 5 dB from 1530 to 1615 nm are obtained with the EDSF, whereas the EDFF yields $F(\lambda_s)$ under 5 dB from 1530 to 1620 nm.

NFs for the multistage configurations of Fig. 13.3 are expressed as follows. When we consider only the signal–spontaneous beat noise component as contributing to the NF for simplicity, the NF of the single-stage EDFA, $F_{1\text{stg}}$, is expressed using equivalent spontaneous emission factor $n_{\text{eq}} = n_{\text{sp}}(G-1)/G$, where n_{sp} is the spontaneous emission factor, as

$$F_{1\text{stg}} = 2n_{\text{eq}}. \tag{13.3a}$$

The NF of the two-stage EDFA, $F_{2\text{stg}}$, is also expressed as [1]

$$F_{2\text{stg}} = 2n_{\text{eq}1} + \frac{2n_{\text{eq}2}}{G_1 L}. \tag{13.3b}$$

Therefore, the difference between the two NFs in dB units, ΔF_{dB}, is

$$\Delta F_{\text{dB}} = 10\log\left(2n_{\text{eq}1} + \frac{2n_{\text{eq}2}}{G_1 L}\right) - 10\log(2n_{\text{eq}}) \cong 10\log\left(1 + \frac{1}{G_1 L}\right). \tag{13.4}$$

This assumes the approximation of $n_{\text{eq}} \cong n_{\text{eq}1} \cong n_{\text{eq}2}$. As an example, ΔF_{dB} is 1 and 0.5 dB, when the gain–loss product $G_1 L$ is set at 6.1 dB and 9.2 dB, respectively. As another example, ΔF_{dB} is 0.41 dB, when $G_1 = 25$ dB, $L = 15$ dB, and $G_2 = 15$ dB. The NF of the three-stage EDFA can also be analyzed in the same way. As an example, ΔF_{dB} is 0.34 dB, when $G_1 = 28$ dB, $L_1 = 15$ dB, $G_2 = 17$ dB, $L_2 = 15$ dB, and $G_2 = 10$ dB.

The above analyses were verified by several experiments [11–14]. The results of two experimental studies [12, 13] are introduced below as examples. The first study reports the results gained from 1.48 μm pumped two-stage EDFAs [12]. A two-stage EDSFA (silica EDFA) generated a 3 dB gain-reduction bandwidth ($\Delta\lambda_{3dB}$) of 52 nm (1556 to 1608 nm) with a peak gain of 19.7 dB and a GEQ peak loss of 17.5 dB. A two-stage EDFFA (fluoride EDFA) generated a $\Delta\lambda_{3dB}$ of 50 nm (1554 to 1604 nm) with a peak gain of 22.4 dB and a GEQ peak loss of 17.0 dB. The GEQs were simple Mach–Zehnder type filters. Therefore, there were some discrepancies, in terms of the loss spectra, between the experimental GEQs and the ideal GEQs. The NFs in $\Delta\lambda_{3dB}$ ranged from 4.4 to 5.7 dB for the EDSFA and EDFFA. The signal output power was 50 mW (17 dBm) for the EDSFA and 40 mW (16 dBm) for the EDFFA when the pump power launched into the output side EDF was 79 mW. However, the second study reports the results gained from 1.48 μm pumped three-stage EDFAs [13]. A three-stage EDSFA generated a $\Delta\lambda_{3dB}$ of 57 nm (1550 to 1607 nm) with a peak gain of 20.9 dB and a GEQ total peak loss of 33 dB. Moreover, a three-stage EDFFA generated a $\Delta\lambda_{3dB}$ of 62 nm (1540.5 to 1602.5 nm) with a peak gain of 22.5 dB and a GEQ total peak loss of 34 dB. Low NFs and high output powers were also confirmed for the EDFAs that were comparable to those of two-stage EDFAs.

Several methods have been employed to enlarge the gain bandwidths of fiber amplifiers. Figure 13.6 compares the gain bands of several types of wideband fiber amplifiers reported to date. The types are (1) the multistage EDFA including one or more gain equalizers [11–14], (2) the two-gain band EDFA [2, 3], (3) the multiwavelength pumped Raman amplifier [17–19], and (4) the hybrid EDFA/Raman amplifier [20]. Three types of EDFAs were used in the wideband fiber amplifiers as shown by EDXFA where X = S, F, and T (S: silica, F: fluoride, T: tellurite) [21]. The gain bands of the SWB-HA (seamless and wideband HA) are shown in detail in Section 13.5. Typical gain bands are as follows.

1. A two-stage EDSFA with a $\Delta\lambda_{3dB}$ of about 47 nm around 1550 nm was reported [11]. The amplifier used two EDSFs and an internal GEQ set between the EDSFs. Another two-stage EDSFA generated a $\Delta\lambda_{3dB}$ of 52 nm (1556 to 1608 nm) as mentioned above [12]. Moreover, a three-stage EDFFA, which used three EDFFs and two internal GEQs set among the EDFFs, generated a $\Delta\lambda_{3dB}$ of 62 nm (1540.5 to 1602.5 nm) [13]. A three-stage EDTFA generated a $\Delta\lambda_{3dB}$ of 76 nm (about 1530 to 1606 nm) [14]. Note that the amplifier used a 0.98 mm pumped EDTF in the first stage and achieved a low NF of 5 dB.
2. A two-gain band EDSFA was reported to generate an aggregate bandwidth of 84.3 nm (~1527 to ~1568 nm and ~1569 to ~1612.5 nm) [3]. The width of the dead band between the two gain bands was only 1 nm or so. Although some quasiseamless two-gain band amplifiers yielded narrow dead bands between the two gain bands, some interference noises around the dead bands must be treated [22]. Note that most of the two-gain band EDFAs used in transmission experiments have dead bandwidths of about 5 to 10 nm; see the examples of [2, 8, 9].
3. Multiwavelength pumped RAs were reported as seamless and wideband fiber amplifiers. A DRA with a $\Delta\lambda_{3dB}$ of about 100 nm (1520 to 1620 nm) with an

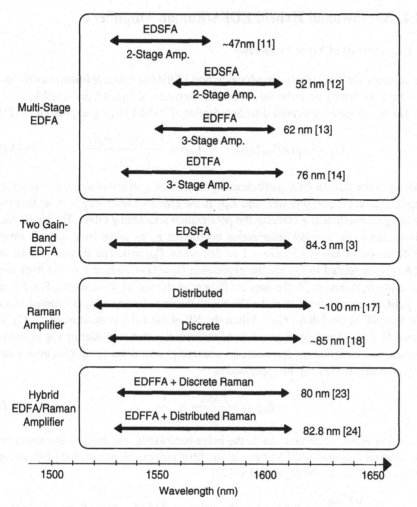

Fig. 13.6. Gain bands of wideband fiber amplifiers. ED(S, F, T)FA: erbium-doped (silica, flouride, tellurite) fiber amplifier.

on-off gain of about 13 dB was reported [17]. Moreover, an LRA was reported to generate a gain bandwidth of about 85 nm (1510 to 1595 nm) with a net gain (amplifier in-out gain) of over 13 dB (on-off Raman gain of over 20 dB) [18].

4. An SWB-HA with a LRA was reported to generate a $\Delta\lambda_{3dB}$ of 80 nm (1530 to 1610 nm) [23]. Moreover, an SWB-HA with a DRA generated a $\Delta\lambda_{3dB}$ of 82.8 nm (1528.8 to 1611.6 nm) [24].

The gains of the wideband fiber amplifiers listed above are, except for the Raman amplifiers, above 20 dB or so. Therefore, the SWB-HAs have the largest seamless bandwidths, about 80 nm, with gain above 20 dB. The bandwidths are comparable to those of two-gain band EDFAs.

13.4. Narrowband Hybrid EDFA/Raman Amplifiers

13.4.1. Analysis of Noise Properties

We consider the NB-HAs (narrowband hybrid EDFA/Raman amplifiers) in this section. First we briefly describe the noise characteristics of an LRA and a DRA.

The on-off gain (or overall distributed gain) of a DRA (G_d) is expressed as [25]

$$G_d = \exp(g P_{\text{pin}} L_{\text{eff}}), \qquad L_{\text{eff}} \equiv \frac{1 - \exp(-\alpha_p l)}{\alpha_p}, \qquad (13.5)$$

where g is the Raman gain coefficient, α_p is the loss coefficient at the pump wavelength, P_{pin} is the pump power, and L_{eff} is the effective fiber length. Note that the Raman gain coefficient g contains the polarization scrambling effect. The coefficient g is related to the parallel-polarization coefficient g_p as given by $g \cong g_p/2$. From the definition of Eq. (13.5), $G_d = 1$ (0 dB) when $P_{\text{pin}} = 0$. The Raman gain of an LRA (G_1) is related to G_d via the expression $G_l = G_d L$, where L is the fiber loss $L = \exp(-\alpha_s l)$, and α_s is the loss coefficient at the signal wavelength. The NF of the fictitious amplifier set after the DRA with the on-off gain of G_d is denoted as the effective NF of the DRA (F_{eff}). When the NF of the LRA is denoted as F, F_{eff} is related to F as shown by $F_{\text{eff}} = FL$. Moreover, F_{eff} can, considering the signal to spontaneous–emission (or spontaneous–scattering) beat noise (s–sp beat noise) and the signal shot noise [26], be expressed as

$$F_{\text{eff}} = \frac{P_{\text{ASE}}}{h\nu \Delta\nu G_d} + \frac{1}{G_d}, \qquad (13.6)$$

where h is Plank's constant, $\Delta\nu$ is the noise bandwidth, and P_{ASE} is the amplified spontaneous emission (ASE) power at the fiber output. Moreover, the propagation equation of P_{ASE} can be expressed as [27]

$$\frac{dP_{\text{ASE}}}{dz} = g P_p \{ P_{\text{ASE}} + 2h\nu_n (N_{\text{phon}} + 1)\Delta\nu \} - \alpha(\nu_n) P_{\text{ASE}}$$

$$N_{\text{phon}} = \frac{1}{\exp\{h(\nu_p - \nu_s)/kT\} - 1}, \qquad (13.7)$$

where N_{phon} is the phonon population number at finite temperature T in Kelvin, ν_s and ν_p are the optical frequencies at the signal and pump wavelengths, respectively, and k is Boltzmann's constant.

Figure 13.7 shows the NF of a backward pumped LRA as a function of G_d. The NFs were calculated by Eq. (13.7). Signal and pump wavelengths were set to 1580 and 1480 nm, respectively. The fiber loss coefficients of the LRA were set to 0.5 dB/km for both signal and pump lights for simplicity. The fiber length of the LRA was set to 5, 7.5, or 10 km, and corresponding fiber losses were 2.5, 3.75, or 5 dB. The NF at $G_d = 10$ (20) dB for the 5, 7.5, or 10 km fiber was 5.8 (5.2), 5.1 (4.7), or 4.4 (4.2) dB, respectively. Therefore, the NFs of the LRA range from 4 to 6 dB in most operating conditions.

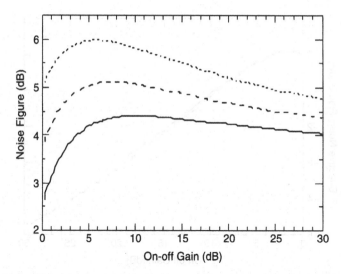

Fig. 13.7. Noise figures as functions of the on-off gain for a discrete Raman amplifier. Fiber loss coefficient is 0.5 dB/km; fiber length is 5 km (solid line), 7.5 km (broken line), or 10 km (dotted line).

F_{eff} of a DRA are shown as a function of G_d in Fig. 13.8 as calculated by Eqs. (13.5) through (13.7). Signal and pump wavelengths were set to 1580 and 1480 nm, respectively. The fiber length of the DRA was set to 100 km. Loss coefficients at the signal and pump wavelengths were set to 0.2, 0.3, or 0.4 dB/km. F_{eff} of the 0.2 dB/km fiber at $G_d = 10$ (20) dB was -1 (-3 dB). The differences in F_{eff} for the three loss coefficients are less than 0.2 dB. Although we can lower F_{eff} by increasing G_d to more than 20 dB as shown in Fig. 13.8, there is a limitation due to the double-Rayleigh scattering noise as shown in Chapter 3.

Next, noise characteristics of NB-HAs are analyzed. We consider an NB-HA with the configuration of Fig. 13.1(c). The NB-HA consists of a DRA with a transmission fiber and an EDFA after the DRA. The NF and effective NF of the NB-HA are denoted as F and F_{eff}, respectively. The on-off gain of the DRA and the loss of the transmission fiber are denoted as G_d and L, respectively. F_{eff} is, from its definition, related to F as shown by $F_{\text{eff}} = FL$. F_{eff} can be expressed as

$$F_{\text{eff}} = \frac{P_{\text{ASE}d}}{h\nu\Delta\nu G_d} + \frac{P_{\text{ASE}l}}{G_d h\nu\Delta\nu G_l} + \frac{1}{G_d G_l} = \frac{P_{\text{ASE}d}}{h\nu\Delta\nu G_d} + \frac{2n_{spl}(G_l - 1)}{G_d G_l} + L,$$

(13.8)

where $G_d G_l L = 1$, $P_{\text{ASE}d}$ is the ASE power from the DRA, and $P_{\text{ASE}l}$ is the ASE power from the EDFA with no input ASE. Moreover, G_l and n_{spl} are the gain and the spontaneous emission factor of the EDFA (lumped amplifier), respectively. The first, second, and third terms in the right-hand side of Eq. (13.8) originate from the s–sp beat noise generated in the DRA, the s–sp beat noise generated in the EDFA, and the signal shot noise, respectively. It is obvious that the EDFA can be replaced by a LRA in the above analysis if needed.

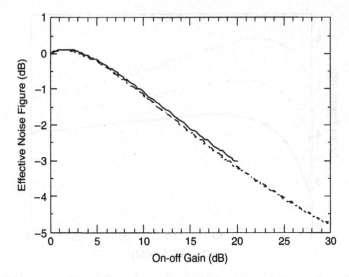

Fig. 13.8. Effective noise figures as functions of the on-off gain for a distributed Raman amplifier with a transmission fiber of 100 km. Fiber loss coefficient is 0.2 dB/km (solid line), 0.3 dB/km (broken line), or 0.4 dB/km (dotted line).

In calculations, the fiber length and the loss coefficient of the transmission fiber were set to 100 km and 0.2 dB/km, respectively, and $2n_{spl}$ was set to 11, 7, or 3 dB. The signal and pump wavelengths were set to 1580 and 1480 nm, respectively. Figure 13.9 shows F_{eff} of the NB-HA as a function of G_d. F_{eff} of the DRA, calculated by Eq. (13.6), is also shown for comparison. F_{eff} of the NB-HA decrease with G_d. For example, F_{eff} decreased in order 11, 7, and 3 dB for G_d values of 0 dB to 2.3, 0.6, and -0.5 dB at $G_d = 10$ dB, respectively. The third term ($L = 0.01$) in the right-hand side of Eq. (13.8) is negligible compared to the sum of the other terms. The second term decreases with G_d. On the other hand, the first term increases with G_d in the range of $G_d = 0$ to 8 dB, and stays around 0.6 or so with G_d in the range of $G_d = 8$ to 20 dB. The second term equals the first term at $G_d \sim 5$, 8, or 12 dB for n_{spl} values of 3, 7, or 11 dB, respectively.

For a nonlinear impairment-limited transmission system, we have to consider an increase in the nonlinear impairment with an increase in path-averaged signal power (ΔP_{ave}) [28]. In the calculation above, ΔP_{ave} is 0.15, 0.6, 1.5, and 1.9 dB for G_d values of 5, 10, 15, and 20 dB, respectively. We can obtain higher transmission performance with higher G_d value up to $G_d = 20$ dB, because the increase in F_{eff} is larger than the increase in ΔP_{ave}.

13.4.2. An Experimental Example

Several experiments have been reported on the NB-HA [4–7], including NB-HAs in the L-band with DSFs. The improvements in the OSNR and the transmission distance with a fixed inline repeater spacing [5] and a repeater spacing upgrade [4] with the

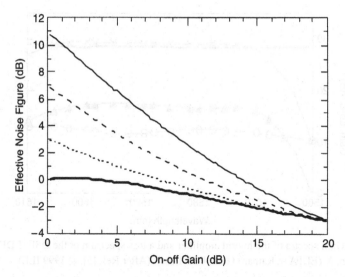

Fig. 13.9. Effective noise figures as functions of the on-off gain for a hybrid EDFA/distributed Raman amplifier with a transmission fiber of 100 km. The effective noise figure for the distributed Raman amplifier is also shown by a thick line. Fiber loss coefficient is 0.2 dB/km. Twice the spontaneous emission factor fo the EDFA is 11 dB (solid line), 7 dB (broken line), or 3 dB (dotted line).

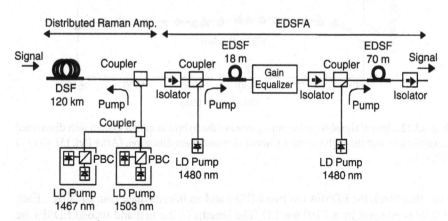

Fig. 13.10. Configuration of a narrowband hybrid EDFA/Raman amplifier (PBC: polarization beam combiner). (After Ref. [5]. © 1999 IEE)

NB-HAs were confirmed. Moreover, reports on NB-HAs in the C-band with DSFs have confirmed the significant suppression of nonlinear impairments and the improvement of the transmission distance [6, 7, 30].

Figure 13.10 shows the configuration of the L-band NB-HA of [5]. The HA consists of a DRA and an EDSFA. The DRA has a 120 km DSF as the gain medium and a pump light source with two LD pumps. Each of the LD pumps has two LDs and a polarization combiner. The wavelengths of the LD pumps are 1467 and 1503 nm. On

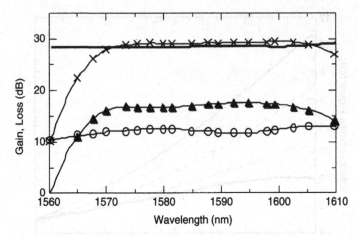

Fig. 13.11. Gain spectra of the hybrid amplifier and a loss spectrum of the DSF (EDFA Gain, Raman Gain, X (EDFA + Raman) Gain, DSF Loss). (After Ref. [5]. © 1999 IEE)

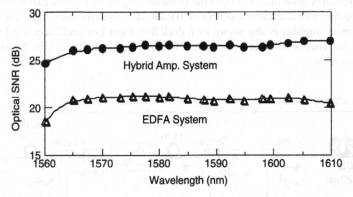

Fig. 13.12. Optical signal-to-noise ratio spectra of the hybrid amplifier system with distributed amplification and the EDFA system without distributed amplification. (After Ref. [5]. © 1999 IEE)

the other hand, the EDSFA has two EDSFs and an intermediate gain equalizer. Each EDSF is pumped by a 1480 nm LD. The lengths of the first and second EDSFs are 18 and 70 m, respectively.

The gain characteristics of the NB-HA are shown in Fig. 13.11. The figure shows the gain spectra of the NB-HA and the loss spectrum of the DSF. The gain spectra are of the EDSFA, the DRA, and the sum of them. The NB-HA gains well compensate the DSF losses over a wide wavelength range. A 0.3 dB gain reduction bandwidth of 30 nm (1572 to 1602 nm) was obtained. Figure 13.12 shows the OSNR spectra of the NB-HA system with distributed amplification and the EDFA system without distributed amplification (DSF + EDFA). The resolution bandwidth of the noise was 0.5 nm and the signal power launched into the DSF was 0 dBm. G_d of the NB-HA was 12.3 dB in the gain bandwidth. The OSNRs of the NB-HA system and the EDFA system were 26.4 ± 0.2 dB and 21.0 ± 0.2 dB, respectively. Therefore, the improvement in OSNRs for the NB-HA system compared to those of the EDFA system is 5.4 dB; this is due to the G_d value (12.3 dB). The NF of the EDFA was about 5 dB. The improvement in OSNRs calculated by Eqs. (13.5) to (13.8) is 5.8 dB, which is close to the measured value of 5.4 dB.

The performance of an NB-HA in a WDM transmission system was tested in an experiment [5]. The amplifier was used as an inline amplifier in a recirculating loop transmission experiment using 2.5 Gb/s × 7 channel WDM signals and the 120 km DSF as the transmission fiber. The average transmission distance of the WDM signals was 570 km for the EDFA system (without distributed amplification) and 2088 km for the NB-HA system (with distributed amplification). Therefore, the improvement in transmission distance was 5.6 dB, which well coincides with the OSNR improvement of 5.4 dB.

13.5. Seamless and Wideband Hybrid EDFA/Raman Amplifiers

13.5.1. Bandwidth Analysis

We can fabricate a wideband hybrid EDFA/Raman amplifier by simply combining, in a serial cascade, a wide and flat gain band EDFA such as multistage EDFAs (Section 13.3) with a multiwavelength pumped flat gain band RA (Chapter 14). However, the bandwidth of the HA is limited by that of the EDFA or the RA. Moreover, each of the EDFA and the RA needs many optical components so cost is high. This section describes SWB-HAs (seamless and wideband hybrid EDFA/Raman amplifiers) that have a simple structure with few optical components and so are cost effective. The EDFA and the RA have opposite gain spectral slopes over a wide wavelength region; the gain bandwidth of the SWB-HA is as large as about 80 nm (1530 to 1610 nm). The 80 nm gain band seamlessly covers the two EDFA gain bands (the C- and L-bands).

Figure 13.13 shows typical gain spectra of the SWB-HA. Figure 13.13(a) shows the gain of an EDFFA as a function of the average fractional population $N_{2\text{ave}}$. Figure 13.13(b) shows the on-off Raman gain of an RA with a dispersion-compensation fiber (DCF) as a function of pump wavelength λ_p. Similar spectra are observed with an EDSFA and an RA with a DSF or a standard SMF. The gain spectrum of the SWB-HA can be synthesized as an appropriate combination of an EDFA gain spectrum and an RA gain spectrum. We can achieve a maximized gain bandwidth with the combination of the EDFA gain spectrum at $N_{2\text{ave}} \sim 0.7$ and the RA gain spectrum at $\lambda_p \sim 1510$ nm.

Figure 13.14 shows examples of SWB-HA gain spectra. Figure 13.14(a) shows the gain spectra of an SWB-HA that consists of an RA pumped at a single wavelength of $\lambda_p = 1510$ nm and an EDFFA at $N_{2\text{ave}} = 0.7$. Moreover, Fig. 13.14(b) shows the gain spectra of an SWB-HA that consists of an RA pumped at two wavelengths of $\lambda_p = 1510$ and 1480 nm, and an EDFFA at $N_{2\text{ave}} = 0.7$. For single-wavelength Raman pumping, the EDFFA and Raman peak gains were both set at 20 dB. For two-wavelength Raman pumping, however, the EDFFA peak gain was set to 20 dB, and the Raman peak gains for $\lambda_p = 1510$ and 1480 nm were set to 22 and 11 dB,

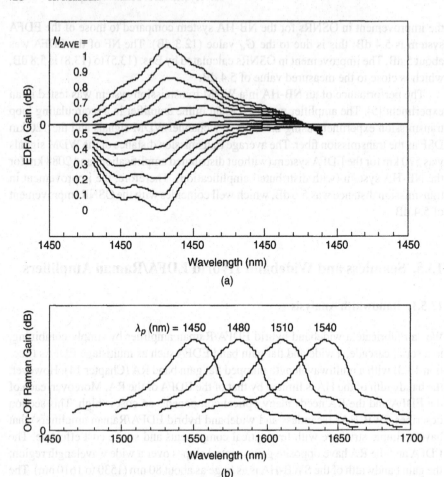

Fig. 13.13. Gain spectra of a flouride EDFA and a Raman amplifier with a dispersion-compensation fiber: (a) EDFFA gain as a function of average fractional population of the upper state; (b) on-off Raman gain as a function of pump wavelength (λp).

respectively. The significantly wider gain bandwidth of the SWB-HA, compared to the individual gain bandwidths of the EDFFA and the RA, was obtained without a gain equalizer by the single-wavelength pumping approach, because the gain spectra of the EDFFA and RA have opposite gain slopes. Moreover, significantly improved gain flatness is obtained by the two-wavelength pumping if the optimum λ_p values are selected. The gain bandwidths obtained with an ideal GEQ for the two cases are both about 90 nm (1530 to 1620nm). The SWB-HA also offers the benefits of low noise and high output power as shown in Fig. 13.2.

The wideband NF spectra of the single-wavelength pumped Raman amplifiers together with their gain spectra are shown in Fig. 13.15. The NF and gain spectra were calculated by Eqs. (13.5) to (13.8). The spectra of the on-off Raman gain G_d and the effective noise figure F_{eff} of the DRA with 80 km DSF are shown in Fig. 13.15(a).

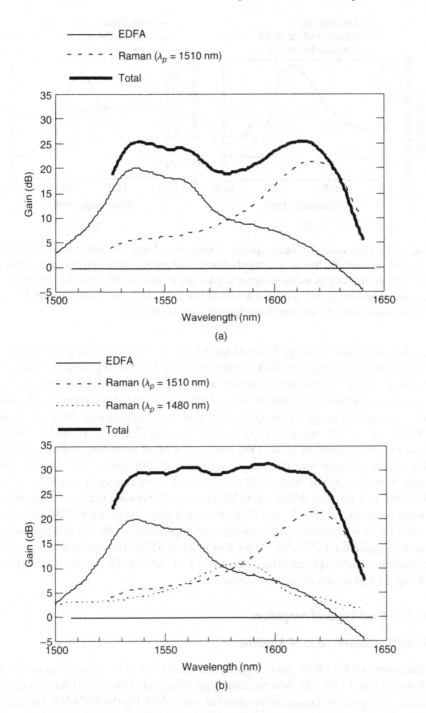

Fig. 13.14. Examples of gain spectra of a hybrid amplifier. λp: Raman pump wavelength; N_{2ave}: average fractional population of the upper state. (a) $N_{2ave} = 0.7$, $\lambda p = 1510$ nm; (b) $N_{2ave} = 0.7$, $\lambda p = 1510$ and 1480 nm.

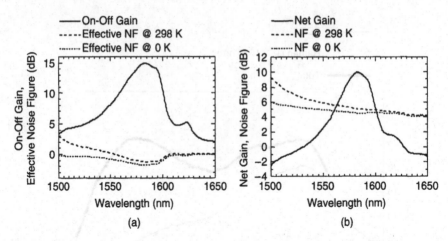

Fig. 13.15. Gain and noise figure spectra of single-wavelength pumped Raman amplifiers: (a) on-off gain and effective noise figure of a distributed Raman amplifier with 80 km dispersion-shifted fiber; (b) net gain and noise figure of a discrete Raman amplifier with 8 km dispersion-compensation fiber (pump wavelength = 1480 nm; gain: solid line; noise figure at 298° K (room temperature): broken line; noise figure at 0° K: dotted line).

On the other hand, the net gain G_l and noise figure F of the LRA (lumped or discrete Raman amplifier) with 8 km DCF are shown in Fig. 13.15(b). F_{eff} and F at the temperatures of 298° K (room temperature) and 0° K are shown for the two amplifiers in the figures. The pump wavelength is set to 1480 nm. Measured absorption coefficients at the signal and pump wavelengths were used in the calculations. Peak G_d of the DRA is set to 15 dB, and peak G_l of the LRA is set to 10 dB. F_{eff} of the DRA at the peak gain wavelength of about 1580 nm is -2.3 dB at room temperature, whereas F_{eff} at a wavelength near the pump wavelength of 1500 nm is 3 dB. On the other hand, F of the LRA at about 1580 nm is 5 dB at room temperature, whereas F at 1500 nm is 9 dB. The difference in NF (F_{eff} or F) between the two temperatures (room temperature, 298° K, and 0° K) for signal wavelengths over 1500 nm is less than 3 dB or so; this reflects the temperature dependence of the phonon population factor N_{phon} in Eq. (13.7). NFs of the four types of SWB-HAs can be determined by considering the NF spectra of the RAs of Fig. 13.15 and the NF spectra of the EDFAs of Fig. 13.5 as shown in the next section.

13.5.2. Four Types of Amplifiers

13.5.2.1. Classification of Amplifiers

Partitioning of the EDFA gain (G_{EDFA}) and the Raman gain (G_{Raman}) in an HA is shown in Fig. 13.16. The total amplifier gain (G_{Hybrid}) is the sum of the two gains: $G_{Hybrid} = G_{EDFA} + G_{Raman}$, where gain has units of dB. For the SWB-HA, G_{Raman} is defined as the peak Raman gain and G_{EDFA} is, in this section, defined at the wavelength that gives the peak Raman gain G_{Raman}. We can classify the SWB-HA into four types according to its G_{Raman} and gain types (distributed or discrete). Table 13.1 shows the

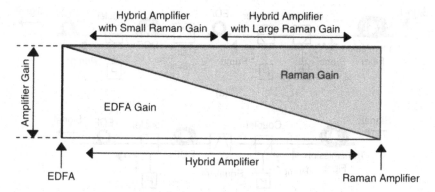

Fig. 13.16. Gain partitioning in a hybrid amplifier.

Table 13.1.

Raman Gain	Distributed	Discrete
Small	Type-1	Type-3
Large	Type-2	Type-4

classification with the four types [20]. The SWB-HA with small (large) distributed Raman gain is denoted as Type-1 (2). On the other hand, the SWB-HA with a small (large) discrete Raman gain is denoted as Type-3 (4). Typical gain examples are $G_{Hybrid} = 25$ dB, $G_{Raman} = 10$ dB, and $G_{EDFA} = 15$ dB, for Types -1 and -3 with small Raman gain, and $G_{Hybrid} = 25$ dB, $G_{Raman} = 20$ dB, and $G_{EDFA} = 5$ dB, for Types -2 and -4 with large Raman gain.

The four types of SWB-HAs have different basic configurations as shown in Fig. 13.17 [20], and thus have different gain, noise, and output characteristics. Amplifier configurations of Types -1 to -4 are shown in Figs. 13.17(a) to 13.17(d), respectively. Optical components such as isolators in the amplifiers are not shown in the figures for simplicity. Note that the EDFs are forward pumped and the DCFs are backward pumped in Fig. 13.17, because this approach is common. However, the opposite pump directions can be employed if needed. The basic amplifier configurations and the amplification characteristics of the four types are described below. Experimental amplifier configurations and their amplification characteristics are described in the following sections.

First, the Type-1 amplifier has a two-stage EDFA with an intermediate GEQ and a DCF. The two-stage EDFA configuration is employed because large EDFA gain is required. The amplifier also has a DRA with a transmission fiber as its gain medium in front of the EDFA. The peak loss of the GEQ is almost equal to that of the wideband two-stage EDFA described in Section 13.3. The effective NF spectrum of the amplifier is determined by both the effective NF spectrum of the DRA and the NF spectrum of the two-stage EDFA. The output power characteristics, however, are determined by the two-stage EDFA.

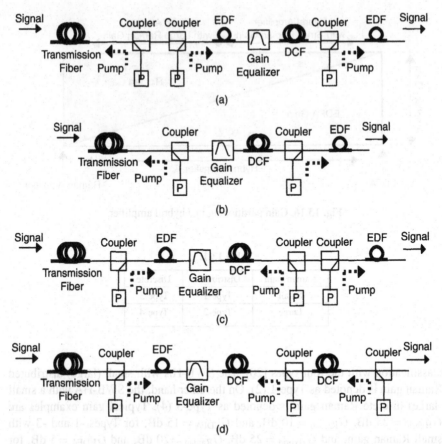

Fig. 13.17. Configurations of the four types of seamless and wideband hybrid amplifiers: (a) Type-1 with small distributed Raman gain; (b) Type-2 with large distributed Raman gain; (c) Type-3 with small discrete Raman gain; (d) Type-4 with large discrete Raman gain; (EDF: erbium-doped fiber amplifier; DCF: dispersion compensation fiber).

Next, the Type-2 amplifier has a single-stage EDFA with a GEQ and a DCF set in front of the EDF in the EDFA. The amplifier also has a DRA with a transmission fiber as its gain medium. The peak loss of the GEQ is small as is expected from the gain spectra of Fig. 13.14. The effective NF spectrum of the amplifier is mainly determined by that of the DRA. However, both the single-stage EDFA and the DRA determine the output power.

The Type-3 amplifier has a two-stage EDFA with intermediate GEQ and DCF. The DCF is pumped and operates as an LRA. The peak loss of the GEQ is large. The NF spectrum of the amplifier is mainly determined by that of the first-stage EDF of the two-stage EDFA, but the output power is determined by the second-stage EDF.

Finally, the Type-4 amplifier has a single-stage EDFA, a two-stage LRA, and an intermediate GEQ. The LRA has two DCFs as its gain media and generates a large Raman gain. The peak loss of the GEQ is small. The NF spectrum of the amplifier is determined by the NF spectra of the EDFA and the LRA.

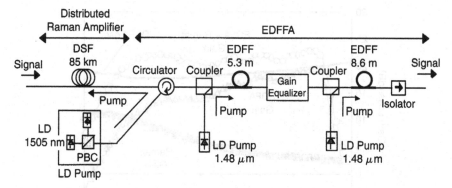

Fig. 13.18. Configuration of a Type-1 seamless and wideband hybrid amplifier (DSF: dispersion-shifted fiber; PBC: polarization beam combiner; EDFF(A): erbium-doped flouride fiber (amplifier)). (After Ref. [31]. © 1998 IEE)

13.5.2.2. Type-1 Amplifier with a Small Distributed Raman Gain

Figure 13.18 shows the experimental configuration of a Type-1 SWB-HA with small distributed Raman gain [31]. The amplifier has a two-stage erbium-doped fluoride fiber amplifier and a DRA. The EDFFA consists of two EDFFs and a GEQ. Each of the EDFFs is forward pumped by a 1.48 μm LD. The first and second EDFFs are 5.3 and 8.6 m long, respectively. The GEQ is a cascade of two Mach–Zehnder filters (GEQ-1 and -2). The free spectral ranges of GEQ-1 and -2 are both 100 nm. The DRA consists of an 85 km DSF as the gain medium and a 1505 nm polarization-multiplexed pump LD source. The pump light from the LD source is launched into the DSF via an optical circulator. The pump power launched into the DSF is 209 mW.

Gain characteristics of the HA are shown in Fig. 13.19. Spectra of the EDFFA gain, the DRA on-off gain G_d, and the HA gain (the sum of the EDFA and Raman gains) are shown together with the loss spectrum of the DSF in Fig. 13.19(a). Losses of GEQ-1 and -2, and the sum of them (GEQ) are shown in Fig. 13.19(b). The peak loss of GEQ is 26 dB at 1560 nm. The net gain is defined as the difference between the HA gain and the DSF loss. The $\Delta\lambda_{3dB}$ value (3 dB gain reduction bandwidth) of the net gain is 75 nm (1531 to 1606 nm) with a peak gain of 22.8 dB. The peak G_d of the DRA is 11.1 dB at 1610 nm.

Figure 13.20 shows the spectral characteristics of the OSNR of the HA. The figure shows measured OSNRs of the HA, measured OSNRs of the transmission line using a 1.48 μm pumped EDFFA, and calculated OSNRs of the transmission line using a discrete amplifier with a noise figure of 3, 6, or 9 dB. The signal power launched into the DSF was set to -14.5 dBm per channel, and the noise bandwidth was set to 0.1 nm. The effective NFs of the HA were 5.5 dB at 1540 nm and 0.7 dB at 1600 nm, whereas the NFs of the EDFFA were 7.5 dB at 1540 nm and 6.7 dB at 1600 nm. Therefore, the improvement in NF between the HA and the EDFFA transmission lines ranged from 2 to 7 dB in the 3 dB gain band. The measured effective NF spectrum of the HA can be well explained by the analyses described in Sections 13.4.1 and 13.5.1 (Figs. 13.9 and 13.15).

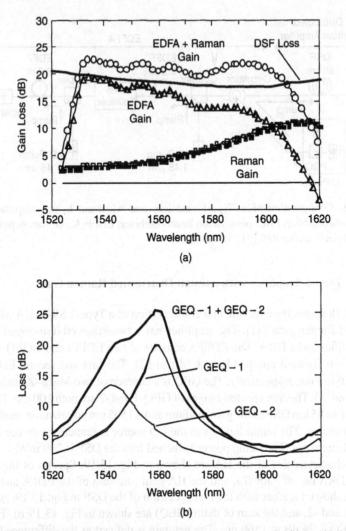

Fig. 13.19. Characteristics of the gain spectra of a Type-1 seamless and wideband hybrid amplifier; (a) Spectra of gains (symbols) and DSF loss (line), EDFA: open triangle, Raman amplifier: filled square, EDFA + Raman amplifier: open circle; (b) Loss spectra of gain equalizers (GEQ-1 and -2); (c) Net gain spectrum. (After Ref. [31]. © 1998 IEE)

A Type-1 SWB-HA using an EDSFA was also fabricated and evaluated [32]. The SWB-HA reported in 1997 was to the author's knowledge, the first hybrid EDFA/Raman amplifier to be described. $\Delta\lambda_{3dB}$ of the net gain was 67 nm (1549 to 1616 nm) with a peak gain of 20 dB. The $\Delta\lambda_{3dB}$ of the SWB-HA using an EDFFA (75 nm) is larger than that of the SWB-HA using an EDSFA (67 nm). This is because there is a smaller difference between the gain near 1530 nm and the gain near 1600 nm in the EDFFA without GEQ than in the EDSFA without GEQ (Fig. 13.4).

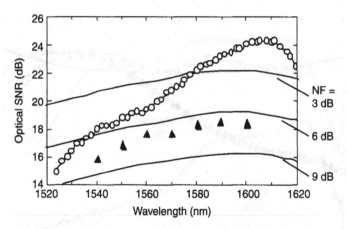

Fig. 13.20. Optical signal-to-noise ratio spectra of a Type-1 seamless and wideband hybrid amplifier (hybrid amplifier, EDFA (measured), EDFA with noise figure of 3, 6, or 9 dB (calculated)). (After Ref. [31]. © 1998 IEE)

13.5.2.3. Type-2 Amplifier with a Large Distributed Raman Gain

A Type-2 SWB-HA with a large peak G_d of 25 dB was fabricated using two 50 km DSFs as the Raman gain media [24]. The detailed configuration of the experimental HA is not shown here for simplicity. Note that the basic configuration is shown in Fig. 13.17(b). The HA has two units in the 100 km transmission line consisting of the two DSFs. The first HA unit has a single stage 1.48 μm pumped EDFFA and a 50 km DSF. The DSF was backward pumped by a 1.51 μm polarization-multiplexed LD module and was forward pumped by a 1.48 μm LD. However, the second HA unit has a single-stage 1.48 μm pumped EDSFA and a 50 km DSF. The DSF was backward pumped by a 1.51 μm polarization-multiplexed LD module. The pump powers launched into the DSFs were about 200 mW for the two 1.51 μm LD modules, and about 100 mW for the 1.48 μm LD. A simple Mach–Zehnder type GEQ with a peak-to-peak loss of about 2 dB was set after the second HA unit if needed.

Figure 13.21 shows gain spectra of the Type-2 HA and loss spectrum of the 100 km transmission line without the GEQ. The peak gain of the EDFA (the sum of the EDFFA and EDSFA) was 14.5 dB at 1530 nm, and the peak G_d of the DRA (the sum of the two DRAs) was 25 dB at 1610 nm. The $\Delta\lambda_{3dB}$ for the net gain was 79.9 nm (1530.6 to 1610.5 nm) without the GEQ and 82.8 nm (1528.8 to 1611.6 nm) with the GEQ. The peak gain of the HA with the GEQ was 22 dB.

Optical SNR spectra for the HA are shown in Fig. 13.22. The figure shows a measured OSNR spectrum of the HA without the GEQ and three calculated OSNR spectra of the transmission line using two discrete amplifiers with the same NF of 3, 5, or 7 dB. The noise bandwidth is 0.1 nm. F_{eff} of the HA was 3 dB at 1530 nm and -2 dB at 1600 nm. The Type-2 HA has lower F_{eff} values than the Type-1 HA (Section 13.5.2.2). This is because it has larger G_d values.

Another Type-2 HA, consisting of a 1497 nm wavelength pumped DRA and a 0.98 μm pumped EDSFA, was reported as an inline amplifier for a submarine

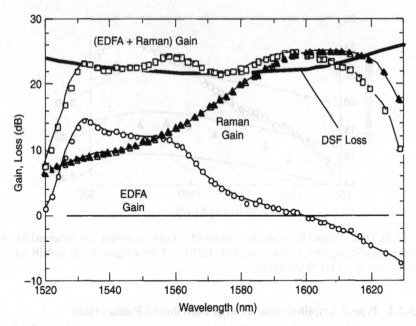

Fig. 13.21. Gain spectra of a Type-2 seamless and wideband hybrid amplifier and loss spectrum of the transmission line with two 50 km DSFs (total length is 100 km). (After Ref. [24]. © 1998 IEE)

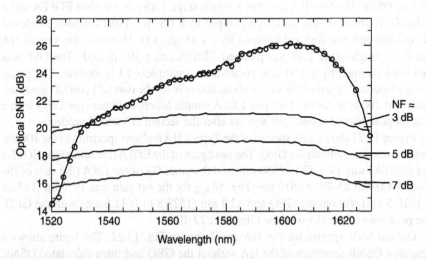

Fig. 13.22. Optical signal-to-noise ratio spectra of a Type-2 seamless and wideband hybrid amplifier (hybrid amplifier (measured), transmission line using two discrete amplifiers with a noise figure of 3, 5, or 7 dB (calculated)). (After Ref. [24]. © 1998 IEE)

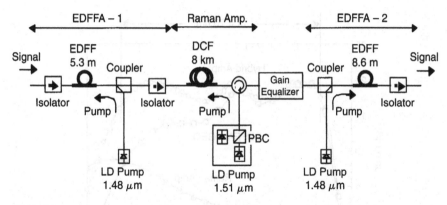

Fig. 13.23. Configuration of a Type-3 seamless and wideband hybrid amplifier (PBC: polarization beam combiner, DCF: dispersion-compensation fiber, EDFF(A): erbium-doped flouride fiber (amplifier)). (After Ref. [34]. © 1998 IEE)

transmission system [33]. The peak gain of the EDSFA was comparable to that of the DRA. A seamless flat gain bandwidth of 80 nm (1527 to 1607 nm) was obtained.

13.5.2.4. Type-3 Amplifier with Small Discrete Raman Gain

The experimental configuration of a Type-3 discrete SWB-HA is shown in Fig. 13.23 [34]. The amplifier has two-stage EDFFAs (EDFFA-1 and -2), an intermediate LRA, and an intermediate GEQ. EDFFA-1 has a 5.3 m EDFF, which is backward pumped by a 1.48 μm LD, and EDFFA-2 has an 8.6 m EDFF, which is forward pumped by a 1.48 μm LD. The LRA uses an 8 km DCF as its gain medium, which is pumped by a 1.51 μm polarization-multiplexed LD module. The 1.51 μm pump power launched into the DCF is 197 mW. The GEQ has the same structure of the GEQ used in the Type-1 HA (Section 13.5.2.2). The peak loss of the GEQ is about 25 dB at 1560 nm.

Figure 13.24 shows the gain spectra of the Type-3 HA. The figure plots the gains of the two-stage EDFFAs including the GEQ loss (EDFFA-1 + EDFFA-2 + GEQ), the LRA, and the sum of them (HA). The peak gain of the two-stage EDFFAs including the GEQ is 19 dB at 1535 nm, and the peak gain of the LRA is 8.9 dB (on-off gain of 12.1 dB) at 1610 nm. Moreover, the peak gain of the HA is 19.8 dB. $\Delta\lambda_{3dB}$ is 75 nm (1531 to 1606 nm).

The NF spectrum of the HA is shown in Fig. 13.25. NFs in the 3 dB gain band range from 4.9 to 7.4 dB at the input point of the HA. The NFs are from 3.8 to 6.3 dB at the input point of the first EDFF when we consider the 1.1 dB insertion loss of the input isolator. The NF spectrum is mainly determined by that of the 1.48 μm pumped EDFF (Fig. 13.5).

13.5.2.5. Type-4 Amplifier with Large Discrete Raman Gain

A Type-4 discrete SWB-HA was fabricated using the basic amplifier configuration of Fig. 13.17(d) [23]. The experimental detailed configuration is not shown here for simplicity. The amplifier had a single-stage EDFFA in the input side of the amplifier,

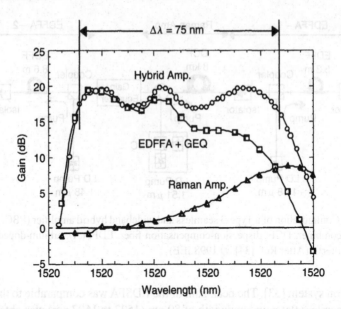

Fig. 13.24. Gain spectra of a Type-3 seamless and wideband hybrid amplifier (Open square: EDFFA + GEQ, EDFFA = EDFFA-1 + EDFFA-2; Open triangle: Raman amplifier; Open circle: hybrid amplifier (EDFFA + GEQ + Raman amp.); Δλ: 3 dB gain reduction bandwidth). (After Ref. [34]. © 1998 IEE)

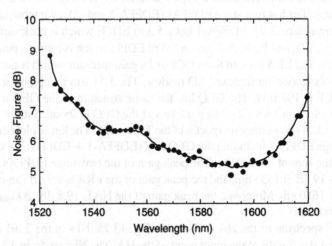

Fig. 13.25. Noise figure spectrum of a Type-3 seamless and wideband hybrid amplifier. (After Ref. [34]. © 1998 IEE)

a two-stage LRA in the output side of the amplifier, and a GEQ downstream of the Raman amplifier. The GEQ can be set between the EDFFA and LRA with negligible variations in the noise and gain characteristics because the peak loss of the GEQ was only 3.5 dB. The EDFFA had a short length (3.3 m) of a EDFF, which was forward pumped by a 1465 nm LD. The LRA had two DCFs as its gain media. The lengths of the first and second DCFs (DSF-1 and -2) were 8.0 and 8.3 km, respectively. DCF-1

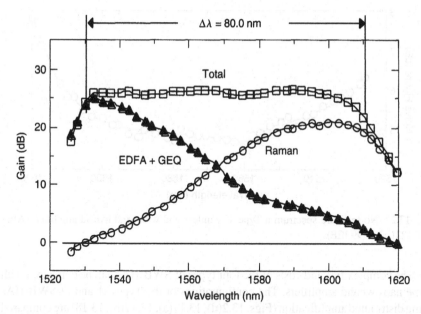

Fig. 13.26. Gain spectra of a Type-4 seamless and wideband hybrid amplifier (GEQ: gain equalizer; filled triangle: EDFA + GEQ; open circle: Raman amplifier; open square: hybrid amplifier (EDFA + GEQ + Raman amplifier); $\Delta\lambda$: 3 dB gain reduction bandwith). (After Ref. [23]. © 1999 IEE)

was pumped by a 1472 nm forward pump (P1) and a 1495 nm backward pump (P2), whereas DCF-2 was pumped by a 1503 nm forward pump (P3) and a 1503 nm backward pump (P4). Each of the pump light sources was a polarization-multiplexed LD module. The pump powers launched into the DCFs for P1 to P4 were 188, 108, 170, and 149 mW, respectively. Therefore, the LRA was bidirectionally pumped with three wavelengths (1472, 1495, and 1503 nm) and with a total power of 615 mW.

Figure 13.26 shows the gain spectra of the HA. The figure plots the gains of the sum of the EDFFA and GEQ, the LRA, and the sum of them (HA). The peak gain of the sum of the EDFFA and GEQ is 25 dB at 1531 nm, and the peak gain of the LRA is 21 dB at 1600 nm. Moreover, the peak gain of the HA is 28.1 dB. $\Delta\lambda_{3dB}$ is 80 nm (1530 to 1610 nm). The NF spectrum of the amplifier is shown in Fig. 13.27. The NFs in the 3 dB gain band are less than 6.0 dB. The NFs at the wavelengths shorter than about 1570 nm are mainly determined by the NFs of the EDFFA, because the EDFFA gains are larger than 10 dB at the wavelengths. On the other hand, the NFs at the wavelengths longer than about 1570 nm are determined by both the NFs of the EDFFA and LRA. A total output power with 15 WDM signals of 13.8 dBm was obtained under the error-free operation.

Each performance of the four types of SWB-HAs described above was tested in a WDM transmission experiment [20, 23, 24, 31, 32, 35, 36]. Each SWB-HA was used as an inline amplifier. There was no power penalty within the experimental accuracy in the bit error rate curves between before and after transmission for all WDM signal channels in the 3 dB gain band.

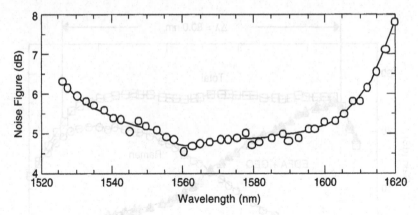

Fig. 13.27. Noise figure spectrum a Type-4 seamless and wideband hybrid amplifier. (After Ref. [23]. © 1999 IEE)

The complexities and costs of the four types of SWB-HAs can be compared with some narrowband amplifiers. The configurations of the Types -1 and -2 SWB-HAs using distributed amplification (Figs. 13.2(d), 13.17(a), 13.17(b), 13.18) are compared with those of the NB-HA (Figs. 13.1(c), 13.10). However, the configurations of the Types -3 and -4 SWB-HAs without using distributed amplification (Figs. 13.2(a), 13.2(b), 13.2(c), 13.17(c), 13.17(d), 13.23) are compared with those of the C- or L-band EDFA (Fig. 13.1(a)). Differences in the number of optical components and the cost between each of the SWB-HAs and the NB-HA or the EDFA are small. However, the gain bandwidths of the SWB-HAs (~80 nm) are about twice those of the NB-HA or the EDFA (~30 to 40 nm). Therefore, the SWB-HAs are cost-effective wideband amplifiers.

13.6. Hybrid TDFA/Raman Amplifiers and Hybrid Tellurite/Silica Raman Amplifiers

Two kinds of HAs other than EDFA/Raman HAs (hybrid EDFA/Raman amplifiers) have been reported. The first kind is an HA using a thulium-doped fiber amplifier (TDFA) [37–39] and an RA (TDFA/Raman HA) in cascade [40–44]. The other kind is an HA using a tellurite Raman amplifier and a silica fiber Raman amplifier (tellurite/silica Raman HA) in cascade [45–48].

The TDFA/Raman HA has some types as in the case of the EDFA/Raman HA. The first type has basic configurations such as Figs. 13.2(a) and 13.2(b). The HA using the TDFA followed by a discrete RA or using the discrete RA followed by a TDFA were reported [40, 41]. As an example, the HA has a TDFA peak gain of 31 dB at 1462 nm and an RA peak gain of 15 dB at 1510 nm [40]. $\Delta\lambda_{3dB}$ of the amplifier is about 50 nm (1460 to 1510 nm) with a peak gain of 25 dB. The bandwidths with over 20 dB gains are about 70 nm (1450 to 1520 nm) [40, 41]. The second type is a narrowband hybrid TDFA/distributed Raman amplifier [42]. The amplifier uses a gain-shifted TDFA with a gain bandwidth of about 30 nm (1480 to 1510 nm).

Finally, the third type is a wideband TDFA/Raman HA using two-stage RAs and an intermediate TDFA [44]. The amplifier has two DCFs as Raman gain media at its input and output sides. $\Delta\lambda_{3dB}$ is 76 nm (1462 to 1538 nm) with a peak gain of 31.5 dB, and the bandwidth with over 20 dB gains is 90 nm (1455 to 1545 nm).

The tellurite/silica Raman HA has two types [45–48]: using a discrete silica RA and using both discrete and distributed silica RAs. A single-wavelength pumped tellurite RA has a two-peak gain spectrum [45], whereas a single-wavelength pumped silica RA has a single-peak gain spectrum as is well known [25]. Moreover, a multiwavelength pumped tellurite RA generated an ultrawide gain bandwidth of 160 nm (1490 to 1650 nm) with gains over 10 dB and NFs under 10 dB [45]. However, the relative gain flatness is a little bit poor (100%). The flatness is defined as $(G_{max} - G_{min})/G_{min}$, where G_{max} and G_{min} are the maximum and minimum gains in dB units, respectively. The first type, tellurite/silica Raman HA, has two-stage tellurite RAs and an intermediate silica RA [46]. The pump wavelengths of the silica RA were chosen so that the gain peaks of the silica RA compensated for the gain dips of the two-stage tellurite RAs. A flattened gain spectrum with the flatness of 52% without gain equalizer was obtained. A seamless gain bandwidth of 135 nm (1497 to 1632 nm) with a minimum gain of 22.8 dB was obtained.

The second type, tellurite/silica Raman HA, has two-stage tellurite RAs, an intermediate discrete silica RA, an intermediate GEQ, and a distributed silica RA with a transmission fiber (80 km standard SMF) in front of the first-stage tellurite RA [47]. A 6 dB gain-reduction bandwidth of the amplifier was 120 nm (1485 to 1605 nm) with a peak gain of 18 dB. The effective NFs in the short wavelength region were about 5 dB. The low effective NFs are important for the wideband and large capacity transmission.

13.7. Conclusions

Technologies on hybrid EDFA/Raman amplifiers, each of which use an EDFA and a DRA or LRA (distributed or discrete fiber Raman amplifier) in cascade, have been described. The HAs (hybrid amplifiers) have two kinds, a narrowband HA (NB-HA) and a seamless and wideband HA (SWB-HA). The typical gain bandwidths $\Delta\lambda$ of the NB-HA were ~30 to 40 nm in the C- or L-band, whereas those of the SWB-HA were ~70 to 80 nm in the C- and L-bands. An NB-HA showed an improvement in the optical SNR of about 5 dB both experimentally and theoretically thanks to the distributed amplification in the amplifier. Inherent spectral characteristics of the EDFA gain and Raman gain were utilized so that wideband and efficient gain equalization were achieved. The large Raman gains compensate for the small EDFA gains in the long-wavelength region, and simultaneously the large EDFA gains also compensate for the small Raman gains in the short-wavelength region. The SWB-HA has four types according to its manner (distributed or discrete) and magnitude (small or large) of Raman amplification. The two types of SWB-HAs with a small peak Raman gain (Types -1 and -3) generated $\Delta\lambda$ of about 75 nm, whereas the two types of SWB-HAs with a large peak Raman gain (Types -2 and -4) generated $\Delta\lambda$ of about 80 nm. Each type of

SWB-HAs had a simple configuration with the number of optical components, which
is comparable with that of the NB-HA. Therefore, the four types of SWB-HAs are
cost-effective wideband amplifiers. The SWB-HAs also showed low noise and high
output power characteristics due to their multistage amplification schemes. The hybrid
amplifiers (both NB- and SWB-HAs) were successfully used in WDM transmission
experiments with error-free operation for all signal channels in their gain bands.

References

[1] E. Desurvire, *Erbium-Doped Fiber Amplifier*, New York: Wiley, 1994.
[2] M. Yamada, H. Ono, T. Kanamori, S. Sudo, and Y. Ohishi, *Electron. Lett.*, 33: 710, 1997.
[3] Y. Sun, J.W. Sulhoff, A.K. Srivastava, A. Abramov, T.A. Strasser, P.F. Wysocki, J.R.
    Pedrazzani, J.B. Judkins, R.P. Espindola, C. Wolf, J.L. Zyskind, A.M. Vengsarker, and J.
    Zhou, *ECOC*, 53, 1998.
[4] A.K. Srivastava, L. Zhang, Y. Sun, J.W. Sulhoff, and C. Wolf: *OFC*, FC2: 53, 1999.
[5] H. Masuda, S. Kawai, and K-I. Suzuki, *Electron. Lett.*, 35: 411, 1999.
[6] P.B. Hansen, A. Stentz, T.N. Nielsen, R. Espindola, L.E. Nelson, and A.A. Abramov,
    OFC, PD8, 1999.
[7] N. Takachio, H. Suzuki, H. Masuda, and M. Koga, *OFC*, PD9, 1999.
[8] Y. Frignac, G. Charlet, W. Idler, R. Discher, P. Tran, S. Lanne, S. Borne, C. Martinelli, G.
    Veith, A. Jourdan, J-P. Hamaide, and S. Bigo, *OFC*, FC5, 2002.
[9] T.N. Nielsen, A.J. Stentz, K. Rottwitt, D.S. Vengsarkar, Z.J. Chen, P.B. Hansen, J.H. Park,
    K.S. Feder, S. Cabot, S. Stult, D.W. Peckham, L. Hsu, C.K. Kan, A.F. Judy, S.Y. Park,
    L.E. Nielson, and L. Gruner-Nielsen, *IEEE Photon. Technol. Lett.*, 12: 1079, 2000.
[10] A. B. Puc, M.W. Chbat, J.D. Henrie, N.A. Weaver, H. Kim, A. Kaminski, A. Rahman,
    and H. Fevrier, *OFC*, PD39, 2001.
[11] P. Wysocki, J.B. Judkins, R.P. Espindola, M. Andrejco, and A.M. Vengsarkar: *IEEE
    Photon. Technol. Lett.*, 9: 1343, 1997.
[12] H. Masuda, S. Kawai, K.-I. Suzuki, K. Aida, *Electron. Lett.*, 33: 1070, 1997.
[13] H. Masuda, S. Kawai, and K. Aida, *Electron. Lett.*, 34: 567, 1998.
[14] H. Ono, A. Mori, K. Shikano, and M. Shimizu, *IEEE Photon. Technol. Lett.*, 14: 1073,
    2002.
[15] H. Ono, M. Yamada, T. Kanamori, S. Sudo, and Y. Ohishi: *J. Lightwave Technol.*, 17:
    490, 1999.
[16] M. Jinno, T. Sakamoto, M. Fukui, S. Aisawa, J. Kani, and K. Oguchi, *OECC*, 16A1: 404,
    1998.
[17] Y. Emori, K. Tanaka, and S. Namiki, *Electron. Lett.*, 35: 1355, 1999.
[18] S.A.E. Lewis, S.V. Chernikov, and J.R. Taylor, *OAA* ThA2: 72, 1999; S.A.E. Lewis, S.V.
    Chernikov, and J.R. Taylor, *Electron. Lett.*, 35: 1761, 1999.
[19] H. Masuda and S. Kawai, *ECOC*, 2: 146, 1999.
[20] H. Masuda, *OFC*, TuA1: 2, 2000.
[21] A. Mori, Y. Ohishi, and S. Sudo, *Electron. Lett.*, 33: 863, 1997.
[22] N. Jolley, R.D. Muro, S. Parry, K. Cordina, and J. Mun, *OAA*, TuD2: 124, 1998.
[23] H. Masuda and S. Kawai, *IEEE Photon. Technol. Lett.*, 11: 647, 1999; H. Masuda, S.
    Kawai, and K. Aida, *OAA*, PD7, 1998.
[24] H. Masuda, S. Kawai, K.-I. Suzuki, and K. Aida, *ECOC*, 51, 1998.
[25] G.P. Agrawal, *Nonlinear Fiber Optics*, San Diego: Academic Press, 1989.

[26] P.B. Hansen, L. Eskildsen, A.J. Stentz, T.A. Strasser, J. Judkins, J.J. DeMarco, R. Pedrazzani, and D.J. DiGiovanni, *IEEE Photon. Technol. Lett.*, 10: 159, 1998.

[27] K. Rottwitt, M. Nissov, and F. Kerfoot, *OFC*, TuG1: 30, 1998; H. Kidolf, K. Rottwitt, M. Nissov, M. Ma, and E. Rabarijaona, *IEEE Photon. Technol. Lett.*, 11: 530, 1999; C.R.S. Fludger, V. Handerek, and R.J. Mears, *OFC*, MA5, 2001.

[28] V. Curri, *NFOEC*, B1.1, 2000.

[29] H. Kawakami, Y. Miyamoto, K. Yonenaga, and H. Toba: *OAA*, ThB5, 110, 1999.

[30] H. Suzuki, J. Kani, H. Masuda, N. Takachio, K. Iwatsuki, Y. Tada, M. Sumida, *IEEE Photon. Technol. Lett.*, 12: 903, 2000.

[31] H. Masuda, and S. Kawai, *IEEE Photon. Technol. Lett.*, 10: 516, 1998.

[32] H. Masuda, K.-I. Suzuki, S. Kawai, and K. Aida, Type-1, 65 nm, *Electron. Lett.*, 33: 753, 1997. H. Masuda, S. Kawai, K.-I. Suzuki, and K. Aida, *OAA*, MC3, 40, 1997.

[33] D.G. Foursa, C.R. Davidson, M. Nissov, M.A. Mills, L. Xu, J.X. Cai, A.N. Pilipetskii, Y. Cai, C. Breverman, R.R. Cordell, T.J. Carvelli, P.C. Corbett, H.D. Kidorf, and N.S. Bergano, *OFC*, FC3, 2002.

[34] S. Kawai, H. Masuda, K.-I. Suzuki, and K. Aida, *Electron. Lett.*, 34: 897, 1998.

[35] S. Kawai, H. Masuda, K.-I. Suzuki, and K. Aida, *OFC*, FC3, 56, 1999.

[36] H. Suzuki, N. Takachio, Y. Hamazumi, H. Masuda, S. Kawai, and K. Araya, *OFC*, ThO4: 221, 1999.

[37] S. Sudo, ed., *Optical Fiber Amplifiers*, Boston: Artech House, 1997.

[38] T. Sakamoto, *OFC*, TuQ1, 2001.

[39] S. Aozasa, H. Masuda, H. Ono, T. Sakamoto, T. Kanamori, Y. Ohishi, and M. Shimizu, *OFC*, PD1, 2001.

[40] J. Kani and M. Jinno, *Electron. Lett.*, 35: 1004, 1999.

[41] J. Masum-Thomas, D. Crippa, and A. Maroney, *OFC*, WDD9, 2001.

[42] K. Fukuchi, T. Kasamatsu, M. Morie, R. Ohhira, T. Ito, K. Sekiya, D. Ogasahara, and T. Ono, *OFC*, PD24, 2001.

[43] Y. Miyamoto, A. Hirano, S. Kuwahara, Y. Tada, H. Masuda, S. Aozasa, K. Murata, and H. Miyazawa, *OAA*, PD6, 2001.

[44] H. Masuda, S. Aozasa, and M. Shimizu, *Electron. Lett.*, 38: 500, 2002.

[45] A. Mori, H. Masuda, K. Shikano, K. Oikawa, K. Kato, and M. Shimizu, *Electron. Lett.*, 37: 1442, 2001.

[46] H. Masuda, A. Mori, K. Shikano, K. Oikawa, K. Kato, and M. Shimizu, *Electron. Lett.*, 38: 867, 2002.

[47] H. Takara, H. Masuda, K. Mori, K. Sato, Y. Inoue, T. Ohara, A. Mori, M. Koutoku, Y. Miyamoto, T. Morioka, and S. Kawanishi, *OFC*, FB1, 2002.

[48] H. Masuda, A. Mori, S. Aozasa, and M. Shimizu, *OAA*, OTuC1, 2002.

Chapter 14

Wideband Raman Amplifiers

Mohammed N. Islam, Carl DeWilde, and Amos Kuditcher

14.1. Introduction

This chapter describes designs and experiments that apply the Raman effect to wideband amplifiers (WBAs). In the context of this chapter, wideband corresponds to a bandwidth of approximately 50 to 100 nm or more. We start by explaining the need for WBAs, and briefly review some of the key enabling technologies for wideband systems. Section 14.2 describes several approaches for WBA, including the erbium-doped fiber amplifier (EDFA) and Raman amplifier combinations as well as all-Raman amplifiers. Section 14.3 summarizes the advantages and challenges of the all-Raman approach, the focus of this chapter. Section 14.4 identifies the key physical principles that need to be considered in the design of all-Raman WBAs. Then, perhaps the most important section of this chapter, Section 14.5 describes engineering design rules for construction of all-Raman WBAs that satisfy gain and noise figure performance requirements of typical long-haul and ultra-long-haul fiber-optic transmission systems. Several WBA experiments that use either EDFA/Raman amplifier combinations or all-Raman amplifiers are illustrated in Section 14.6, and exemplary wideband system experiments are described in Section 14.7. Finally, we summarize and conclude the chapter in Section 14.8.

14.1.1. Paths to High-Capacity Systems

As demand for throughput of fiber-optic networks continues to grow, the need to expand network capacity continues to increase, fueled primarily by the growth of the Internet. In response to the increasing demand for throughput, significant research and development effort has been directed at technologies for accessing more of the intrinsic capacity of optical fiber. The capacity of optical fiber is the product of channel count and bit rate per channel or, alternately, the product of spectral width and spectral efficiency. For capacity expansion to benefit network operators, upgraded equipment must be provided at low ownership and operating cost. A common cost metric is the cost of transmitting unit capacity over a unit distance (measured in units of $/(Gb/s km)) [39]. Cost reduction then entails minimizing equipment cost and maximizing performance. Reducing the cost metric, either by minimizing equipment cost or by maximizing capacity and reach, ensures cost savings are provided without sacrificing performance. Raman amplification in wideband systems enables cost reduction through optimization of all three factors of the cost metric, namely, equipment cost, capacity, and reach. First, a single Raman amplifier requires fewer components than splitband amplifiers, thereby reducing equipment cost [2]. Second, the wide bandwidth achievable with all-Raman amplification can provide substantial capacity allowing, for example, a bandwidth of 100 nm and a channel count of 240 with a channel spacing of 50 GHz. Finally, Raman amplification including distributed Raman amplification provides a lower noise figure than purely discrete amplification, thereby maximizing system reach (typically 1500 km or more, depending on the number of channels required).

There are several ways of increasing fiber capacity: increasing bit rate, channel density, available spectrum, or spectral efficiency. Figure 14.1 schematically illustrates the evolution of these approaches. In the early years of communication networks, the most common means of increasing capacity was to increase time-division-multiplexing bit rates. For example, bit rates increased from 64 kb/s (the base rate of a voice signal) through 1.5 Mb/s (so-called T1), 45 Mb/s (so-called T3), 155 Mb/s (so-called OC-3), and on up to about 2.5 Gb/s (so-called OC-48). In recent years, a channel speed of 10 Gb/s has become common, and products supporting 40 Gb/s channel speeds have become available, although deployment has been hampered to date by prohibitive costs associated with the need for polarization-mode

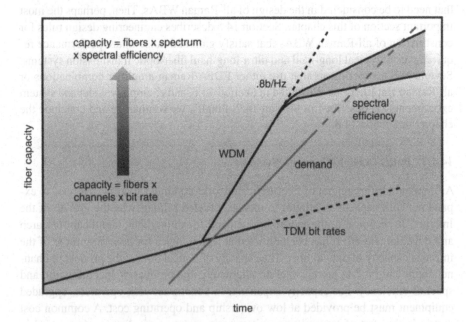

Fig. 14.1. Evolution of optical transmission capacity.

dispersion compensation, fine-tuned dispersion and dispersion slope-compensation, and overcoming other transmission impairments.

Since the first deployment of wavelength-division-multiplexed (WDM) systems, the number of WDM channels has continued to increase. The number of channels in commercially available WDM systems has increased to about 240 and the bit rate to 10 Gb/s, corresponding to a single fiber capacity of 2.4 Tb/s. Back in 1995, it was common to use a bit rate of 2.5 Gb/s on channels spaced by 200 GHz, which, at the time, yielded four to eight channels. Channel spacing was limited by the accuracy and stability of the transmitting lasers and of the multiplexing and demultiplexing components. By 1999, systems featuring 10 Gb/s channels on 100 GHz channel spacing had become commonplace [3], and over the next three years 50 GHz spacing was introduced. Up to this point, the spectral extent of the signal comprising any channel fits well within the bandwidth available to that channel, as long as a fairly simple modulation scheme is used. Therefore, total capacity for a single fiber can be calculated by summing the bit rate on each of the channels over the number of channels available within the total bandwidth delivered by the amplifier or amplifiers.

Any further reduction in channel spacing or increase in bit rate at the same channel spacing requires that careful consideration be given to spectral efficiency. As a result, increasing bandwidth capacity is no longer a matter of just decreasing channel spacing, but requires more sophisticated and costlier modulation techniques. At that point an increase in bit rate without a change in modulation technique requires a proportional increase in channel spacing and, consequently, a proportional decrease in the total number of channels. Overall capacity remains the same and bandwidth capacity for a single fiber, which is determined by multiplying available channels within a given spectrum with the bit rate on each channel, no longer increases as bit rate increases or channel spacing decreases. Only an increase in available spectrum or an increase in spectral efficiency will lead to a higher bandwidth capacity. It is expected that the cost of increasing spectral efficiency above about 0.8 bps/Hz will grow substantially, by reason of a combination of factors: complicated transmitters and receivers, active laser stabilization, as well as passive component stabilization.

Since the commercial deployment of optical amplifiers, there has also been a significant increase in available bandwidth. The first systems used only the flat gain section of an EDFA, corresponding to 10 to 15 nm. With the introduction of static gain-flattening filters, gain bandwidths in commercial systems increased to 32 nm, with laboratory experiments reporting as much as 40 nm. With the introduction of the long-wavelength L-band by the end of the 1990s, bandwidth increased to as much as 80 nm in EDFA-based systems utilizing a splitband configuration. In 2002, 100 nm bandwidth commercial systems using all-Raman amplification became available, and laboratory experiments demonstrated amplifiers with bandwidth greater than 130 nm. In comparing the three techniques for increasing capacity, the path followed in commercial systems has proved to be the most economical mature technology. Although combinations of all of the approaches can be used, the lowest cost approach for the next several years is expected to be bandwidth expansion.

14.1.2. Enabling Technologies for Wideband Systems

Several technologies are required to realize practical and economical wideband systems. Consider, for example, the key components for 100 nm bandwidth systems. (Smaller bandwidth systems still require the same components or subsystems, but with less stringent performance requirements.) First, wideband amplifiers are required. Various approaches for WBAs are described in the next section, and the remainder of the chapter is focused on the all-Raman amplifier approach. Second, broadband components such as multiplexers and demultiplexers are required. The passband for couplers and other passive components needs to be larger than the actual bandwidth being used by the channels because in a cascaded system a spectrally dependent transfer function decreases in bandwidth as a cascade of components is traversed.

A third component that is critical for wideband systems is gain equalizers and gain-flattening filters. Again, in a system consisting of a cascade of spans (e.g., a 1500 km system may have 19 spans each 80 km long), any nonuniformity in the gain spectra of the amplifiers is accentuated as more and more amplifiers are traversed. A gain variation of 0.5 dB can grow to well over 5 dB at the end of a system, resulting in substantial variation in signal strength between channels, which can require a large dynamic range at the receivers and lead to unwanted gain saturation, crosstalk between channels, and substantial variation in bit error rate performance for different channels. From a design perspective, the worst-performing channel limits the overall system performance. Therefore, the best design achieves roughly equally performing channels. Gain equalizers and gain-flattening filters are typically utilized to equalize channels. In EDFAs, a bandwidth greater than 10 to 15 nm can only be obtained with gain equalizing devices. In Raman amplifiers, gain shape can be more flat, reducing constraints on gain equalization. However, gain equalization in Raman amplifiers is still beneficial and is especially useful in wideband amplifiers. For uniform signal traffic and time-invariant components, a static gain equalizer can be utilized; if, on the other hand, traffic or component characteristics vary with time as may be the case in a system that utilizes dynamic optical add-drop multiplexers or optical cross-connects, then a dynamic gain equalizer may be required. Various devices have been used for gain equalization. Static gain equalizers have been realized with waveguide notch filters, Mach–Zender interferometers, dielectric filters, and fiber Bragg gratings. Dynamic gain equalizers have been realized with microelectromechanical systems (MEMS), liquid crystal, or thermooptic switch arrays integrated with a dispersive element such as a diffraction grating. Acoustooptic filters have also been used for dynamic gain equalization.

Another essential-component for wideband systems operating at 10 Gb/s or higher bit rates comprises dispersion-compensating and dispersion slope-compensating modules. Dispersion-compensating fiber (DCF), chirped fiber Bragg gratings, higher-order mode fibers, and virtually imaged phased array microoptic devices have been used for dispersion compensation, but DCF is the most widely deployed. As system bandwidth increases, it is not adequate to match the dispersion magnitude at the center of the band, as is commonly done in systems with less than 30 nm bandwidth. The dispersion slope of the compensator must also be matched to the dispersion slope of the transmission fiber in order to achieve compensation over a wide spectral band. The quantity to be matched is then the relative dispersion slope $RDS = S/D$, the ratio of dispersion slope S to dispersion D. Manufacturers of dispersion-compensating fiber are aware of the need for dispersion slope-compensation, and slope-compensating fibers are now becoming available for most deployed transmission fiber types.

14.2. Alternate Approaches for Implementing Wideband Amplifiers

Wideband amplification has been demonstrated with EDFAs, EDFA/Raman amplifier hybrids, and all-Raman amplifiers. Figure 14.2 shows applicable wavelength bands for various amplifier types [4]. Silica-based EDFAs have been most successful to date and can provide amplification over the conventional C-band (~1530 to 1565 nm) and the long-wavelength L-band (~1570 to 1605 nm). Bandwidths greater than 80 nm have been demonstrated in splitband configurations employing parallel C- and L-band EDFAs, and a continuous EDFA gain band greater than 70 nm has been demonstrated by using new glass compositions such as tellurite. Other rare earth dopants such as thulium can be used to provide gain in the short-wavelength S-band or in the L+ band (Chapter 10, Section 10.2). The spectral location of the gain band in Raman amplification only depends on the spectral location of the pump band, so Raman amplification can provide gain anywhere in the low-loss telecommunications window between 1450 and 1650 nm. Likewise, semiconductor optical amplifiers can be used, but a separate amplifier is required for every 30 nm band.

14.2.1. EDFAs in Various Glass Compositions

The most common glass composition for EDFAs comprises silica doped with aluminum, germanium, and erbium. For typical inversion levels of 40 to 60%, the flat

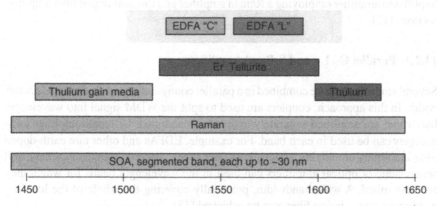

Fig. 14.2. Gain bands of optical fiber amplifiers [4].

gain band of silica-based EDFAs (EDSFAs) extends from about 1540 to 1560 nm, corresponding to the central portion of the C-band. Using gain-equalizing filters, the bandwidth of C-band amplifiers can be extended to about 40 nm. Further bandwidth extension by means of gain-equalizing filters is hampered by very low gain near the ends of the C-band. Therefore, gain shifting has been used to extend operation to the 1560 to 1620 nm (L-band) wavelength range and, recently, the 1450 to 1520 nm (S-band) range. Long lengths of erbium-doped fiber together with low population inversion rates, in the neighborhood of 20 to 40%, shift the gain band to long wavelengths, yielding bandwidths of 43 nm for L-band EDSFAs [5]. Suppressing the principal spontaneous emission peak at 1530 nm with distributed filtering shifts the EDFA gain band to the S-band [6]. By combining gain-flattening and gain-shifting techniques, a composite bandwidth of 84.3 nm was demonstrated for a splitband amplifier employing parallel C- and L-band EDSFAs [5].

Ground state absorption on the blue end and excited state absorption on the red end of the erbium emission spectrum limit the aggregate bandwidth that can be obtained with EDSFAs. However, host glasses with a broader erbium emission spectrum can be used to achieve a wider bandwidth. For example, erbium-doped fluoride fiber amplifiers can achieve a wider flat gain band, and bandwidth can be further extended with the aid of gain-flattening filters [7]. A gain-flattened tellurite-based EDFA has been demonstrated to yield a continuous bandwidth greater than 70 nm [8]. In addition, antimony–silicate multicomponent silica glasses have been shown to extend the bandwidth of L-band EDFAs to 60 nm, with which the composite bandwidth of a splitband EDFA could be increased to 90 nm [9].

14.2.2. Other Dopings

Other dopants can be used to modify the characteristics of EDFAs or to achieve amplification in wavelength regions not accessible to conventional EDFAs. For example, the bandwidth of an L-band EDFA is extended to 50 nm when a phosphorous-doped fiber is utilized [10], and thulium-doped fiber amplifiers have been developed for S-band amplification [11]. A bandwidth of approximately 70 nm can be achieved for a splitband amplifier employing a Raman amplifier and thulium-doped fiber amplifier sections [12].

14.2.3. Parallel C-, L-, and S-Band Amplifiers

Several amplifiers can be combined in a parallel configuration to achieve a wide bandwidth. In this approach, couplers are used to split the WDM signal into wavelength bands that are amplified separately and then recombined. Different amplifier technologies can be used in each band. For example, EDFAs and other rare earth-doped fiber amplifiers employing various host fiber types, discrete Raman amplifiers, and semiconductor optical amplifiers can used in the wavelength bands for which they are best suited. A wide bandwidth, potentially covering the whole of the low-loss window of transmission fiber, can be achieved [13].

14.2.4. Hybrid Raman/EDFA

Hybrid amplification employing an EDFA and a Raman amplifier in series provides another approach for extending the bandwidth of EDFAs. The EDFA provides gain in the C-band whereas the Raman amplifier operates in the L-band. By incorporating gain-flattening filters, continuous bandwidths ranging from 50 to 80 nm have been achieved [14].

14.2.5. All-Raman Amplifiers

Raman amplification in silica fiber can yield a continuous bandwidth in excess of 100 nm, limited only by the Raman frequency shift in silica [15]. Wider bandwidths can be achieved in fiber types with a larger frequency shift. A Raman amplifier using tellurite fiber as gain medium was shown to yield a bandwidth of 160 nm [16]. The Raman frequency spectrum in phosphorous-doped fiber includes peaks up to 40 THz frequency shift, and phosphorous-doped fiber can potentially provide a bandwidth of 300 nm in all-Raman amplifiers [17].

Raman amplification provides gain over a single continuous band. This characteristic is advantageous because a single-band amplifier requires fewer components than one employing multiple amplifiers in a splitband configuration, and guard bands are not required for the couplers (c.f. Fig. 1.15 in Chapter 1) [2]. Monitoring and management of the gain profile is less complex over a single band because only a pair of monitor taps, one at the input and one at the output of the amplifier, is required. In addition, the dispersion-compensation module as well as much of the control circuitry can be consolidated, thereby reducing the cost and size of the amplifier. In comparison, a splitband architecture employing separate amplifiers for each subband requires separate dispersion-compensation modules, control circuitry, and monitoring of each subband amplifier, thereby increasing the cost and footprint of the amplifier.

14.3. Advantages and Challenges of All-Raman Wideband Amplifiers

Current trends suggest extensive deployment of Raman amplification in the future for a number of reasons. First, distributed Raman amplification alone or in combination with lumped Raman amplification can improve the noise figure and reduce system penalties arising from fiber nonlinearities. In optically amplified transmission systems, the principal limitations usually arise from amplifier noise and fiber nonlinearities. In particular, the range of allowed signal levels is determined by two factors: the minimum signal level is determined by the signal-to-noise ratio (SNR) required to achieve a prescribed bit error rate; and the maximum signal level is determined by impairments arising from nonlinear propagation effects in optical fiber. Raman amplification eases the constraints arising from both of these limitations and improves performance. As shown in Chapter 1, Fig. 1.18, distributed Raman amplification reduces the excursion of the signal power level in comparison to purely lumped

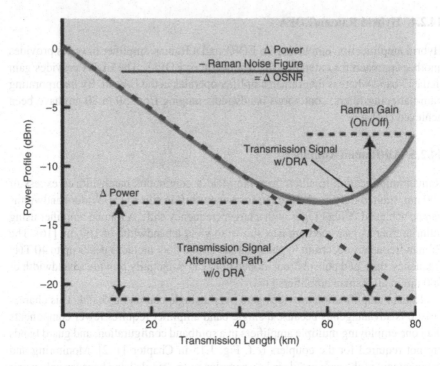

Fig. 14.3. Variation of signal power with distance in an optical fiber with and without DRA. The minimum signal power is higher by a factor of ΔPower with Raman amplification than without amplification and optical signal-to-noise ratio is improved in the Raman amplified case by ΔPower divided by the optical noise figure.

amplification. The low signal power swing reduces signal impairments arising from fiber nonlinearity, and the higher minimum signal level reduces SNR degradation. Figure 14.3 illustrates the variation of signal power with distance along a distributed Raman amplifier (DRA) and along an unamplified span of equal length. Improvement in optical SNR (OSNR) in the distributed Raman amplifier depends on the ratio ΔPower of the minimum signal power with Raman amplification to the minimum signal power without Raman amplification. The SNR improvement is given by ΔPower divided by the noise figure of the Raman amplifier. This improvement can be as much as 7 dB for an 80 km fiber span.

A second advantage of Raman amplification is that gain can be obtained in any spectral region and over a wide spectral band. The location and width of the gain band are determined by the pump wavelength location and distribution. The available gain band for Raman amplification can be compared to the gain band for EDFAs in Fig. 14.2. Although conventional EDFAs operate over the C-band (~1525 to 1565 nm) and the long-wavelength L-band (~1565 to 1605 nm), Raman amplification can be obtained anywhere within the telecommunication band, typically defined as 1300 to 1650 nm. Raman amplifiers can easily achieve 100 nm of continuous bandwidth, and laboratory experiments have shown much larger bandwidths in a single amplifier. In present generation Raman amplifiers, several pump laser diodes operating at different wavelengths are used. As a rule of thumb, the Raman gain bandwidth for each pump laser is about 20 nm. Therefore, by using approximately five properly spaced pump wavelengths, a gain bandwidth of 100 nm can be achieved. Note that the location and number of pump wavelengths are determined not only by the gain bandwidth, but also by gain-flatness requirements: the flatter the gain spectrum required, the greater the number of wavelengths required.

A third advantage of Raman amplification is that gain and dispersion-compensation can be integrated in a lumped amplifier. Such an amplifier can have higher power conversion efficiencies at high channel counts than EDFAs. Most systems operating at line rates of 10 Gb/s (OC-192) or higher require dispersion compensation, commonly implemented using dispersion-compensating fiber. Dispersion-compensating fiber turns out to be an efficient fiber type for Raman amplification: the Raman gain coefficient can be as much as an order of magnitude larger in DCF than in standard single-mode fiber (Chapter 1, Fig. 1.10). Hence, gain and dispersion-compensation can be integrated in the same unit. Moreover, the efficiency of Raman amplification in DCF can exceed that of EDFAs pumped at 1480 nm for channel counts exceeding about 200 (c.f., Chapter 1, Fig. 1.16).

Beyond the advantages of Raman amplification enumerated above, additional features of Raman amplification enable simple practical implementation in deployed fiber-optic systems. Some of these features include:

- All-silica components: all Raman amplifier components can be based on fused-silica, and all amplifier parts can be spliced into existing systems;
- Simple channel equalization: because gain shape can be smooth in a Raman amplifier, gain equalization can be considerably simplified;
- Simple adjustment of gain flatness: adjustment of gain flatness can be made by adjusting pump power levels [18], an important feature in, for example, a transmission system using optical add-drop multiplexers or optical crossconnects;
- Large gain and power levels: because Raman amplifiers are relatively weakly saturated even at high powers, gain and output power levels can be increased by simply increasing pump power, and output powers in the watt range have been reported; and
- Large power dynamic range for flat gain operation: unlike an EDFA where the gain shape changes as the pump power is changed (i.e., as inversion level is changed), the gain shape of a Raman amplifier can be maintained flat over a large dynamic range of pump power and amplifier gain.

Despite a significant list of desirable features, Raman amplification is beset by numerous challenges that dictate amplifier design and architecture. Gain flatness of a wideband Raman amplifier can be significantly affected by pump–pump interactions as well as interband and intraband stimulated Raman scattering (i.e., transfer of energy from short wavelength signals to longer wavelength signals through the Raman effect). In addition, as the signal band in a wideband Raman amplifier approaches the longest wavelength pump band, the noise figure of the amplifier increases because of thermal noise as further described below (see also Chapter 8). Furthermore, the electrical noise figure can be poorer than the optical noise figure in a Raman amplifier. Because the upper state lifetime associated with Raman amplification is short and long fiber lengths are typically used in Raman amplifiers, additional noise sources contribute to the electrical noise figure. Multipath interference (MPI) arising from double-Rayleigh scattering, coupling of pump fluctuations to signals, and pump-mediated signal crosstalk increase the electrical noise figure with minimal effect on the optical noise figure as usually determined from amplified spontaneous emission (ASE) measurements. The next section examines the physical origins of these limitations. Subsequent sections illustrate amplifier design approaches to mitigate the limitations and present exemplary implementations.

14.4. Physical Origin of All-Raman Wideband Amplifier Limitations

Several physical principles underlie the design of wideband Raman amplifiers. The first to be considered is the interaction between different pump waves and different signal waves through the Raman effect itself. Consider the Raman gain curve in fused silica, as shown in Chapter 1, Fig. 1.1. The gain curve is continuous from the pump frequency to a frequency shift as great as 40 THz away from the pump frequency with a significant Stokes peak near 13 THz. In general, for any reference frequency, energy is lost to lower frequency waves and gained from higher frequency waves. This is true for Raman interaction between any set of waves and gives rise to signal–signal as well as pump–pump power exchange. Therefore, in addition to energy transfer from pump to signal waves, energy is transferred from short-wavelength to long-wavelength pumps with the result that lower launched powers are usually required for long-wavelength pumps and higher powers are needed at short pump wavelengths to achieve a flat gain shape at a prescribed gain level. Figure 14.4 schematically exhibits a pump launch power distribution taking into account pump–pump interaction to achieve comparable gain at long and short signal wavelengths. Note that the shortest wavelength pumps have the highest power at the pump input end of the fiber, because they not only provide gain to signal waves, but also supply energy to the longer wavelength pumps.

To model the Raman interaction between waves propagating on an optical fiber, it is customary to use a set of coupled mode equations. For copropagating steady-state pump and signal waves, the coupled mode equations describing stimulated Raman scattering take the form

$$\frac{dP_s}{dz} = g P_p P_s - \alpha_s P_s$$

$$\frac{dP_p}{dz} = -\frac{v_p}{v_s} g P_p P_s - \alpha_p P_p,$$

where P_p and P_s are the respective pump and signal powers, α_p and α_s are the fiber loss coefficients at pump and signal frequencies v_p and v_s ($v_p > v_s$), and g is the Raman gain coefficient. Accompanying the scattering process are vibrational excitations that

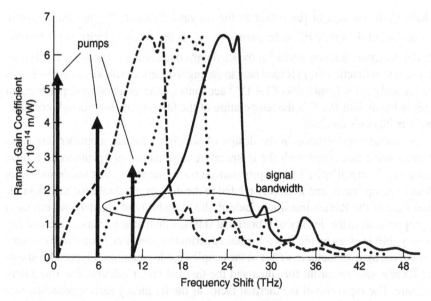

Fig. 14.4. Multipump Raman amplification. Gain bands from different pump wavelengths combine to yield a broad net gain band. The long-wavelength pumps are launched with lower power than the short-wavelength pumps to achieve comparable gain across the signal band.

make up for the energy difference between the signal and the pump and represent a loss mechanism for pump waves. The loss of pump energy to vibrations of the host material during the scattering process is accounted for by the factor $-v_p/v_s$ in the foregoing equations. For multiple pump and signal waves, P_p and P_s are summed over all pump and signal waves. For waves with frequency lower than the signal frequency under consideration, the terms of the sum include the energy loss factor.

For an accurate model useful for predicting the properties of Raman amplifiers, it is necessary to include noise terms describing processes originating from spontaneous Raman scattering (the source of ASE in Raman amplification) and Rayleigh scattering. In particular, the temperature dependence of spontaneous Raman scattering needs to be included to accurately predict the noise figure of wideband Raman amplifiers. Furthermore, dispersion must be accounted for when time-dependent waves (e.g., modulated signals) are considered. For quasicontinuous wave signals, phenomenological propagation terms in the coupled mode equations suffice [19]. A more complete set of coupled mode equations then takes the form [20, 21]

$$\frac{\partial P_j^{\pm}(z)}{\partial z} \pm \frac{1}{V_g} \frac{\partial P_j^{\pm}(z)}{\partial t} = -\alpha_j P_j + S\alpha_j^R P_j^{\mp}$$

$$+ \sum_k \Theta(k-j)(gP_kP_j + 2hv_jg[n(v_k - v_j) + 1]P_kB_o)$$

$$- \Theta(j-k)\frac{v_j}{v_k}(gP_kP_j + 2hv_kg[n(v_k - v_j) + 1]P_kB_o),$$

where P_j is the sum of the power in the forward direction P_j^+ and the power in the backward direction P_j^- at frequency v_j, V_g is the group velocity at v_j, S is the Rayleigh capture fraction, and α_j^R is the Rayleigh contribution to loss at v_j. The Heaviside unit step function $\Theta(j)$ is used here to distinguish between Stokes and anti-Stokes waves, and $n(v) = (\exp[hv/k_b T] - 1)^{-1}$ accounts for the temperature dependence of ASE in bandwidth B_0. T is the temperature of the fiber, k_B is Boltzmann's constant, and h is Planck's constant.

A second consideration in the design of wideband Raman amplifiers concerns thermal noise associated with the temperature dependence of spontaneous Raman scattering. Thermal noise is most pronounced for signal waves that reside spectrally close to pump waves and is closely related to the density of vibrational modes in the final state of the Raman transition. Indeed, the probability of spontaneous emission is proportional to the density of vibrational states, which at any temperature and frequency shift is given by the Bose–Einstein distribution function, hence the functional form of the thermal factor $n(v) + 1$ in the coupled mode equations. Figure 14.5 shows the Raman gain spectrum for silica and the thermal factor calculated at room temperature. The rapid rise of the thermal factor as the frequency shift approaches zero results in an increase in spontaneous emission near the pump wavelength [22], which leads to a higher noise figure for signals with wavelength close to the pump wavelength. The increase in noise for short-wavelength signals close to the pump band increases with maximum gain corresponding to the longest wavelength pumps [23]. In wideband amplifiers, pump–pump stimulated Raman scattering can lead to signif-

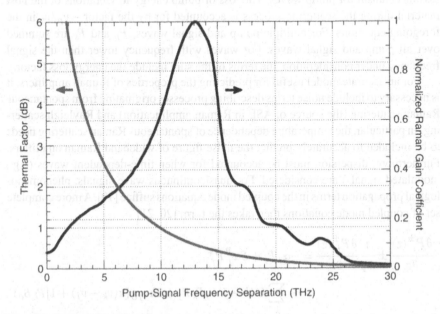

Fig. 14.5. Room-temperature thermal factor and Raman gain as a function of pump–signal frequency separation.

icant power at the longest pump wavelength, thereby significantly increasing thermal noise. Note that the temperature dependence of Raman scattering only affects the noise performance of Raman amplifiers; gain is largely temperature independent.

Multipath interference due to double-Rayleigh scattering (DRS) can also be important. Rayleigh scattering corresponds to an elastic scattering event that occurs when propagating waves encounter subwavelength density variations in the fiber core. The fiber core captures a fraction of the scattered radiation, and scattered waves may undergo a second scattering event. The forward component of the twice-scattered radiation, having been recaptured by the fiber core and constituting delayed copies of the original signal, beats with the signal at the receiver and gives rise to inband noise. Although the recaptured fraction is small, the Rayleigh scattered radiation is amplified as it propagates along the fiber. Therefore, MPI noise due to DRS strongly depends on amplifier properties: it increases with gain and fiber length [24] and dominates noise performance for pump-on/pump-off gains in excess of about 15 dB. Because the delayed signal copies reside in the same spectral band as the primary signal, DRS crosstalk cannot be filtered out in the same manner as ASE and can lead to severe degradation of the electrical noise figure and system quality factor Q. To ensure Q penalties less than 1 dB_{20}, DRS crosstalk less than -15 dB is required, thereby limiting the gain achievable in single-stage Raman amplifiers.

A further consideration concerns the short upper state lifetime associated with Raman amplification, which leads to a number of requirements on pump lasers and amplifier architecture. Because the response time associated with the Raman effect is very short, on the order of femtoseconds, transfer of pump intensity fluctuations to the signals can occur. The signal degradation that results depends on the spectrum of pump noise, fiber properties, and amplifier architecture [25]. In addition, the signals also impress patterns on the pumps through gain saturation. The signal patterns impressed on the pump waves constitute fluctuations of pump intensity that can in turn be transferred to other signals, leading to pump-mediated crosstalk between signal waves. As with pump–signal noise transfer, the penalty arising from pump-mediated signal crosstalk depends on the noise frequency spectrum, fiber properties, and pumping configuration [26].

To counter the deleterious effects of the short upper state lifetime, purely counter-propagating (backward pumping, where the pump and signal waves are launched into opposite ends of the fiber) geometry can be used. In this case, the transit time through the fiber introduces an effective upper state lifetime over which pump fluctuations are averaged. In cases where copropagation of pumps and signals (forward pumping) is desired, pump lasers with very low relative intensity noise (RIN) are required to minimize signal degradation from pump fluctuations. In both forward and backward pumping, the gain fiber behaves as a low-pass filter, but with properties that depend on the pumping configuration. For forward pumping, the 3 dB frequency of the filter is inversely proportional to fiber dispersion and frequency shift and proportional to fiber loss, whereas the 3 dB frequency is proportional to loss and group velocity for backward pumping [25]. The cut-off frequency is generally higher for forward pumping than for backward pumping, indicating that lower RIN pump lasers are required for forward than for backward pumping to achieve low noise transfer penalties. In practical terms, laser diodes that are wavelength stabilized with an external cavity and a fiber Bragg grating have a RIN of approximately -140 dB/Hz, and Fabry–Perot lasers have a RIN of approximately -160 dB/Hz. Therefore, if forward pumping is desired, Fabry–Perot laser diodes typically need to be used. Although less noisy, Fabry–Perot lasers are not as stable in wavelength as grating-stabilized lasers.

A final consideration in the design of wideband Raman amplifiers is connected with the location of the zero dispersion wavelength of the gain fiber relative to the pump and signal bands. This is typically not a problem in DCF, but can pose a problem in the DRA section that uses the transmission fiber as the gain medium because four-wave-mixing can be phase-matched near the zero dispersion wavelength. Four-wave-mixing leads to generation of sidebands, which have three effects: the sidebands deplete the pump waves, effectively increasing pump attenuation; new sidebands on the pump waves can lead to gain spectrum distortions, which may require additional gain equalization; and the sidebands can fall within signal bands, leading to unwanted crosstalk. The crosstalk arising from the third of these effects can lead to severe SNR degradation. The magnitude of the crosstalk depends on the relative propagation directions of pump and signal waves. In forward pumping, the sidebands copropagate and directly interfere with signals. In backward pumping, the sidebands do not directly interfere with signals. However, Rayleigh scattering reflects the interfering waves, and the fiber core captures a fraction of the scattered sidebands propagating in the signal direction. In both cases, Raman amplification enhances the crosstalk level, with the result that severe SNR degradation can occur [27].

14.5. Design of Wideband Raman Amplifiers

This section describes design techniques to overcome the physical limitations described in the previous section. As can be seen by the end of the section, each of the limitations can be mitigated and relatively simple implementation of wideband Raman amplifiers can be realized.

14.5.1. Electrical Noise Figure

As mentioned in the previous section, it is important to account for the electrical noise figure as well as the optical noise figure in Raman amplifiers. The optical noise figure primarily depends on signal–spontaneous emission beating, whereas additional effects such as MPI from DRS and pump–signal RIN transfer contribute to the electrical noise figure. The electrical noise figure is, therefore, generally higher than the optical noise figure if the additional effects contribute significantly to amplifier noise performance. However, amplifier architectures exist that enable reduction of the additional noise contributed by DRS and pump–signal RIN transfer.

Because of MPI penalties from DRS, wideband Raman amplifiers are typically sectioned into multiple stages, where each stage is separated by an isolator. For an amplifier comprising discrete and distributed stages, the MPI penalty is usually larger in the discrete stages. This is because discrete stages are usually constructed of DCF, which has a higher gain coefficient and a higher DRS coefficient than transmission fiber. Although the MPI penalty needs to be computed for each particular design, one rule of thumb is that it is difficult to simultaneously obtain pump-on/pump-off gain of more than 15 dB and DRS crosstalk of less than -15 dB (corresponding to a Q penalty of less than 1 dB$_{20}$) in a single stage. By using isolators between stages, the buildup of DRS is minimized, and high-gain, low MPI amplifiers can be constructed.

Reduction in pump–signal RIN transfer and other noise sources associated with the short upper state lifetime of Raman amplifiers is achieved by employing backward pumping, which introduces an effective lifetime equal to the transit time through the fiber. Because amplification mostly occurs near the signal egress end of the fiber in backward pumping, the resulting noise figure is not optimal; however, backward pumping also reduces dependence of amplifier gain on the state of polarization of the signal. Where forward pumping is desired (as in bidirectional systems, e.g.), low-RIN pump lasers need to be used.

14.5.2. Pump–Pump Interactions

Stimulated Raman scattering in the gain fiber leads to pump–pump power exchange, which must be properly accounted for to obtain a desired gain profile. If equal frequency spacing and equal amplitude pumps are used in a wideband Raman amplifier, the resulting gain increases with wavelength (c.f. Fig. 14.6) [20]. Depending on the power and bandwidth of the signal bands, there can also be transfer of energy from short- to long-wavelength signals, further accentuating the gain tilt.

Several techniques can be used to achieve flat-gain spectra in the presence of pump–pump interactions. The basic idea is to increase the spectral density at short

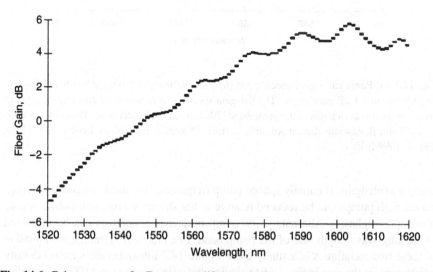

Fig. 14.6. Gain spectrum of a Raman amplifier employing a uniform pump power spectral density [20]. Source: H. Kidorf et al: "Pump Interactions in a 100-nm Bandwidth Raman Amplifier," IEEE Photonics Technology Letters, Vol. 11, pg 530 (© 1999 IEEE)

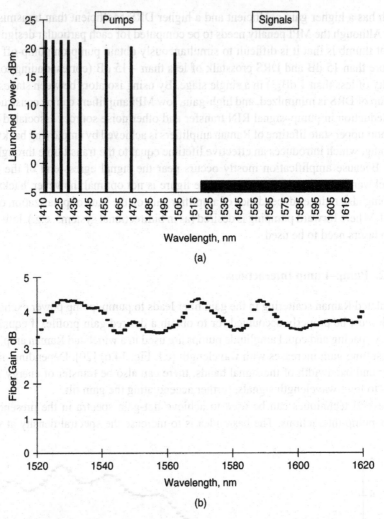

Fig. 14.7. (a) Pump and signal spectra; and (b) gain spectrum of a 100 nm bandwidth Raman amplifier with 1.1 dB gain ripple. The flat-gain spectrum is achieved by concentrating pump power at short-wavelengths of the pump band [20]. Source: H. Kidorf et al: "Pump Interactions in a 100-nm Bandwidth Raman Amplifier," IEEE Photonics Technology Letters, Vol. 11, pg 530 (© 1999 IEEE)

pump wavelengths. If equally spaced pump frequencies are used, the power of long-wavelength pumps can be reduced relative to the shorter wavelength pumps (pump power preemphasis). Alternately, the pump frequency separation can be increased to the long-wavelength side of the pump band. In a practical system, a combination of these two techniques is actually used. Figure 14.7 illustrates the spectral density adjustment for the case in Fig. 14.6 to obtain a flat-gain spectrum in a 100 nm amplifier [20]. Although there is some decrease in pump power at long-wavelengths, in this case the spectral density at the long-wavelength end of the pump band is primarily reduced by spacing the long-wavelength pumps farther apart.

Another completely different approach to reducing the effects of pump–pump interaction is to arrange for the pumps not to encounter each other altogether. This can be simply achieved by segregating the pumps in separate amplifier stages such that strongly interacting pumps are launched in different stages [28]. Another approach involves time modulation of the pumps. If the pumps are time modulated and then multiplexed or timed so that they do not overlap, then they are not simultaneously in the gain fiber and do not interact. A particular realization of this approach has been demonstrated by Fludger et al. [29], where four pump wavelengths were separated into two bands, each pair was modulated, and the relative phase was selected so that the most strongly interacting bands did not coincide within the gain fiber. This scheme reduces interaction between the pumps, thereby reducing depletion and increasing penetration into the fiber of the short-wavelength pumps. The result is amplification of short-wavelength signals deeper into the gain fiber and reduction of the short-wavelength noise figure. The reverse occurs for long-wavelength signals, inasmuch as reduction of interaction between the pump bands decreases the gain experienced by the long-wavelength pumps as well as the distance along the fiber over which long-wavelength signals experience gain.

Gain slope in a multistage configuration, as shown in Fig. 14.8 for a two-stage amplifier, can also be used to reduce pump–pump interactions. In the example shown in Fig. 14.8, gain in the first stage increases with wavelength and decreases with wavelength in the second stage. The gain slopes in the two stages are complementary so the net gain is substantially flat. Although noise performance may be degraded (see next section), higher power conversion efficiencies may be realized by such a scheme. This gain tilt allocation, in which gain increases with wavelength in the first stage, allows use of fewer pump wavelengths and lower pump powers because of energy transfer from short to long pump wavelengths through the Raman effect. The long-wavelength pumps provide gain in the first stage, and then the shorter pump wavelengths are introduced with more power in the second stage. In other words, by selecting a pump wavelength distribution that leads to a sloped up gain followed by a sloped down gain, the effect of pump–pump interactions is actually reduced.

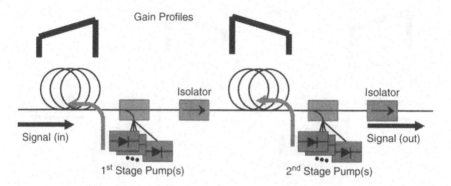

Fig. 14.8. Block diagram of a two-stage Raman amplifier with complementary gain spectra. The gain of the first stage increases with wavelength, and that of the second stage decreases as wavelength increases in such a manner that the composite gain shape is flat.

14.5.3. Thermally Induced Phonon Noise

Thermal noise is particularly important in wideband Raman amplifiers. The peak of the Raman gain curve at 13.2 THz corresponds to a separation of approximately 100 nm between pump and signals near 1550 nm. Thus, for a 100 nm bandwidth amplifier with gain band near 1550 nm, the short-wavelength end of the signal band is close to the long-wavelength pumps, and the contribution of thermal noise to the short-wavelength noise figure can be substantial when the maximum gain contributed by the long-wavelength pumps is large [23].

To mitigate the effect of the thermally induced noise, the short-wavelength signals can be amplified before they encounter significant noise. In other words, if the signals susceptible to thermal noise are amplified before encountering noise from the long-wavelength pumps, the signal-to-noise ratio remains high throughout the amplifier because the signal strength remains high. Because the most vulnerable signals are those closest to the pumps in wavelength (i.e., the shortest wavelength signals), gain for the shortest wavelength signals can be made high at the input of the amplifier. However, as discussed earlier, there is also a need to equalize gain shape across the signal band to obtain roughly equivalent behavior for all channels. A method for accommodating both of these requirements is to use a multistage amplifier, each with a gain slope.

The concept of using gain tilt in multiple stages to mitigate the noise limitation is illustrated in Fig. 14.9, showing a two-stage Raman amplifier and gain profiles associated with each stage. Gain in the first stage decreases from short to long-wavelengths, and the second stage has an approximately complementary gain spectrum. Thus short-wavelength signals are substantially amplified in the first stage before encountering the long-wavelength pumps responsible for thermal noise, and the net gain shape after the two stages is substantially flat.

Signal preemphasis, discussed in Section 14.5.7, can also be used to mitigate the effects of thermal noise. As with the gain tilt scheme previously described, the strength of short-wavelength signals is made high to achieve a high signal-to-noise ratio. Thus short-wavelength signals are launched with a higher power than long-

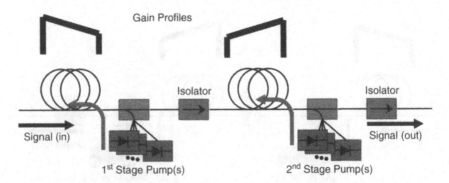

Fig. 14.9. Block diagram of a two-stage Raman amplifier with complementary gain spectra. The gain of the first stage decreases as wavelength increases, and that of the second stage increases with wavelength in such a manner that the composite gain shape is flat.

wavelength signals. In this manner, the signal-to-noise ratio at short-wavelengths can be made comparable to that at long-wavelengths.

14.5.4. Zero Dispersion Wavelength

Four-wave-mixing impairments in Raman amplifiers are most pronounced in dispersion-shifted fibers for which the zero dispersion wavelength falls within the pump band or between the pump and signal bands. In fibers for which the dispersion wavelength falls within the pump band, four-wave-mixing between pump waves can generate sidebands within the signal band and severely degrade OSNR for some channels. In fibers for which the zero dispersion wavelength falls between the pump and signal bands, nondegenerate four-wave-mixing between pump and signal waves generates sidebands that fall in the signal band. In both cases, the power in the sidebands depends on how close the longest wavelength pumps are to the zero dispersion wavelength. This is because phase matching over long lengths of fiber, necessary for significant four-wave-mixing to occur, is easily achieved when one of the pump wavelengths is near to or coincides with the zero dispersion wavelength of the fiber.

Four-wave-mixing presents a crosstalk problem when the frequencies of the sidebands coincide with channel frequencies. For example, in single-band transmission using Raman amplification and only C- or L-band signals, pump–pump four-wave-mixing sidebands usually have wavelengths too short to fall within the signal band. On the other hand, in wideband systems where both C- and L-band transmission are used, the longest wavelength pumps are typically close enough to the signal band that four-wave-mixing sidebands fall within the signal band and can interfere with some channels [27]. In that case, several methods can be used to mitigate the effects of four-wave-mixing. One technique is simply to delete those channels that coincide with sidebands, although that approach reduces overall capacity. Four-wave-mixing efficiency is strongly dependent on phase matching, therefore employing fiber with high local dispersion tends to reduce the crosstalk level. However, in many cases, the DRA section within which four-wave-mixing is troublesome employs preexisting transmission fiber, and the associated dispersion properties are predetermined. Increasing pump frequency spacing, particularly for pump wavelengths close to the zero dispersion wavelength, can also be used. Increasing pump frequency separation increases dispersive walk-off between the pump waves, reducing phase matching and four-wave-mixing efficiency. In addition, reducing the power of long-wavelength pumps is also effective. Four-wave-mixing efficiency is at least a quadratic function of pump intensity. Raman amplification further increases efficiency so the functional dependence of sideband intensity on pump intensity can be stronger than quadratic [30]. Therefore, decreasing pump power can significantly reduce four-wave-mixing. In reducing the power at long pump wavelengths, gain and gain flatness can be kept unchanged by maintaining the spectral power density of the long-wavelength pumps.

Another approach to mitigation of four-wave-mixing employs pump amplitude modulation to prevent interaction between the pumps involved in pump–pump four-wave-mixing. The offending pumps are time modulated and multiplexed in such a manner that different pump wavelengths occupy separate time slots. For example,

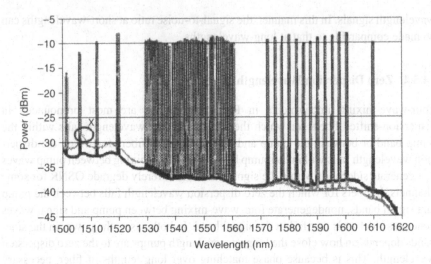

Fig. 14.10. Signal spectra with (black curve) and without (gray curve) pump modulation. The marked band on the spectrum obtained without pump modulation is a four-wave-mixing sideband arising from pump–pump interactions. The sideband is suppressed in the modulated case [29]. Source: C.R.S. Fludger et al: "Ultra-broadband high performance distributed Raman amplifier employing pump modulation," OFC Technical Digest Postconference Edition Vol. 70, pg 183 (© 2002 OSA)

Fig. 14.10 depicts the signal spectrum for modulated and nonmodulated pump waves [29]. Pump–pump four-wave-mixing sidebands appear near the short-wavelength end of the spectrum in the nonmodulated pump case, shown in gray on the figure. The sidebands disappear when the pumps are time modulated and a phase delay of π is introduced between interacting pumps.

14.5.5. Pump Power Adjustment to Control Gain Shape

The behavior of Raman amplifiers differs significantly from rare earth-doped amplifiers such as EDFAs with regard to the change in pump power and the corresponding noise figure shape. In an EDFA, changing the pump power has the effect of changing the level of inversion. The noise figure in an EDFA is minimum for a completely inverted amplifier, and the noise figure increases as the level of inversion is decreased (i.e., the spontaneous emission factor increases). On the other hand, in a Raman amplifier the gain shape and noise figure shape as a function of frequency can be controlled by the pump powers in the various pump wavelengths.

Management of Raman gain tilt is required in almost all wideband amplified systems. Although present in any DWDM system, Raman gain tilt is most problematic in systems spanning near or above 100 nm of gain bandwidth. This tilt management is less complicated with an all-Raman system as monitoring of the signals can occur across the entire band at just two positions: immediately before and directly after amplification (Fig. 14.11). Dynamic adjustment of any measured tilt (or other gain affecting event, such as channel drops or additions) can be made through the straight-

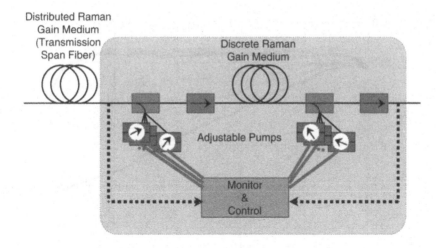

Fig. 14.11. Raman amplifier configuration including monitoring and control functionality.

forward adjustment of the Raman pump power levels. The very nature of Raman amplification permits this ability to "shape" the gain spectrum. Monitoring and adjustment of the banded system is more intricate, as a number of monitoring points need to be used together with the different gain adjustments for the discrete amplifiers. This is further complicated by the fact that the greatest amount of Raman gain tilt will be most prevalent between the bands (intraband) and careful coordination of all adjustments, regardless of band, needs to take place.

Figure 14.12 illustrates gain tilt control obtained simply by varying pump power levels in a Raman amplifier pumped with 12 wavelengths [31]. Decreasing the power of the low-frequency pumps and adjusting the other pump powers results in the tilted gain shapes shown in the figure. Another example of gain tilt control is depicted in Fig. 14.13 [18]. The figure shows results of simulations for flat (curve (a)), linear (curve (b)), and nonlinear (curve (c)) gain profiles in a Raman amplifier with a bandwidth of 35 nm. All gain profiles shown are achieved simply by varying the pump power levels. A further example of gain tilt control through pump power adjustment is given in Chapter 10, Section 10.5.

14.5.6. Signal Preemphasis to Equalize Channel Performance .

In designing multispan amplified transmission links, it is usual to select signal launch powers that assure that the worst performing channel achieves a target SNR dictated by bit error rate requirements. The noise figure of EDFAs is spectrally flat to a large extent. Therefore, in an EDFA-based system with negligible gain ripple, a uniform signal power spectrum yields comparable SNR for all channels. Where gain ripple is not negligible but accumulated gain variations are low (up to about 10 dB), preemphasis of the input signal spectrum has been used to obtain uniform SNR at the output of point-to-point transmission systems [32]. Total input power is redistributed among the WDM channels in such a manner as to maintain the composite output power and

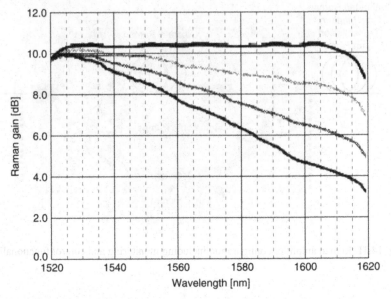

Fig. 14.12. Gain tilt control by pump power adjustment in a Raman amplifier comprising 25 km long dispersion-shifted fiber and a 12-wavelength pump module [31]. Source: S. Namiki and Y. Emori: "Ultrabroad-band Raman Amplifiers Pumped and Gain-Equalized by Wavelength-Division-Multiplexed High-Power Laser Diodes," IEEE Journal of Select Topics in Quantum Electronics, Vol 7, pg 3 (© 2001 IEEE)

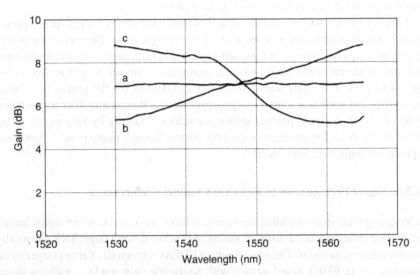

Fig. 14.13. (a) Flat, (b) tilted linear, and (c) nonlinear gain profiles obtained by pump power control [18].

to yield equal SNR for all channels. The SNR of the worst channels is improved, and a uniform bit error rate performance is obtained across the entire signal band.

The noise figure in Raman amplifiers is wavelength-dependent even when gain ripple is negligible. Noise figure variations in Raman amplifiers arise from a number of sources. Stimulated Raman scattering transfers energy from short- to long-wavelength signals, constitutes an additional loss mechanism at short-wavelengths, and results in a larger noise figure at short-wavelengths than at long-wavelengths. In addition, thermal noise in spectral bands close to pump wavelengths leads to a higher noise figure for short-wavelength signals. Furthermore, the quartic frequency-dependence of Rayleigh scattering results in a higher double-Rayleigh crosstalk for short-wavelength signals than for longer wavelength signals and a correspondingly higher electrical noise figure. Therefore, a uniform signal spectrum does not yield comparable SNR performance for all channels even when gain variations are negligible. Although power variations at the output may be small, there may still be significant variation of SNR of the channels because of the noise figure variations. A nonuniform SNR at the output translates into substantial bit error rate variation among the channels.

Input signal preemphasis can again be used to equalize channel performance in Raman amplifiers. However, unlike the case of EDFA-based systems where preemphasis is applied in such a manner as to maintain the composite signal power at a prescribed level, Raman amplification permits preemphasis to be applied to lower the composite signal power while equalizing output SNR for all channels. The launch power for channels with the worst noise figure is selected to achieve the target SNR as is usually done in systems with uniform signal launch power. A higher output SNR would be then obtained for channels with better noise figure if equal launch power were used for all channels. To achieve equal output SNR, the launch power of channels with lower noise figure is reduced. The pump powers may need to be adjusted to maintain the gain of the amplifier. An iterative adjustment of signal and pump powers results in a signal spectrum that mirrors the noise figure spectrum and yields a flat SNR spectrum. Depending on the saturation level of the Raman amplifier, reduction of the composite signal power through preemphasis of launch signal powers based on the magnitude of the corresponding noise figure can reduce the total pump power.

As an illustration, consider a simulation of a two-stage Raman amplifier comprising an 80 km long distributed section and a dispersion-compensating discrete section. Net gain through the two stages is nominally 0 dB with a flat-gain spectrum over a bandwidth of 100 nm in each stage. Figure 14.14 depicts the signal and Fig. 14.15 the OSNR spectra for launch signal powers preemphasized according to the shape of the noise figure spectrum. Also shown in the figures are spectra for a three-stage Raman amplifier comprising an 80 km long distributed section and a two-stage discrete section using 0 dBm and −2.26 dBm per channel (no preemphasis). Net gain is again nominally 0 dB over 100 nm of bandwidth. Using complementary gain slopes as described in Section 14.5.3, a nearly flat OSNR is obtained across the entire 100 nm wide spectrum, as shown in Fig. 14.15 for launch powers of 0 dBm and −2.26 dBm per channel. However, OSNR variations greater than 2 dB remain. The OSNR spectrum obtained for input signal preemphasis applied in conformity to the noise figure

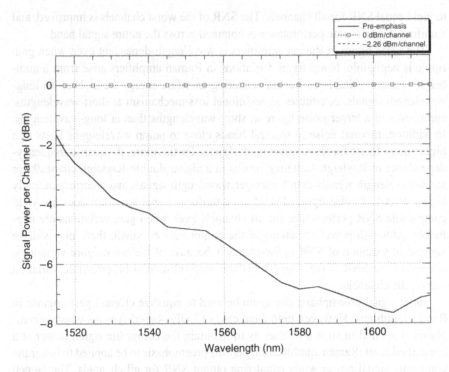

Fig. 14.14. Signal power spectra for Raman amplifiers with (solid curve) and without (dashed curves) input signal preemphasis. The shape of the input signal power spectrum for the preemphasized case is that of the noise figure of the amplifier.

spectrum is substantially flat compared to the case of the uniform signal spectrum, as attested by the spectra in Fig. 14.15. The small slope of the OSNR spectrum evident in the figure for the preemphasized case arises from neglecting a wavelength-dependent factor in the SNR calculation.

14.5.7. Other Enhancements

The foregoing techniques for wideband Raman amplification are by no means exhaustive, and more features and techniques can be expected to emerge as research and development efforts continue to be directed at Raman amplification. A number of developments in all-Raman amplification that have not been treated in detail here include: forward pumping, higher-order pumping, and time-multiplexed or frequency-swept pumping.

Most Raman amplifiers use a counterpropagating pump configuration to avoid coupling of pump fluctuations to the signal. In addition, by using the counterpropagating geometry, other effects such as pump-mediated crosstalk, polarization dependence of gain, and four-wave-mixing between pumps are also reduced. However, forward pumping improves the noise figure and enables implementation of Raman amplified bidirectional systems. Forward pumping improves the noise figure because the signal is amplified at the signal input end of the fiber before it is significantly

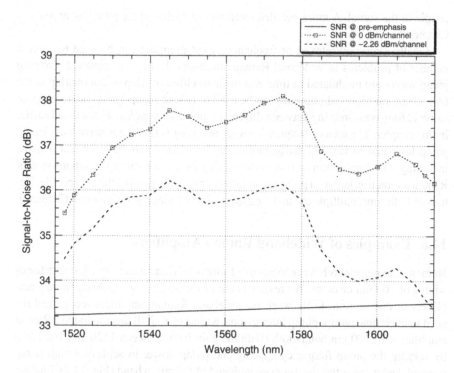

Fig. 14.15. Output optical signal-to-noise ratios for Raman amplifiers with (solid curve) and without (dashed curves) signal preemphasis.

attenuated, thereby maintaining a high signal-to-noise ratio throughout the fiber. As mentioned earlier, forward pumping requires the use of low-RIN pump lasers such as Fabry–Perot laser diodes. More detail on forward pumping is provided in Chapter 9, and some of the effects of the fast gain dynamics are covered in Chapter 8.

One way to obtain some of the benefits of forward pumping without the direct coupling of the pump fluctuations to the signal is to use so-called higher-order pumping. In higher-order pumping, pump waves at one or more Raman frequency shift above the primary pump frequency are used. The higher-order pumps amplify the primary pump waves and can reduce the amplifier noise figure. A copropagating higher-order pump can be used to pump a counterpropagating pump, thereby amplifying the counterpropagating pump at the signal input end of the fiber. Thus the signal gain is increased at the beginning of the fiber. In general, the copropagating pump is of shorter wavelength than the counterpropagating pump, and the counterpropagating pump is primarily responsible for the gain to the signal wavelengths. This approach has several limitations, however. First, the process tends to be somewhat inefficient, because one pump amplifies the other pump before it supplies energy to the signal. Also, the pumps are at the two ends of the fiber, and are therefore maximally attenuated by the propagation loss. Second, the copropagating pump can still provide gain to the signal directly through the high-frequency features on the gain spectrum (c.f., Fig. 1.1 in Chapter 1 for frequencies above 13 THz), and pump fluctuations can still couple to the signal. A more detailed treatment of higher-order pumping appears in Chapter 9.

Finally, time-multiplexed or frequency-swept pumping can be used to solve a number of problems in wideband Raman amplifiers. In time-multiplexed pumping pump waves are modulated in time and made to either overlap or not overlap in the fiber. Time-multiplexed pumping can be used to reduce pump–pump interactions or to avoid four-wave-mixing between different pump wavelengths, as discussed earlier in this chapter. The idea of frequency-swept pumping is to use a tunable wavelength pump laser and to vary the frequency of the laser. This may have several benefits including: (1) suppression of four-wave-mixing between pumps; (2) suppression of Raman interaction between pumps; and (3) very precise control of the gain profile. Details of both time-multiplexed and frequency-swept pumping are given in Chapter 3.

14.6. Examples of Wideband Raman Amplifiers

Numerous experiments have demonstrated wideband Raman amplification with bandwidths of 100 nm or more. We review some of the results on wideband Raman amplifiers in this section. Early work on wideband Raman amplifiers recognized the need to overcome pump–pump interactions. Kidorf et al. [20] demonstrated a Raman amplifier with 100 nm bandwidth (with the gain band between 1520 and 1620 nm) by varying the pump frequency spacing and pump power to achieve a high pump spectral density near the short-wavelength end of the pump band (Fig. 14.7). Fludger et al. [29] reduced the pump–pump interactions by time-multiplexing the strongly interacting pumps, by this means demonstrating a 100 nm amplifier with a gain band between 1510 and 1610 nm.

For a fixed bandwidth, the gain spectrum of a wideband Raman amplifier becomes increasingly flat as the number of pump wavelengths increases. An example of using closely spaced pumps is provided by Namiki and Emori [31], who demonstrate gain flatness of about 0.1 dB over the wavelength range between 1527 and 1607 nm in 25 km of dispersion-shifted fiber by using 12 pump wavelengths (Chapter 1, Fig. 1.6). A few pump wavelengths together with gain-flattening filters can also be used to achieve a flat-gain shape. The trade-off to be considered in electing to use a large number of pump wavelengths or a smaller number together with gain-flattening filters concerns cost, ease of implementation, and flexibility. For example, a gain flatness similar to that reported in [31] can be achieved with five or six pump wavelengths and a gain-flattening filter. Although a large number of pump wavelengths permits adjustment of gain shape in response to changes in channel loading or other dynamic effects, using fewer pump lasers and a gain-flattening filter may be cost effective because amplifier cost is often dominated by pump cost.

Although commercially available wideband Raman amplifiers have been limited to a bandwidth of 100 nm to date, laboratory experiments have demonstrated amplifiers with much greater bandwidths. For example, Naito et al. [33] have demonstrated a Raman amplifier with a bandwidth of 136.6 nm by using a pump and signal wavelength interleaving technique. The technique intermixes a few of the pump wavelengths into the signal bands with a guard band surrounding the pumps. The broadband DRA

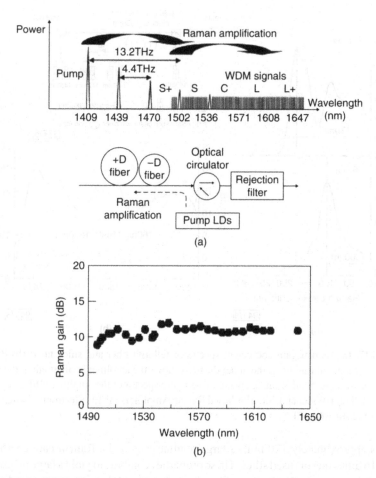

Fig. 14.16. (a) Schematic spectrum showing pump and signal wavelength allocation (upper diagram) and experimental configuration (lower diagram) for pump–signal wavelength interleaved Raman amplification; (b) gain profile of single-stage DRA [33]. Source: T. Naito et al: "A broadband distributed Raman amplifier for bandwidths beyond 100 nm" OFC Technical Digest, Postconference Edition Vol. 70, page 116 (© 2002 OSA)

scheme is shown in Fig. 14.16(a), where the signal and pump wavelength allocation is shown as well as the circulator that is used to inject the pump waves. The gain fiber consists of 28 km of positive dispersion fiber and 12 km of negative dispersion fiber. The pump wavelengths used are 1409, 1439, 1470, 1502, and 1536 nm. The total pump power is set at 480 mW. Figure 14.16(b) shows the Raman gain profile for a single-stage DRA. The transmission bandwidth is 3.7 nm in the S+ band between 1496 and 1499.7 nm, 30.1 nm in the S-band between 1504.2 and 1534.3 nm, and 102.8 nm in the C- and L-bands between 1537.2 and 1640 nm. The sum of WDM signal bandwidths is 136.6 nm, and the total width of guard bands is 7.4 nm.

Mori et al. [16] have studied Raman amplification in tellurite fiber. They show that tellurite fiber, a nonsilica fiber, has a large intrinsic bandwidth (Stokes shift of gain

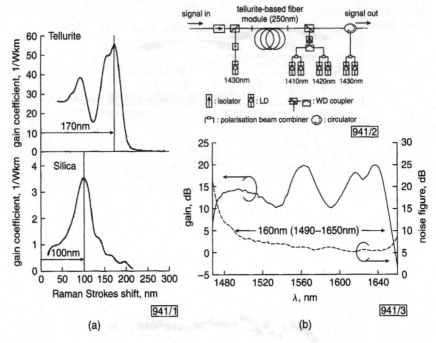

Fig. 14.17. (a) Raman gain coefficient spectra of tellurite fiber and silica fiber; (b) Raman amplification in tellurite fiber: the upper diagram shows the amplifier configuration; the lower diagram shows the small-signal gain and noise figure spectra of the amplifier [16]. Source: A. Mori et al "Ultra-Wideband Tellurite-based Raman Amplifiers" IEEE Electronics Letters, Vol. 37, pg 1442 (© 2001 IEEE)

peak is approximately 170 nm), a large nonlinearity, and a Raman gain coefficient about 16 times that of fused silica. These properties combine to yield a large bandwidth in short lengths of tellurite gain fiber. Experiments with four pump wavelengths yielded a bandwidth of 160 nm (Fig. 14.17) [16]. Figure 14.17 shows the Raman gain spectrum in tellurite fiber along with that of silica fiber. It can be seen that the gain spectrum of tellurite fiber includes two prominent peaks that straddle the main peak in the spectrum for silica. This fortuitous conjunction of spectral maxima and minima enables construction of flat-gain amplifiers by cascading tellurite and silica fiber amplifier stages. In experiments discussed next, a bandwidth of 120 nm could be obtained using silica and tellurite gain fibers.

Takara et al. [34] demonstrated a multistage all-Raman amplifier with a 3.5 dB gain bandwidth of 120 nm (between 1485 and 1605 nm) and equivalent noise figure between 4 to 10 dB within that range. The multistage amplifier configuration, consisting of a DRA in standard single-mode fiber followed by a three-stage discrete amplifier in silica and tellurite fiber, is shown in Fig. 14.18 along with gain and noise figure spectra. In particular, the large bandwidth is achieved by using lumped Raman amplification in tellurite fiber, lumped Raman amplification in DCF, distributed Raman amplification in the transmission fiber, and gain equalization filters. Using this amplifier configuration, 313 channels at 10 Gb/s are transmitted over a distance of 160 km.

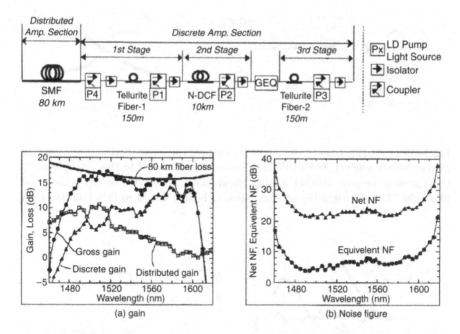

Fig. 14.18. Configuration, gain spectrum, and noise figure spectrum of inline hybrid Raman amplifier (N-DCF: negative-slope dispersion-compensating fiber. GEQ: gain equalizer) [34]. Source: H. Takahara et al: "Ultra-wideband tellurite/silica fiber Raman amplifier and supercontinuum lightwave source for 124-nm seamless bandwidth DWDM transmission," OFC Technical Digest, Postconference Edition Vol. 70, pg FB1-1 (© 2002 OSA)

14.7. Wideband Raman Amplified Systems

Raman amplification has been successfully used in high-performance transmission systems in conjunction with EDFAs and other high-performance techniques to extend the reach and/or capacity of fiber-optic transmission systems. High-performance transmission systems utilizing all-Raman amplification have also been demonstrated. The wide bandwidth, low noise figure, and distributed gain of Raman amplification are critical to achieving high performance.

As an example of bandwidth and reach extension using Raman amplification, Foursa et al. [35] demonstrated 2.56 Tb/s (256 channels at 10 Gb/s per channel) transmission over 11,000 km using hybrid Raman/EDFAs with 80 nm of continuous bandwidth. There have been several earlier reports of bandwidth extension by addition of L-band EDFAs in parallel with C-band EDFAs. Parallel C- and L-band EDFAs are limited by inferior noise performance compared to C-band EDFAs because of additional loss from couplers used to split and recombine the subband signals. These limitations were overcome in [35] by using a hybrid amplifier comprising an erbium-doped fluoride fiber amplifier in series with a Raman amplifier (R-EDFA) that provided a continuous bandwidth over the C and L. Ultra-long-distance transmission was achieved by using state-of-the-art gain equalization techniques as well as concatenated Reed–Solomon forward error correction (FEC). The configuration

474   M.N. Islam et al.

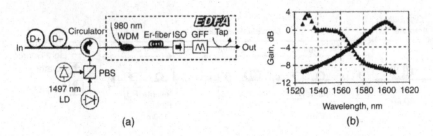

(a)   (b)

Fig. 14.19. (a) Amplifier configuration and (b) gain spectra of Raman (circles) and EDFA (triangles) sections of a hybrid DRA/EDFA [35]. Source: D.G. Foursa: "2.56 Tb/s (256/spl times/10 Gb/s) transmission over 11,000 km using hybrid Raman/EDFAs with 80 nm of continuous bandwidth," OFC Technical Digest, Postconference Edition Vol. 70, pg FC3-1 (© 2002 OSA)

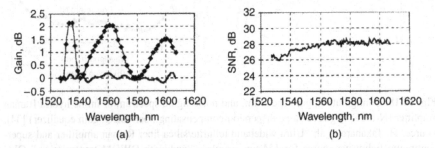

(a)   (b)

Fig. 14.20. (a) Hybrid Raman-EDFA gain shapes showing equalized (solid line) and unequalized (diamonds) gain spectra; (b) measured SNR of a chain of 13 hybrid Raman/EDFAs [35]. Source: D.G. Foursa: "2.56 Tb/s (256/spl times/10 Gb/s) transmission over 11,000 km using hybrid Raman/EDFAs with 80 nm of continuous bandwidth," OFC Technical Digest, Postconference Edition Vol. 70, pg FC3-1 (© 2002 OSA)

for the R-EDFA is illustrated in Fig. 14.19(a), which shows unpolarized backward pumping in a positive dispersion/negative dispersion $(+D/ - D)$ fiber combination. Figure 14.19(b) shows that the gain spectra of the Raman and EDFA sections complement each other and provide a wide continuous bandwidth with minimum gain ripple. Typical unequalized and equalized gain shapes are shown in Fig. 14.20(a). To evaluate the performance of the amplifier, a 525 km chain of 13 R-EDFAs was constructed. The SNR measured at the output end of the chain is illustrated in Fig. 14.20(b).

In another example, transmission of 10.2 Tb/s (256 channels at 42.7 Gb/s per channel) over three 100 km spans of nonzero dispersion-shifted fiber was demonstrated by Frignac et al. [36]. A number of techniques were used to achieve this ultra-high-capacity transmission. First, hybrid Raman/EDFA amplifiers with improved noise performance achieved by second-order pumping were used. Second, polarization-multiplexing was used: two sets of 128 WDM channels were multiplexed in the polarization domain (with adjacent channels orthogonally polarized to reduce interaction or crosstalk between channels). Third, as is becoming more common in almost all system experiments, FEC was used to improve the BER at the output. Finally, vestigial sideband (VSB) coding was used, which in combination with polarization-

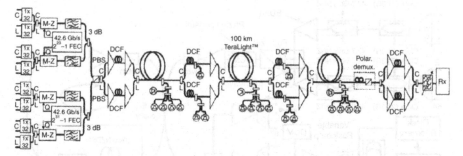

Fig. 14.21. Experimental configuration for transmission of 256 × 42.7 Gb/s channels over three 100 km laps of nonzero dispersion-shifted fiber [36]. Source: Y. Frignac et al "Transmission of 256 wavelength- division and polarization-division-multiplexed channels at 42.7Gb/s (10.2Tb/s capacity) over 3/spl times/100km of TeraLight/spl trade/ fiber," OFC Technical Digest, Postconference Edition Vol. 70, pg FC5-1 (© 2002 OSA)

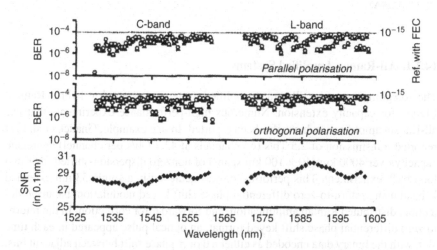

Fig. 14.22. Summary of experimental results for transmission of 256 × 42.7 Gb/s channels over three laps of 100 km nonzero dispersion-shifted fiber showing bit error rate spectra for parallel (top) and orthogonal (center) polarized signals and SNR at output after FEC (bottom) [36]. Source: Y. Frignac et al "Transmission of 256 wavelength- division and polarization-division-multiplexed channels at 42.7Gb/s (10.2Tb/s capacity) over 3/spl times/100km of TeraLight/spl trade/ fiber," OFC Technical Digest, Postconference Edition Vol. 70, pg FC5-1 (© 2002 OSA)

multiplexing allowed a high spectral density of 0.64 bps/Hz to be realized. The experimental configuration is shown in Fig. 14.21, where the WDM transmitter consists of two sets of 128 DFB lasers covering the C- and L-bands. As shown in the figure, a split-band configuration is used for the amplifiers, with 8 nm of guard band centered around 1565 nm between the C- and L-bands. The experimental results are summarized in Fig. 14.22. The top curves correspond to the bit error rate (BER) for parallel and orthogonal polarizations, and the bottom curve shows the SNR at the output after FEC.

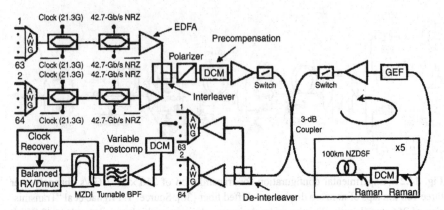

Fig. 14.23. Experimental setup for transmission of 64 × 42.7 Gb/s channels over 40 laps of 100 km nonzero dispersion-shifted fiber [37]. Source: A.H. Gnauck et al: "2.5 Tb/s (64/spl times/42.7 Gb/s) transmission over 40/spl times/100 km NZDSF using RZ-DPSK format and ' all-Raman-amplified spans," OFC Technical Digest, Postconference Edition Vol. 70, pg FC2-1 (© 2002 OSA)

14.7.1. All-Raman Amplified Systems

The foregoing examples illustrate amplified systems that employ hybrid Raman/ EDFAs for capacity extension. Numerous high-performance experiments that use all-Raman amplification have also been reported. In one example, Gnauck et al. [37] reported transmission of 2.5 Tb/s (64 channels at 42.7 Gb/s per channel) aggregate capacity over 4000 km (40 × 100 km spans) of nonzero dispersion-shifted fiber in a loop-back experiment. This performance was achieved in a single 53 nm extended L-band using return-to-zero differential phase shift keyed modulation, balanced detection, distributed Raman amplification, and forward error correction. In the return-to-zero differential phase shift keyed format, an optical pulse appeared in each time slot, with the binary data encoded as either a 0 or π phase shift between adjacent bits. In principle, this coding scheme can reduce the SNR required to achieve a given BER by 3 dB, thereby increasing system margin. The experimental setup for the loop-back experiment is shown in Fig. 14.23. The amplifier is a DRA followed by a discrete Raman amplifier implemented in a dispersion-compensating module. The experimental results are shown in Fig. 14.24, where the SNR without FEC of the 64 channels is shown. After FEC, an uncorrected BER of 2.4×10^{-4} corresponds to a corrected BER of less than 10^{-12}.

In another experiment, Sugahara et al. [38] demonstrated transmission of 32 × 42.7 Gb/s DWDM signals over a distance of 6050 km using a Raman-amplified quadruple-hybrid span configuration. Transmission of the DWDM signals, which were spaced by 100 GHz, was accomplished through suppression of fiber nonlinearity effects by distributed all-Raman amplification in the quadruple-hybrid span configuration of Fig. 14.25. The dispersion map consisted of lengths of low-loss, pure silica core fibers (PSCFs) interleaved with DCF. The experimental configura-

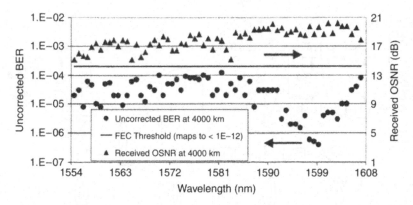

Fig. 14.24. Experimental results for transmission of 64 × 42.7 Gb/s channels over 40 laps of 100 km nonzero dispersion-shifted fiber. Optical signal-to-noise ratio (triangles) and uncorrected bit error rate (circles) are shown [37]. Source: A.H. Gnauck et al: "2.5 Tb/s (64/spl times/42.7 Gb/s) transmission over 40/spl times/100 km NZDSF using RZ-DPSK format and all- Raman-amplified spans," OFC Technical Digest, Postconference Edition Vol. 70, pg FC2-1 (© 2002 OSA)

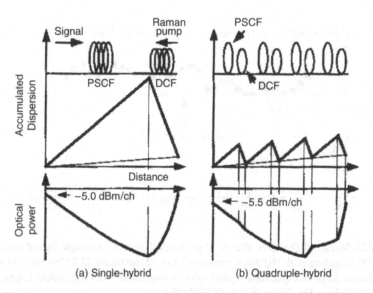

Fig. 14.25. Dispersion maps and power profiles for (a) single hybrid and (b) quadruple-hybrid span configurations (PSCF: pure silica core fiber; DCF: dispersion-compensating fiber) [38]. Source: H. Sugahara et al: "6,050km transmission of 32 /spl times/ 42.7 Gb/s DWDM signals using Raman-amplified quadruple-hybrid span configuration," OFC Technical Digest, Postconference Edition Vol. 70, pg FC6-1 (© 2002 OSA)

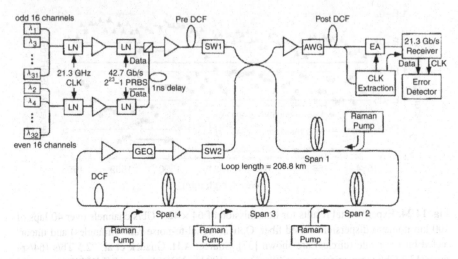

Fig. 14.26. Experimental configuration for Raman-amplified transmission using quadruple-hybrid spans [38]. Source: H. Sugahara et al: "6,050km transmission of 32 /spl times/ 42.7 Gb/s DWDM signals using Raman-amplified quadruple-hybrid span configuration," OFC Technical Digest, Postconference Edition Vol. 70, pg FC6-1 (© 2002 OSA)

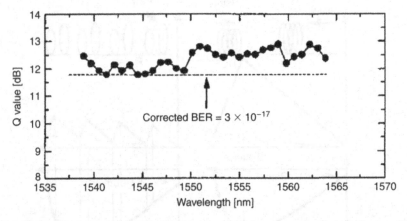

Fig. 14.27. Measured Q-values after 6050 km transmission on quadruple-hybrid spans [38]. Source: H. Sugahara et al: "6,050km transmission of 32 /spl times/ 42.7 Gb/s DWDM signals using Raman-amplified quadruple-hybrid span configuration," OFC Technical Digest, Postconference Edition Vol. 70, pg FC6-1 (© 2002 OSA)

tion is illustrated in Fig. 14.26, showing the distributed scheme used. Figure 14.27 shows the measured Q-values after 6050 km transmission. The Q-values are all above 11.7 dB, corresponding to a BER below 3×10^{-17} with FEC.

As another example, a 1.6 Tb/s system (40×42.7 Gb/s) transmission over 3600 km of dispersion-managed fiber was accomplished using all-Raman amplification in 100 km terrestrial spans using a loop-back configuration [39]. The long transmission distance and high capacity were achieved with a carrier-suppressed

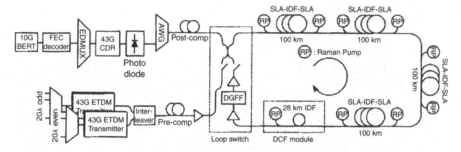

Fig. 14.28. Configuration of all-Raman amplified loop experiment for 1.6 Tbit/s transmission [39]. Source: F. Liu et al: "1.6 Tbit/s (40/spl times/42.7 Gbit/s) transmission over 3600 km UltraWave/spl trade/ fiber with all-Raman amplified 100 km terrestrial spans using ETDM transmitter and receiver," OFC Technical Digest, Postconference Edition Vol. 70, page FC7-1 (© 2002 OSA)

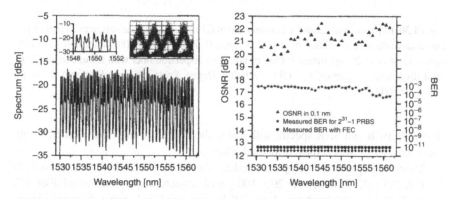

Fig. 14.29. BER with and without FEC and optical signal-to-noise ratio after 3600 km transmission [39]. Source: F. Liu et al: "1.6 Tbit/s (40/spl times/42.7 Gbit/s) transmission over 3600 km UltraWave/spl trade/ fiber with all-Raman amplified 100 km terrestrial spans using ETDM transmitter and receiver," OFC Technical Digest, Postconference Edition Vol. 70, page FC7-1 (© 2002 OSA)

return-to-zero (CS-RZ) modulation format, distributed Raman amplification, and new fiber types. Unlike many 40 Gb/s systems that use optical time-division-multiplexing, these results were obtained using an electronic time-division-multiplexed (ETDM) transmitter and ETDM receiver. Even without using polarization interleaving, a high spectral density (40 Gb/s on 100 GHz channel spacing) was achieved because of this combination of techniques. The loop-back experimental configuration is shown in Fig. 14.28, depicting the 100 km spans comprising 35.5 km of large effective area fiber in series with 29 km of perfectly slope-matched dispersion-compensating fiber and an additional 35.5 km of large effective area fiber. Each link was also bidirectionally pumped, with a forward gain of ~7.5 dB and a backward gain of ~18.5 dB. In addition, fixed gain-flattening filters were used after each amplifier, and a dynamic gain-flattening filter was used in one location of the loop. Figure 14.29

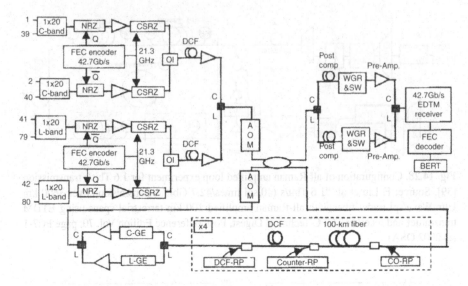

Fig. 14.30. Experimental setup for transmission of 80 × 42.7 Gb/s channels over 20 laps of 100 km nonzero dispersion-shifted fiber [40]. Source: B. Zhu: "3.2Tb/s (80 /spl times/ 42.7Gb/s) transmission over 20 /spl times/ 100km of non-zero dispersion fiber with simultaneous C+L-band dispersion compensation," OFC Technical Digest, Postconference Edition Vol. 70, pg FC8-1 (© 2002 OSA)

shows the BER and SNR spectra with and without FEC. Using FEC, a bit error rate lower than 10^{-11} was achieved for each channel.

A similar experiment was performed by Zhu et al. [40], where transmission of 3.2 Tb/s (80 × 42.7 Gb/s) over 20 × 100 km of nonzero dispersion-shifted fiber was accomplished in a recirculating loop. All-Raman amplified spans, dispersion slope matched DCF modules covering a 75 nm band, CS-RZ modulation format, and FEC were employed. The low dispersion slope of the transmission fiber enables C- and L-band dispersion and dispersion slope-compensation of each span in a single DCF module. Also, the location of the zero dispersion wavelength of the transmission fiber at <400 nm enables forward and backward pumping of the fiber spans for flat-gain and noise figure across the entire C- and L-bands by eliminating pump–pump and pump–signal four-wave-mixing. The experimental setup for the recirculating loop is shown in Fig. 14.30, where the all-Raman amplifier is shown to be a bidirectionally pumped DRA in the 100 km transmission spans and a backward pumped discrete Raman amplifier in the DCF. Figure 14.31 shows the measured pump-on/pump-off gains used in the experiment where each span is pumped to transparency. In general, the SNRs at short wavelengths are lower than at long wavelengths because the shorter wavelength pumps amplify the longer wavelength pumps through the Raman effect. Therefore, the use of higher copumped Raman gain at short wavelengths and the complementary slope for the counter-pumped Raman gain provides both flat noise figure and gain. Gain in the DCF is only used to compensate for losses in the DCF module and other components. With FEC, the worst-case BER corresponds to a corrected BER below 10^{-13} at the receiver. This experiment illustrates the importance of both dispersion

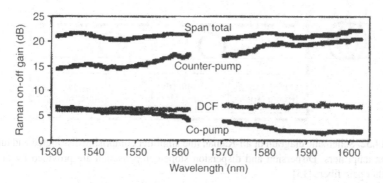

Fig. 14.31. Measured on-off Raman gains in all-Raman amplified loop experiment [40]. Source: B. Zhu: "3.2Tb/s (80 /spl times/ 42.7Gb/s) transmission over 20 /spl times/ 100km of non-zero dispersion fiber with simultaneous C+L-band dispersion compensation," OFC Technical Digest, Postconference Edition Vol. 70, pg FC8-1 (© 2002 OSA)

and dispersion slope matching, as well as proper use of complementary gain slopes to obtain flat-gain and noise figure spectra.

14.7.2. Ultra-Long-Haul WDM Systems Based on Wideband Raman Amplification

Telecommunications operators are shifting from demand for "capacity at any price" to demand for "capacity at the lowest price" [39]. Thus the focus of equipment suppliers is now to achieve large throughputs on a single strand of optical fiber over long distances at the lowest cost. The effective metric for long-haul and ultra-long-haul WDM transmission systems thus changes from just the product of total throughput and maximum unregenerated distance (measured in units of Gb/s km), to cost per unit bandwidth transmitted over unit distance ($/(Gb/s km)). Raman amplifier technology not only provides the basis for low-noise amplification of a very large continuous bandwidth, but also contributes to a substantial reduction in transmission cost by enabling high-capacity WDM system design utilizing low-cost terminal components.

A single, large transmission band is desirable because it minimizes amplifier cost by lowering the total number of passive components and pump sources, reduces the complexity of amplifier configurations, simplifies amplifier management, and minimizes the overall terminal cost by simplifying the multiplexer/demultiplexer structure (c.f., Chapter 1, Fig. 1.15) [2]. An amplifier structure capable of providing gain over a single ~100 nm wide band, with performance and total transmission cost comparable to those of traditional C-band (~35 nm) EDFAs with DRA and dispersion-compensation, can yield almost a threefold reduction of the economic metric. By allowing gain in any wavelength window with simple tailoring of the gain bandwidth, Raman amplification presents a reliable and low-cost solution to extending bandwidth.

An amplifier structure combining both lumped Raman and distributed amplification can give significant advantages of reach and capacity while dramatically lowering the overall cost of the system. Such a design has been implemented and the

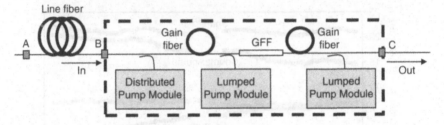

Fig. 14.32. Schematic diagram of all-Raman line amplifier comprising distributed and lumped Raman amplifiers. Dispersion and dispersion slope-compensation are provided by discrete amplifier gain fibers [15].

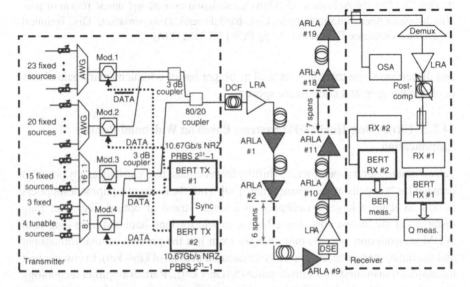

Fig. 14.33. Experimental setup for ultra-long-haul transmission of 67×10.7 Gb/s channels over 1565 km of standard single-mode fiber [15].

feasibility of transmitting 240 OC-192 channels over 1565 km standard single-mode fiber has been demonstrated [15]. In this all-Raman experiment, a wideband 100 nm Raman amplifier is used both for the booster amplifier and the inline amplifier (ILA). The booster amplifier is placed after the transmitter and consists of an LRA (lumped Raman amplifier). The ILAs use a hybrid distributed/lumped Raman amplifier in a three-stage configuration (Fig. 14.32). Dispersion and dispersion slope-compensation are provided by the dispersion-compensation fiber used as gain medium in the LRAs, and a static gain-flattening filter is used to obtain a substantially flat-gain spectrum over 100 nm. Each LRA is composed of two stages separated by isolators to reduce double-Rayleigh scattering, so that the resulting amplifier has a low MPI level.

The experimental setup for the ultra-long-haul transmission experiment is shown in Fig. 14.33. For this example, the transmission of 67×10.7 Gb/s channels spanning >97 nm (from 1519.9 to 1617.0 nm) was achieved over 19 spans of average loss

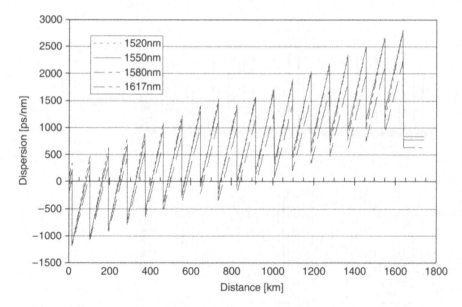

Fig. 14.34. Dispersion map for transmission experiment utilizing 100 nm bandwidth all-Raman line amplifiers. A common precompensation of ~−1000 ps/nm is provided across the full bandwidth. Banded postcompensation of 0, −300, −300, and 0 ps/nm is provided in four subbands [15].

19.7 dB. The total distance was 1565 km, and the average span length was 82.4 km. Using tunable sources, each measured channel was set at the center of a group of five 50 GHz-spaced channels, all NRZ-modulated with 2^{31} to 1 pseudorandom bit sequence. The overall system performance was measured using a bit error rate tester (BERT).

The dispersion map for the experiment is illustrated in Fig. 14.34. For all of the channels there is a common precompensation of −1000 ps/nm, which is provided by the DCF after the transmitter block and before the booster amplifier shown in Fig. 14.33. Because the dispersion slope match is not perfect, the continuous band is divided at the receiver end into four subbands, which receive respective postcompensation of 0, −300, −300, 0 ps/nm. Better slope matching of the DCF to the transmission fiber would reduce the need for postcompensation.

Input and output spectra, measured with a resolution of 0.1 nm on an optical spectrum analyzer, are illustrated in Fig. 14.35. Whereas the input spectrum is more or less flat, the output shows some accumulation of ripple after the 19 amplifier spans. The amplifier output power per channel is 0 dBm (1mW), and the minimum optical SNR is 18 dB (measured from ASE floor to the peak of the signal). On the right side of the figure, an expanded view is shown of several channels at the short-wavelength, midrange, and long-wavelength side of the band.

The transmission performance measured with the BERT is shown in Fig. 14.36. The end-of-life BER target is 10^{-15}, which corresponds to a quality factor Q of $12.5\,\mathrm{dB}_{20}$ (i.e., $20\log(Q) = 12.5$). However, an additional 3 dB of end-of-life margin

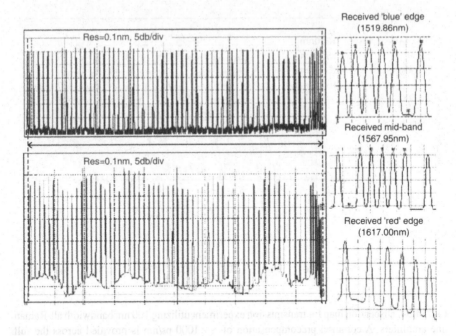

Fig. 14.35. Input (top left) and output (bottom left) signal spectra in the 100 nm bandwidth all-Raman amplified transmission experiment. The insets on the right show output spectra near the short-wavelength (top), central (middle), and the long-wavelength portions of the signal band [15].

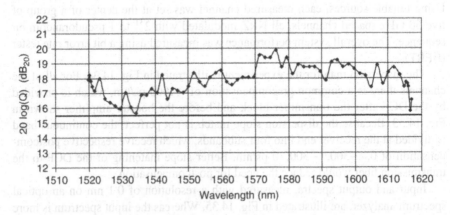

Fig. 14.36. Experimental Q value spectrum in the 100 nm bandwidth all-Raman amplified transmission experiment. The solid horizontal line indicates the target end-of-life Q [15].

needs to be allocated for loss and other impairments. Hence the target end-of-life Q is 15.5 dB_{20}. Figure 14.36 shows the measured Q across the ~100 nm band by using the tunable lasers to obtain five consecutive 50 GHz-spaced channels for each measurement. All channels showed a performance exceeding the end-of-life target of 10^{-15} BER, given a 5.5 dB coding gain from FEC (RS 255/239).

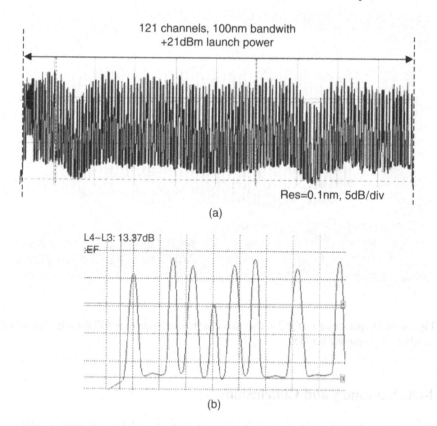

Fig. 14.37. (a) Signal spectrum after transmission of 121 channels over 1565 km of standard single-mode fiber; (b) detail of signal spectrum showing system margin for 50 GHz channel spacing at 0 dBm/channel. The composite launch power is 21 dBm [15].

Although this exemplary experiment uses only 67 channels, newer results do show the performance of the system up to the full load of 240 channels. For example, Fig. 14.37 shows the output spectrum after 1565 km for transmission of 121 channels over a 100 nm bandwidth with a composite launch power of +21 dBm. As the lower curve shows, the system margin is tested by using the tunable lasers to create five channels at 50 GHz channel spacing with the output from each amplifier set at 0 dBm per channel. All channels are decoded error-free with FEC (RS 255/239). Margin measurements on all channels from 6.3 to 7.5 dB were obtained by input channel deemphasis from the nominal amplifier output channel power (0 dBm). Figure 14.38 shows the output spectrum for 240 channels after eight spans (~650 km total length) for a substantially flat input spectrum. Each ILA now has an output power of +24 dBm, corresponding to 0 dBm per channel for the 240 channels. With the use of dynamic gain equalizers placed periodically in the span (i.e., one or two in the entire 19-span link), transmission of all 240 channels over a distance above 1500 km can be achieved.

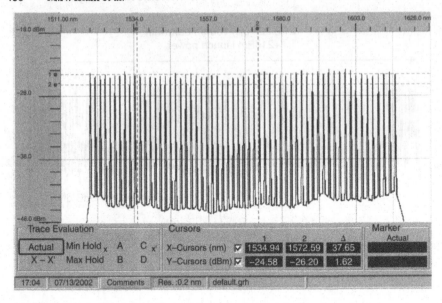

Fig. 14.38. Output spectrum for 240 channels after transmission over 650 km (eight spans) of standard single-mode fiber [15].

14.8. Summary and Conclusions

This chapter described design and implementation of wideband Raman amplifiers. All-Raman amplification enables the lowest cost and smallest footprint system for a number of reasons. First, an all-Raman amplifier requires fewer components than the alternative splitband approach. Second, broadband gain (100 nm or more) allows use of less dense channel spacing (e.g., 10 Gb/s on 50 GHz channel spacing) and standard nonreturn-to-zero modulation, thereby lowering transponder cost. Third, although most long-haul and ultra-long-haul amplified systems are limited by SNR or nonlinearity, the use of distributed Raman amplification offers significant improvement in noise figure and mitigation of nonlinearity compared with discrete amplification as offered by EDFAs.

To realize wideband systems, a number of enabling technologies are required. First among these is the wideband amplifier itself. Approaches for constructing wideband amplifiers include: EDFAs in various glass compositions, other rare earth dopings, splitband S/C/L amplifiers, hybrid Raman/EDFA amplifiers, and all-Raman amplifiers. Second, broadband components are required, such as multiplexers and demultiplexers. Third, wideband systems require gain equalizers and gain-flattening filters, which may be static or dynamic. Finally, for operation at 10 Gb/s dispersion-compensation as well as dispersion slope-compensation is required periodically in the system.

Raman amplification provides a simple single platform for long-haul and ultra-long-haul fiber-optic transmission systems. Distributed Raman amplification alone or in combination with lumped Raman amplification improves the noise figure and nonlinear penalty performance. Also, gain in Raman amplification is obtainable in practically any wavelength band and can be over a wideband, determined primarily by the pump wavelength and power distribution. In addition, lumped Raman amplification can integrate dispersion-compensation by using DCF as the gain fiber. Other advantages of Raman amplification include simpler equalization, simple adjustment of gain flatness, and large power dynamic range for flat-gain operation. Despite this significant list of advantages, a number of challenges exist for Raman amplification, including: pump–pump interactions, interband and intraband Raman gain tilt, noise arising from thermally induced phonons near the pump wavelengths, MPI from double-Rayleigh scattering, coupling of pump fluctuations to the signal, and pump-mediated signal crosstalk.

Fortunately, design techniques exist for overcoming all of these physical limitations, thus allowing for the relatively simple implementation of 100 nm Raman amplifiers. Section 14.5 on design of wideband Raman amplifiers provides in detail the engineering design rules used to determine the architecture of wideband Raman amplifiers. Because of MPI penalties from DRS, wideband Raman amplifiers are typically sectioned into multiple stages, where each stage is separated by an isolator. Also, the easiest way to deal with the short upper state lifetime of the Raman effect is to introduce an effective lifetime equal to the transit time by making the pump and signal counterpropagating.

For wideband Raman amplifiers, several physical effects become particularly important. The Raman effect leads to power exchange between the pumps and between the signals (interband and intraband). To overcome the pump–pump interactions, the basic idea is to increase the pump spectral density at shorter wavelengths. In addition, thermal noise is a significant limitation for wideband Raman amplifiers. To mitigate the effect of thermal noise at room temperature, one strategy is to amplify the vulnerable signals first. In a hybrid amplifier comprising a DRA followed by a lumped Raman amplifier, this technique leads to gain slopes that are complementary in the two stages with the DRA gain increasing with frequency.

Several unique features of Raman amplification also lead to simplified system implementations. For example, in a Raman amplifier the gain shape as a function of frequency can be controlled by varying the pump powers. This leads to a simple system where the signal input and output spectra can be monitored and used to control the pump powers applied to the Raman amplifier.

Section 14.6 illustrates various wideband Raman amplifiers that have been reported in the literature. Although commercially available wideband Raman amplifiers have been limited to a bandwidth of 100 nm to date (i.e., the separation of the pump and signal at roughly the peak of the gain band), laboratory experiments have shown amplifiers with much larger bandwidths. Bandwidths greater than 100 nm are usually achieved with such special techniques as new glass compositions or wavelength guard bands around the pump wavelengths.

Finally, Section 14.7 describes successful application of wideband Raman amplification in high-performance transmission systems. For example, a hybrid Raman amplifier structure, combining both LRA and DRA, can give significant advantages of reach and capacity while dramatically lowering the overall cost of the system. Such a design has been implemented and the transmission feasibility of 240 OC-192 channels over 1565 km standard single-mode fiber has been demonstrated. The BER target is 10^{-15} and a 3 dB of end-of-life margin for loss and other impairments leads to a target Q of 15.5 dB_{20}. In the systems experiment all channels showed a performance exceeding the end-of-life target, given a 5.5 dB coding gain from FEC (RS 255/239).

Although the history of Raman amplification in fibers dates back to the 1970s, only in recent years has the technology matured to the point that practical systems can be built and commercialized. Engineering design rules have been developed to address almost all physical limitations of Raman amplification, and advances in pump lasers, fiber quality, and passive components have made Raman amplifiers economical.

Raman amplification enables other system cost reductions, although amplifiers account for about 20% of the total loaded system cost in fiber-optic transmission systems. For example, due to the lower noise figure and reduced nonlinear penalty with DRAs, reach can be 1500 km or more, thus reducing the cost associated with channel-by-channel regeneration. Because of the broad bandwidth enabled by Raman amplification, several terabit-per-second capacities can be achieved on a single strand of fiber without exceeding a spectral density of 0.2 bps/Hz and using standard NRZ modulation format. Hence, transponder cost can be low for wideband Raman systems. Moreover, because a single wideband amplifier can be used instead of three amplifiers in a splitband configuration for a 100 nm bandwidth, the amplifier cost is reduced by eliminating band couplers and by consolidating dispersion-compensation modules and monitoring and control circuits. With fewer parts, the footprint for all-Raman amplification can be lowered, thereby also reducing the operating expense to carriers. For all of these reasons, Raman amplification can be expected to be dominant in long-haul and ultra-long-haul systems in the near future.

References

[1] S. Lin, *Lightwave* 19: 3, 75, 2002.

[2] M. Islam and M. Nietubyc, *WDM Solutions*, 4: 51, 2000.

[3] R. C. Alferness, H. Kogelnik, and T. H. Wood, *Bell Labs Tech. J.*, 5: 188, 2000.

[4] A. Hadjifotiou. In *Optical Amplifiers and their Applications*, OSA Technical Digest, Washington DC: Optical Society of America, 2, 2000.

[5] Y. Sun, A. K. Srivastava, J. Zhou, and J. W. Sulhoff, *Bell Labs Tech. J.* 4: 187, 1999.

[6] H. Ono, M. Yamada, and M. Shimizu, *Electron. Lett.*, 38: 1084, 2002.

[7] T. Sugawa, T. Komukai, and Y. Miyajima, *IEEE Photon. Technol. Lett.*, 2: 475, 1990.

[8] M. Yamada, A. Mori, K. Kobayashi, H. Ono, T. Kanamori, K. Oikawa, Y. Nishida, and Y. Ohishi, *IEEE Photon. Technol. Lett.*, 10: 1244, 1998.

[9] J. D. Minelly and A. J. E. Ellison. In *Proceedings of the Conference on Lasers and Electro-Optics*, OSA Technical Digest, Washington DC: Optical Society of America, 352, 2001.

[10] S. Tanaka, K. Imai, T. Yazaki, and H. Tanaka. In *Proceedings of OSA Trends in Optics and Photonics (TOPS) Vol. 70, Optical Fiber Communication Conference*, Technical Digest, Postconference Edition, Washington DC: Optical Society of America, 458, 2002.

[11] T. Sakamoto. In *Proceedings of OSA Trends in Optics and Photonics (TOPS) Vol. 54, Optical Fiber Communication Conference*, Technical Digest, Postconference Edition, Washington DC: Optical Society of America, TuQ1-1, 2001.

[12] J. Masum-Thomas, D. Crippa, and A. Maroney. In *Proceedings of OSA Trends in Optics and Photonics (TOPS) Vol. 54, Optical Fiber Communication Conference*, Technical Digest, Postconference Edition, Washington DC: Optical Society of America, WDD9-1, 2001.

[13] D. Bayart, P. Baniel, A. Bergonzo, J.-Y. Boniort, P. Bousselet, L. Gasca, D. Hamoir, F. Leplingard, A. Le Sauze, P. Nouchi, F. Roy, and P. Sillard, *Electron. Lett.*, 36: 1569, 2000.

[14] H. Masuda and S. Kawai, *IEEE Photon. Technol. Lett.*, 11: 647, 1999.

[15] M. W. Chbat and H. A. Fevrier. In *Proceedings of the 28th European Conference on Optical Communication* (COM, University of Denmark, Copenhagen, Denmark), 2002; M. W. Chbat. In *OSA Technical Digest Series, Conference on Lasers and Electro-Optics*, Washington DC: Optical Society of America, 430, 2002.

[16] A. Mori, H. Masuda, K. Shikano, K. Oikawa, K. Kato, and M. Shimizu, *Electron. Lett.*, 37: 1442, 2001.

.[17] T. Yagi. In *Proceedings of OSA Trends in Optics and Photonics (TOPS) Vol. 77, Optical Amplifiers and Their Applications*, OSA Technical Digest, Postconference Edition, Washington DC: Optical Society of America, OMC1-1, 2002.

[18] S. G. Grubb, R. Zanoni, and T. D. Stephens, Patent Number 6115174, United States Patent Office, 2000.

[19] G. P. Agrawal, *Nonlinear Fiber Optics*, New York: Academic, 1995.

[20] H. Kidorf, K. Rottwitt, M. Nissov, M. Ma, and E. Rabarijaona, *IEEE Photon. Technol. Lett.*, 11: 530, 1999.

[21] C.-J. Chen and W. S. Wong, *Electron. Lett.*, 37: 371, 2001.

[22] S. A. E. Lewis, S. V. Chernikov, and J. R. Taylor, *Opt. Lett.*, 24: 1823, 1999.

[23] K. Rottwitt, J. Bromage, D. Mei, and A. Stentz. In *Proceedings of the 26th European Conference on Optical Communication* Berlin: VDE Verlag, vol. 2, 67, 2000.

[24] M. Nissov, K. Rottwitt, H. D. Kidorf, and M. X. Ma, *Electron. Lett.*, 35: 997, 1999.

[25] C. R. S. Fludger, V. Handerek, and R. J. Mears, *IEEE J. Lightwave Technol.*, 19: 1140, 2001; 20: 316, 2002.

[26] F. Forghieri, R. W. Tkach, and A. R. Chraplyvy. In *Optical Fiber Communication Vol. 4*, Technical Digest Series, Conference Edition, Washington DC: Optical Society of America, 294, 1994.

[27] R. E. Neuhauser, P. M. Krummrich, H. Bock, and C. Glingener. In *Proceedings of the Optical Fiber Communication Conference*, OSA Technical Digest, Washington DC: Optical Society of America, MA4-1, 2001.

[28] S. A. E. Lewis, S. V. Chernikov, and J. R. Taylor, *Electron. Lett.*, 35: 1761, 1999.

[29] C. R. S. Fludger, V. Handerek, N. Jolley, and R. J. Mears. In *Proceedings of OSA Trends in Optics and Photonics (TOPS) Vol. 70, Optical Fiber Communication Conference*, Technical Digest, Postconference Edition, Washington DC: Optical Society of America, 183, 2002.

[30] J. Bromage, P. J. Winzer, L. E. Nelson, and C. J. McKinstrie. In *Proceedings of OSA Trends in Optics and Photonics (TOPS) Vol. 77, Optical Amplifiers and Their Applications*, OSA Technical Digest, Post Conference Edition, Washington DC: Optical Society of America, OWA5-1, 2002.

[31] S. Namiki and Y. Emori, *IEEE J. Select. Topics Quantum Electron.*, 7: 3, 2001.

[32] A. R. Chraplyvy, R. W. Tkach, K. C. Reichmann, P. D. Magill, and J. A. Nagel, *IEEE Photon. Technol. Lett.*, 5: 428, 1993.

[33] T. Naito, T. Tanaka, K. Torii, N. Shimojoh, and H. Nakamoto. In *Proceedings of OSA Trends in Optics and Photonics (TOPS) Vol. 70, Optical Fiber Communication Conference*, Technical Digest, Postconference Edition, Washington DC: Optical Society of America, 116, 2002.

[34] H. Takara, H. Masuda, K. Mori, K. Sato, Y. Inoue, T. Ohara, A. Mori, M. Kotoku, Y. Miyamoto, T. Morioka, and S. Kawanishi. In *Proceedings of OSA Trends in Optics and Photonics (TOPS) Vol. 70, Optical Fiber Communication Conference*, Technical Digest, Postconference Edition, Washington DC: Optical Society of America, FB1-1, 2002.

[35] D. G. Foursa, C. R. Davidson, M. Nissov, M. A. Mills, L. Xu, J. X. Cai, A. N. Pilipetskii, Y. Cai,C. Breverman, R. R. Cordell, T. J. Carvelli, P. C. Corbett, H. D. Kidorf, and N. S. Bergano. In *Proceedings of OSA Trends in Optics and Photonics (TOPS) Vol. 70, Optical Fiber Communication Conference*, Technical Digest, Postconference Edition, Washington DC: Optical Society of America, FC3-1, 2002.

[36] Y. Frignac, G. Charlet, W. Idler, R. Dischler, T. Tran, S. Lanne, S. Borne, C. Martinelli, G. Veith, A. Jourdan, J.-P. Hamaide, and S. Bigo. In *Proceedings of OSA Trends in Optics and Photonics (TOPS) Vol. 70, Optical Fiber Communication Conference*, Technical Digest, Postconference Edition, Washington DC: Optical Society of America, FC5-1, 2002.

[37] A. H. Gnauck, G. Raybon, S. Chandrasekhar, J. Leuthold, C. Doerr, L. Stulz, A. Agarwal, S. Banerjee, D. Grosz, S. Hunsche, A. Kung, A. Marhelyuk, D. Maywar, M. Movassaghi, X. Liu, C. Xu, X. Wei, and D. M. Gill. In *Proceedings of OSA Trends in Optics and Photonics (TOPS) Vol. 70, Optical Fiber Communication Conference*, Technical Digest, Postconference Edition, Washington DC: Optical Society of America, FC2-1, 2002.

[38] H. Sugahara, K. Fukuchi, A. Tanaka, Y. Inada, and T. Ono. In *Proceedings of OSA Trends in Optics and Photonics (TOPS) Vol. 70, Optical Fiber Communication Conference*, Technical Digest, Postconference Edition, Washington DC: Optical Society of America, FC6-1, 2002.

[39] F. Liu, J. Bennike, S. Dey, C. Rasmussen, B. Mikkelsen, P. Mamyshev, D. Gapontsev, and I. Ivshin. In *Proceedings of OSA Trends in Optics and Photonics (TOPS) Vol. 70, Optical Fiber Communication Conference*, Technical Digest, Postconference Edition, Washington DC: Optical Society of America, FC7-1, 2002.

[40] B. Zhu, L. Leng, L. E. Nelson, L. Gruner-Nielsen, Y. Qian, J. Bromage, S. Stulz, S. Kado, Y. Emori, S. Namiki, P. Gaarde, A. Judy, B. Palsdottir, and R. L. Lingle, Jr. In *Proceedings of OSA Trends in Optics and Photonics (TOPS) Vol. 70, Optical Fiber Communication Conference*, Technical Digest, Postconference Edition, Washington DC: Optical Society of America, FC8-1, 2002.

Chapter 15

Multiple Path Interference and Its Impact on System Design

J. Bromage, P.J. Winzer, and R.-J. Essiambre

15.1. Introduction

Lightwave communication systems carry information that is encoded onto the intensity, phase, or polarization of light from one point to another along an optical path. When designing such systems, many mechanisms that degrade the transfer of information must be taken into account. Until the late 1990s, the main causes of signal degradation in transmission were *fiber nonlinearity* and *amplified spontaneous emission* (ASE) from optical amplifiers. More recently, however, a third type of system degradation, involving the unwanted beating of the signal with a number of weak *interferers*, has become increasingly important. With reference to Fig. 15.1(a), such interferers can result from imperfect extinction of the drop-signal in optical cross-connects and add-drop multiplexers, which are both key elements for flexible and transparent optical network architectures [1, 2]. Also, single-Rayleigh backscattering in bidirectional transmission systems [3, 4] can lead to unwanted interferers at the receiver. Although these two examples involve interferers that are independent of the main signal, the important class of *multiple-path interference* (MPI) involves interferers that are delayed replicas[1] of the main signal. In the case of MPI, additional (unwanted) optical paths, with losses orders of magnitude greater than the main path, lead to interfering signals at the receiver, and can have a significant impact on system performance. With reference to Figs. 15.1(b) and 15.1(c), MPI is encountered for

- discrete reflections within or surrounding optical amplifiers [5],
- double-Rayleigh scattering in optical amplifiers [6, 7],
- double-Rayleigh scattering in the transmission span [8], or
- unwanted transverse mode mixing in higher-order mode dispersion compensators [9].

---

[1] Note that for path-dependent signal distortion, as found in highly nonlinear or dispersive systems, MPI may lead to interference between fields that have significantly different spectra.

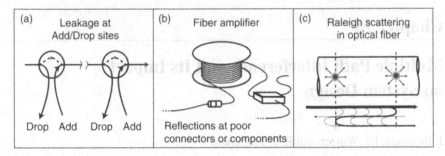

Fig. 15.1. Schematic examples of possible sources of inband crosstalk: (a) discrete reflection sites at connectors or components on either side of an optical amplifier; (b) double-Rayleigh scattering within a fiber; (c) poor extinction of add-drop multiplexers.

MPI has become increasingly relevant with the advent of optical amplifiers, in particular, of distributed Raman amplification [10–12]. Lightwave systems incorporating optical amplifiers can transmit information much farther before electronic regeneration is necessary. Such increased network transparency, therefore, places stricter requirements on system components, or the accumulated MPI may reach unacceptable levels [13, 7, 14]. Furthermore, optical amplification can exacerbate MPI by providing gain for paths that would otherwise have too much attenuation to be significant [5, 15]. The classic example is that of two weak partial reflectors, spaced by some distance along a fiber link. If an amplifier is placed between the reflectors, the double-pass gain may be enough to offset the weak reflections, increasing the MPI. (In the most extreme case, the fiber may even lase.) Even if care is taken to eliminate discrete reflections, unavoidable Rayleigh backscatter in sufficiently long fibers may cause similar effects [8, 16, 17].

15.1.1. Inband Crosstalk and MPI

Inband crosstalk occurs in general when a signal beats with other interfering fields, producing beat frequency components that fall within the receiver's bandwidth and add to other sources of receiver noise. This process is visualized in Fig. 15.2. The beating between signal (black) and interferer (gray) at difference frequency δf falls within the receiver's electrical bandwidth B_e, thus producing inband crosstalk. Note that larger beat frequencies (e.g., Δf) are also produced. These, however, exceed B_e and thus do not lead to inband crosstalk, even though the two interfering fields occupy the same optical frequency band. This latter process is referred to as *out-of-band crosstalk*. As discussed along with Fig. 15.1, inband crosstalk can either be caused by interferers originating from separate transmitters, or by MPI, where the signal beats with delayed replicas of itself that have reached the receiver by one or more additional paths.

The negative impact of MPI on communication systems can also be understood as a consequence of the conversion of phase fluctuations to intensity fluctuations

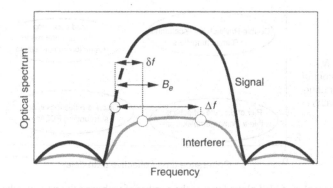

Fig. 15.2. The beating between signal (black) and interferer (gray) at difference frequency δf falls within the receiver's electrical bandwidth B_e, thus producing inband crosstalk. Beating at Δf exceeds B_e and is thus called out-of-band crosstalk.

by interference. As an illustration, consider continuous-wave (CW) light passing through a simple two-path interferometer. The intensity of light at an output port depends critically on the phase relationship between the electric fields from each path, set by the optical path difference. If the source of light is purely monochromatic and the optical path difference is static, this phase relationship is fixed. Therefore, the interferometer output has constant intensity, fixed between a maximum value (constructive interference) and a minimum value (destructive interference). A more realistic case, however, is when the CW source emits light with a finite linewidth, often the result of time-dependent random phase fluctuations [18]. Now the phase *difference* between the interfering fields also fluctuates in time, producing intensity fluctuations. In other words, the interferometer converts phase noise to intensity noise [19, 20]. This intensity noise can mask information-carrying modulation to the point where information is lost.

15.1.2. Classification of MPI

Sources of MPI can be classified using two parameters. The first parameter is N, the number of paths between the transmitter and receiver. N can vary from two (e.g., Mach–Zehnder-like interferometers) to infinity (e.g., double-Rayleigh backscattering distributed along a fiber amplifier). The other parameter describes the mutual coherence[2] of the interfering fields. For MPI one can consider the ratio of the coherence time of the source to the path delay. If many paths exist, as in the case of double-Rayleigh backscattering (DRB), there will not be a clearly defined delay time. Instead one can consider an effective delay time of the dominant paths that produce the maximum beat noise.

---

[2] Mutual coherence [21] describes the amount by which two optical fields are deterministically related to each other. Two fields are mutually coherent if amplitude and phase of one field can be predicted by knowledge of amplitude and phase of the other field.

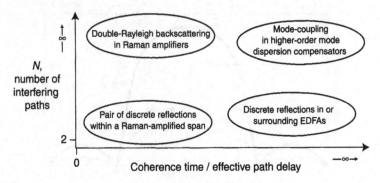

Fig. 15.3. Examples of MPI plotted versus the number of paths and the ratio of coherence time to path delay.

These two parameters are useful, because two MPI-producing devices that have very different forms and functions can have similar MPI properties. If they have the same number of paths and mutual coherence properties for the interfering fields, the degradation caused for the same level of MPI will be identical. Examples of MPI for different numbers of paths and different coherence properties are shown in Fig. 15.3.

Although much of this chapter applies to all sources of inband crosstalk mentioned along with Fig. 15.1, we mainly concentrate on MPI produced inside Raman amplifiers by discrete reflections or by unavoidable Rayleigh backscattering in the fiber. As Raman amplification can be used in many different parts of lightwave systems, and in many different types of optical fibers, an understanding of this noise source is crucial when designing and characterizing Raman-amplified systems.

In this chapter we first present some of the main physical concepts that describe MPI and are used to quantify it. Then we discuss the impact the resulting intensity noise has on system performance and receiver design. Finally we show how one optimally designs Raman amplifiers to take into account MPI from double-Rayleigh backscattering.

15.2. Properties of MPI

In this section we examine some important properties of intensity noise resulting from multiple path interference. First we consider the intensity noise in the *frequency domain*. Starting with expressions for the interfering optical fields, we derive an expression for the spectrum of intensity noise for the case of two paths (see Section 15.2.1). Next, the common case of sources with Lorentzian lineshapes is treated in Section 15.2.1.2. In Section 15.2.1.3 these results are generalized to include multiple interfering fields.

In Section 15.2.2, we examine the intensity fluctuations in the *time domain*. Here we compare the probability density functions for the intensity noise from different numbers of paths. This analysis points out important differences between cases of MPI involving a few paths to cases where there are a large or infinite number of paths.

15.2.1. Intensity Noise Spectra

15.2.1.1. Two Interfering Fields

The objective of this section is to derive a general expression for the spectrum of the intensity noise created by interference between optical fields from two paths. The *noise power spectral density* is valuable for understanding how MPI leads to inband crosstalk as well as how levels of MPI can be measured (see Section 15.4). As a result of this analysis, we show that the spectral shape of the beat noise depends on the optical spectra and mutual coherence properties of the interfering optical fields. The mathematical treatment used for this simple case of two interfering fields also serves as an introduction to the analysis of the more complex case of double-Rayleigh-induced MPI (see Section 15.3.3), which basically uses the same approach.

First consider the following expression for the optical field of a laser at the signal frequency f_o,

$$\overrightarrow{\boldsymbol{E}}_s(t) = \mathrm{Re}[\overrightarrow{\boldsymbol{\varepsilon}}_s(t)e^{-j2\pi f_o t}], \tag{15.1}$$

where Re denotes the real part and $\overrightarrow{\boldsymbol{\varepsilon}}_s(t)$ is the stochastic complex amplitude vector given by

$$\overrightarrow{\boldsymbol{\varepsilon}}_s(t) = \overrightarrow{p}_s \boldsymbol{\varepsilon}_s(t) = \overrightarrow{p}_s \sqrt{I_s} e^{j\boldsymbol{\phi}(t)}. \tag{15.2}$$

The quantity $\sqrt{I_s}$ is the field amplitude, $\overrightarrow{p}_s$ is the field polarization vector, and $\boldsymbol{\phi}(t)$ is the random instantaneous phase of the field's complex envelope. Throughout this chapter we use bold print (e.g., $\boldsymbol{X}$) to indicate a random quantity. For simplicity we assume that the source has negligible amplitude noise. Therefore the linewidth of the source is only due to phase noise, which is typically the case for semiconductor lasers. The instantaneous optical intensity is $I_s = |\overrightarrow{\boldsymbol{\varepsilon}}_s(t)|^2$.

Consider the case of two-path interference, shown in Fig. 15.4. Added to the signal field $\overrightarrow{\boldsymbol{\varepsilon}}_s$ is the doubly reflected field $\overrightarrow{\boldsymbol{\varepsilon}}_r$, given by

$$\overrightarrow{\boldsymbol{\varepsilon}}_r(t) = \overrightarrow{p}_r \sqrt{R_1 R_2 \underset{\rightarrow}{G}_{12} \underset{\leftarrow}{G}_{12}} \boldsymbol{\varepsilon}_s(t - T_d)e^{j2\pi f_o T_d}, \tag{15.3}$$

where $\overrightarrow{p}_r$ is the polarization vector, T_d is the round-trip delay, and R_i is the intensity reflectivity at position z_i ($i = 1, 2$). The quantity $\underset{\rightarrow}{G}_{12}$ is the intensity gain for light

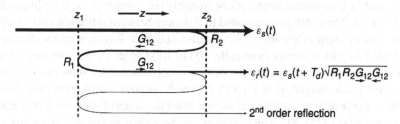

Fig. 15.4. MPI caused by two discrete reflections at z_1 and z_2. The variable $\boldsymbol{\varepsilon}_s$ is the output field amplitude from the straight-through path; $\boldsymbol{\varepsilon}_r$ is from the two-reflection path with round-trip delay T_d. R_1 and R_2 are the intensity reflectivities of the mirrors and $\underset{\rightarrow}{G}_{12} \underset{\leftarrow}{G}_{12}$ is the round-trip intensity gain between them. If $R_1 R_2 \underset{\rightarrow}{G}_{12} \underset{\leftarrow}{G}_{12} \ll 1$, higher-order paths may be neglected.

propagating in the $+z$-direction, from z_1 to z_2; $\underset{\leftarrow}{G}_{12}$ is the intensity gain for light propagating in the $-z$-direction, from z_2 to z_1. This notation allows for cases in which the gain is nonreciprocal, that is, for cases where the gain experienced by light passing between two points depends on the direction of propagation. An important case of a nonreciprocal gain is, for example, when an isolator is located between z_1 and z_2 [22, 23]. Here we assume that $R_1 R_2 \underset{\rightarrow}{G}_{12} \underset{\leftarrow}{G}_{12} \ll 1$ so that higher-order reflections can be neglected. This is a realistic assumption because, as we show in Section 15.5.2.3, significant penalties already result when this term is as small as 0.003 or -25 dB. The instantaneous intensity at the output is

$$I(t) = I_s + I_r + \Delta I(t, T_d), \tag{15.4}$$

where ΔI is the time-dependent intensity fluctuation given by

$$\Delta I(t, T_d) = (\vec{p}_s \cdot \vec{p}_r^{\,*})\sqrt{R_1 R_2 \underset{\rightarrow}{G}_{12} \underset{\leftarrow}{G}_{12}}\, \boldsymbol{\varepsilon}_s^*(t)\boldsymbol{\varepsilon}_s(t - T_d) + c.c. \tag{15.5}$$

(The common notation $c.c.$ stands for the complex conjugate of the expression preceding it.) Here we treat the worst case where the polarization states of $\vec{\boldsymbol{\varepsilon}}_s$ and $\vec{\boldsymbol{\varepsilon}}_r$ are identical ($\vec{p}_s \cdot \vec{p}_r^{\,*} = 1$), producing maximum intensity noise. To obtain an expression for the noise spectrum, we first calculate the autocorrelation function of the noise. In general, this is given by

$$\Gamma_{\boldsymbol{X}}(t, \tau) = \langle \boldsymbol{X}^*(t)\boldsymbol{X}(t + \tau)\rangle, \tag{15.6}$$

where $\langle \cdot \rangle$ denotes an ensemble average. For a wide-sense stationary process (WSS) [24], $\Gamma_{\boldsymbol{X}}$ only depends on τ and not on t. The autocorrelation function of the WSS intensity fluctuations is

$$\Gamma_{\Delta I}(\tau) = R_1 R_2 \underset{\rightarrow}{G}_{12} \underset{\leftarrow}{G}_{12} \langle [\boldsymbol{\varepsilon}_s^*(t)\boldsymbol{\varepsilon}_s(t - T_d) + c.c.][\boldsymbol{\varepsilon}_s^*(t + \tau)\boldsymbol{\varepsilon}_s(t - T_d + \tau) + c.c.]\rangle. \tag{15.7}$$

To proceed further, we need to evaluate the ensemble averages using statistical information about the phase fluctuations and therefore the optical spectrum of the laser light. However, if we consider the case where the delay T_d is much greater than the coherence time of the source, we can obtain a general expression for the noise spectral density that is valid for any lineshape. The coherence time of signal lasers used in fiber telecommunications (roughly the inverse of the laser linewidth) is 0.1 to 1 μs. These delays correspond to distances between reflection sites of 10 to 100 m. (As we show in Section 15.3.2, significant double-Rayleigh backscattering in distributed amplifiers comes from paths 100 to 1000 longer than this.) In the limit of large delay, $\boldsymbol{\varepsilon}_s(t)$ and $\boldsymbol{\varepsilon}_s(t - T_d)$ are uncorrelated. This allows us to reduce the intensity autocorrelation in Eq. (15.7), which contains fourth-order moments of the form $\langle \boldsymbol{\varepsilon}_1 \boldsymbol{\varepsilon}_2^* \boldsymbol{\varepsilon}_3 \boldsymbol{\varepsilon}_4^* \rangle$, to a product of second-order moments of the form $\langle \boldsymbol{\varepsilon}_1 \boldsymbol{\varepsilon}_2^* \rangle \langle \boldsymbol{\varepsilon}_3 \boldsymbol{\varepsilon}_4^* \rangle$ and $\langle \boldsymbol{\varepsilon}_1 \boldsymbol{\varepsilon}_2 \rangle \langle \boldsymbol{\varepsilon}_3^* \boldsymbol{\varepsilon}_4^* \rangle$ [25]. The latter product can be shown to vanish.[3] If we use our

---

[3] This can be seen by writing $\phi(t) + \phi(t + \tau) = 2\phi(t) + (\phi(t + \tau) - \phi(t)) = 2\phi(t) + \Delta\phi(t, \tau)$. Because for a random walk process, $\phi(t)$ and $\Delta\phi(t, \tau)$ are statistically independent, we are left with $\langle e^{j2\phi(t)}\rangle\langle e^{j\Delta\phi(t,\tau)}\rangle$. The first term $\langle e^{j2\phi(t)}\rangle = 0$ as the phase at time t is a random variable that is uniformly distributed between 0 and 2π.

assumption of wide-sense stationarity we obtain

$$\Gamma_{\Delta I}(\tau) = 2R_1 R_2 \underset{\rightarrow}{G}_{12} \underset{\leftarrow}{G}_{12} |\Gamma_{\boldsymbol{\varepsilon}_s}(\tau)|^2, \tag{15.8}$$

that is, the *intensity* autocorrelation expressed in terms of the *field* autocorrelation. According to the Wiener-Khintchine theorem, the power spectral density of a random process is derived from the Fourier transform of the autocorrelation [24]. Therefore we can express the intensity noise power spectral density as

$$
\begin{aligned}
S_{\Delta I}(f) &= 2\mathcal{F}\{\Gamma_{\Delta I}(\tau)\}, \\
&= 4R_1 R_2 \underset{\rightarrow}{G}_{12} \underset{\leftarrow}{G}_{12} \mathcal{F}\{|\Gamma_{\boldsymbol{\varepsilon}_s}(\tau)|^2\}, \\
&= 4R_1 R_2 \underset{\rightarrow}{G}_{12} \underset{\leftarrow}{G}_{12} [S_{\boldsymbol{\varepsilon}_s}(f) * S_{\boldsymbol{\varepsilon}_s}(f)], \tag{15.9}
\end{aligned}
$$

where $*$ denotes a convolution. This shows that the intensity noise spectrum depends on the autocorrelation of the field spectrum $S_{\boldsymbol{\varepsilon}_s}(f)$. ($\mathcal{F}$ denotes the Fourier transform.) Here $S_{\Delta I}(f)$ is the *single-sided* spectral density, which is nonzero for $f \in [0, \infty)$.[4] The self-homodyne technique for measuring laser linewidths [26] makes use of this relation between the spectrum of the intensity noise and the spectrum of the optical field, which is also a critical part of one of the MPI measurement approaches discussed in Section 15.4.

An intuitive explanation for this result goes as follows. Consider decomposing the two interfering fields $\boldsymbol{\varepsilon}_s(t)$ and $\boldsymbol{\varepsilon}_r(t)$ into quasimonochromatic spectral components. Although the two fields $\boldsymbol{\varepsilon}_s$ and $\boldsymbol{\varepsilon}_r$ are mutually incoherent, any two of their spectral components will have a high degree of mutual coherence as a consequence of their narrow bandwidth. Therefore interference between these components will produce a strong RF intensity modulation at their difference frequency. Such modulation is produced by all possible pairs of spectral components, which explains why the spectral width of the $S_{\Delta I}$ is closely tied to the spectral width of the source. The fact that the two beams are mutually incoherent means that all the RF components simply add on an intensity basis, resulting in the autocorrelation of the source spectrum shown in Eq. (15.9).

15.2.1.2. Two Fields with Lorentzian Lineshapes

Consider the case of a Lorentzian laser lineshape, typical for single-mode semiconductor lasers. The field spectrum is [26]

$$S_{\boldsymbol{\varepsilon}_s}(f - f_o) = \frac{2\overline{P}_0}{\pi \Delta f} \left(\frac{1}{1 + \left(\dfrac{f - f_o}{\Delta f/2}\right)^2} \right), \tag{15.10}$$

where $\overline{P}_0$ is the average optical power, f_o is the laser center frequency, and Δf is the full-width at half-maximum laser linewidth. From Eq. (15.9), we can calculate the

---

[4] This leads to an additional factor of two in Eq. (15.9) as the single-sided spectral density is a factor of two larger than the *double-sided* spectral density, defined as nonzero for $f \in (-\infty, \infty)$.

(single-sided) intensity noise spectrum. It is convenient to use a normalized quantity, the *relative intensity noise* (RIN), defined as[5]

$$\text{RIN}(f) = \frac{S_{\Delta I}(f)}{\overline{P}_0^2} = \frac{4R_1R_2 \underrightarrow{G}_{12} \underleftarrow{G}_{12}}{\pi \Delta f} \left(\frac{1}{1 + \left(\dfrac{f}{\Delta f} \right)^2} \right). \tag{15.11}$$

Because of the Lorentzian laser lineshape, the noise spectrum also has a Lorentzian lineshape[6] with a *half-width* at half-maximum linewidth of Δf. Furthermore, for this case note that the total noise power, calculated by integrating $S_{\Delta I}(f)$ over $f \in [0, \infty)$, is independent of Δf.

For a Lorentzian lineshape, closed-form expressions can be derived for arbitrary delays between the two paths [27], not only the incoherent case considered above where $T_d \gg 1/\Delta f$. In general, RIN is given by

$$\text{RIN}(f) = b \frac{4R_1R_2 \underrightarrow{G}_{12} \underleftarrow{G}_{12}}{\pi \Delta f} \left(\frac{1}{1 + \left(\dfrac{f}{\Delta f} \right)^2} \right)$$

$$\times [\sin^2(2\pi f_o T_d)(1 + e^{-4\pi \Delta f T_d} - 2e^{-2\pi \Delta f T_d} \cos(2\pi f T_d))$$

$$+ \cos^2(2\pi f_o T_d)(1 - e^{-4\pi \Delta f T_d} - 2e^{-2\pi \Delta f T_d} \frac{\Delta f}{f} \sin(2\pi f T_d))].$$

$$\tag{15.12}$$

In contrast to Eq. (15.11), we see the mean optical frequency of the source f_o appearing in the expression for the RIN. This is because when T_d is less than the coherence time, the noise power spectral density depends on the average phase difference between the mutually coherent fields. When T_d is such that $2\pi f_o T_d = (n + 1/2)\pi$ where n is any integer, RIN(f) integrated over $f \in [0, \infty)$ is maximum, and therefore there is the maximum amount of intensity noise. In this case, the interferometer is said to be "in-quadrature."

Figure 15.5 shows RIN spectra, calculated for a Lorentzian source with $\Delta f = 2$ MHz and three different values of $\Delta f T_d$. For all three cases the delay is such that the interferometer is in-quadrature, $R_1 R_2 \underrightarrow{G}_{12} \underleftarrow{G}_{12} = -40$ dB, and the fields are partially mutually coherent. However, there are clear differences between the spectra; for both $\Delta f T_d = 0.2$ and 0.02 we can see well-defined minima, located at harmonics of $1/T_d$. This occurs when the delay between the two paths is an integer multiple of the period of phase noise components. Then light from the two paths has phase fluctuations at these frequencies that are in-phase and so the phase *difference* between the optical fields varies less, resulting in less intensity noise. On the other hand, for $\Delta f T_d = 2$, we cannot resolve the oscillations, and so the RIN spectrum appears to be identical to the case of complete mutual incoherence ($\Delta f T_d \to \infty$); that is, the

---

[5] Note that RIN is used here as a spectral density and *not* as the dimensionless quantity corresponding to the integral of RIN(f) over frequency which is also used in the literature.

[6] The convolution of two Lorentzians is itself a Lorentzian.

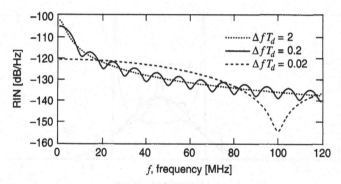

Fig. 15.5. Calculated RIN for different values of $\Delta f T_d$. The laser has a Lorentzian lineshape with a full-width half-maximum, $\Delta f = 2$ MHz.

RIN spectrum is an autocorrelation of the Lorentzian source spectrum. At the other extreme, the fields are completely mutually *coherent* if $\Delta f T_d = 0$, (i.e., the source is monochromatic) and therefore the $RIN(f) = 0 \text{ Hz}^{-1}$. In this chapter we are primarily concerned with MPI in Raman amplifiers for which the degree of mutual coherence is low, and so MPI between fields with a high degree of mutual coherence is not considered further.

15.2.1.3. Multiple Interfering Fields

The result for the RIN spectrum shown in Eq. (15.11) can be applied to multiple paths $(N > 2)$ if the path difference between any pair of paths is longer than the coherence length of the source [28]. Consider a total of M reflection sites, with any pair specified by reflectivity coefficients R_i and R_j and with gain G_{ij} between them. This leads to N paths where $N = M(M-1)/2 + 1$ (including the "straight-through" path). Generalizing Eq. (15.11) gives [28]

$$\text{RIN}(f) = \frac{4}{\pi} \left(\frac{1}{1 + \left(\dfrac{f}{\Delta f}\right)^2} \right) \sum_{i=2}^{N} \sum_{j=1}^{i-1} R_i R_j \, \overrightarrow{G}_{ij} \, \overleftarrow{G}_{ij}. \tag{15.13}$$

Note that for multiple incoherent paths, the frequency dependence of the intensity noise does not depend on the number of interfering paths.

15.2.2. Intensity Distribution for N-Path Interference

In this section we discuss the *time-domain* view of intensity fluctuations produced by MPI. In general, the intensity produced by N interfering fields is given as a function of time by

$$I(t) = \left| \overrightarrow{\varepsilon_s}(t) + \sum_{i=1}^{N-1} \overrightarrow{\varepsilon_i}(t) \right|^2, \tag{15.14}$$

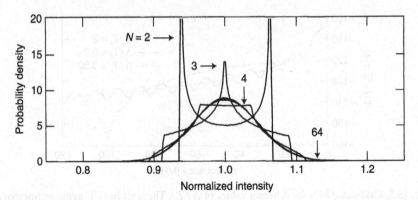

Fig. 15.6. Numerical simulations of the probability density functions for the normalized intensity due to incoherent MPI. The number of paths (N) is 2, 3, 4, 8, 16, 32, and 64. The crosstalk ratio (R_C) is -30 dB for all cases.

where $\vec{\varepsilon}_s(t)$ is the main component of the signal. Each delayed, weakly interfering component $\vec{\varepsilon}_i(t)$ results from one of $N-1$ paths. We now consider the worst case of aligned polarizations,

$$I(t) = \left| \sqrt{I_s}e^{j\phi_s(t)} + \sum_{i=1}^{N-1} \sqrt{I_i}e^{j\phi_i(t)} \right|^2, \qquad (15.15)$$

where I_i is the intensity and $\phi_i(t)$ is the time-dependent phase of light from the ith path. If the fields are all mutually incoherent, the difference between any pair of phases is uniformly distributed between 0 and 2π.

A statistical quantity relevant for predicting errors caused by fluctuations of $I(t)$ is the *probability density function* (PDF). Figure 15.6 shows the resulting intensity PDF for incoherent interference between N paths. Shown are PDFs resulting from different numbers of paths, ranging from $N=2$ to 64, which were calculated numerically from an ensemble of 10^9 realizations for each case. We can express the amount of interference in terms of a crosstalk ratio for the average intensities R_C where

$$R_C = \frac{\sum_{i=1}^{N-1} I_i}{I_s}. \qquad (15.16)$$

Thus R_C is the ratio of average intensities from all the delayed paths to the intensity from the main signal path. In Fig. 15.6, $R_C = -30$ dB for all cases. Also, for a given value of N, it was assumed that I_i was the same for each path, and so $R_C = (N-1)I_i/I_s$.

As can be seen in Fig. 15.6, for the case of $N=2$, the PDF has clearly defined boundaries [29, 1, 30]. These correspond to cases of complete destructive or constructive interference. For $R_C = -30$ dB, the maximum and minimum possible values for

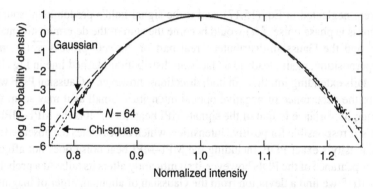

Fig. 15.7. Probability density function for 64 mutually incoherent paths for $R_C = -30$ dB, calculated using an ensemble of 10^9 realizations. Note the asymmetry of the PDF. Also shown is the Gaussian and chi-square PDF.

the total intensity are $(1 \pm \sqrt{0.001})^2 I_s$; values outside this range are not possible. However, as more paths are added, the PDF extends beyond these boundaries, even though R_C is still -30 dB. The PDF spreads because R_C depends on the total *average* intensity of all the mutually incoherent delayed paths, but if at some instant all these field components interfere constructively, the instantaneous intensity can be significantly larger. Simply, the average intensity of $N-1$ delayed paths with equal I_i is $\langle I \rangle = (N-1)I_i$, whereas the maximum instantaneous intensity, $I_{max} = (N-1)^2 I_i$, is larger than the average by a factor of $N-1$ as it includes all beat terms. Therefore the total intensity ranges over $I = I_s(1 \pm \sqrt{(N-1)R_c})^2$. So, although in the frequency domain the number of incoherent paths does not affect the MPI noise spectrum, in the time domain PDF one can clearly see a dependence on the number of interfering paths.

In the limit of large N, the PDF becomes Gaussian-like. Figure 15.7 shows the PDF for 64 interfering paths on a logarithmic scale along with a Gaussian fit. Note that unlike the Gaussian, the PDF is not symmetric about the mean intensity, but is skewed to higher intensities. The exact form of this skewed PDF is called *chi-square* (χ^2) *distribution* (shown dotted in Fig. 15.7), and is also encountered in systems degraded solely by amplified spontaneous emission (see, e.g., [31]), where the random ASE optical field takes the role of the interferers' fields. Intuitively, the skewness results from the fact that the optical intensity, by definition, is a strictly positive quantity (cf. Eq. (15.14)). In analogy to the two-path interference process (Eq. (15.4)), the total intensity $I(t)$ (Eq. (15.14)) can be written as

$$I(t) = |\vec{\varepsilon_s}(t)|^2 + |\vec{n}(t)|^2 + 2\text{Re}[\vec{\varepsilon_s}(t) \cdot \vec{n}(t)], \qquad (15.17)$$

where $\vec{n}(t)$ denotes the sum of all interferers' optical fields. The second and third contributions to Eq. (15.17) can, respectively, be associated with the beating of MPI with itself (MPI–MPI beat noise) and of the signal with MPI (signal–MPI beat noise). In the case of a large number of interferers, $\vec{n}(t)$ takes on circularly symmetric complex Gaussian (ccG) statistics due to the central limit theorem. If MPI–MPI beat noise were neglected in Eq. (15.17), and if the signal field's predominant source of randomness is phase noise, $I(t)$ would become the sum of the deterministic quantity $|\vec{\varepsilon}_s(t)|^2$ and the Gaussian-distributed[7] real part of the ccG process $\vec{\varepsilon}_s(t) \cdot \vec{n}(t)$. This superposition, again, leads to a Gaussian distribution (dashed line in Fig. 15.7). With its tails extending infinitely in both directions, however, a Gaussian PDF would allow for the occurrence of negative optical intensities. Small as it may seem when comparing its variance to that of the signal–MPI beat noise,[8] it is the MPI–MPI beat noise that is responsible for positive intensities, which necessitates the observed skew in the intensity's exact PDF. Including the MPI–MPI beat noise does not affect the overall appearance of the PDF, however, it significantly alters its tails. At a probability level of 10^{-6} we find a deviation from the Gaussian of about an order of magnitude. Approximating the exact PDF by a Gaussian may therefore yield slightly pessimistic bit error predictions [32, 30].

15.2.3. Section Summary

In this section we have presented descriptions of the intensity fluctuation caused by MPI-induced beat noise. First, in a frequency domain description, we calculated the noise power spectral density of the beat noise for the case of interference between two fields, showing that it can be related to the autocorrelation of the *field* spectrum if the fields are mutually incoherent. We presented results for the case of Lorentzian source optical spectra and showed how these results can be generalized for multiple interfering fields where the number of paths $N > 2$. In the time domain, we calculated the intensity probability density function, showing that its form depends on the number of interfering fields. For the case of many paths ($N \gtrsim 16$), the PDF looks Gaussian-like, but is more accurately represented by a chi-squared distribution.

15.3. Properties of Rayleigh Scattering

In this section we describe the nature of Rayleigh scattering, focusing on properties that are relevant for our treatment of MPI resulting from double-Rayleigh backscattering.

15.3.1. Dependence on Fiber Properties

Rayleigh scattering of light is elastic scattering caused by small-scale inhomogeneities of the refractive index of the media. In an ideal medium that is perfectly homogeneous, secondary radiation from material dipoles that are driven by the incident optical field cancels in all but the propagation direction of the incident field. Inhomogeneities

---

[7] Uniformly distributed phase noise on the signal field does not alter these statistics.

[8] The ratio of signal–MPI beat noise to MPI–MPI beat noise amounts to 2000 in our example, as can be evaluated with the help of Eqs. (15.68) and (15.69).

cause imperfect cancellation and so light is scattered in other directions. The scale over which inhomogeneities are correlated is much smaller than the wavelength of the light, leading to the well-known wavelength dependence of $1/\lambda^4$ [24].

Multiple path interference in fibers results from Rayleigh *backscattering*, where a fraction of the incident light is scattered in the opposite direction. The fraction of light that is backscattered depends on two main factors: the composition of the glass and the waveguide properties of the fiber. The amount of scattered optical power per unit length is simply the loss coefficient due to Rayleigh scattering α_R. For modern fibers this is usually the dominant loss mechanism, but other mechanisms such as OH-ion absorption, bend losses, and loss from draw-induced fluctuations of the core [33] may also be significant depending on the fiber type and wavelengths of interest.

The recapture fraction S determines the fraction of scattered optical power that is captured in a guided mode, propagating in the backscatter direction. To a first approximation, we can assume that light scatters equally from all points in a fiber, both radial and axial. Then for Gaussian transverse modes $[I(r) \sim \exp(-r^2/w^2)]$, the recapture fraction is [34]

$$S = \frac{3}{4n^2k^2w^2} = \frac{3\pi}{2n^2k^2A_{\text{eff}}}, \tag{15.18}$$

where $k = 2\pi/\lambda$. From this expression we see that the larger the effective area A_{eff}, the smaller the amount of captured backscattered light. A more thorough treatment, which takes into account the radial dependence of Rayleigh scattering and the exact mode profiles, can be done by integrating the overlap of an optical mode with the radial composition profile of a fiber [35–37]. This approach can be used, for example, where fiber index profiles are produced using dopants with significantly different Rayleigh scattering properties from the silica host, or for fibers with modes that are far from Gaussian.

The product of the Rayleigh scattering loss and recapture fraction $\alpha_R S$ represents the amount of Rayleigh backscattered power per unit length of the fiber, often referred to as the Rayleigh backscatter coefficient. This parameter is important for predicting the amount of DRB-induced MPI in fiber amplifiers and, in general, may be z-dependent (e.g., if multiple fiber types are used within an amplifier).

15.3.2. Rayleigh Scattered Power Within Fiber Amplifiers

In this section we examine the average optical powers of the single-Rayleigh backscattered (SRB) and double-Rayleigh backscattered light inside a fiber amplifier. We present differential equations that can be used to predict these optical powers at any point in the amplifier, given the gain, loss, and Rayleigh backscattering properties. Then, using these equations, we identify the most significant DRB paths for DRB-induced MPI in Raman amplifiers.

Consider a fiber amplifier extending from $z = 0$ to $z = L$. The net gain per unit length is $\underset{\rightarrow}{g}(z)$ for light propagating in the forward direction (increasing z) and $\underset{\leftarrow}{g}(z)$

for the backward direction (decreasing z). The Rayleigh backscattering coefficient per unit length is $\alpha_R S$. Here we consider the optical powers:

1. $\overline{P}_s(z)$, the average forward propagating signal,
2. $P_{SRB}(z)$, the single-Rayleigh backscattered light, and
3. $P_{DRB}(z)$, the double-Rayleigh backscattered light.

Both SRB and DRB are stationary stochastic processes and so the ensemble-averaged powers P_{SRB} and P_{DRB} are constant in time (see Section 15.3.3.2). These optical powers are governed by the following coupled differential equations.

$$d_z \overline{P}_s = \underset{\rightarrow}{g}(z)\overline{P}_s(z) \tag{15.19}$$

$$-d_z P_{SRB} = \underset{\leftarrow}{g}(z)P_{SRB}(z) + \alpha_R S \overline{P}_s(z) \tag{15.20}$$

$$d_z P_{DRB} = \underset{\rightarrow}{g}(z)P_{DRB}(z) + \alpha_R S P_{SRB}(z), \tag{15.21}$$

where d_z is the operator d/dz. Higher-order scattering may be ignored if the combination of gain and backscattering is sufficiently weak. This is a realistic assumption because significant DRB-induced beat noise occurs when the ratio of P_{DRB} to $\overline{P}_s$ is only -25 dB (see Section 15.5.2.3). The boundary values for the scattered fields are $P_{SRB}(L) = 0$, and $P_{DRB}(0) = 0$. By eliminating P_{SRB}, we can solve for P_{DRB} in terms of $\overline{P}_s$ at the output of the amplifier $z = L$. The ratio of these powers is

$$\frac{P_{DRB}(L)}{\overline{P}_s(L)} = \int_0^L dz_2 \int_0^{z_2} dz_1 (\alpha_R S)^2 \underset{\rightarrow}{G}(z_1, z_2) \underset{\leftarrow}{G}(z_1, z_2), \tag{15.22}$$

where $\underset{\rightarrow}{G}(z_1, z_2)$ is the intensity gain going forward from z_1 to z_2, and $\underset{\leftarrow}{G}(z_1, z_2)$ is the gain going backwards from z_2 to z_1. This notation allows Eq. (15.22) to be used for arbitrary nonreciprocal gain distributions, such as occur when isolators are positioned within the amplifier. For the amplifier described by Eqs. (15.19) to (15.21), $\underset{\rightarrow}{G}(z_1, z_2) = \exp\left(\int_{z_1}^{z_2} \underset{\rightarrow}{g}(z')dz'\right)$ and $\underset{\leftarrow}{G}(z_1, z_2) = \exp\left(-\int_{z_2}^{z_1} \underset{\leftarrow}{g}(z')dz'\right)$.

In Raman amplifiers, the gain distribution along the fiber depends on the pump distribution. Consider a simple amplifier formed from a single fiber so that $\alpha_R S$ does not depend on z. When this fiber is counter-pumped using one pump wavelength, the small-signal[9] pump power distribution is $P_p(z) = P_p(L) \exp[-\alpha_p(L-z)]$ where α_p is the loss at the pump wavelength. $\underset{\rightarrow}{g}(z)$ and $\underset{\leftarrow}{g}(z)$ are given by $C_R P_p(z) - \alpha_s$, where C_R is the Raman gain efficiency of the fiber $[(W \cdot km)^{-1}]$ and α_s is the signal loss coefficient $[km^{-1}]$. Hence we can write the round-trip gain term in the integrand of Eq. (15.22) as

$$\underset{\rightarrow}{G}(z_1, z_2)\underset{\leftarrow}{G}(z_1, z_2) = \exp[2C_R P_p(L) \exp(-\alpha_p L)(e^{\alpha_p z_2} - e^{\alpha_p z_1})/\alpha_p - \alpha_s(z_2 - z_1)]. \tag{15.23}$$

---

[9] "Small-signal" means the power of the signals is sufficiently weak that that pump power can be assumed to be undepleted by the signals.

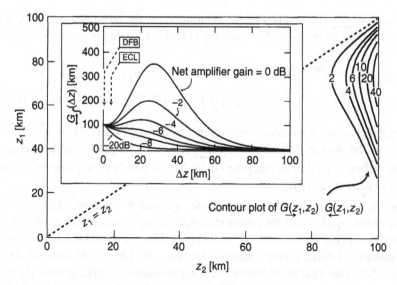

Fig. 15.8. Small-signal round-trip gain between reflection sites z_1 and z_2 in a 100-km span that is counter-pumped to transparency. Inset shows the round-trip gain integrated over z_1, and plotted as a function of $\Delta z = z_2 - z_1$ for net gains ranging from -20 dB (no Raman pumping) to 0 dB (pumped to transparency). Dashed arrows indicate Δz values that equal *half* the coherence lengths of two common sources.

This is the round-trip gain between Rayleigh backscatter sites at z_1 and z_2, where $z_1 \leq z_2$. By examining the round-trip gain for all possible pairs of z_1 and z_2 we can find the paths that are most important. This is shown in Fig. 15.8, which is a contour plot of $\underset{\rightarrow}{G}(z_1, z_2)\underset{\leftarrow}{G}(z_1, z_2)$ versus z_1 and z_2 for an amplifier that is pumped to transparency ($\underset{\rightarrow}{G}(0, L) = 0$ dB). Other parameters are $L = 100$ km, $\alpha_s = 0.2$ dB/km, $\alpha_p = 0.25$ dB/km, and $C_R P_p(L) = 0.265$ km^{-1}.

First note that the round-trip gain is only significant within the region of high pump power, roughly a distance of $1/\alpha_p$ (with α_p in km^{-1}) from the pump input at $z = L$. Secondly, the round-trip gain is not significant if $z_1 = z_2$, marked by the dashed line. The separations of z_1 and z_2 that give the largest contributions to Eq. (15.22) can be found using a new variable $\Delta z = z_2 - z_1$, which is the distance between Rayleigh backscatter sites. Then the integrated round-trip gain can be rewritten as

$$\int_0^L dz_2 \int_0^{z_2} dz_1 \underset{\rightarrow}{G}(z_1, z_2)\underset{\leftarrow}{G}(z_1, z_2)$$

$$= \int_0^L d(\Delta z) \int_0^{L-\Delta z} dz_1 \underset{\rightarrow}{G}(z_1, z_1 + \Delta z)\underset{\leftarrow}{G}(z_1, z_1 + \Delta z),$$

$$= \int_0^L d(\Delta z) \underset{\leftrightarrow}{G}_f(\Delta z). \tag{15.24}$$

The variable $\underset{\leftrightarrow}{G}_f(\Delta z)$ represents the contribution to the integrated round-trip gain from all paths *with the same* Δz. From Eq. (15.23), $\underset{\leftrightarrow}{G}_f(\Delta z)$ is given by

$$\underset{\leftrightarrow}{G}_f(\Delta z) = e^{2\alpha_s \Delta z} \big[Ei(2C_R P_p(L)(1 - e^{-\alpha_p \Delta z})/\alpha_p)$$

$$- Ei(2C_R P_p(L)e^{-\alpha_p L}(e^{\alpha_p \Delta z} - 1)/\alpha_p)\big]/\alpha_p, \quad (15.25)$$

where $Ei(z)$ is the exponential integral function defined by $Ei(z) = -\int_{-z}^{\infty} e^{-t}/t \, dt$ for $z > 0$. The inset in Fig. 15.8 shows $\underset{\leftrightarrow}{G}_f(\Delta z)$ for a range of amplifier net gains corresponding to no Raman pumping ($\underset{\rightarrow}{G}(0, L) = -20$ dB) up to pumping to transparency ($\underset{\rightarrow}{G}(0, L) = 0$ dB).

In Fig. 15.8, the dashed arrows indicate values of Δz corresponding to round-trip distances equal to half the coherence length of two types of sources: distributed feedback lasers (DFB) and external-cavity lasers (ECL). Typical DFB linewidths are approximately 1 MHz, giving coherence lengths in the fiber on the order of 200 m, whereas ECLs have narrower linewidths of approximately 25 kHz, giving coherence lengths on the order of 8 km. For ECL sources and no Raman pumping, 60% of the contribution to P_{DRB} comes from round-trip distances that are greater than the source coherence length. When the span is pumped to transparency, this fraction increases to 96%. For DFB sources, the shorter coherence length means that more than 98% of the contribution to P_{DRB} comes from round-trip distances greater than the coherence length, even without Raman pumping. Therefore, for most relevant situations, the signal and DRB light are *mutually incoherent* when they interfere to produce intensity noise.

15.3.3. Spectral Properties of Rayleigh Backscattered Light

The spectral properties of DRB light play an important role, both in the quantification of DRB-induced MPI and in assessing its impact on system performance. In this section, we present results showing that the intensity noise spectrum produced by DRB-induced MPI can be identical to that of incoherent two-path interference (treated in Section 15.2.1.1) [38, 39]. Therefore, when considering DRB-induced beat noise in the frequency domain, DRB light can be treated as having the same optical spectrum as the source.

Consider a fiber span from $z = 0$ to L as shown in Fig. 15.9. In addition to light propagating straight through the fiber (with optical field $\overrightarrow{\mathbf{\varepsilon}}_s(L, t)$) there will be light at the output from double-Rayleigh scattering within the span. To obtain an expression for the optical field of the DRB light we first note that the field at z_1 that is single-Rayleigh backscattered from an element ΔL at z_2 is given by

$$\Delta \overrightarrow{\mathbf{\varepsilon}}_{SRB}(z_1, z_2, t) = \mathbb{J}_{21} \, \Delta\boldsymbol{\rho}(z_2)\sqrt{\underset{\leftrightarrow}{G}(z_1, z_2)} \, \overrightarrow{\mathbf{\varepsilon}}_s(z_2, t - (z_2 - z_1)/v_g)e^{-j\beta(z_1 - z_2)},$$

$$(15.26)$$

where $\mathbb{J}_{21}$ is the Jones matrix describing the polarization evolution as a result of

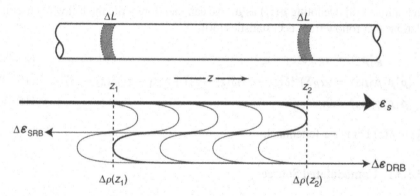

Fig. 15.9. Schematic showing DRB within a fiber amplifier. Highlighted is the path showing Rayleigh backscattering in a length ΔL at z_2 with (field) reflectivity $\Delta\boldsymbol{\rho}(z_1)$, followed by another similar scattering event at z_1.

propagation from z_2 to z_1. The *field* Rayleigh backscatter coefficient of a short section of fiber $\Delta\boldsymbol{\rho}(z_i)$ has circular complex Gaussian statistics (ccG) when ΔL is shorter than the source coherence length but longer than the correlation distance of refractive index fluctuations [38]. The variable v_g is the group velocity, $\underset{\leftarrow}{G}(z_1, z_2)$ is the intensity gain when propagating backwards from z_2 to z_1, and β is the propagation constant of the optical mode.

A contribution to the total DRB field from one particular path, shown in Fig. 15.9, comes from backscattering in two short sections of fiber of length ΔL at positions z_1 and z_2. This contribution to the total field at $z = L$ is

$$\Delta\overrightarrow{\boldsymbol{\varepsilon}}_{\text{DRB}}(L, z_1, z_2, t) = \mathbb{J}_{12}\mathbb{J}_{21}\,\Delta\boldsymbol{\rho}(z_1)\Delta\boldsymbol{\rho}(z_2)\sqrt{\underset{\rightarrow}{G}(z_1, z_2)\underset{\leftarrow}{G}(z_1, z_2)}$$

$$\times \overrightarrow{\boldsymbol{\varepsilon}}_s(L, t - 2(z_2 - z_1)/v_g)e^{j2\beta(z_2 - z_1)}, \tag{15.27}$$

where $\underset{\rightarrow}{G}(z_1, z_2)$ is the intensity gain when propagating forwards from z_1 to z_2. The round-trip delay is $2(z_2 - z_1)/v_g$. This treatment neglects spectral distortion of both the signal and DRB fields on propagating through the fiber. The total double-Rayleigh field is given by integrating over all possible z_1 and z_2 as

$$\overrightarrow{\boldsymbol{\varepsilon}}_{\text{DRB}}(L, t) = \int_0^L dz_2 \int_0^{z_2} dz_1 \mathbb{J}_{12}\mathbb{J}_{21}\boldsymbol{\rho}(z_1)\boldsymbol{\rho}(z_2)\sqrt{\underset{\rightarrow}{G}(z_1, z_2)\underset{\leftarrow}{G}(z_1, z_2)}$$

$$\times \overrightarrow{\boldsymbol{\varepsilon}}_s(L, t - 2(z_2 - z_1)/v_g)e^{j2\beta(z_2 - z_1)}. \tag{15.28}$$

Here $\Delta\boldsymbol{\rho}(z_i)$ has been replaced by a differential Rayleigh backscatter coefficient $\boldsymbol{\rho}(z_i) = \lim_{\Delta L \to 0}[\Delta\boldsymbol{\rho}(z_i)/\Delta L]$. We can continue to assume ccG statistics for this differential coefficient because scales over which there are changes in field amplitude and polarization are much larger than the correlation distance of refractive index fluctuations [38]. Denoting $\boldsymbol{\rho}(z_i)$ as $\boldsymbol{\rho}_i$, we can therefore write the following useful-relations that follow from ccG statistics [40].

$$\langle \boldsymbol{\rho}_1^* \boldsymbol{\rho}_2 \rangle = \alpha_R S \, \delta(z_1 - z_2) \tag{15.29}$$

$$\langle \boldsymbol{\rho}_1^* \boldsymbol{\rho}_2^* \boldsymbol{\rho}_3 \boldsymbol{\rho}_4 \rangle = (\alpha_R S)^2 [\delta(z_1 - z_3)\delta(z_2 - z_4) + \delta(z_1 - z_4)\delta(z_2 - z_3)] \tag{15.30}$$

$$\langle \boldsymbol{\rho}_1 \boldsymbol{\rho}_2 \boldsymbol{\rho}_3 \boldsymbol{\rho}_4 \rangle = 0, \tag{15.31}$$

where $\delta(z)$ Dirac's δ-functional.

15.3.3.1. Unmodulated Source

As an intermediate step in obtaining the spectrum of DRB-induced beat noise, we first calculate the autocorrelation of intensity fluctuations $\Gamma_{\Delta I}(\tau)$, following the approach used in Section 15.2.1. We can obtain a closed-form expression assuming that the DRB field is completely polarized with the same state of polarization as the signal[10] and the source field to be WSS with a correlation length much shorter than paths that give significant contributions[11] to $\vec{\boldsymbol{\varepsilon}}_r(L, t)$. As shown in Appendix A15.A, we get

$$\Gamma_{\Delta I}(\tau) = \int_0^L dz_2 \int_0^{z_2} dz_1 \, 2(\alpha_R S)^2 \, \underset{\rightarrow}{G}(z_1, z_2) \underset{\leftarrow}{G}(z_1, z_2) |\Gamma_{\boldsymbol{\varepsilon}_s}(\tau)|^2 \tag{15.32}$$

for the autocorrelation of the intensity fluctuations ΔI. Note once again that $\Gamma_{\Delta I}(\tau)$ can be expressed in terms of the magnitude-squared field autocorrelation function $|\Gamma_{\boldsymbol{\varepsilon}_s}(\tau)|^2$. Fourier transforming Eq. (15.32) gives the (single-sided) noise power spectral density for DRB-induced MPI beat noise

$$S_{\Delta I}(f) = \int_0^L dz_2 \int_0^{z_2} dz_1 \, 4(\alpha_R S)^2 \, \underset{\rightarrow}{G}(z_1, z_2) \underset{\leftarrow}{G}(z_1, z_2) [S_{\boldsymbol{\varepsilon}_s}(f) * S_{\boldsymbol{\varepsilon}_s}(f)]. \tag{15.33}$$

Note the similarity between this equation and Eq. (15.9). This shows that DRB-induced beat noise has a noise power spectrum with the *same frequency dependence* as the self-homodyne spectrum of two-path incoherent interference.

As a simple example, consider the case of a Lorentzian source with linewidth Δf, propagating in a hypothetical distributed amplifier where gain is distributed *uniformly* along the length of the fiber and is independent of direction of propagation (i.e., reciprocal). Thus $\underset{\rightarrow}{G}(z_1, z_2) = \underset{\leftarrow}{G}(z_1, z_2) = \exp[g(z_2 - z_1)]$, where g is a z-independent coefficient that specifies the gain per unit length. Then, if $\alpha_R S$ is also uniform, the RIN from double-Rayleigh MPI is

$$\text{RIN}(f) = \frac{4(\alpha_R S)^2}{\pi \Delta f} \left(\frac{e^{2gL} - 2gL - 1}{4g^2} \right) \frac{1}{1 + (\frac{f}{\Delta f})^2}. \tag{15.34}$$

---

[10] This restriction is lifted in Section 15.3.4.

[11] This assumption is discussed in Section 15.3.2 along with Fig. 15.8.

In some situations, the optical spectrum of double-Rayleigh scattered light may be significantly different from the source spectrum. For example, this can occur when the single-backscattered light copropagates with a noisy pump laser. If the dispersion properties of the fiber are such that the walk-off between the pump and scattered signal light is small (as is the case when the fiber's zero dispersion wavelength falls directly between pump and signal), high-bandwidth pump noise amplitude-modulates the single-scattered signal light, and can significantly broaden its spectrum. For certain systems this can be beneficial as it reduces the spectral overlap between the signal and double-Rayleigh scattered light [41] .

15.3.3.2. Modulated Source

To assess the impact of DRB on system performance in Section 15.5 we will have to know the autocorrelation of the DRB optical field $\boldsymbol{\varepsilon}_{\mathrm{DRB}}(t)$ for data-modulated optical signals,

$$\Gamma_{\boldsymbol{\varepsilon}_{\mathrm{DRB}}}(t, t+\tau) = \langle \boldsymbol{\varepsilon}_{\mathrm{DRB}}^*(t) \boldsymbol{\varepsilon}_{\mathrm{DRB}}(t+\tau) \rangle. \tag{15.35}$$

To derive an expression for $\Gamma_{\boldsymbol{\varepsilon}_{\mathrm{DRB}}}(t, t+\tau)$, we start by using the scalar version of Eq. (15.28), assuming that the state of polarization does not change during propagation ($\mathbb{J}_{12} = \mathbb{J}_{21} = \mathbb{I}$), and we neglect laser phase noise; that is, we assume a purely deterministic signal field,[12] $\boldsymbol{\varepsilon}_s(t) = \varepsilon_s(t)$. Proceeding similarly to the derivation of Eq. (15.32), we arrive at

$$\Gamma_{\boldsymbol{\varepsilon}_{\mathrm{DRB}}}(t, t+\tau) = (\alpha_R S)^2 \int_0^L dz_2 \int_0^{z_2} dz_1 \, \varepsilon_s^*(L, t - 2[z_2 - z_1]/v_g)$$

$$\times \varepsilon_s(L, t + \tau - 2[z_2 - z_1]/v_g)$$

$$\times \underset{\rightarrow}{G}(z_1, z_2) \underset{\leftarrow}{G}(z_1, z_2). \tag{15.36}$$

Substituting $\xi = 2(z_2 - z_1)/v_g$, we can rewrite this expression as

$$\Gamma_{\boldsymbol{\varepsilon}_{\mathrm{DRB}}}(t, t+\tau) = (\alpha_R S)^2 \frac{v_g}{2} \int_0^L dz_2 \int_0^{2z_2/v_g} d\xi \varepsilon_s^*(L, t - \xi) \varepsilon_s(L, t + \tau - \xi)$$

$$\times \underset{\rightarrow}{G}(z_2 - v_g \xi/2, z_2) \underset{\leftarrow}{G}(z_2 - v_g \xi/2, z_2). \tag{15.37}$$

The ξ-integration can then be interpreted [39] as the convolution of the functions $x(t; \tau) = \varepsilon_s^*(L, t) \varepsilon_s(L, t+\tau)$ and $y(t; z_2) = \underset{\rightarrow}{G}(z_2 - v_g t/2, z_2) \underset{\leftarrow}{G}(z_2 - v_g t/2, z_2)$ rect$(t, [0, 2z_2/v_g])$, with τ and z_2 acting as parameters, and the windowing function defined as rect$(t, [0, 2z_2/v_g]) = 1$ in $[0, 2z_2/v_g]$ and 0 elsewhere. For double-Rayleigh backscatter in optical fibers as the source of MPI, the function $y(t, z_2)$ has low-pass characteristics, with cut-off frequencies much lower than those of typical

---

[12] Performing the more complicated derivations with phase noise included, it turns out that the result is unaffected by phase noise for realistic data modulation bandwidths that exceed the laser linewidth.

data signals.[13] Therefore the convolution becomes time-independent and can be replaced by the product of the time average of $x(t; \tau)$ and the integral over $y(t; z_2)$,

$$
x(t; \tau) * y(t; z_2) \approx \underbrace{\lim_{T \to \infty} \frac{1}{T} \int_{-T/2}^{T/2} \varepsilon_s^*(L, t) \varepsilon_s(L, t + \tau) dt}_{\Gamma_{\varepsilon_s}(\tau)}
$$

$$
\times \int_0^{2z_2/v_g} \underset{\rightarrow}{G}(z_2 - v_g t/2, z_2) \underset{\leftarrow}{G}(z_2 - v_g t/2, z_2) dt. \quad (15.38)
$$

The first term on the right-hand side of this equation is identified as the temporal autocorrelation $\Gamma_{\varepsilon_s}(\tau)$ of the deterministic signal field, and the DRB field's autocorrelation becomes

$$
\Gamma_{\mathbf{s}_{\mathrm{DRB}}}(t, t + \tau) = \Gamma_{\mathbf{s}_{\mathrm{DRB}}}(\tau) =
$$

$$
\Gamma_{\varepsilon_s}(\tau) \cdot (\alpha_R S)^2 \frac{v_g}{2} \int_0^L dz_2 \int_0^{2z_2/v_g} d\xi \, \underset{\rightarrow}{G}(z_2 - v_g \xi/2, z_2) \underset{\leftarrow}{G}(z_2 - v_g \xi/2, z_2).
$$

$$
(15.39)
$$

Several interesting properties of DRB can be seen from Eq. (15.39). First, because the DRB autocorrelation depends on the time difference τ only, the stochastic process is (at least wide-sense) stationary [42], with power

$$
P_{\mathrm{DRB}} = \Gamma_{\mathbf{s}_{\mathrm{DRB}}}(0) =
$$

$$
= \overline{P}_s (\alpha_R S)^2 \frac{v_g}{2} \int_0^L dz_2 \int_0^{2z_2/v_g} d\xi \, \underset{\rightarrow}{G}(z_2 - v_g \xi/2, z_2) \underset{\leftarrow}{G}(z_2 - v_g \xi/2, z_2),
$$

$$
(15.40)
$$

where $\overline{P}_s = \Gamma_{\varepsilon_s}(0)$ is the average signal power.

Second, owing to the Fourier relationship between $\Gamma_{\varepsilon_s}(\tau)$ and the signal's power spectrum,

$$
\mathcal{F}\{\Gamma_{\varepsilon_s}(\tau)\} = \lim_{T \to \infty} \frac{1}{T} \int_{-T/2}^{T/2} \Gamma_{\varepsilon_s}(\tau) \exp[-j2\pi f\tau] d\tau = |\mathcal{E}_s(f)|^2, \quad (15.41)
$$

with

$$
\mathcal{F}\{\varepsilon_s(t)\} = \lim_{T \to \infty} \frac{1}{T} \int_{-T/2}^{T/2} \varepsilon_s(t) \exp[-j2\pi ft] dt = \mathcal{E}_s(f), \quad (15.42)
$$

---

[13] Assuming a typical exponential gain/loss profile $\underset{\rightarrow}{G}(z_1, z_2) = \underset{\leftarrow}{G}(z_1, z_2) = \exp[g|z_2 - z_1|]$ with $-2.3 \cdot 10^{-4} m^{-1} \lesssim g \lesssim 0.06 m^{-1}$, corresponding to the extreme cases of 1 dB/km fiber loss (lower limit) and pumping a high Raman efficiency fiber with 10 W of pump power (upper limit), we arrive at time constants on the order of 100 ns and above. Concerning the windowing function, a 10 ns wide window corresponds to the backscatter originating from only the last meter of optical fiber, which can be safely neglected in the scattering integral.

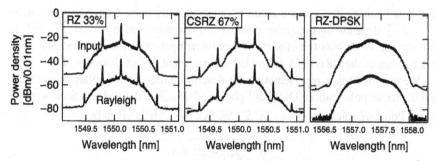

Fig. 15.10. Measured Rayleigh backscattered spectra (lower traces) and spectra entering the fiber (upper traces) for typical 40-Gbit/s amplitude and phase modulated optical data signals. Note that Rayleigh scattering preserves the full information content of the signals' spectral magnitudes. For formats with isolated spectral components, spontaneous Brillouin scattering lines are additionally observed in the Rayleigh scattered spectra to both sides of these discrete tones.

the magnitude of the DRB power spectrum can be obtained from the modulated signal's power spectrum. Rayleigh backscattering thus preserves the full information content of the data signal's spectral magnitude, independent of whether the signal is phase or amplitude modulated. To illustrate this point, Fig. 15.10 shows measurements of optical data spectra taken at 40 Gbit/s for on-off keying (OOK) signals (33% duty cycle return-to-zero (RZ) and 67% duty cycle carrier-suppressed return-to-zero (CSRZ) formats), and for phase-coded signals (33% duty cycle return-to-zero differential phase shift keying (RZ-DPSK) format). The upper traces give the signals' spectra at the fiber input, and the lower traces refer to the single-Rayleigh backscattered signals. Note that the spectral shapes are accurately preserved by Rayleigh scattering, irrespective of whether amplitude or phase is modulated. For formats that have isolated spectral tones (33% RZ and 67% CSRZ), it is also possible to see *spontaneous* Brillouin scattering lines positioned 10 GHz above and below the discrete spectral tones.[14]

Knowing that the DRB process is stationary with power spectral density $|\mathcal{E}_s(f)|^2$, it is also possible to generate realizations in the *frequency domain* for use in Monte Carlo simulations [43]. Thereby, the magnitudes of the Fourier components are weighted by $|\mathcal{E}_s(f)|^2$, and their phases are chosen to be independent[15] random variables, uniformly distributed between 0 and 2π.

15.3.4. State and Degree of Polarization

Thus far we have ignored the way the polarization state of light can change when propagating in an optical fiber. For simplicity, we have assumed that the Jones matrices affecting the DRB field ($\mathbb{J}_{12}$ and $\mathbb{J}_{21}$) are the identity matrix $\mathbb{I}$. However, we must

---

[14] The absence of *stimulated* Brillouin scattering (SBS) was verified by varying the optical power going into the fiber.

[15] The statistical independence of any two spectral components is a direct consequence of stationarity.

take polarization effects into account because intensity noise from beating between $\vec{\varepsilon_1}$ and $\vec{\varepsilon_2}$ depends on the scalar product of their polarization vectors $\vec{p_1} \cdot \vec{p_2}^*$. In this section we discuss the impact of fiber birefringence on the state and degree of polarization of the DRB field, following the treatment of van Deventer [44].

We use the Stokes representation for describing the state of polarization (SOP) and degree of polarization (DOP) of light [26]. For a SOP defined by a Stokes vector on the Poincaré sphere as $\vec{S} = (S_0, S_1, S_2, S_3)$, the DOP is given by

$$DOP = \frac{\sqrt{S_1^2 + S_2^2 + S_3^2}}{S_0}. \tag{15.43}$$

The component S_0 represents the average power, which we can use to normalize all components of $\vec{S}$ so that $S_0 = 1$. The evolution of the Stokes vector on the Poincaré sphere is determined by Müller matrices. Two types are needed here: the matrix M_t describes transmission through fiber and M_r describes a reflection. (Here we are ignoring polarization-dependent gain or loss.) These matrices are given by

$$M_t = \begin{pmatrix} 1 & 0 & 0 & 0 \\ 0 & m_1 & m_2 & m_3 \\ 0 & m_4 & m_5 & m_6 \\ 0 & m_7 & m_8 & m_9 \end{pmatrix} \quad \text{and} \quad M_r = \begin{pmatrix} 1 & 0 & 0 & 0 \\ 0 & 1 & 0 & 0 \\ 0 & 0 & 1 & 0 \\ 0 & 0 & 0 & -1 \end{pmatrix}. \tag{15.44}$$

The transmission matrix M_t is an orthogonal matrix ($M_t^T = M_t^{-1}$) with $\det(M_t) = 1$. The elements of the reflection matrix M_r show that for a (lossless) reflection, only the handedness of the state of polarization is changed; that is, $S_3 \rightarrow -S_3$. This point is key to understanding the state and degree of polarization of double-Rayleigh scattered light.

First consider the case of single-Rayleigh backscattered light. Light first propagates forward through the fiber, is reflected at some point, and then propagates backward through the same section of fiber. This path can be described by M_{SRB}, given by

$$M_{SRB} = M_t^T \cdot M_r \cdot M_t = \mathbb{I} - 2 \begin{pmatrix} 1 & 0 & 0 & 0 \\ 0 & m_7^2 & m_7 m_8 & m_7 m_9 \\ 0 & m_7 m_8 & m_8^2 & m_8 m_9 \\ 0 & m_7 m_9 & m_8 m_9 & m_9^2 \end{pmatrix}, \tag{15.45}$$

where $\mathbb{I}$ is the identity matrix. The total Rayleigh backscattered field is the sum of contributions from all points in the fiber. Significant contributions come from points within the effective length of the fiber, which for Rayleigh backscattering in passive fiber is $(1 - e^{-2\alpha L})/2\alpha$ [34]. If the coherence length of the source is much smaller than this, the Stokes parameters are additive and the sum over all contributions depends only on the average of the matrices from each reflection point.

In [44], results are given for fiber with varying degrees of birefringence: zero, low, and high. Here we present results for low birefringence fiber as this is the most relevant for telecommunications applications where polarization beat lengths are typically on the order of 10 m [45]. For fiber where the beat length is much less than the effective length, Müller matrices for transmission through sections of fiber are random orthogonal matrices [46]. From $M_t^T \cdot M_t = \mathbb{I}$ it follows that $m_1^2 + m_2^2 + m_3^2 = 1, m_4^2 + m_5^2 + m_6^2 = 1$, and $m_7^2 + m_8^2 + m_9^2 = 1$. As different matrix elements are uncorrelated and have the same variance, we have $\langle m_i^2 \rangle = 1/3$, and $\langle m_i m_j \rangle = 0$ for $i \neq j$. Therefore, for low-birefringence fiber,

$$\langle M_{SRB} \rangle = \begin{pmatrix} 1 & 0 & 0 & 0 \\ 0 & \frac{1}{3} & 0 & 0 \\ 0 & 0 & \frac{1}{3} & 0 \\ 0 & 0 & 0 & \frac{1}{3} \end{pmatrix}. \tag{15.46}$$

Here we see two important results. First, because the Müller matrix is diagonal, the SOP for backscattered light is the same as the input SOP. Second, the DOP is reduced by a factor of three; for completely polarized input light (DOP = 1), the DOP for the backscattered light is 1/3.

Brinkmeyer [47] gives the following explanation for the depolarizing mechanism. As signal light propagates down a low birefringence fiber, the SOP evolves, sampling all points on the Poincaré sphere equally. If the light is linearly polarized when reflected ($S_3 = 0$), the SOP is preserved, but if it is circularly polarized ($S_1 = S_2 = 0$), the handedness is changed from left- to right-handed or vice versa, to give the orthogonal polarization. (See the sign of the elements of M_r, Eq. (15.44).) It is this change in the polarization state that depolarizes the Rayleigh backscattered light. However, as the polarization vector is more likely to cross the equator of the Poincaré sphere (linearly polarized) than to arrive at the poles (circularly polarized), it is more likely that the original state of polarization is preserved on reflection. Therefore the Rayleigh backscattered light is not completely depolarized, but is partially polarized with the same SOP as the source.

Next we apply these results to the case of DRB. We assume again that both the source coherence length and the fiber beat length are much less than the effective length over which significant contributions originate (see Section 15.3.2). Then the average Müller matrix for DRB light is given by

$$\langle M_{DRB} \rangle = \begin{pmatrix} 1 & 0 & 0 & 0 \\ 0 & \frac{1}{9} & 0 & 0 \\ 0 & 0 & \frac{1}{9} & 0 \\ 0 & 0 & 0 & \frac{1}{9} \end{pmatrix}; \tag{15.47}$$

that is, the SOP is the same as the output signal light, and the DOP is 1/9.

This depolarizing effect means that there is less intensity noise from DRB-induced MPI than was derived by assuming no change in the polarization along the fiber. We can weight the previous results (Eqs. (15.32) to (15.34), (15.39), (15.40)) with a depolarizing factor to take this into account. If there were complete depolarization (DOP = 0), we could decompose the double-Rayleigh light into two uncorrelated halves, one with the same polarization as the signal and the other with the orthogonal polarization. In this case, the factor would be $1/2$. However, as the DOP equals $1/9$ and the polarized component has the same SOP as the signal light, the correct factor is $1/9$ plus half of the remaining $8/9$ that is depolarized. Thus the total factor, as shown in [44], is $(1/9) + (1/2) \cdot (8/9) = (5/9)$.

15.3.5. Section Summary

In this section we presented some of the properties of Rayleigh scattering in single-mode optical fibers that are relevant for DRB-induced MPI. We first discussed the fiber properties (such as composition and waveguide design) that dictate the amount of Rayleigh backscattered light. Next, we presented differential equations showing the coupling between the average signal power and the single- and double-Rayleigh backscattered power at any point in a fiber. Using these equations, we showed that the most significant paths for DRB-induced MPI in Raman amplifiers have a length of several kilometers, much longer than a typical source coherence length. Next we analyzed the spectral properties of Rayleigh backscattered light, with and without signal modulation, showing that the spectrum of the intensity fluctuations due to DRB is closely related to the source spectrum, analogous to the case of self-homodyning. Finally, we presented a polarization analysis that shows how DRB light is depolarized if the fiber is birefringent.

15.4. Measurement of MPI Levels

In this section we describe ways to quantify the level of multiple path interference. Then we discuss two techniques that are used to measure the level of MPI in fiber amplifiers.

15.4.1. Definition of Optical Signal-to-Noise Ratio and Crosstalk Ratio for MPI

Two types of quantities are used to describe the level of MPI. One type is the optical signal-to-noise ratio due to MPI ($OSNR_{MPI}$) which is the ratio of average signal power $\overline{P}_s$ to the average MPI power P_{MPI} (including all polarizations),

$$OSNR_{MPI} = \frac{\overline{P}_s}{P_{MPI}}. \tag{15.48}$$

This is similar to the commonly used definition of OSNR due to ASE,

$$OSNR_{ASE} = \frac{\overline{P}_s}{2N_{ASE}B_{ref}}, \tag{15.49}$$

where N_{ASE} is the ASE power spectral density per (polarization) mode, and B_{ref} is the reference optical bandwidth, typically 12.5 GHz. The other type is a crosstalk ratio R_C, which is simply the inverse of $OSNR_{MPI}$,

$$R_C = \frac{P_{MPI}}{\overline{P}_s} = \frac{1}{OSNR_{MPI}}. \tag{15.50}$$

The crosstalk ratio is predominantly used when describing the performance of a single component, whereas $OSNR_{MPI}$ is generally favored when considering the cumulative effect of many components or amplifiers in a system [14].

Care must be taken when using these quantities to state assumptions for the SOP and DOP for the MPI light. Here we distinguish between definitions that use the total MPI power in *all* polarizations, and definitions that only use MPI power with the *same* polarization as the signal. For the latter we add a "P" superscript. For example, $OSNR_{MPI}^P$ and R_C^P are the ratio of the signal power to the MPI power with the same polarization as the signal P_{MPI}^P,

$$OSNR_{MPI}^P = \frac{1}{R_C^P} = \frac{\overline{P}_s}{P_{MPI}^P}. \tag{15.51}$$

As MPI only results in intensity noise from beating with the signal when there is some degree of polarization overlap, $OSNR_{MPI}^P$ and R_C^P are the more relevant quantities. To highlight the symmetry between ASE and MPI-induced beat noises, we also define $OSNR_{ASE}^P$ as

$$OSNR_{ASE}^P = \frac{\overline{P}_s}{N_{ASE}B_{ref}}. \tag{15.52}$$

Example 15.1 (Two Discrete Reflectors). Consider the case of gain between two discrete reflectors, R_1 and R_2, shown in Fig. 15.4. For sufficiently weak reflections, $R_1 R_2 \underset{\rightarrow}{G}_{12} \underset{\leftarrow}{G}_{12} \ll 1$, so we only need to consider the path that double-passes the gain medium. The crosstalk ratio R_C for this device is then simply given by $R_C = R_1 R_2 \underset{\rightarrow}{G}_{12} \underset{\leftarrow}{G}_{12}$. On the other hand, the polarization-dependent crosstalk ratio R_C^P depends on the polarization overlap between the main signal and the delayed component. If they are both completely polarized with the same SOP, $R_C^P = R_C$, but if they are orthogonal, $R_C^P = 0$.

Example 15.2 (Double-Rayleigh Backscatter). Another useful example is the case of distributed reflections from DRB throughout a fiber amplifier. Consider the amplifier described in Section 15.3.2, which has a length L, and gain and Rayleigh backscattering per unit length at position z are given by $g(z)$ and $\alpha_R S$, respectively. From Eqs. (15.22) and (15.50) we get

$$R_C = \frac{P_{DRB}(L)}{\overline{P}_s(L)} = \int_0^L dz_2 \int_0^{z_2} dz_1 (\alpha_R S)^2 \underset{\rightarrow}{G}(z_1, z_2) \underset{\leftarrow}{G}(z_1, z_2). \tag{15.53}$$

A couple of useful analytic results can be obtained from Eq. (15.53). For the simple case of uniform gain and backscattering per unit length, $\alpha_R S$ and g are z-independent constants. Then R_C is given by

$$R_C = (\alpha_R S)^2 \left(\frac{e^{2gL} - 2gL - 1}{4g^2} \right). \tag{15.54}$$

Note that this result appears as a factor in Eq. (15.34) for the RIN from such an amplifier. This connection between RIN and R_C (or OSNR$_{MPI}$) forms the basis of the electrical beat-noise measurement technique, described in Section 15.4.2. Using the results of Section 15.3.4, we can take into account depolarization effects in the fiber, and write

$$R_C^P = \frac{5}{9} R_C = \frac{5}{9} (\alpha_R S)^2 \left(\frac{e^{2gL} - 2gL - 1}{4g^2} \right). \tag{15.55}$$

In Raman amplifiers, the gain distribution along the fiber depends on the pump distribution. For an amplifier pumped with a single wavelength, $g(z)$ is given by $C_R P_p(z) - \alpha_s$, where C_R is the Raman gain efficiency of the fiber [(W · km)$^{-1}$], $P_p(z)$ is the pump power distribution, and α_s is the signal loss coefficient [km^{-1}]. If we assume the pump and signal loss are the same ($\alpha_p = \alpha_s = \alpha$), in the small-signal limit Eq. (15.53) can be solved, giving

$$R_C^P = \frac{5(\alpha_R S)^2}{72 (\alpha\theta)^2} e^{-(\alpha L - 2\theta e^{-\alpha L})} \left[4\theta^2 (e^{\alpha L} + 2\theta) \{ Ei(2\theta) - Ei(2\theta e^{-\alpha L}) \} \right.$$
$$\left. + (1 + 2\theta)(e^{\alpha L + 2\theta e^{-\alpha L}} - e^{\alpha L + 2\theta} - 2\theta e^{2\theta} + 2\theta e^{\alpha L + 2\theta e^{-\alpha L}}) \right], \tag{15.56}$$

where we used a dimensionless parameter $\theta = C_R P_p(L)/\alpha$ to simplify notation, and $Ei(z)$ is the exponential integral function defined by $Ei(z) = - \int_{-z}^{\infty} e^{-t}/t \, dt$ for $z > 0$.

15.4.2. Electrical Beat-Noise Measurement Technique

In this section we describe a technique to measure the amount of DRB-induced MPI produced by fiber amplifiers. The goal is to measure R_C (or equivalently, its inverse, OSNR$_{MPI}$) at the output of the amplifier. Unfortunately, one cannot simply use an optical spectrum analyzer to measure OSNR$_{MPI}$ as is routinely done when measuring the ratio of signal light to ASE. This is because the DRB light responsible for MPI-induced beat noise has the same spectrum and state of polarization as the signal light (see Sections 15.3.3 and 15.3.4) and so the DRB light cannot be distinguished from the signal.[16] A different approach must be used.

This section presents an overview of an electrical beat-noise measurement technique [48, 43, 49], so called because this approach measures the level of MPI by quantifying beat noise added to the electrical output of a photodetector. From this, one can infer the power ratio of the signal light to double-Rayleigh light. Because no interference noise is produced by the polarization component of double-Rayleigh light that is orthogonal to the signal polarization state, this technique measures the ratio of signal light to DRB light *with the same polarization state*. Therefore, using the notation of the previous section, the quantity that is measured by this technique is R_C^P (or, equivalently, OSNR$_{MPI}^P$).

---

[16] In principle, one can use a polarization nulling technique, analogous to that used for measuring OSNR$_{ASE}$ [26]. By placing high-extinction polarization-selective components before the optical spectrum analyzer, one can measure the component of the DRB power that is orthogonally polarized to the signal.

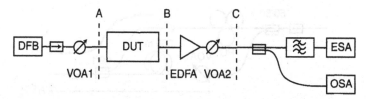

Fig. 15.11. Setup for measuring MPI in the device-under-test using the electrical beat-noise measurement technique. RIN measurements are made using the ESA (which consists of a square-law photodetector, low-noise RF amplifier, optical power meter, and an electrical spectrum analyzer). An optical spectrum analyzer measures the $OSNR_{ASE}$ at point C to calculate the RIN contribution from signal–ASE beating (RIN_{s-ASE}).

A typical measurement setup is shown in Fig. 15.11 where the amplifier in question is the device-under-test (DUT) between points A and B. The device is probed using a low-noise CW source at the signal wavelength. Its linewidth needs to be large enough that its coherence length is much shorter than significant round-trip paths in the amplifier. This ensures that the DRB light is incoherent with respect to the straight-through signal light, greatly simplifying the analysis of the noise power spectrum (see Section 15.2.1). For Raman amplifiers made from several kilometers of fiber, linewidths greater than 1 MHz suffice.[17] Therefore, distributed feedback semiconductor lasers with RIN levels below -155 dB/Hz and 1 to 10 MHz linewidths are ideal. MPI generated in shorter amplifiers, such as EDFAs made from a few meters of erbium-doped fiber, can also be measured if the linewidth of the source is sufficiently broad [50].

After passing through the device-under-test, the signal may need to be optically amplified to a level suitable for the RIN-measurement apparatus. This apparatus, labeled ESA, consists of a high-bandwidth photodetector, optical power meter, low-noise electrical preamplifier, and electrical spectrum analyzer [51, 26]. An optical spectrum analyzer (OSA) may be used to measure the signal-to-ASE ratio at point C of Fig. 15.11 to determine the noise contribution from signal–ASE beating [43, 49]. The bandpass filter (< 1 nm bandwidth) before the ESA reduces noise from ASE–ASE beating to insignificance [26].

The relative intensity noise RIN_{total} measured by the ESA comes from a variety of sources, one of which is the RIN_{MPI} we want to measure. Other noise contributions must therefore be subtracted from RIN_{total}. These contributions are RIN_{s-ASE} from signal–ASE beating, RIN_{DFB} from the excess noise of the DFB measured at point A, RIN_{shot} from shot noise, and RIN_{th} from thermal noise in the ESA electronic components.

Figure 15.12 shows results demonstrating this subtraction technique. For this case, two-path MPI was produced using a Mach–Zehnder interferometer (Fig. 15.12(a)). A 500 m path difference between the arms was chosen to ensure incoherent interference for a 2.5 MHz DFB linewidth. By adjusting a polarization controller (PC) in one arm

---

[17] The linewidth of the source used to measure RIN from DRB is sufficiently large if the DRB RIN spectral shape shown in Fig. 15.13(b) matches the incoherent self-homodyne RIN spectra of Fig. 15.5.

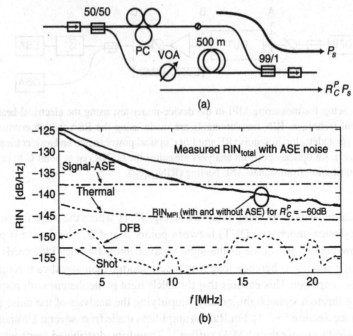

Fig. 15.12. (a) Mach–Zehnder interferometer used to test and calibrate electrical beat-noise measurements of MPI; (b) MPI measurement on interferometer ($R_C^P = -60$ dB) demonstrating the RIN subtraction technique. RIN spectra are shown, measured either with or without signal–ASE beat-noise. RIN$_{MPI}$ is the same for both cases.

to maximize the intensity noise, and varying a variable optical attenuation (VOA) in the other arm, values of R_C^P ranging from -20 to -60 dB could be produced. To determine R_C^P, we simply measured the ratio of optical powers at the output when each arm of the interferometer was blocked in turn.

Figure 15.12(b) shows RIN$_{MPI}$ curves between 2 and 22 MHz for $R_C^P = -60$ dB. First the RIN was measured without the EDFA and VOA2, and RIN$_{MPI}$ was obtained. Next the EDFA and VOA2 (see Fig. 15.11) were added and configured to increase RIN$_{s-ASE}$. This noise contribution, labeled "Signal–ASE," was calculated from the measured OSNR$_{ASE}$ at point C using RIN$_{s-ASE}(f) = 2/($OSNR$_{ASE} B_{ref})$, where B_{ref} is the reference optical bandwidth (in Hz) used to define OSNR$_{ASE}$. (RIN measurements were calibrated using a RIN transfer standard [51, 26] formed from a broadband light-emitting diode, an EDFA, and a 1.6 nm bandpass filter.) The fact that both RIN$_{MPI}$ curves lie on top of each other shows the accuracy of the RIN subtraction technique even for very low levels of R_C^P, which was only -60 dB in this demonstration.

The same interferometer can be used to calibrate RIN$_{MPI}$ measurements made on fiber amplifiers. The key point in this calibration technique is that the RIN$_{MPI}$ traces for incoherent interference do not depend on the number of interfering paths (see Section 15.3.3). Therefore, if the RIN$_{MPI}$ curves of an amplifier and interferometer

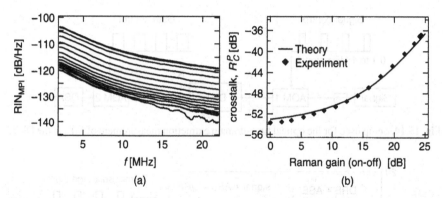

Fig. 15.13. (a) Family of RIN_{MPI} curves measured when the on-off Raman gain was varied from 0 dB (pumps off) to 24.3 dB; (b) crosstalk ratio R_C^P plotted as a function of gain. The markers show the measured values, and the line shows the result of Eq. (15.56).

are identical, the values of R_C^P for both devices are also identical. Then R_C^P can be obtained by simply measuring the power ratio between the arms of the interferometer.

Figure 15.13 shows the results of measurements made on a distributed Raman amplifier consisting of 100 km of nonzero dispersion fiber that had a Rayleigh backscatter coefficient of 100×10^{-6} km^{-1} at 1550 nm. The amplifier was counter-pumped by multiplexed semiconductor diodes at 1435 and 1450 nm. We used a 1550 nm DFB with a linewidth of 2.5 MHz as the signal. Figure 15.13(a) shows RIN_{MPI} curves measured for Raman gains varying from 24.3 dB (the top curve) to 0 dB (the bottom curve, which is for the pumps turned off). Each RIN_{MPI} curve is converted (using the interferometer calibration technique) to a corresponding R_C^P value, plotted versus on-off Raman gain in Fig. 15.13(b). The theoretical curve is from Eq. (15.56). For small on-off gains, the contribution to R_C^P is dominated by double-Rayleigh scattering within the entire 100 km of fiber, and so increases in gain do not increase R_C^P significantly. For large gain values, however, double-Rayleigh scattering within the effective length of the amplifier dominates, and so R_C^P increases more dramatically as the gain is increased: a 1 dB increase in gain produces a 1.2 dB increase in R_C^P (or, equivalently, a 1.2 dB *decrease* in $\text{OSNR}_{\text{MPI}}^P$.)

15.4.3. Optical Time-Domain Extinction Technique

More recently, an optical time-domain extinction technique has been demonstrated for quantifying MPI in fiber amplifiers [52]. This technique measures the ratio of signal light to double-Rayleigh light using a time-domain extinction approach similar to that used in EDFA noise figure measurements [26].

A schematic of the setup is shown in Fig. 15.14. Light from a CW source at the wavelength of interest first passes through an acoustooptic modulator (AOM 1) used as a switch to produce pulses with a high extinction ratio. Next these pulses pass through the DUT, through another high-extinction switch (AOM 2), and are then

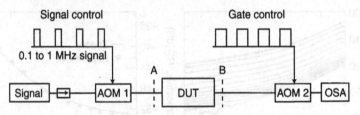

Fig. 15.14. Setup used for the optical time-domain extinction measurement of MPI in the DUT.

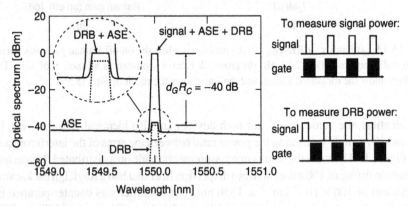

Fig. 15.15. Calculated OSA spectra (0.1 nm resolution bandwidth) for an optical time-domain extinction measurement of MPI. The signal and gate pulse trains for both parts of the measurement are shown. When the gate pulses are out of phase with the signal pulses, the signal light is blocked so only ASE and DRB light are detected.

measured using an OSA. The pulse trains used to control AOM 1 and AOM 2 have the same repetition rate, but their duty cycles may be different. The combination of AOM 2 and the OSA forms a detector that is both gated in time and wavelength sensitive. Therefore the spectrum recorded on the OSA depends on the relative phase of the pulse trains that control the signal and gate. If the phase of the gate pulses is set to block the straight-through signal pulses, the OSA trace will only show light leaving the amplifier out of phase with the signal pulses. This light includes double-Rayleigh "echoes," which originate from signal pulses that are delayed inside the amplifier. The other type of light is ASE from the amplifier, but since it is usually broadband, it can easily be distinguished from the narrowband double-Rayleigh light.

To make a MPI measurement, two OSA traces are needed (see the schematic in Fig. 15.15). First, the signal power is measured from a trace with the gate set in phase with the signal pulses. Next the double-Rayleigh power is determined from a trace with the gate out of phase and the ASE noise floor subtracted by interpolation. A calibration factor is needed to account for the duty cycle of the signal pulses (d_S) and the gate pulses (d_G), where $d_S + d_G < 1$ to ensure the signal can be blocked by the gate. (The duty cycle is defined as the time for which the AOM passes light divided by the repetition period.) If the average power levels measured on the OSA are P'_s and P'_{DRB} for the signal and double-Rayleigh light, respectively, R_C is given by

$$R_C = \frac{1}{d} \frac{P'_{DRB}}{P'_s}, \qquad (15.57)$$

where d is the larger of the two duty cycles d_S and d_G. Note that, for the setup described above, the measurement is not polarization sensitive and so all polarization components of double-Raleigh light are measured. Thus, unlike the electrical beat-noise measurement approach, this approach cannot measure R_C^P unless polarization-selective components are added before the OSA [53].

Several factors must be considered for an accurate measurement. First of all, the probe signal must be weak enough that it does not change the gain of the amplifier when it is turned off, or complicated transient gain effects may distort the measurement [54]. Secondly, the extinction ratio of the AOM switches must be high enough to prevent leakage of the signal when the gate is set to block it. Typically, a single switch is formed from a pair of cascaded AOMs to provide greater than 90 dB extinction. Thirdly, care must be taken when subtracting the ASE noise floor from the double-Rayleigh signal, especially for devices with low levels of MPI, as this can be the largest source of error. Finally, the choice of modulation rate depends on the duration of the double-Rayleigh echo [55]; this rate must be high enough that the double-Rayleigh light is not time-dependent. Then, the double-Rayleigh power measured on the OSA is proportional to the double-Rayleigh power leaving the amplifier multiplied by the gate duty cycle d_G. Rates on the order of 1 MHz are suitable for amplifiers made from many kilometers of fiber.

Typically, this technique is sensitive enough to measure R_C down to -50 dB when ASE is also present, which is not as low as can be measured with the electrical beat-noise technique (-60 dB). Also, only devices with sufficient time delay between the signal pulses and the double-Rayleigh "echo" can be characterized. But, because the time domain extinction technique does not require a low-noise signal source with an appropriate linewidth, any standard tunable laser can be used. This flexibility makes it easier to characterize the MPI properties of fiber amplifiers over their entire gain spectrum [56].

15.4.4. Section Summary

In this section we presented ways to quantify and measure the level of MPI. Two types of quantities, inversely proportional to each other, were defined: an optical signal-to-noise ratio ($OSNR_{MPI}$) and a crosstalk ratio (R_C). We also defined measures that take into account the state and degree of polarization of the MPI-induced light, $OSNR_{MPI}^P$ and R_C^P. This extension is crucial, because only the MPI copolarized with the signal produces beat noise. Next we discussed two techniques used to measure the level of MPI in either the electrical domain (by measuring the noise spectrum of beat noise) or in the optical domain (by measuring the optical spectrum of gated echoes produced by MPI).

15.5. Impact of MPI on Beat-Noise Limited Receivers

The most widely used class of optical receivers employed today uses optically pream-plified direct detection in order to reduce the impact of electronics noise (most notably, thermal noise) to insignificance. Thus higher receiver sensitivities can be obtained than those achieved in practice for pin or avalanche photodiode (APD) receivers. In an optically preamplified receiver, the dominating noise sources originate from beating between the signal and ASE as well as from beating of ASE with itself [57]. Therefore this class of receivers is frequently termed *beat-noise limited*. In optical communication systems using discrete inline optical amplifiers, ASE is added at each amplifier site [58]. If distributed amplification is employed, ASE is generated contin-uously along the transmission path [59]. Because ASE is present in both propagation directions, originally counterpropagating ASE that is Rayleigh backscattered and am-plified together with the signal can also be observed at the receiver [60, 61]. In the presence of MPI, the beating of the signal and MPI (as well as the beating of ASE with MPI and of MPI with itself) can give rise to appreciable levels of MPI-induced beat noise, which additionally impairs detection.

To properly design a receiver in the presence of both ASE-induced beat noise and MPI-induced beat noise, one has to take into account the *combined* influence of both beat noise sources. This problem has been tackled in different ways: Wan and Con-radi [17, 62, 63] derive important statistical properties of the DRB process, including the variance of signal–DRB beat noise. However, because they consider optical in-tensity fluctuations rather than fluctuations of the photocurrent, their analyses only apply to the case of infinitely broadband electrical receiver front ends. For the ASE-induced beat noise terms, they use the well-known approximations of Olsson [57], which assume rectangular optical and electrical receive filters as well as constant optical signal power levels (i.e., no isolated "1"-bits or "0"-bits, and no pulsed signal formats). Along these lines, other groups [30, 43, 49, 64–68] have theoretically and experimentally studied detection degraded by both MPI and ASE for different num-bers of interfering signals as well as for DRB. The discussions on the impact of MPI and ASE on receiver performance in this section are mainly based on [69], which generalizes the work of Wan and Conradi by providing a theoretical analysis of beat noise due to MPI and ASE that takes into account the effect of modulation formats as well as of optical and electrical filtering. A specialization for Gaussian optical pulses and Gaussian filter shapes allows for closed-form expressions of the noise variances.

15.5.1. General Expressions for Beat Noise Variances

Figure 15.16 shows the basic setup of a direct detection optical receiver in a wavelength-division-multiplexed (WDM) transmission scenario impaired by ASE and MPI. At the receiver input, we find a superposition of the optical signal field $\varepsilon_s(t)$, the ASE field $\varepsilon_{ASE}(t)$, and the MPI field $\varepsilon_{MPI}(t)$. The optical spectra of the three field contributions are also represented in the figure. While ASE is a white pro-cess (i.e., the ASE field's power spectral density is constant over the frequency range of interest), MPI (and DRB as a special case) basically has the same spectral shape as

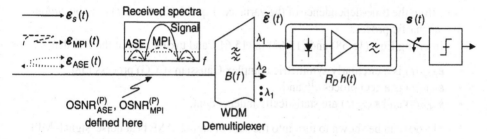

Fig. 15.16. Setup of a direct detection optical receiver impaired by ASE and MPI.

the signal (see Sections 15.1 and 15.3.3.2). The optical signal-to-noise ratios for ASE (OSNR$_{ASE}$ and OSNR$_{ASE}^P$) and MPI (OSNR$_{MPI}$ and OSNR$_{MPI}^P$) are defined in front of the optical filter at the receiver input according to Section 15.4.1. The WDM de-multiplexer at the receiver input both serves to separate the WDM channels at center wavelengths λ_i, and acts as a narrowband optical filter to suppress out-of-band ASE. The demultiplexing filter has a field transfer function $B(f)$ (impulse response $b(t)$). The filtered optical field after the demultiplexer thus reads

$$\widetilde{\boldsymbol{\varepsilon}}(t) = \left[\varepsilon_s(t) + \boldsymbol{\varepsilon}_{MPI}(t) + \boldsymbol{\varepsilon}_{ASE}(t) \right] * b(t) = \widetilde{\varepsilon}_s(t) + \widetilde{\boldsymbol{\varepsilon}}_{MPI}(t) + \widetilde{\boldsymbol{\varepsilon}}_{ASE}(t), \quad (15.58)$$

where $*$ denotes a convolution. When referred to *after* the optical demultiplexing filter, optical fields and powers are written with a tilde. Photodetection is followed by electrical amplification and low-pass filtering. The overall optoelectronic conversion factor is denoted R_D ([A/W] or [V/W]). The impulse response of the entire opto-electronic detection chain is $h(t)$ (transfer characteristics $H(f)$), and is considered normalized to $H(0) = \int h(t)dt = 1$. The resulting electrical signal $\boldsymbol{s}(t)$ is then applied to a sampling-and-decision gate to restore the desired digital information.

Observing that both $\boldsymbol{\varepsilon}_{ASE}(t)$ and $\boldsymbol{\varepsilon}_{MPI}(t)$ are statistically independent, stationary, zero-mean stochastic processes, it can readily be shown [40] that the ensemble average of $\boldsymbol{s}(t)$ reads

$$\langle \boldsymbol{s}(t) \rangle = R_D(|(\varepsilon_s * b)(t)|^2 * h(t) + \widetilde{P}_{ASE} + \widetilde{P}_{MPI}), \quad (15.59)$$

where $\widetilde{P}_{ASE}$ and $\widetilde{P}_{MPI}$ denote the ASE and MPI powers at the photodetector. Assuming realistically that shot noise is negligible compared to the beat noise terms for preamplified receivers [31], the variance $\sigma^2(t)$ of $\boldsymbol{s}(t)$ can be shown to be [40]

$$\sigma^2(t) = R_D^2 \iint_{-\infty}^{\infty} C_{\widetilde{\boldsymbol{P}}}(\tau, \tau')h(t-\tau)h(t-\tau')d\tau d\tau', \quad (15.60)$$

where

$$C_{\widetilde{\boldsymbol{P}}}(\tau, \tau') = \langle \widetilde{\boldsymbol{P}}(\tau)\widetilde{\boldsymbol{P}}(\tau') \rangle - \langle \widetilde{\boldsymbol{P}}(\tau) \rangle \langle \widetilde{\boldsymbol{P}}(\tau') \rangle \quad (15.61)$$

is the autocovariance function of the total optical power at the detector,

$$\widetilde{\boldsymbol{P}}(t) = |\widetilde{\boldsymbol{\varepsilon}}(t)|^2. \quad (15.62)$$

Note from the time-dependence of the variance, Eq. (15.60), that $s(t)$ is a *nonstation-ary* stochastic process.

Inserting Eqs. (15.58) and (15.62) in (15.61), and taking note of the fact that

- $\pmb{\varepsilon}_{ASE}(t)$ is a circularly symmetric complex Gaussian (CCG) process [58],
- $\pmb{\varepsilon}_{MPI}(t)$ is a ccG process,[18] and
- $\pmb{\varepsilon}_{MPI}(t)$ and $\pmb{\varepsilon}_{ASE}(t)$ are statistically independent,

Eq. (15.60) can be shown to turn into the sum of signal–ASE beat noise, signal–MPI beat noise, ASE–ASE beat noise, MPI–MPI beat noise, and ASE–MPI beat noise,

$$\sigma^2(t) = \sigma^2_{s\text{-ASE}}(t) + \sigma^2_{s\text{-MPI}}(t) + \sigma^2_{ASE\text{-ASE}} + \sigma^2_{ASE\text{-MPI}} + \sigma^2_{MPI\text{-MPI}}. \qquad (15.63)$$

Introducing the autocorrelation of the stationary ASE and MPI fields, $\Gamma_{\widetilde{\pmb{\varepsilon}}_N}(\tau, \tau') = \Gamma_{\widetilde{\pmb{\varepsilon}}_N}(\tau - \tau') = \langle \widetilde{\pmb{\varepsilon}}^*_N(\tau)\widetilde{\pmb{\varepsilon}}_N(\tau')\rangle$, with N standing for either ASE or MPI, the variances of the beat noise terms involving the signal are

$$\sigma^2_{s-N}(t) = 2R^2_D \, \mathrm{Re}\left\{ \iint_{-\infty}^{\infty} \widetilde{\varepsilon}_s(\tau)\widetilde{\varepsilon}^*_s(\tau')\Gamma_{\widetilde{\pmb{\varepsilon}}_N}(\tau - \tau')h(t - \tau)h(t - \tau')d\tau d\tau' \right\},$$

$$(15.64)$$

the N–N beat noise terms are

$$\sigma^2_{N-N} = R^2_D \iint_{-\infty}^{\infty} |\Gamma_{\widetilde{\pmb{\varepsilon}}_N}(\tau - \tau')|^2 h(\tau)h(\tau')d\tau d\tau', \qquad (15.65)$$

and the ASE–MPI beat noise term is

$$\sigma^2_{ASE\text{-MPI}} = 2R^2_D \, \mathrm{Re}\left\{ \iint_{-\infty}^{\infty} \Gamma_{\widetilde{\pmb{\varepsilon}}_{ASE}}(\tau - \tau')\Gamma^*_{\widetilde{\pmb{\varepsilon}}_{MPI}}(\tau - \tau')h(\tau)h(\tau')d\tau d\tau' \right\}.$$

$$(15.66)$$

Note that these equations represent a generalization of the ones derived for ASE only (see, e.g., [70–72]). By viewing the sum of ASE and MPI as a single source of noise, it is also possible to compress the five beat noise terms in Eq. (15.63) to just two terms (one that involves the signal, and one that does not). This is readily done by replacing $\widetilde{\pmb{\varepsilon}}_N$ in Eqs. (15.64) and (15.65) by $\widetilde{\pmb{\varepsilon}}_{ASE} + \widetilde{\pmb{\varepsilon}}_{MPI}$. This approach can be advantageous for numerical simulations, however, it does not allow for straightforward statements on the relative impact of ASE and MPI.

As evident from Eq. (15.64), the signal-induced beat noise variances depend on the combined influence of the filtered signal waveform $\widetilde{\varepsilon}_s(t)$, on the detector impulse response $h(t)$, and on the spectra of the optically filtered ASE and MPI, reflected in the correlation functions $\Gamma_{\widetilde{\pmb{\varepsilon}}_{ASE}}(\tau, \tau')$ and $\Gamma_{\widetilde{\pmb{\varepsilon}}_{MPI}}(\tau, \tau')$. Assuming no electrical filtering

---

[18] For a sufficiently large number of discrete interferers, this property follows from the central limit theorem. For DRB, it also follows from the statistical properties of $\rho(z)$, Eqs. (15.29) to (15.31).

(i.e., infinitely broadband receiver electronics, $h(t) = \delta(t)$, with $\delta(t)$ being Dirac's δ-functional), we arrive at

$$\langle s(t) \rangle = R_D \widetilde{P}_s(t) \tag{15.67}$$

for the signal, and

$$\sigma_{s-N}^2(t) = 2R_D^2 \, \widetilde{P}_s(t) \widetilde{P}_N^P \tag{15.68}$$

and

$$\sigma_{N-N}^2 = R_D^2 \, \widetilde{P}_N^2, \tag{15.69}$$

for the beat noise variances, consistent with the simplified model of [63]. Here $\widetilde{P}_s(t)$ and $\widetilde{P}_N(t)$ denote the optical signal power and the ASE or MPI power at the detector; $\widetilde{P}_N^P(t)$ is the ASE or MPI power copolarized with the signal.

15.5.2. From Beat Noise to Receiver Performance

The quantity of ultimate interest when assessing the performance of optical receivers is their sensitivity, which for beat noise limited receivers is best specified in terms of a *required OSNR* [73], that is, an OSNR value at the optical receiver input that guarantees a certain bit error ratio (BER) at the output of the sampling and decision device. To arrive at a required OSNR by means of analysis or simulation, we need to know the signal waveform at the decision gate (including deterministic signal distortions that can lead to intersymbol interference (ISI)), as well as the probability distribution of the noise that is corrupting detection. It has been shown (see, e.g., [31, 74]) that replacing the exact (non-Gaussian) probability density of the decision variable $s(t)$ by a Gaussian density results in highly accurate predictions of receiver sensitivity, provided that the decision threshold is assumed optimum.[19] Because deviations from the optimum threshold cannot be modeled appropriately using Gaussian statistics, we assume throughout this section that the receiver adaptively adjusts its decision threshold for minimum BER, which is also the most relevant situation encountered in practice.

The Gaussian distribution is determined solely by its mean and variance. Thus these two parameters are sufficient to assess receiver performance, and we arrive at a closed-form expression for the BER,

$$\text{BER} = 0.5 \, \text{erfc}[Q/\sqrt{2}], \tag{15.70}$$

where $\text{erfc}[x] = (2/\sqrt{\pi}) \int_x^\infty \exp(-\xi^2) d\xi$ denotes the complementary error function, and the Q-factor is defined as [76]

$$Q = \frac{\langle s_1 \rangle - \langle s_0 \rangle}{\sigma_1 + \sigma_0}. \tag{15.71}$$

---

[19] Note that this statement is only true for formats that are received using intensity modulation at a single photodetector. For example, it does *not* apply for balanced detection of differential phase shift keying (DPSK) [75].

Here $\langle s_1 \rangle$ and $\langle s_0 \rangle$ denote the noise-free electrical signal values at the optimum sampling instant. In the presence of bit-pattern-dependent deterministic signal distortions (in particular, in the presence of ISI), taking the worst (lowest) 1-bit and the worst (highest) 0-bit contained in the bit sequence usually yields the most accurate results.[20] In Eq. (15.71), the signal variances corresponding to the signal values $\langle s_1 \rangle$ and $\langle s_0 \rangle$ are denoted σ_1^2 and σ_0^2. Although based on a number of approximations, the Q-factor provides valuable analytical insight in the noise performance of optical receivers.

15.5.2.1. The Q-Factor in the Presence of ASE and MPI

If signal-induced beat noise dominates all other noise terms, the Q-factor can be approximated by

$$Q = \frac{\langle s_1 \rangle - \langle s_0 \rangle}{\sqrt{\sigma_{1,s-\text{ASE}}^2 + \sigma_{1,s-\text{MPI}}^2} + \sqrt{\sigma_{0,s-\text{ASE}}^2 + \sigma_{0,s-\text{MPI}}^2}}$$

$$= \left(f_s(1 - \sqrt{r}) \right) \bigg/ \left(\sqrt{1+r} \sqrt{\frac{f_{s-\text{ASE}}^2}{\text{OSNR}_{\text{ASE}}^P} + \frac{f_{s-\text{MPI}}^2}{\text{OSNR}_{\text{MPI}}^P}} \right). \tag{15.72}$$

To arrive at this expression, we have factored out the absolute optical power according to

$$\langle s_1 \rangle = f_s \overline{P}_s / (1+r) \tag{15.73}$$

$$\sigma_{1,s-\text{ASE}}^2 = f_{s-\text{ASE}}^2 \overline{P}_s N_{\text{ASE}} B_{\text{ref}} / (1+r) \tag{15.74}$$

$$\sigma_{1,s-\text{MPI}}^2 = f_{s-\text{MPI}}^2 \overline{P}_s P_{\text{MPI}}^P / (1+r), \tag{15.75}$$

and included the *inverse* extinction ratio of the optical signal in front of the demultiplexer,

$$r = P_{s,0} / P_{s,1}. \tag{15.76}$$

The 0-bit mean and beat noise variances thus read

$$\langle s_0 \rangle = r \langle s_1 \rangle \quad \text{and} \tag{15.77}$$

$$\sigma_{0,s-\text{N}}^2 = r \sigma_{1,s-\text{N}}^2. \tag{15.78}$$

In Eqs. (15.73) through (15.75), $\overline{P}_s$ is the average optical signal power, leading to a 1-bit peak power of

$$P_{s,1} = \hat{a} \overline{P}_s / (1+r), \tag{15.79}$$

---

[20] If only a few bits within a long bit pattern are highly distorted, as can be the case for distortions brought by, for example, intrachannel four-wave mixing, taking the worst 1- and 0-bits will result in noticeable underestimates of the Q-factor [77].

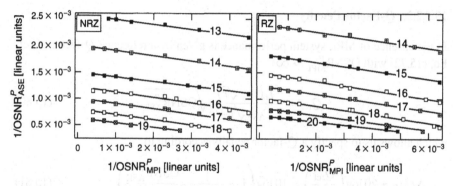

Fig. 15.17. The experimental verification of Eq. (15.72) reveals excellent agreement with theory [78].

with $\hat{a}$ characterizing the shape of the optical pulses. (For OOK with rectangular pulses of duty cycle d, we have $\hat{a} = 2/d$.) In what follows, quantities with a hat refer to the peak of the information-carrying data pulses. The symbols $N_{ASE}B_{ref}$ and P_{MPI}^{P} in Eqs. (15.74) and (15.75) denote the power of the ASE and MPI components that are *copolarized* with the signal, the former measured in a reference optical bandwidth B_{ref} (typ. 12.5 GHz). All powers are measured *in front of* the optical demultiplexer filter, as indicated in Fig. 15.16. The constant f_s reflects the attenuating influence of optical and electrical filtering on the peak of the optical pulses, and f_{s-ASE}^{2} and f_{s-MPI}^{2} represent the influence of filtering on the beat noise variances at the pulse peak.[21] In order to introduce the effect of finite extinction ratios, it has to be assumed that the influence of optical and electrical filtering is identical for 1-bits and 0-bits, that is, that ISI at the decision gate is negligible. Otherwise, the expressions for 1-bit and 0-bit means and variances would not be as straightforward as Eqs. (15.73) through (15.75) and (15.77) and (15.78).

According to Eq. (15.72), the Q-factor for a system limited by both ASE and MPI can be accurately approximated by a function involving the sum of the inverse OSNRs. The excellent validity of this relationship has been verified experimentally at 10 Gbit/s for NRZ and 50% duty cycle RZ on-off keying [78], and is shown in Fig. 15.17. The figure shows contours of constant Q-factor (expressed in dB units) as a function of $OSNR_{ASE}^{P}$ and $OSNR_{MPI}^{P}$ together with linear fits. We now proceed to calculate the

- MPI-induced Q-factor penalty ΔQ, that is, the decrease in Q-factor when MPI is added to a system that is originally limited by signal–ASE beat noise only, and the

- OSNR-penalty $\Delta OSNR$, that is, the increase in $OSNR_{ASE}$ that is required when MPI is added, in order to maintain the same Q-factor obtained without MPI.

---

[21] Note that f_s, f_{s-ASE}, and f_{s-MPI} can substantially differ from one another.

15.5.2.2. Q-Factor Penalty

In the absence of MPI, system performance is given by a reference Q-factor of (cf. Eq. (15.72) with $\mathrm{OSNR}_{\mathrm{MPI}}^{P} \to \infty$)

$$Q_{\mathrm{ref}} = \frac{f_s (1 - \sqrt{r})}{f_{s-\mathrm{ASE}}\sqrt{1+r}} \sqrt{\mathrm{OSNR}_{\mathrm{ASE}}^{P}}, \tag{15.80}$$

which allows us to specify a Q-factor penalty according to

$$\Delta Q_{dB} = 20 \log\left(\frac{Q_{\mathrm{ref}}}{Q}\right) = 10 \log\left(1 + \frac{1}{F_{\mathrm{OSNR}}} \cdot \frac{\mathrm{OSNR}_{\mathrm{ASE}}^{P}}{\mathrm{OSNR}_{\mathrm{MPI}}^{P}}\right) \tag{15.81}$$

$$= 10 \log\left(1 + \frac{1}{F_Q} \cdot \frac{Q_{\mathrm{ref}}^2}{\mathrm{OSNR}_{\mathrm{MPI}}^{P}} \frac{1+r}{(1-\sqrt{r})^2}\right). \tag{15.82}$$

Equation (15.81) expresses the Q-factor penalty in terms of the two optical signal-to-noise ratios $\mathrm{OSNR}_{\mathrm{ASE}}^{P}$ and $\mathrm{OSNR}_{\mathrm{MPI}}^{P}$, using the *OSNR-based MPI tolerance factor*

$$F_{\mathrm{OSNR}} = f_{s-\mathrm{ASE}}^2 / f_{s-\mathrm{MPI}}^2, \tag{15.83}$$

whereas Eq. (15.82) quantifies the Q-factor penalty for a given reference Q_{ref}, using the *Q-factor-based MPI tolerance factor*

$$F_Q = f_s^2 / f_{s-\mathrm{MPI}}^2. \tag{15.84}$$

Both F_{OSNR} and F_Q indicate the *tolerance* of a system to degradations by MPI; that is, the larger these quantities, the more robust the system will be to impairments by MPI. Note also from comparing Eq. (15.83) with (15.74) and (15.75) that F_{OSNR} gives the ratio of signal–ASE beat noise variance to signal–MPI beat noise variance for equal copolarized OSNRs ($\mathrm{OSNR}_{\mathrm{ASE}}^{P} = \mathrm{OSNR}_{\mathrm{MPI}}^{P}$).

If we neglect the influence of electronic filtering (i.e., we set $h(t) = \delta(t)$ and use Eqs. (15.67) and (15.68)), and if we take note of the fact that $\mathrm{OSNR}_{\mathrm{MPI}}^{P}$ does not change upon transition through optical filters, which is true if signal and MPI have the same spectral shape, we can approximate the two MPI tolerance factors by

$$F_{\mathrm{OSNR}} \approx \frac{B_o P_{\mathrm{MPI}}^{P}}{B_{\mathrm{ref}} \widetilde{P}_{\mathrm{MPI}}^{P}} \tag{15.85}$$

and

$$F_Q \approx \frac{\widetilde{a}}{2}, \tag{15.86}$$

where $B_o = \widetilde{P}_{\mathrm{ASE}}^{P} / N_{\mathrm{ASE}}$ denotes the equivalent square bandwidth of the demultiplexer filter, and $\widetilde{a}$ characterizes the optical pulse shape after the demultiplexer (cf. Eq. (15.79)). Note from Eq. (15.85) that the tolerance to MPI for *fixed* OSNRs increases linearly with the data rate, which is a consequence of the bit-rate-independent reference bandwidth entering the definition of $OSNR_{ASE}$ (see also Section 15.5.2.4). Also broadband optical signaling, resulting in a reduction of the optically filtered MPI power $\widetilde{P}^P_{MPI}$ for fixed optical demultiplexer bandwidths, increases the tolerance to MPI (see Section 15.5.3.1).

For perfect signal extinction ratio ($r = 0$), $\widehat{a}$ denotes the ratio of peak to average optical power after the demultiplexer, and we arrive at

$$F_Q \approx 0.5\widehat{\overline{P}}/\overline{P}, \tag{15.87}$$

which, again, implies that broadband signaling is beneficial for a receiver's robustness to MPI. Disregarding a slightly different conversion of the Q-factor to dB units, and assuming unpolarized MPI ($OSNR_{MPI}=OSNR^P_{MPI}/2$), we then arrive at the results of Fludger and Mears, who also showed good experimental agreement in a 40 Gbit/s CSRZ experiment [100].

15.5.2.3. OSNR Penalty

Setting $Q = Q_{ref}$ (using Eqs. (15.72) and (15.80)), we readily find for the OSNR penalty,

$$\Delta OSNR_{ASE,\,dB} = 10\log\left(\frac{OSNR_{ASE}}{OSNR_{ASE,\,ref}}\right) = 10\log\left(\frac{OSNR^P_{ASE}}{OSNR^P_{ASE,\,ref}}\right)$$

$$= -10\log\left(1 - \frac{1}{F_{OSNR}}\cdot\frac{OSNR^P_{ASE}}{OSNR^P_{MPI}}\right) \tag{15.88}$$

$$= -10\log\left(1 - \frac{1}{F_Q}\cdot\frac{Q^2_{ref}}{OSNR^P_{MPI}}\frac{1+r}{(1-\sqrt{r})^2}\right), \tag{15.89}$$

with the MPI tolerance factors defined in Eqs. (15.83) and (15.84). Neglecting, again, the influence of electronic filtering, we have $F_Q \approx \widehat{a}/2$ (see Eq. (15.86)), which, inserted into Eq. (15.89), directly reproduces the results of Liu et al. [65].

Figure 15.18 shows measurements (circles) and theoretical predictions (solid line) of the OSNR penalty due to MPI using Eq. (15.89) under the approximation (15.86). The measurement results are taken from Kim et al. [79], and apply for 10 Gbit/s NRZ modulation with a measured extinction ratio of $r = -13$ dB and $Q_{ref} = 6$, corresponding to BER$= 10^{-9}$. The dashed line shows the OSNR penalty that would have been predicted for perfect extinction ($r = 0$). It can be clearly seen that neglecting the extinction ratio can lead to a significant underestimation of the performance degradation due to MPI, as pointed out in [64–66]. The dotted line shows the Q-factor penalty according to Eq. (15.82) under the approximation (15.86) for reference. The Q-factor penalty is seen to be lower than the OSNR penalty. For $OSNR^P_{MPI} \gtrsim 25$ dB, the two penalties closely match each other. This can be understood by comparing Eqs. (15.82) and (15.89) and taking note of the Taylor expansion $1/(1 - x^2) \approx 1 + x^2$ around $x = 0$.

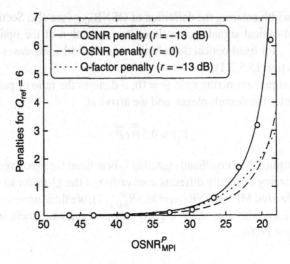

Fig. 15.18. Measurements (circles) and theoretical prediction (solid line) of OSNR penalty due to MPI for $r = -13$ dB. The dashed line gives the predicted OSNR penalty for perfect extinction ($r = 0$), and the dotted line shows the Q-factor penalty. The experimental data are taken from [79].

A third penalty parameter occasionally used in the literature is the *power penalty*. This quantity specifies the required increase in received optical signal power to maintain a given reference Q-factor. When using power penalties in the context of optically preamplified receivers, it is implicitly assumed that ASE generated by the receiver's optical preamplifier dominates any ASE originating from other sources (e.g., from a saturated booster amplifier at the transmitter). If the level of MPI at the receiver input is referred to the increased signal power that, in the presence of MPI, guarantees detection with Q_{ref}, the power penalty due to MPI can be shown to be identical to the OSNR penalty $\Delta \mathrm{OSNR}_{ASE}$ derived in this section.

15.5.2.4. Bit Rate Scaling

Equations (15.82), (15.84), (15.86), and (15.89) reveal that neither the Q-factor penalty nor the OSNR penalty due to MPI are data-rate-dependent, if the penalties are measured with reference to a fixed (data-rate-independent) reference Q-factor to guarantee a certain (data-rate-independent) BER.

On the other hand, Eq. (15.83) and, more explicitly, Eq. (15.85) predict a linear *increase* in tolerance to MPI with data rate, if two systems at different data rates are compared on a *constant-OSNR* basis. The higher MPI tolerance at increased data rates is reflected by the linear dependence of F_{OSNR} on the demultiplexer bandwidth B_o, which scales linearly with the data rate for fixed spectral efficiencies. The physical origin of this behavior resides in the fact that the ASE power spectral density (and thus OSNR_{ASE}) is defined in a fixed bandwidth B_{ref}, and is thus independent of the bit rate.

Because ASE is more broadband than B_{ref}, increasing the receiver filter bandwidths when transitioning to higher data rates leads to *more* signal–ASE beat noise within the receiver. In contrast, the MPI power spectral density has to decrease linearly with bit rate to maintain a constant level of MPI (i.e., constant $OSNR_{MPI}$). This leads to a *bit-rate-independent* signal–MPI beat noise. For fixed OSNRs, higher bit rate systems therefore suffer more from signal–ASE beat noise than lower bit rate systems. The impact of changes in per channel launch powers, leading to scalings in OSNR with bit rate, as well as the influence of fiber nonlinearities on the bit rate scaling of MPI degradations is covered in Section 15.6.

15.5.2.5. Impact of Signal Quality on Degradations due to MPI

The results of Sections 15.5.2.2 and 15.5.2.3 show that the system penalty caused by MPI strongly depends on the *quality* of the optical signal prior to adding MPI, as has been pointed out by several groups [43, 64–67]. Most importantly,

- a large amount of ASE (or, equivalently, a low reference Q-factor, as found in systems using forward error correction, FEC) mitigates the impact of MPI, whereas
- a poor signal extinction ratio dramatically reduces a system's robustness to MPI, and
- a high peak-to-average optical power ratio after the demultiplexer helps to reduce the impact of MPI. This statement, which implies that RZ using short pulses is more robust towards MPI than high duty cycle RZ or even NRZ, is discussed in more depth in the next section.

We explicitly note at this point that the above predictions on the impact of MPI and ASE on receiver performance critically rely on the validity of replacing the exact (non-Gaussian) statistics of the signal at the decision gate by *Gaussian* probability densities. Special caution has to be exerted whenever this assumption breaks down, for example, for balanced detection of DPSK, for fixed-threshold optical receivers, or for situations where only a few interferers produce inband crosstalk, as can be the case in optical networks incorporating add-drop multiplexers with imperfect suppression of the drop channels (see Sections 15.1 and 15.2.2).

Furthermore, if the influence of limited bandwidth electrical filtering in the receiver is to be included to overcome the assumption of infinitely broadband electronics, one has to evaluate F_{OSNR} or F_Q based on the exact expressions (15.59) and (15.64). Although this has to be done numerically in general, closed-form solutions can be given under the assumption of Gaussian pulse shapes and Gaussian (optical and electrical) filter characteristics. This is demonstrated in the following section.

15.5.3. Special Case: Gaussian Pulse Shapes and Filter Characteristics

To better understand and interpret the relative importance of the beat noise due to ASE and MPI as a function of signal pulsewidth, demultiplexer optical bandwidth,

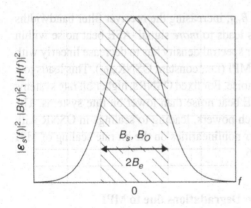

Fig. 15.19. Definition of equivalent square bandwidths for the optical signal spectrum $\mathcal{E}_s(f)$, the optical filter characteristics $B(f)$, and the electrical filter characteristics $H(f)$.

and electrical receiver bandwidth, we analytically solve Eq. (15.64), which is possible by assuming first-order Gaussian functions for all pulse and filter shapes involved,

$$\varepsilon_p(t) = \sqrt{\widehat{P}_s} \exp[-2\pi B_s^2 t^2] \tag{15.90}$$

$$b(t) = \sqrt{2}B_o \exp[-2\pi B_o^2 t^2] \quad \text{and} \tag{15.91}$$

$$h(t) = 2\sqrt{2}B_e \exp[-8\pi B_e^2 t^2], \tag{15.92}$$

where $\widehat{P}_s$ denotes the peak optical pulse power *before* demultiplexing, and B_s, B_o, and B_e are the optical signal, demultiplexer, and electrical equivalent square bandwidths, respectively. With reference to Fig. 15.19, these are defined as the double-sided (B_s and B_o) and single-sided (B_e) bandwidths that define rectangles of the same area as the respective squared spectra. For the optical and electrical filters, the equivalent square bandwidths are related to the 3 dB bandwidths by $B_{o,3dB} = 2B_o\sqrt{\ln(2)/\pi}$ and $B_{e,3dB} = 2B_e\sqrt{\ln(2)/\pi}$.

With these assumptions, the ASE and MPI autocorrelation functions $\Gamma_{\mathcal{E}_{ASE}}(\tau, \tau')$ and $\Gamma_{\mathcal{E}_{MPI}}(\tau, \tau')$ appearing in Eq. (15.64) can be readily evaluated. The ASE autocorrelation is obtained from the realistic assumption that the ASE process is white with a power spectral density N_{ASE} per polarization mode over the optical filter bandwidth, that is, assuming $\Gamma_{\mathcal{E}_{ASE}}(\tau, \tau') = N_{ASE}\,\delta(\tau - \tau')$. This readily leads to

$$\Gamma_{\mathcal{E}_{ASE}}(\tau, \tau') = \left\langle \int_{-\infty}^{\infty} \boldsymbol{\varepsilon}_{ASE}^{\star}(\xi)b^{\star}(\tau - \xi)d\xi \int_{-\infty}^{\infty} \boldsymbol{\varepsilon}_{ASE}(\eta)b(\tau' - \eta)d\eta \right\rangle$$

$$= \iint_{-\infty}^{\infty} \langle \boldsymbol{\varepsilon}_{ASE}^{\star}(\xi)\boldsymbol{\varepsilon}_{ASE}(\eta)\rangle b^{\star}(\tau - \xi)b(\tau' - \eta)d\xi\,d\eta$$

$$= N_{ASE} \int_{-\infty}^{\infty} b^{\star}(\tau - \xi)b(\tau' - \xi)d\xi = N_{ASE}B_o \exp[-\pi B_o^2(\tau - \tau')^2],$$

$$\tag{15.93}$$

which in the spectral domain simply reads $N_{ASE}|B(f)|^2$: ASE is spectrally shaped by the demultiplexer filter, as expected. To obtain a suitable expression for $\Gamma_{\mathcal{E}_{MPI}}(\tau, \tau')$, which relies on finding an analytically tractable approximation to the signal autocorrelation $\Gamma_{\tilde{\varepsilon}_s}(\tau - \tau')$ (see Eq. (15.39)), we use the method detailed in Appendix A15.B to arrive at

$$\Gamma_{\tilde{\varepsilon}_s}(\tau, \tau') = P_{\text{MPI}}^P \sqrt{1/(1 + B_s^2/B_o^2)} \exp[-\pi B_s^2/(1 + B_s^2/B_o^2)(\tau - \tau')^2], \quad (15.94)$$

where P_{MPI}^P is the copolarized optical MPI power measured *in front of* the WDM demultiplexer. The difference between the two autocorrelations (15.93) and (15.94) is caused by the different spectral distributions of ASE and DRB; while the ASE spectrum is only shaped by the optical demultiplexing filter, the spectrum of DRB matches that of the signal. This difference fundamentally distinguishes signal–ASE beat noise from signal–DRB beat noise.

Substituting Eqs. (15.90) and (15.92) through (15.94) into Eq. (15.64), we finally arrive at

$$\sigma_{1,s-\text{ASE}}^2 = \frac{4R_D^2 N_{\text{ASE}} \widehat{P}_s B_e}{[1 + B_s^2/B_o^2 + B_s^2/(4B_e^2)]^{1/2}[1 + 4B_e^2/B_o^2(1 + B_s^2/B_o^2) + 2B_s^2/B_o^2]^{1/2}},$$
$$(15.95)$$

$$\sigma_{1,s-\text{MPI}}^2 = \frac{2R_D^2 P_{\text{MPI}}^P \widehat{P}_s}{[1 + B_s^2/B_o^2]^{1/2}[1 + B_s^2/B_o^2 + B_s^2/(2B_e^2)]^{1/2}[1 + B_s^2/B_o^2 + B_s^2/(4B_e^2)]^{1/2}},$$
$$(15.96)$$

for the beat noise variances at the peak of an isolated electrical pulse. The average electrical signal at that point then becomes (cf. Eq.(15.59))

$$\langle \mathfrak{s}_1 \rangle = R_D \widehat{P}_s/[(1 + B_s^2/B_o^2)(1 + B_s^2/B_o^2 + B_s^2/(2B_e^2))]^{1/2}. \quad (15.97)$$

In the limit of broadband optical filtering and quasi CW signaling ($B_s \ll B_e \ll B_o$), Eq. (15.95) simplifies to the well-known expression [57]

$$\sigma_{1,s-\text{ASE}}^2 \approx 4R_D^2 N_{\text{ASE}} \widehat{P}_s B_e, \quad (15.98)$$

whereas Eq. (15.96) becomes [63]

$$\sigma_{1,s-\text{MPI}}^2 \approx 2R_D^2 P_{\text{MPI}}^P \widehat{P}_s, \quad (15.99)$$

independent of any bandwidth and in agreement with Eq. (15.68). The peak electrical signal in this case is simply $\langle \mathfrak{s}_1 \rangle \approx R_D \widehat{P}_s$.

In Section 15.6.1.4, when calculating the noise figure of a Raman amplified transmission span, we also need an expression for the shot noise variance $\sigma_{\text{shot}}^2(t)$. This quantity can be generally written as [80]

$$\sigma_{\text{shot}}^2(t) = R_D e(|\varepsilon_s|^2 * h)(t), \quad (15.100)$$

with $e = 1.602 \cdot 10^{-19} As$ denoting the elementary charge. Using the Gaussian pulse and filter model (Eqs. (15.90) through (15.92)), this equation becomes

$$\sigma_{\text{shot}}^2(t) = 2R_D e \widehat{P}_s B_e \frac{1}{[1 + B_s^2/B_o^2]^{1/2}[1 + B_s^2/B_o^2 + B_s^2/(4B_e^2)]^{1/2}}, \quad (15.101)$$

with the well-known approximation

$$\sigma_{\text{shot}}^2 \approx 2R_D e \widehat{P}_s B_e \quad (15.102)$$

for large filters and quasi CW signaling.

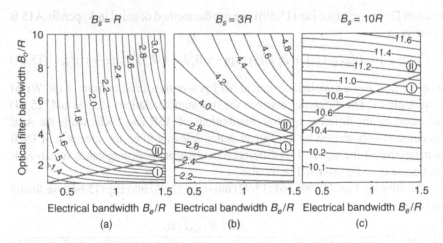

Fig. 15.20. MPI tolerance factor F_{OSNR} as a function of optical and electrical receiver bandwidths, both normalized to the data rate R. The reference bandwidth for defining $OSNR_{ASE}$ is assumed equal to the data rate ($R = B_{ref}$). (a), (b), and (c) correspond to signal bandwidths of $B_s = 1R, 3R$, and $10R$. Gray lines give boundaries between the two asymptotic regions discussed in the text.

15.5.3.1. Impact of MPI for given OSNRs

We next use Eqs. (15.95) and (15.96) together with (15.74), (15.75), and (15.83) to calculate the MPI tolerance factor F_{OSNR}, which can be interpreted as the ratio of signal–ASE beat noise variance to signal–MPI beat noise variance for $OSNR_{MPI}^P = OSNR_{ASE}^P$,

$$F_{OSNR} = 2B_e/R \frac{[1 + B_s^2/B_o^2]^{1/2}[1 + B_s^2/B_o^2 + B_s^2/(2B_e^2)]^{1/2}}{[1 + 4B_e^2/B_o^2(1 + B_s^2/B_o^2) + 2B_s^2/B_o^2]^{1/2}}. \tag{15.103}$$

The results are displayed in Fig. 15.20 for a signal bandwidth of (a) $B_s = R$, (b) $3R$, and (c) $10R$. For the sake of convenience, we assume that the bit rate R equals the reference bandwidth B_{ref} appearing in the definition of $OSNR_{ASE}^P$. If R deviates from B_{ref}, the values for F_{OSNR} shown in Fig. 15.20 have to be rescaled by R/B_{ref}.

It becomes obvious from Fig. 15.20 that a system's tolerance to MPI generally depends on the *signal bandwidth* (i.e., the pulse duration used for RZ signaling) as well as on both *optical and electrical* receiver bandwidths. The tendency for broadband optical signals to result in a higher tolerance to MPI (higher values of F_{OSNR}) lines up with the more simplified notion that the tolerance to MPI depends on the ratio of peak to average optical power, which is higher for lower RZ duty cycles (cf. Eq. (15.87)). Note that broadening the spectrum by means of short pulses is only one way of diminishing the effect of MPI. Another way, which is not captured by the simplified model treated here, is to impose a controlled amount of phase modulation (chirp) on the signal to broaden the spectrum, as proposed by different groups [81–85]. The benefit of using broad signal spectra can be explained as follows.[22] For sufficiently

---

[22] This can also be seen from Eqs. (15.95) and (15.96), when taking note of the fact that the peak optical power is proportional to the average optical power and the signal bandwidth.

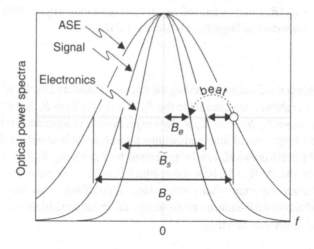

Fig. 15.21. Visualization of the beating process. Only those spectral ASE components whose beat frequencies with the optically filtered signal fall into the receiver electronics' bandwidth produce signal–ASE beat noise.

large signal bandwidths and constant $OSNR_{ASE}$, the signal field's power spectral density has to scale inversely with the signal bandwidth to maintain constant average optical power (and thus constant $OSNR_{ASE}$). Hence the signal–ASE beat noise term decreases linearly with signal bandwidth. The signal–MPI beat noise term, on the other hand, decreases *quadratically* with signal bandwidth, because both the signal power spectral density *and* the MPI power spectral density are inversely proportional to the signal bandwidth for constant $OSNR_{MPI}$.

Two asymptotic regions can be identified in Figs. (15.20)(a) through (c), qualitatively distinguished by contour lines running perpendicular to the B_o-axis (regions I) and contour lines running in parallel to the B_o-axis (regions II). The two regions are separated by gray lines in Fig. 15.20. With reference to Fig. 15.21, these lines indicate, for each electrical receiver bandwidth, the largest possible optical demultiplexer bandwidth B_o that lets pass spectral ASE components which produce a beating with the signal at a beat frequency ΔB that still lies within the detection bandwidth B_e; that is, $\Delta B = B_o/2 - \widetilde{B}_s/2 = B_e$. ($\widetilde{B}_s = B_s B_o/\sqrt{B_s^2 + B_o^2}$ denotes the bandwidth of the filtered optical signal.)

Region I

For demultiplexer bandwidths below the boundary indicated by the gray lines, decreasing B_o reduces the signal–ASE beat noise, thus reducing F_{OSNR} and emphasizing the influence of MPI on system noise. The reduction of F_{OSNR} is more significant for narrower signal spectra (long RZ pulses or even NRZ), where the spectral distribution of MPI significantly differs from that of ASE; for signal spectra significantly broader than the optical demultiplexer bandwidth (short RZ pulses or highly chirped signals), the spectral distribution of MPI becomes constant over the optical filter bandwidth, and is thus similar to ASE (see Eq. (15.94)). In the limit $B_s \gg B_o$ we find $F_{OSNR} \to B_s/R$, independent of any filter bandwidth: signal–MPI beat noise becomes less important at large B_s, as discussed above.

Region II

For demultiplexer bandwidths exceeding the boundary indicated by the gray lines, the contour lines eventually turn parallel to the B_o-axis; in the limit $B_o \gg \{B_e, B_s\}$ we finally find $F_{OSNR} \to 2\sqrt{B_e^2 + B_s^2/2}/R$. This expression is independent of B_o, because for sufficiently large optical filter bandwidths no additional beat noise falling within the electrical filter bandwidth can be generated by increasing B_o (see Fig. 15.21). Note, however, that ASE–ASE beat noise, which is not considered in our model, will eventually become important when going to large optical filter bandwidths, which will lead to an additional ASE-induced performance degradation (and thus to an increase in the system's tolerance to MPI).

15.5.3.2. Tolerance to MPI for Fixed Q-Factors

We now evaluate the MPI tolerance factor F_Q (see Eq. (15.84)), which is used to assess the MPI-induced system degradation on a receiver initially limited by signal–ASE beat noise and specified in terms of a reference Q-factor Q_{ref} (see Eqs. (15.82) and (15.89)). Using Eqs. (15.96) and (15.97), and taking note of the relation $\widehat{P}_s/\overline{P}_s = 4B_s/R$, we arrive at

$$F_Q = 2B_s/R \cdot \frac{[1 + B_s^2/B_o^2 + B_s^2/(4B_e^2)]^{1/2}}{[1 + B_s^2/B_o^2]^{1/2}[1 + B_s^2/B_o^2 + B_s^2/(2B_e^2)]^{1/2}}. \qquad (15.104)$$

As expected from our previous discussion along with Eq. (15.87), for $B_e \to \infty$, this expression turns into

$$F_Q \approx 2B_s/R \frac{1}{\sqrt{1 + B_s^2/B_o^2}} = 0.5\widehat{P}_s/\overline{P}_s. \qquad (15.105)$$

Figure 15.22 shows an evaluation of the MPI tolerance factor F_Q, Eq. (15.104), for the same parameters used in Fig. 15.20, that is, (a) $B_s = R$, (b) $B_s = 3R$, and (c) $B_s = 10R$). The curves show the same qualitative behavior as those for the tolerance factor F_{OSNR} shown in Fig. 15.20. In particular, it is evident from Fig. 15.22 that

- the impact of MPI on receiver performance initially decreases with increasing signal bandwidth. In the limit of large signal bandwidths ($B_s \to \infty$), F_Q approaches the value $2B_o/R\sqrt{(1 + B_o/(4B_e)^2)(1 + B_o/(2B_e)^2)}$. This, again, illustrates that RZ signaling is more robust to MPI than NRZ signaling, which has been experimentally confirmed by several groups [43, 78]. Also,
- the receiver degradation due to MPI depends on the electrical receiver bandwidth in general. The tolerance to MPI is reduced when going to lower receiver bandwidths.

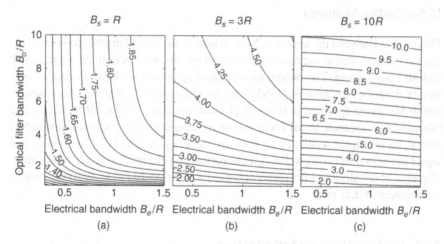

Fig. 15.22. MPI tolerance factor F_Q as a function of optical and electrical receiver bandwidths normalized to data rate R for (a) $B_s = R$, (b) $B_s = 3R$, and (c) $B_s = 10R$.

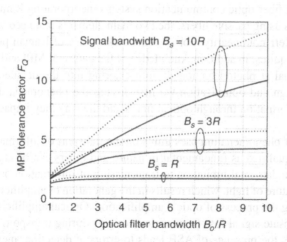

Fig. 15.23. Comparison between simple approximation to F_Q (dashed; see Eq. (15.105) for infinite electrical receiver bandwidths) and exact result using Gaussian pulses and Gaussian optical and electrical filters (solid; $B_e = 0.7R$) for $B_s = R, 3R,$ and $10R$. The approximation underestimates the impact of MPI.

Finally, Fig. 15.23 shows the quality of the approximation given in Eq. (15.105), obtained by neglecting the influence of electrical filtering, for the three signal bandwidths $B_s = R, 3R,$ and $10R$. The dashed lines represent the approximation to F_Q, Eq. (15.105), whereas the solid curves are section lines from Fig. 15.22 at a typical electrical receiver bandwidth of $B_e = 0.7R$, and thus follow Eq. (15.104). It can be seen that the approximation overestimates the receiver's tolerance to MPI (i.e., it underestimates the impact of MPI), with increasing deviation for large signal bandwidths.

15.5.4. Section Summary

This section analyzed the impact of MPI on beat-noise limited digital receivers. General formulae as well as various convenient approximations for beat noise due to ASE and MPI were given. The overall receiver performance was shown to depend on a weighted sum of the inverse OSNRs due to ASE and MPI. Receiver degradations due to MPI were specified in terms of Q-factor penalties and OSNR penalties. The tolerance of a receiver to MPI was shown to depend on the employed modulation format, on the signal quality at the receiver input (e.g., amount of ASE loading and signal extinction ratio), as well as on the optical and electrical receiver bandwidths. The amount of tolerable MPI power is typically some 20 dB below the signal power, independent of data rate.

15.6. System Design Optimization

In this section, we use the knowledge developed in previous sections to optimize the performance of fiber-optic communication systems incorporating Raman amplification. In systems using passive fibers, the two main elements that need to be balanced are ASE and Kerr nonlinearity. For systems making use of Raman pumping, such optimization requires, in addition, knowledge of the nature of MPI and the way MPI affects the signal at photodetection. Before entering into the detailed calculations on system design and optimization we first discuss the phenomena detrimental to transport of information through optical fibers and the way they impact system performance.

A variety of phenomena imposes limits on the transport of information, some of which can be qualified as fundamental in nature and others can be argued as being fortuitous. A fundamental limitation to transmission of information originates from the quantum nature of light, which results in the generation of amplified spontaneous emission during the process of optical amplification. Optical amplification is necessary to compensate signal loss from various origins during transport. As discussed in Section 15.5, the presence of ASE leads to errors at detection, measured by the BER. Because the BER depends on the optical signal-to-noise ratio, it can generally be improved by increasing the optical signal power. However, when the ASE sources are distributed throughout the transmission line, increasing the OSNR at the end of the line requires increasing signal power throughout the line which induces nonlinear effects in the optical fibers.

Because of the very high field confinement in the transverse direction, the intensities reached inside optical fibers quickly reach values at which the signal induces changes in the refractive index of the fiber core. For silica-core optical fibers, the index change is virtually instantaneous and does not depend on the polarization of the signal. It is referred to as Kerr nonlinearity [86]. For systems designed with realistic fiber types and signal powers, the source of signal distortions from Kerr nonlinearity on intensity-modulated formats originates almost solely from nonlinear interactions between the signals themselves (as opposed to interactions between signal and noise).

Interactions among signals are deterministic and, in the regime of powers accessible to current fiber-optic communication systems, do not initiate chaotic transmission behavior. Consequently, distortions imprinted on the signal by Kerr nonlinearity do not exhibit the stochasticity of ASE or MPI and can clearly be compensated for [87]. Nevertheless, we consider here that such distortions are not compensated for because it is technologically difficult and expensive to correct the deterministic signal distortions from Kerr nonlinearity at the receiver. To achieve optimal performance, fiber-optic communication systems are generally designed to operate at the maximum possible signal power at which any further power increase results in signal distortions from Kerr nonlinearity that outweighs the benefits from the improvement in OSNR. Consequently, the optimization of system performance involves a constant balancing of the effects of Kerr nonlinearity and sources of noise, such as ASE.

The amount of ASE an amplifier generates is not a critical function of the signal power going through the amplifier, but depends primarily on how much gain it provides. The impact of the ASE generated by an amplifier on the OSNR at the amplifier output depends directly on the signal input power to the amplifier. Consequently, in a chain of amplifiers, the maximum adverse effect of ASE on OSNR takes place in amplifiers following locations of lowest signal powers in the line.[23] Because pumping stations are generally the most expensive element in a transmission line, one wants to minimize their number by increasing the length of the passive transmission fibers. As a result, the passive transmission fibers are generally the most lossy elements in a line and the signal power in the line is at its lowest at their output. Therefore, the low signal power at the end of a passive transmission fiber is usually the origin of the largest OSNR degradations in a transmission line. Thus one can efficiently avert this OSNR degradation by preventing the signal power from dropping as low as it does when the transmission fiber is left passive. Such improvement in OSNR can be accomplished by converting the passive transmission fiber into an active fiber, for instance, by either using light doping with erbium atoms [88] or by using Raman pumping [89]. These active transmission fibers are called distributed fiber amplifiers (DFAs). The term *distributed amplifier* is used to describe amplifiers of physical length comparable to important characteristic transmission lengths, such as the loss length or the nonlinear length [86].

Because of their long lengths and their nonuniform gain distribution, DFAs are more likely to suffer from the generation of DRB than passive fiber spans. As discussed in Section 15.3.3, DRB produces a large number of low-power replicas of the signal with random delays and phases that propagate along with the signal. As a result, we are confronted with MPI from a large number of interferers. Thus the PDF for the composite intensity becomes Gaussian-like (see Section 15.2.2), which justifies the use of variances only (see Section 15.5.2).

The goal when designing a transmission system is the maximization of system performance by minimizing BER at the electrical decision gate (see Fig. 15.16). To minimize BER, one simultaneously has to reduce the effect of beat noise resulting

---

[23] Other sources of loss are, for instance, connectors, splices, optical filters, dispersion-compensators, crossconnects, add-drop multiplexers, and variable optical attenuators controlling the input power to a device.

from all sources of noise added to the signal (see Section 15.5.2) and prevent buildup of signal distortions from fiber dispersion and Kerr nonlinearity. The widespread availability of dispersion-compensating elements allows dispersion to be tailored to virtually any desired value, leaving Kerr nonlinearity the major source of signal distortion to be considered. In designing Raman-pumped systems it is then essential to balance the effects of three phenomena limiting system performance: ASE, DRB, and Kerr nonlinearity.

15.6.1. Raman Amplification and System Design

Raman amplification can be used to provide a large range of gain values for any fiber type and in any spectral window (or band). Raman amplification can also be applied to any particular fiber, regardless of its location or function in a transmission line. Moreover, Raman gain can be obtained by pumping a given fiber in the same direction as the signal propagation (copumping), in the opposite direction to the signal (counter-pumping), or Raman pumping through both fiber ends (bidirectional pumping). Even though all these Raman pumping configurations can in principle be used to improve system performance, it is generally desirable to reduce the number of Raman pumping sites and Raman pumps to minimize system cost.

The possible utilizations of Raman amplification in systems are represented in Fig. 15.24, depicting a typical span for four different transmission lines. Fig-

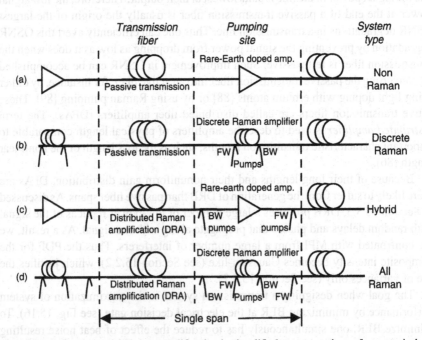

Fig. 15.24. Different use of Raman amplification in simplified representations of a transmission line and corresponding system types: (a) non-Raman; (b) discrete Raman; (c) hybrid; and (d) all-Raman (FW: forward, BW: backward).

ure 15.24(a) represents a transmission line that does not use Raman amplification. Optical amplification occurs periodically at pumping sites following each passive transmission fiber. Such discrete amplification is generally accomplished using a type of rare earth-doped amplifier. By the end of 2002, commercial systems almost universally use a single type of rare earth-doped amplifier, the erbium-doped fiber amplifier (EDFA). One possible application of Raman amplification is to replace discrete rare earth-doped amplifiers in bands where their effectiveness remains uncertain (Fig. 15.24(b)). These bands are essentially those outside the EDFA traditional C- and L-bands (1530 to 1570 nm and 1570 to 1610 nm, respectively), and have been labeled S^+-, S-, and L^+-bands (1450 to 1490 nm, 1490 to 1530 nm, and 1610-1650 nm, respectively) [90]. A second way of using Raman amplification is pumping transmission fibers (Fig. 15.24(c)), providing gain in any desired band. As mentioned earlier, providing distributed Raman amplification (DRA) in the transmission fiber is the most beneficial use of Raman amplification to reduce the buildup of ASE in the transmission line. In such a configuration, DRA is generally complemented with discrete amplification based on rare earth-doped amplifiers (generally EDFAs) at pumping sites. Such discrete amplification is usually necessary to make up for the lack of sufficient gain of DRAs in the presence of loss from various optical components present at pumping sites. The system configuration depicted in Fig. 15.24(c) is referred to as a hybrid-span system or simply hybrid system. Lastly, if DRA is used and all sources of gain in the transmission line come from Raman amplification, such a system is referred to as an all-Raman system (Fig. 15.24(d)). We chose to focus our attention on hybrid systems because they are simpler to analyze than all-Raman systems.

To determine the optimal use of Raman pumping in a hybrid system, one should consider Raman pumping transmission fibers by both ends and varying the amount of gain provided by pumping in each direction. The various pumping configurations of a hybrid system will result in different amounts of ASE, DRB, and distortions from Kerr nonlinearity. In the following sections, we describe a procedure developed for finding the optimum Raman pumping configuration that maximizes system performance. The optimum Raman pumping configuration is expressed in terms of a percentage of forward pumping and net Raman gain.

15.6.1.1. Amplified Spontaneous Emission

Any optical amplification process is simultaneously accompanied by spontaneous emission. Spontaneous emission corresponds to the spontaneous creation of photons through the decay of an electron from an excited atomic state to a lower energy state. This emission can occur at any time and location in an amplifier. Moreover, transitions can cover a large range of photon energies corresponding to all possible atomic level transitions. Because spontaneous emission occurs anywhere in a medium, spontaneous emission is amplified from the moment of its creation to the output end of an amplifier. It follows that the amplified signal comes with ASE that corrupts the information carried by the signal. In the case of a bidirectionally-pumped DRA with a single Raman pump on each side, the spectral density of the optical noise power N_{ASE} in each state of polarization is given by (undepleted pump approximation and assuming that the fiber is at 0 Kelvin) [59, 91],

$$N_{ASE} = hf \int_0^L C_R(\lambda_s, \lambda_p)[P_b e^{-\alpha_p(L-z)} + P_f e^{-\alpha_p z}] \underset{\rightarrow}{G}(z, L)dz, \quad (15.106)$$

where hf is the energy of a single photon at the wavelength at which ASE is to be evaluated, and

$$\underset{\rightarrow}{G}(z_1, z_2) = \underset{\rightarrow}{T}_F(z_1, z_2) \underset{\rightarrow}{G}_R(z_1, z_2) \quad (15.107)$$

is the net gain at the signal wavelength for signals propagating from z_1 to z_2 in the forward direction. The transmission $\underset{\rightarrow}{T}_F(z_1, z_2) = \exp[-\alpha_s(z_2 - z_1)]$ is the passive fiber transmission at the signal wavelength for signals propagating from z_1 to z_2 in the forward direction, and

$$\underset{\rightarrow}{G}_R(z_1, z_2) = \exp\left\{\frac{C_R(\lambda_s, \lambda_p)}{\alpha_p}[P_b(e^{-\alpha_p(L-z_2)} - e^{-\alpha_p(L-z_1)}) + P_f(e^{-\alpha_p z_1} - e^{-\alpha_p z_2})]\right\},$$

$$(15.108)$$

is the Raman gain at the signal wavelength for signals propagating from z_1 to z_2 in the forward direction. The DRA net gain G is given by $G_R T_F$, where $G_R = \underset{\rightarrow}{G}_R(0, L)$ is the Raman on-off gain at the signal wavelength for forward propagating signals and $T_F = \underset{\rightarrow}{T}_F(0, L)$ is the passive fiber transmission (≤ 1) at the signal wavelength for forward propagating signals. The various gain parameters are depicted in Fig. 15.25. In Eqs. (15.106) and (15.108), $C_R(\lambda_s, \lambda_p)$ is the Raman gain efficiency (gain factor

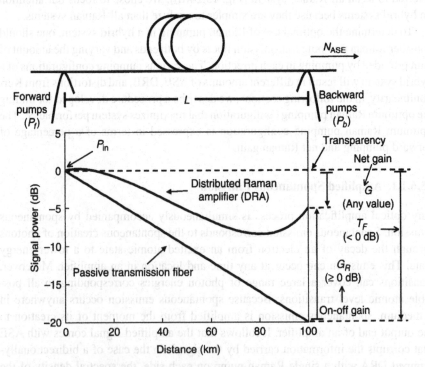

Fig. 15.25. Power evolution for a passive and a bidirectionally Raman-pumped transmission fiber and definition of the corresponding Raman gain parameters.

per unit of length and power) for signal and pump wavelengths λ_s and λ_p, respectively, α_p and α_s are the loss coefficients at the pump and signal wavelengths, and P_b and P_f are the backward and forward pump powers at their respective launch points on either ends of the fiber. The expressions "copumping" and "counter-pumping" are also used to describe forward and backward pumpings, respectively. The impact of nonzero fiber temperature on ASE generation has been neglected in Eq. (15.106) but can be incorporated without too much difficulty [92]. Under the assumption of perfectly depolarized pumps, the ASE from Raman amplification is randomly polarized so that the total spectral density of ASE in all polarization states is given by $2N_{\text{ASE}}$.

As for any amplifier, the spectral density of ASE at the output of a DRA depends on the degree of forward to total pump power and on the net gain. Figure 15.26(a) shows the spectral density of ASE per polarization state N_{ASE} at the output of a DRA for various values of net gain and Fig. 15.26(b) shows the OSNR_{ASE} (see Eq. (15.49)) considering only the ASE levels given in Fig. 15.26(a). For Fig. 15.26(b), the input power to the DRA, $\overline{P}_{\text{in}}$, is 0 dBm. The fiber considered is of the nonzero dispersion fiber (NZDF) type with the parameters: $L = 100$ km, $\alpha_s = 0.21$ dB/km, $\alpha_p = 0.26$ dB/km, $A_{\text{eff}} = 55$ μm^2, and $C_R(\lambda_s, \lambda_p) = 0.68$ (W km)$^{-1}$. One observes in Fig. 15.26(a) that the level of ASE is lower for forward pumping than for backward pumping. This can be understood by the difference in the net gain experienced by the ASE from the moment of its generation to the output end of the fiber (see Eq. (15.106)). The ASE generated near the fiber input by forward pumping experiences the loss of the full length of the fiber in addition of the Raman on-off gain. For backward pumping, the ASE generation occurs mainly near the fiber output end and thus experiences only a fraction of the fiber loss. As a result, forward pumping is more efficient to reduce ASE.

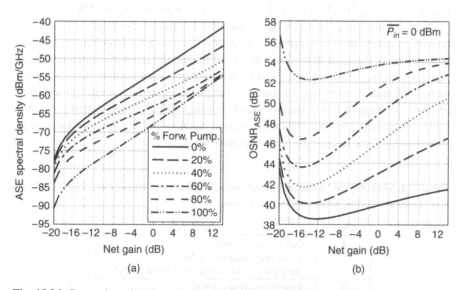

Fig. 15.26. Generation of ASE at the output of a distributed Raman amplifier and the corresponding OSNR_{ASE} as a function of net gain: (a) spectral density of ASE per polarization state (N_{ASE}); (b) OSNR_{ASE} for $\overline{P}_{\text{in}} = 0$ dBm. OSNR_{ASE} is calculated considering that the DRA is the only source of ASE. Fiber parameters are listed in main text.

Accordingly, as seen in Fig. 15.26(b), the OSNR$_{ASE}$ is higher for forward pumping than backward pumping. Note that the OSNR$_{ASE}$ increase observed in Fig. 15.26(b) for low net gain comes from the fact that only ASE from the DRA is considered when plotting OSNR$_{ASE}$. For such low net gain, an optical amplifier following the fiber span is necessary to bring back the signal to an appropriate power level for launching in the next span or photodetection. Such an amplifier then becomes the main source of OSNR$_{ASE}$ degradation. For instance, for the conditions of Fig. 15.26, a 3 dB noise figure amplifier following the fiber span would reduce the OSNR$_{ASE}$ to $\tilde{3}4$ dB after amplification to bring back $\overline{P}_{in}$ to 0 dBm.

15.6.1.2. Fiber Nonlinearity

Most fiber-optic communication systems are limited in signal power by the fiber Kerr nonlinearity, especially for long-reach systems or systems with high fiber span loss. These systems are generally designed to operate at the maximum signal power level that leads to a controlled amount of tolerable waveform distortions.

One can generally separate the effects of Kerr nonlinearity into two classes (see Table 15.1). The first class (Class I) involves continuous waveform distortions without energy transfer between interacting fields. The interacting fields correspond to individual channels for nonlinear WDM interactions [86] and individual pulses for intrachannel nonlinear interactions in high-speed pseudolinear transmission [93]. The main interactions in nonlinearity of Class I are self-phase modulation (SPM) [86], cross-phase modulation (XPM) [86], and intrachannel cross-phase modulation (IXPM) [94]. The second class (Class II) of Kerr nonlinear interactions involves energy transfer between interacting fields. The most important interactions belonging to this class are four-wave mixing (FWM) [86] and intrachannel four-wave mixing (IFWM) [94].

For systems limited by SPM, XPM, and IXPM, different signal power evolutions on a transmission fiber of given constant dispersion (along the fiber length) produce approximately the same distortions on signals having the same integrated nonlinear phase. One should note, however, that for this to be, one must optimize the dispersion map of the transmission line for each Raman pumping configuration. Optimization of dispersion maps refers to the process of choosing the value of dispersion compensation for individual dispersion-compensation modules and their precise location in a transmission line to optimize system performance.

Table 15.1. Separation of Nonlinear Interactions in Kerr Media in Two Classes, I and II, Which Have Different Scaling Laws with Aignal Powers.

Class	Nonlinear Effect
I	Self-Phase Modulation (SPM) Cross-phase Modulation (XPM) Intrachannel Cross-Phase Modulation (IXPM)
II	Four-Wave Mixing (FWM) Intrachannel Four-Wave Mixing (IFWM)

The integrated nonlinear phase of a signal power evolution $P(z)$ over a fiber segment of length L is given by

$$\phi_{NL}(L) = \int_0^L \gamma(z)\, P(z)\, dz, \qquad (15.109)$$

where $\gamma(z) = 2\pi n_2 f_0/(c_0 A_{eff})$ is the fiber nonlinear coefficient, n_2 the nonlinear refractive index coefficient, f_0 the central optical frequency of the signal, and A_{eff} is the fiber effective area [86]. Assuming that $\gamma(z)$ does not vary appreciably over the entire fiber length (which is approximately the case for fiber spans made of fibers of a single type), the ratio of the integrated nonlinear phase in a DFA, $\phi_{DFA}(L)$, to that of a passive fiber, $\phi_{Pas}(L)$, at fixed input power is simply given by the ratio of average signal powers,

$$R_{NL} = \frac{\phi_{DFA}(L)}{\phi_{Pas}(L)} = \frac{\int_0^L \overline{P}_{in}\, \underset{\rightarrow}{G}(0,z)\, dz}{\int_0^L \overline{P}_{in}\, \exp(-\alpha_s z)\, dz} = \frac{\alpha_s}{1 - \exp(-\alpha_s L)} \int_0^L \underset{\rightarrow}{G}(0,z)\, dz.$$

$$(15.110)$$

The nonlinear ratio R_{NL} represents the increase in Kerr nonlinearity of a DFA relative to a passive fiber span for systems limited by nonlinearity of Class I. This nonlinear ratio can be used to rescale signal power to equalize the effects of Kerr nonlinearity of Class I.

For a system in which transmission is limited by the nonlinear interactions of Class II (i.e., FWM and IFWM), the scaling of nonlinear interactions is no longer proportional to the signal power. The origin of this fundamental difference between the two classes lies in the scaling laws of any four-wave mixing process with signal power. In its most simplified form (ignoring, for instance, signal depletion [86]) the power generated during a FWM process is proportional to the product of the three signal powers involved in the FWM process. The amplitude of the beating on the electrical signal produced by this FWM is proportional to the ratio of FWM power to the undistorted signal power. Assuming all neighboring WDM channels have the same average power $\overline{P}_s$, then FWM power is proportional to $\overline{P}_s^3$ and the ratio of FWM to signal that determines signal distortions is proportional to $\overline{P}_s^2$. The same scaling laws also apply to IFWM, which corresponds to FWM interactions between neighboring overlapping pulses from the same channel [93]. Under the assumption that all pulses within a channel have the same shapes and powers, the ratio of powers of the "shadow" pulses (spurious pulses created in the empty bit slot as a result of IFWM) [93] to the undistorted pulse power is also proportional to $\overline{P}_s^2$, where $\overline{P}_s^2$ is average power of a given channel. Note that for IFWM the scaling law applies to each individual channel independently of the power of neighboring channels. The establishment of a nonlinear ratio R_{NL} for Class II nonlinearity requires, in general, knowledge of the specific FWM or IFWM processes involved in the nonlinear interaction. We leave this problem of forming R_{NL} for nonlinearity of Class II for further studies.

In bidirectionally-pumped DRAs, even with identical net gains, the signal power evolution depends strongly on the ratio of bidirectional pumping. This is illustrated

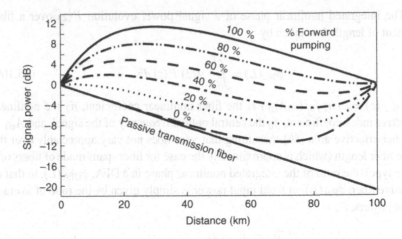

Fig. 15.27. Different signal power evolutions for different ratios of forward to total pump powers in a bidirectionally-pumped fiber. Power evolution in a passive transmission fiber is shown for comparison. Different ratios of bidirectional pumpings lead to different levels of fiber nonlinearity (see Fig. 15.28).

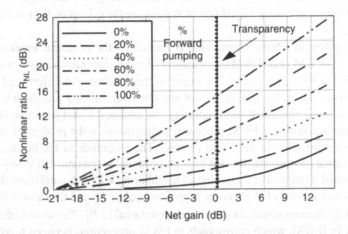

Fig. 15.28. Nonlinearity ratio R_{NL} (in dB) of Kerr nonlinearity in a bidirectionally Raman pumped 100-km-long transmission fiber for Class I nonlinearity (see Table 15.1).

in Fig. 15.27, which shows the signal power evolution for different percentages of forward pump power to total pump powers for a DRA pumped to transparency. Thus R_{NL} depends strongly on the percentage of forward pump power to total pump power. Figure 15.28 shows how R_{NL} scales for a 100-km-long Raman-pumped transmission fiber as a function of net gain and percentage of forward pumping. The fiber loss coefficients at the signal and pump wavelengths α_s and α_p are 0.21 dB and 0.26 dB, respectively. The passive span loss is 21 dB. For 100% backward pumping (0% forward pumping), it is only when the fiber starts to approach transparency (21 dB on-off

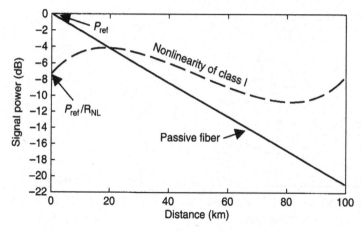

Fig. 15.29. Different signal power evolutions inside a transmission fiber leading to identical fiber nonlinearities in a DRA as in a passive transmission fiber. Solid curve: signal power in passive fiber. Dashed curve: same nonlinearity as passive fiber for Class I nonlinearity. The DRA is transparent and bidirectionally-pumped equally on both sides (50% forward pumping). Other parameters are identical to Fig. 15.28.

gain or 0 dB net gain) that the nonlinear ratio R_{NL} starts to increase significantly. As soon as forward pumping (copumping) is used, a few dBs of on-off gain immediately translates into an increase in nonlinearity ratio. As a result, ensuring constant nonlinearity over one fiber span, the signal launch power must be adjusted for different pump configurations. To illustrate this, Fig. 15.29 shows the signal power evolution for a passive transmission fiber with input power $\overline{P}_{ref}$ and for a bidirectionally-pumped DRA where the input signal power is reduced from $\overline{P}_{ref}$ to $\overline{P}_{ref}/R_{NL}$ to make the effects of Kerr nonlinearity identical in both systems.

One should point out that the nonlinear ratio R_{NL} is a measure of the averaged effects of Kerr nonlinearity of Class I. However, for very large changes in power evolution relative to the reference case, the nonlinear ratio R_{NL} may lose some accuracy. Such a case is exemplified by an extreme case where a DFA would have a constant signal power evolution along the fiber length. For identical values of R_{NL}, such a DFA would partially suppress some specific nonlinear interactions. In the case of dispersion-compensated solitons, for instance, constant signal power evolution suppresses the effect of XPM in symmetrized collisions [95, 96]. On the other hand, partial soliton collisions are less affected by having constant power evolution [95, 96].

15.6.1.3. Double-Rayleigh Backscattering

Double-Rayleigh backscattering of a signal can be significantly enhanced in DFAs because of the nonuniformity of the gain. The total DRB signal power in all polarization states P_{DRB} in a DFA of length L is given by (undepleted pump approximation) [7],

$$P_{DRB} = \overline{P}_{in}\, G(\alpha_R S)^2 \int_0^L \int_z^L \underset{\rightarrow}{G}(z,\zeta)\, \underset{\leftarrow}{G}(z,\zeta)\, d\zeta\, dz, \tag{15.111}$$

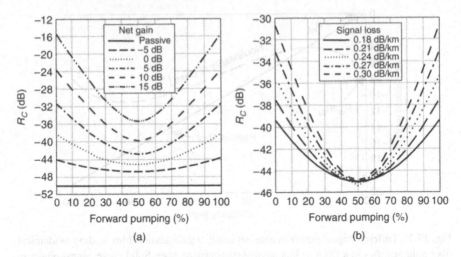

Fig. 15.30. Dependence of the R_C crosstalk ratio as a function of percentage of forward pumping in a 100-km-long bidirectionally-pumped DRA for (a) various levels of net gains and (b) various span losses. The loss at pump wavelength α_p is set 0.05 dB/km higher than the loss at signal wavelength α_s. Other fiber parameters are described in main text.

where $\alpha_R S$ is the Rayleigh backscatter coefficient, $\overline{P}_{in}$ is the input signal power to the DFA, and $\underset{\leftarrow}{G}(0, z)$ is the Raman net gain at the signal wavelength propagating in the backward direction from z to the fiber input end ($z = 0$). Note that different forms of Eq. (15.111) have been derived previously in this chapter (see Eqs. (15.22), (15.40), and (15.53)). It is worth pointing out that DRB is not randomly polarized, but, as described in Section 15.3.4, has a degree of polarization 5/9, copolarized with the signal [97].

To illustrate the effects of bidirectional pumping on DRB generation, we show in Fig. 15.30 how the crosstalk ratio R_C ($= P_{DRB}/\overline{P}_{in} G$, see Eq. (15.50)) is affected by bidirectional pumping of a DRA. The fiber parameters are identical to those of Section 15.6.1.1 (i.e., $L = 100$ km, $\alpha_s = 0.21$ dB/km, $\alpha_p = 0.26$ dB/km, $A_{eff} = 55\mu m^2$, and $C_R(\lambda_s, \lambda_p) = 0.68$ (W km)$^{-1}$) with the Rayleigh backscatter coefficient $\alpha_R S = 1.03 \times 10^{-4}$ km^{-1}.

As seen in Fig. 15.30(a), the crosstalk ratio increases rapidly with increasing Raman net gain except for symmetric bidirectional pumping (50% forward pumping) where the crosstalk ratio does not increase as rapidly with net gain increase. The impact of various span losses on the crosstalk ratio is shown in Fig. 15.30(b). As span losses increase, the crosstalk ratio increases only when the bidirectional Raman pumping is asymmetric (dominated by forward or backward pumping). As the bidirectional pumping becomes symmetric, the crosstalk ratio exhibits little dependence on the span loss for fixed fiber length. Clearly, symmetric bidirectional pumping significantly decreases the generation of DRB (keeping a low R_C), especially for pumping near transparency and above and for large span losses.

The values of the R_C crosstalk ratio plotted in Fig. 15.30 ignore the effects of pump depletion. However, the impact of pump depletion on R_C has been experimentally investigated [49] and it was found that the values of the R_C crosstalk ratio do not depend strongly on pump depletion even for pumps depleted by up to 6 dB where a reduction of only 0.3 dB in R_C was observed [49].

15.6.1.4. Noise Figure

A quantity commonly used to characterize noise performance in optical amplifiers is their noise figure, defined as [58, 98],

$$NF \equiv \frac{SNR_{in}}{SNR_{out}}, \tag{15.112}$$

where SNR_{in} is the electrical signal-to-noise ratio SNR at the input to the amplifiers where the only source of detection noise is the fundamentally unavoidable shot noise. It is given by

$$SNR_{in} \equiv \frac{\langle s_1 \rangle^2}{\sigma_1^2} = \frac{R_D^2 P_{in}^2}{\sigma_{shot}^2}, \tag{15.113}$$

where $\langle s_1 \rangle$ and σ_1 are the electrical signal current and current variance at the 1-bit peak. $\langle s_1 \rangle$ corresponds to $R_D P_{in}$ (see Eq. (15.67)) for the input signal to the amplifier, and σ_1 corresponds to σ_{shot} for a shot noise limited signal (see Section 15.5.3), respectively. The SNR at the output of the amplifier SNR_{out} is obtained by considering all sources of noise in the amplifier in addition to shot noise. For an amplifier limited by signal–ASE and signal–DRB beat noises, the output SNR is given by

$$SNR_{out} = \frac{R_D^2 P_{out}^2}{\sigma_{s-ASE}^2 + \sigma_{s-DRB}^2 + G \sigma_{shot}^2}, \tag{15.114}$$

where $P_{out} = G P_{in}$ is the output signal variance from the amplifier of gain G σ_{s-ASE}^2 and σ_{s-DRB}^2 are beat noise variances for ASE and DRB signal–ASE and signal–DRB beat noises, respectively (Section 15.5.1), and $G \sigma_{shot}^2$ is the shot noise variance at the amplifier's output. In the presence of these types of beat noise, the noise figure of an amplifier is then given by

$$NF = \frac{1}{G} \left(\frac{\sigma_{s-ASE}^2}{G \sigma_{shot}^2} + \frac{\sigma_{s-DRB}^2}{G \sigma_{shot}^2} + 1 \right). \tag{15.115}$$

In the limit of large optical filter bandwidths, the ratio of the signal–ASE beat noise to shot noise variances $\sigma_{s-ASE}^2 / \sigma_{shot}^2$ is given by $2 N_{ASE} / hf$ [57] where N_{ASE} is the spectral density of the optical ASE power copolarized with the signal. The contribution from the signal–DRB beat noise is more difficult to evaluate, as it involves beating of a colored noise with the signal. A detailed calculation of this type of beat noise is presented in Section 15.5 and in [69]. In the framework of Gaussian pulse and filter shapes and in the limit of large optical filter bandwidths, the ratio of the variances is given by (see Eqs. (15.95) and (15.96))

$$\frac{\sigma^2_{s-DRB}}{\sigma^2_{shot}} = \frac{5/9 \; P_{DRB}}{hf \; (B_e^2 + B_s^2/2)^{1/2}}, \tag{15.116}$$

where $5/9 \; P_{DRB}$ is the DRB power copolarized with the signal (see Section 15.3.4 and [97]), B_e is the equivalent square bandwidth of the electrical filter, and B_s is the equivalent square bandwidth of the optical signal (see Fig. 15.19). Note that in Eq. (15.116), the ratio of signal–DRB beat noise to shot noise decreases as either the signal bandwidth or receiver's electrical bandwidth increases [81, 99]. From the above results, one can explicitly write the noise figure NF_{DFA} of a DFA as

$$NF_{DFA} = \frac{1}{G} \left(\frac{2 \, N_{ASE}}{h \, f} + \frac{5/9 \; P_{DRB}}{hf \; (B_e^2 + B_s^2/2)^{1/2}} + 1 \right). \tag{15.117}$$

Both N_{ASE} and P_{DRB} refer to quantities evaluated at the amplifier's output. Note that the noise figure incorporating double-Rayleigh scattering now depends on signal power through $P_{DRB} = R_C \overline{P}_{out}$.

Optical amplifiers are frequently concatenated. The chain of amplifiers itself can be considered as a single amplifier having its own noise figure. In the case of concatenation of two amplifiers, the first one labeled 1 with noise figure NF_1 followed by a second labeled 2 and with noise figure NF_2, the noise figure NF_{12} of the amplifier's pair is given by [58, 98]

$$NF_{12} = NF_1 + \frac{NF_2 - 1}{G_1}, \tag{15.118}$$

where G_1 is the net gain of the first amplifier. The noise figure for concatenation of more than two amplifiers can be obtained by using Eq. (15.118) recursively.

15.6.1.5. Output Signal-to-Noise Ratios

To quantify the benefits of using DRAs instead of passive fiber spans in terms of an improvement of electrical SNR at the receiver, we first need to evaluate the ratio of SNRs at the end of a transmission line operating under these two different conditions (see Fig. 15.31). Under the first operating condition (Fig. 15.31(a)), each fiber span remains passive, and the optimal input signal power $\overline{P}_{ref}$ is adjusted to minimize BER. Note that this optimum signal power typically leads to an eye closure at the end of the line in the range of 10 to 30% relative to the line input. The noise figure of the entire span (including the passive fiber and all other elements of the span) is labeled NF_{Pas}, and is evaluated for the span input signal power $\overline{P}_{ref}$. Under the second operating condition, using DRAs (Fig. 15.31(b)), the signal power at the input of each DRA has to be reduced by a nonlinear ratio R_{NL} relative to $\overline{P}_{ref}$ to ensure similar fiber nonlinearities (Section 15.6.1.2). Similarly to NF_{Pas}, the DRA span noise figure is evaluated at the input signal power $\overline{P}_{ref}/R_{NL}$ and is labeled NF_{Act}. The SNR at the output of a transmission line is obtained by generalizing Eq. (15.114) to take

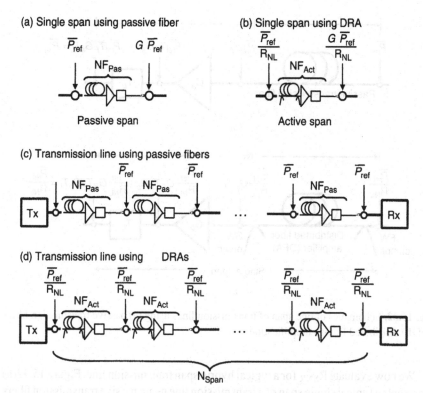

Fig. 15.31. Schematics of two operating conditions of a pair of amplifiers and transmission lines. In (a) and (b), two amplifiers have identical gains but different noise figures and operate with different input signal powers. These amplifiers can be used to represent individual spans in transmission lines using (c) passive fibers and (d) DRAs.

into account the accumulation of signal–ASE and signal–DRB variances in a chain of N_{Span} identical spans, and is given by,

$$\text{SNR}_{\text{out}} = \frac{R_D^2 P_{\text{out}}^2}{N_{\text{Span}} (\sigma_{s-\text{ASE}}^2 + \sigma_{s-\text{DRB}}^2) + G \sigma_{\text{shot}}^2}. \tag{15.119}$$

Applying Eq. (15.119) to the two operating conditions of Fig. 15.31 we can obtain the ratio of output SNRs, R_{SNR}, at the end of a transmission line,

$$R_{\text{SNR}} \equiv \frac{\text{SNR}_{\text{out}}^{\text{Act}}}{\text{SNR}_{\text{out}}^{\text{Pas}}} = \frac{1}{R_{\text{NL}}} \frac{N_{\text{Span}}(\text{NF}_{\text{Pas}} - 1) + 1}{N_{\text{Span}}(\text{NF}_{\text{Act}} - 1) + 1}. \tag{15.120}$$

Note that for a single span, R_{SNR} reduces to the simple expression $\text{NF}_{\text{Pas}}/(R_{\text{NL}}\text{NF}_{\text{Act}})$ corresponding to the ratio of the active and passive span noise figures weighted by the nonlinear ratio (see Fig. 15.31). One should point out that in systems not producing signal distortions from fiber nonlinearities, Eq. (15.120) also applies if one sets the nonlinearity ratio R_{NL} to 1.

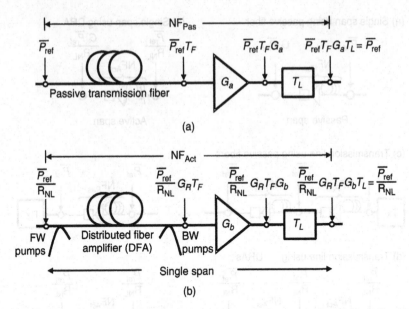

Fig. 15.32. Schematics of one span of transmission lines using (a) passive transmission fibers and (b) distributed Raman amplification.

We now evaluate R_{SNR} for a typical hybrid span transmission line. Figure 15.32(a) represents a typical single span of a transmission line using passive transmission fibers and Fig. 15.32(b) applies to a hybrid span transmission line using DRA. The fiber nonlinearity is made equal in both lines by reducing the signal input power to the DFA by the nonlinear ratio R_{NL}. A discrete lossy element of transmission T_L has been inserted after the discrete amplifier to account for the loss of optical elements such as gain equalizers, add-drop multiplexers, and crossconnects. By using Eq. (15.118) for the concatenation of noise figures of a chain of amplifiers, one can derive the noise figures for a passive span (NF$_{Pas}$) and for a hybrid span (NF$_{Act}$) including all elements of each span. These noise figures for the entire spans are

$$\text{NF}_{\text{Pas}} = \frac{\text{NF}_a}{T_F} + 1 - T_L \tag{15.121}$$

and

$$\text{NF}_{\text{Act}} = \text{NF}_{\text{DFA}} + \frac{\text{NF}_b - 1}{G_R T_F} + 1 - T_L, \tag{15.122}$$

where NF$_a$ and NF$_b$ are the noise figures of the discrete amplifiers in passive and hybrid span transmission lines, respectively. Assuming that the discrete amplifiers of Fig. 15.32 are fully inverted (3-dB noise figures when operated at high gain), one can easily show that

$$R_{\text{SNR}} = \frac{1}{R_{\text{NL}}} \frac{N_{\text{Span}} (2/T_F - 2T_L) + 1}{N_{\text{Span}} [\text{NF}_{\text{DFA}} + 1/(G_R T_F) - 2T_L] + 1}, \tag{15.123}$$

where DRB has been neglected in the calculation of NF_{Pas} as R_C is typically lower than -50 dB for passive fiber spans. Note that our choice of fully inverted amplifiers for transmission lines using passive transmission fibers in Eq. (15.123) corresponds to the best possible scenario for these systems. More realistic optical amplifiers incorporating potentially many functionalities will most likely have higher noise figures that typically range between 4 and 10 dB. One should emphasize that the benefits of using DRA in systems with such high noise figure amplifiers are larger than in systems with ideal amplifiers.

The parameter R_{SNR} is our final measure of system performance. Our interpretation of this factor is as follows: The factor R_{SNR} is the improvement of signal-to-noise ratios at the end of a transmission line using distributed Raman amplification relative to a line using passive transmission fibers and ideal optical amplifiers. This SNR ratio is obtained using different signal input powers to the spans for the two lines corresponding to equal average effects of fiber nonlinearities. Therefore the factor R_{SNR} gives the amount of additional dBs of SNR that become available by making use of distributed Raman amplification in a passive transmission line.

15.6.1.6. Results and Discussion

We illustrate the technique described above by considering the optimization of a hybrid Raman-pumped system limited by the nonlinearity of Class I (SPM, XPM, or IXPM). This example encompasses such systems as long-haul WDM systems operating at 10 Gb/s over NZDF and standard single-mode fibers (SSMFs) or at 40 Gb/s over NZDF.

The DRA is made of a NZDF operating at 1550 nm without any nonreciprocal element present in the fiber path. Under such conditions, forward and backward gains $\underset{\rightarrow}{G}$ and $\underset{\leftarrow}{G}$ are identical. The fiber parameters are identical to those in Section 15.6.1.3, except for $L = 80$ km. The nonlinear coefficient $\gamma = 1.67$ W^{-1} km^{-1}.

Figure 15.33 shows isocontours of SNR ratios, R_{SNR} (in dB), calculated using Eq. (15.123) as a function of the DRA net gain and the percentage of forward pumping. An improvement in SNR from DRA translates into a positive value for R_{SNR} (in dB). Figures 15.33(a) and 15.33(b) are isocontours when either ASE or DRB noise sources are considered in the DRAs, respectively. We considered system lengths of 40 spans. For the beat noise calculation from DRB we used an electrical receiver bandwidth $B_e = 5$ GHz and an optical signal bandwidth $B_s = 9.4$ GHz corresponding to the equivalent square optical bandwidth of a 50% duty cycle return-to-zero (RZ) format. The DRB power in Eq. (15.111) is calculated using $\overline{P}_{in} = \overline{P}_{ref}/R_{NL}$, where $\overline{P}_{ref} = 0.5$ mW (-3 dBm). The lossy element has 10 dB of loss ($T_L = 0.1$). It is worth mentioning that for up to 10 dB of loss, the presence of this lossy element does not have a significant impact on either the value of R_{SNR} or the optimum pumping scheme configuration. Figure 15.33(c) shows the isocontours when both sources of noise are simultaneously included in the calculation of R_{SNR}. One sees in Fig. 15.33(a) that when ASE is the only source of noise in the DRA, the optimum pumping scheme corresponds to 30% forward Raman pumping power and the optimum net gain is a few

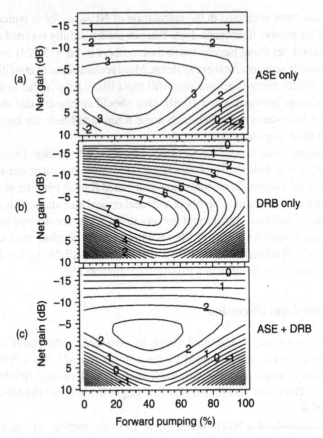

Fig. 15.33. Ratio of output SNRs, R_{SNR} (in dB): (a) considering ASE only in the DRA; (b) considering DRB only in the DRA; and (c) with both sources of optical noise in the DRA included.

dBs above transparency. When DRB is the only source of noise (Fig. 15.33(b)), the optimum pumping scheme corresponds to 35% forward pumping and a slightly negative net gain (-2 dB). When both noise sources of DRAs are included (Fig. 15.33(c)), the optimum Raman pumping scheme is 40% forward pumping with a net gain of -3 dB. At high net gain, generation of DRB in the DRA degrades the SNR whereas for low net gain the limitations come from the generation of ASE. One should note that generation of DRB can be strongly reduced by inserting isolators in the fiber span while generation of ASE remains unaffected. Thus limitations from ASE seen in Fig. 15.33 can be interpreted as an ultimate limit to improvement in system performance from DRA.

More generally, the optimal Raman pumping configuration depends on the length of each fiber span and type(s) of fiber used, their loss and backscatter parameters, the noise figure of the discrete amplifiers, the class of nonlinearity limiting transmission, the data rate, and filter design. Moreover, to achieve broadband gain flatness, a few Raman pumps are generally required and the procedure presented in this section needs to be generalized to accommodate multiple pumps. Optimization of Raman pumping should be applied to each system that one wishes to optimize.

15.6.1.7. Bit-Rate Scaling Laws for System Design

Sections 15.6.1.1 to 15.6.1.3 described the three important parameters determining system performance in transmission lines using Raman amplification. These impairments can be summarized by the three parameters: $OSNR_{ASE}$, the signal power per channel $\overline{P}_s$, and $OSNR_{MPI}$. Scaling of system performance with bit rate per channel R for *fixed* received $OSNR_{ASE}$ and *fixed* received $OSNR_{MPI}$ have been described in Section 15.5.2.4. In this section, we first consider the effect of bit rate scaling on MPI requirements for a transmission line at *fixed* received $OSNR_{ASE}$ and *fixed* received $OSNR_{MPI}$. We then discuss the effects of varying the signal power $\overline{P}_s$.

A transmission line using Raman amplification designed to minimize the effects of optical noise sources will generate levels of ASE and MPI that produce comparable signal degradation at the receiver for a given bit rate R (degradations can be expressed, for instance, by Eqs. (15.81), (15.82), (15.88), and (15.89)). Let's call $OSNR_{ASE}$ and $OSNR_{MPI}$ the corresponding OSNR levels at this optimum operating point. If one wants to increase the bit rate n-fold at *fixed* signal power then, as discussed in Section 15.5.2.4, the required $OSNR_{ASE}$ increases n-fold and the required $OSNR_{MPI}$ remains unchanged. Expressed differently, the electrical bandwidths B_e and B_s are naturally scaled up n-fold with the bit rate, reducing n-fold the relative impact of the MPI (or DRB) term in the noise figure given in Eq. (15.117). Consequently, the $OSNR_{ASE}$ at the bit rate $n\,R$ will be insufficient to provide the same BER as for bit rate R. The situation can be corrected, for instance, by operating at higher net gain for the DRA which will increase $OSNR_{ASE}$, at the expense of $OSNR_{MPI}$ (see Fig. 15.33(b)). A new optimum balance between the effects of ASE and MPI on detection will be established. Another alternative to recover system performance is to simply scale back system reach to increase $OSNR_{ASE}$ and to allow relaxed specifications on sources of MPI. Additional margin on MPI specifications especially becomes of value when sources of MPI with high variability are present in a system. An example of a source of MPI that is difficult to predict accurately is the MPI produced by discrete reflections by damaged or dirty connectors in the transmission path.

We now additionally consider the effect of varying the signal power $\overline{P}_s$. For a given transmission line layout, the level of ASE is fixed and $OSNR_{ASE}$ is determined solely by the signal power $\overline{P}_s$. As pointed out earlier, the required $OSNR_{MPI}$ does not change with $\overline{P}_s$ or bit rate whereas the required $OSNR_{ASE}$ increases n-fold with bit rate and $\overline{P}_s$. As a result, including both the effects of ASE and MPI, an n-fold increase in bit rate R followed by a corresponding n-fold increase in $\overline{P}_s$ leads to $OSNR_{ASE}$ and $OSNR_{MPI}$ that leaves the BER unaffected (assuming identical signal degradations from fiber nonlinearities). It is interesting to note that for a fixed signal spectral density ($= R/\Delta f$, where Δf is the channel spacing), the n-fold scaling for $\overline{P}_s$ with R for constant BER is automatically fulfilled with an optical amplifier with constant *total* output power.

Whether it is possible to raise n-fold the signal power $\overline{P}_s$ in a transmission line depends essentially on fiber nonlinearity. The maximum value of $\overline{P}_s$ that can transmit with limited degradation from nonlinear signal distortions depends on a large number of system parameters such as fiber types, dispersion maps, modulation formats, and so on. It is beyond the scope of this chapter to determine how $\overline{P}_s$ scales with bit rate. Nevertheless, as can be assessed from the published literature (see [93] and references therein), we can identify some cases where it is more likely to observe a significant increase in the maximum value of $\overline{P}_s$ as the bit rate R increases. An example of such a case is for a spectral efficiency of 0.4 bits/s/Hz where systems operating at 40 Gb/s (100 GHz channel spacing) have generally higher $\overline{P}_s$ than systems operating at 10 Gb/s (25 GHz spacing). For such spectral efficiency, just from the transmission and detection standpoints, it may become advantageous to use 40 Gb/s per channel instead of 10 Gb/s if, for the particular system design of interest, the maximum value of $\overline{P}_s$ at 40 Gb/s exceeds by more than 6 dB the maximum $\overline{P}_s$ at 10 Gb/s.

15.6.2. Section Summary

In this section we derived a procedure to evaluate the ratio of SNRs at the end of a transmission line using passive transmission fibers and DRAs (hybrid spans) keeping constant the effects of fiber nonlinearity. Two classes of fiber nonlinearity with their respective scaling laws were introduced to accurately describe the effect of various nonlinear interactions. Using the procedure described in this section, we explored bidirectional Raman pumping of a hybrid span transmission line and showed that there exist an optimum net gain and percentage of forward pumping that maximize the benefits of distributed Raman amplification in communications systems.

A15.A. Derivation of Autocorrelation of DRB-Induced Intensity Noise

This appendix sketches important steps in the derivation of the autocorrelation $\Gamma_{\Delta I}(\tau)$ of the DRB-induced intensity fluctuations $\Delta I(t)$, Eq. (15.32). The mathematical steps involved are also needed in the derivation of the DRB field autocorrelation $\Gamma_{\boldsymbol{\varepsilon}_{DRB}}(t, t+\tau)$, Eq. (15.39).

We start by writing the autocorrelation of the intensity fluctuations $\Delta I(t)$ in terms of the involved scalar optical fields,

$$\Gamma_{\Delta I}(t, t+\tau) = \left\langle \left[\boldsymbol{\varepsilon}_s(t)\boldsymbol{\varepsilon}_{DRB}(t) + \boldsymbol{\varepsilon}_s^{\star}(t)\boldsymbol{\varepsilon}_{DRB}^{\star}(t) \right]^{\star} \right.$$
$$\left. \times \left[\boldsymbol{\varepsilon}_s(t+\tau)\boldsymbol{\varepsilon}_{DRB}(t+\tau) + \boldsymbol{\varepsilon}_s^{\star}(t+\tau)\boldsymbol{\varepsilon}_{DRB}^{\star}(t+\tau) \right] \right\rangle, \quad (A15.1)$$

where we set $\boldsymbol{\varepsilon}_s(L, t) = \boldsymbol{\varepsilon}_s(t)$ and $\boldsymbol{\varepsilon}_{DRB}(L, t) = \boldsymbol{\varepsilon}_{DRB}(t)$ for notational convenience. Next we perform the multiplication, and substitute the scalar version of the integral expression in Eq. (15.28) for $\boldsymbol{\varepsilon}_{DRB}(t)$ (i.e., we ignore the matrices $\mathbb{J}_{12}$ and $\mathbb{J}_{21}$). Then we interchange the order of integration and ensemble averaging. The result is a sum of four terms, each containing a fourfold integration. Because the differential Rayleigh backscatter coefficient $\rho(z)$ is statistically independent of any random fluctuations of $\boldsymbol{\varepsilon}_s(t)$, we can separate the average over $\rho(z)$ and over $\boldsymbol{\varepsilon}_s(t)$, and are left with ensemble averages of the form (15.30) and (15.31) within the integrals. Due to Eq. (15.31), two of the four integral expressions vanish. The remaining two fourfold integrals are complex conjugated to each other, which turns their sum into a real part,

$$
\begin{aligned}
\Gamma_{\Delta I}(t, t+\tau) = 2\mathrm{Re}\Bigg\{ &\int_0^L dz_2 \int_0^L dz_4 \int_0^{z_2} dz_1 \int_0^{z_4} dz_3 \langle \rho^*(z_1)\rho^*(z_2)\rho(z_3)\rho(z_4)\rangle \\
&\times \sqrt{\underrightarrow{G}(z_1, z_2)\,\underleftarrow{G}(z_1, z_2)} \sqrt{\underrightarrow{G}(z_3, z_4)\,\underleftarrow{G}(z_3, z_4)} \\
&\times \langle \boldsymbol{\varepsilon}_s^*(t)\boldsymbol{\varepsilon}_s^*(t - 2(z_2-z_1)/v_g)\boldsymbol{\varepsilon}_s(t+\tau)\boldsymbol{\varepsilon}_s(t+\tau - 2(z_4-z_3)/v_g)\rangle \\
&\times e^{-j2\beta(z_2-z_1)}e^{j2\beta(z_4-z_3)}\Bigg\}.
\end{aligned}
$$

$$\text{(A15.2)}$$

Inserting the correlation relation (15.30) for $\rho(z)$, the fourfold integral in Eq. (A15.2) splits into a sum of two terms, $\Gamma_{\Delta I}(t, t+\tau) = 2\mathrm{Re}\{F_1(t, t+\tau) + F_2(t, t+\tau)\}$, with

$$
F_1(t, t+\tau) = \int_0^L dz_2 \int_0^L dz_4 \int_0^{z_2} dz_1 \int_0^{z_4} dz_3\, f(z_1, z_2, z_3, z_4; t, t+\tau)\delta(z_1 - z_4)\delta(z_2 - z_3)
$$

$$\text{(A15.3)}$$

$$
F_2(t, t+\tau) = \int_0^L dz_2 \int_0^L dz_4 \int_0^{z_2} dz_1 \int_0^{z_4} dz_3\, f(z_1, z_2, z_3, z_4; t, t+\tau)\delta(z_1 - z_3)\delta(z_2 - z_4),
$$

$$\text{(A15.4)}$$

where $f(z_1, z_2, z_3, z_4; t, t+\tau) = (\alpha_R S)^2 \sqrt{\underrightarrow{G}(z_1, z_2)\,\underleftarrow{G}(z_1, z_2)} \sqrt{\underrightarrow{G}(z_3, z_4)\,\underleftarrow{G}(z_3, z_4)}$
$\langle \boldsymbol{\varepsilon}_s^*(t)\boldsymbol{\varepsilon}_s^*(t - 2(z_2-z_1)/v_g)\boldsymbol{\varepsilon}_s(t+\tau)\boldsymbol{\varepsilon}_s(t+\tau - 2(z_4-z_3)/v_g)\rangle e^{-j2\beta(z_2-z_1)}e^{j2\beta(z_4-z_3)}$.

We first show that $F_1(t, t+\tau) \equiv 0$. The innermost integral (over z_3) results in replacing $f(z_1, z_2, z_3, z_4; t, t+\tau)$ by $f(z_1, z_2, z_2, z_4; t, t+\tau)$ if the Dirac functional can take action, that is, if $z_2 - z_3 = 0$ lies within the range of integration. This is only the case if $z_2 < z_4$. Otherwise the integration result is zero. Next we integrate over z_1, which results in replacing $f(z_1, z_2, z_2, z_4; t, t+\tau)$ by $f(z_4, z_2, z_2, z_4; t, t+\tau)$, if $z_1 - z_4 = 0$ lies within the range of integration. This is only the case if $z_2 > z_4$. Therefore, the integrals over z_1 and z_3 are mutually exclusive, and $F_1(t, t+\tau) \equiv 0$, *independent* of the nature of $f(z_1, z_2, z_3, z_4; t, t+\tau)$.

Next we simplify $F_2(t, t+\tau) \equiv 0$: By the same argument as before, the integral over z_3 results in replacing $f(z_1, z_2, z_3, z_4; t, t+\tau)$ by $f(z_1, z_2, z_1, z_4; t, t+\tau)$ if $z_4 > z_1$. Otherwise the integration result is zero. A subsequent integration over z_4 results in replacing $f(z_1, z_2, z_1, z_4; t, t+\tau)$ by $f(z_1, z_2, z_1, z_2; t, t+\tau)$, and in replacing the condition $z_4 > z_1$ by $z_2 > z_1$. Because z_4 covers the entire range of z_2, the Dirac functional always takes action here. The remaining expression reads

$$
F_2(t, t+\tau) = \int_0^L dz_2 \int_0^{z_2} dz_1\, f(z_1, z_2, z_1, z_2; t, t+\tau). \qquad \text{(A15.5)}
$$

The necessary condition $z_2 > z_1$ for a nonvanishing z_3-integration is always satisfied here, therefore this is the final result. Assuming further that the correlation length of the stationary source is much shorter than paths that lead to significant contributions to the DRB field, we can write $\langle \boldsymbol{\varepsilon}_s^*(t)\boldsymbol{\varepsilon}_s^*(t - 2(z_2 - z_1)/v_g)\boldsymbol{\varepsilon}_s(t + \tau)r\varepsilon_s(t + \tau - 2(z_2 - z_1)/v_g)\rangle = \langle \boldsymbol{\varepsilon}_s^*(t)\boldsymbol{\varepsilon}_s(t + \tau)\rangle\langle \boldsymbol{\varepsilon}_s^*(t - 2(z_2 - z_1)/v_g)\boldsymbol{\varepsilon}_s(t + \tau - 2(z_2 - z_1)/v_g)\rangle = |\Gamma_{\boldsymbol{\varepsilon}_s}(\tau)|^2$, and directly arrive at Eq. (15.32).

A15.B. Derivation of MPI Field Autocorrelation

This appendix describes a way to calculate the autocorrelation and spectrum of a digitally modulated optical sequence, given the signaling constellation and the pulse waveform. The approximations are tailored to meet the needs of signal–MPI beat noise calculations.

We first write the digitally modulated, optically filtered data signal as

$$\widetilde{\boldsymbol{\varepsilon}}_s(t) = \sum_{k=-\infty}^{\infty} \boldsymbol{a}_k \widetilde{\varepsilon}_p(t - k/R), \tag{A15.6}$$

where $\boldsymbol{a}_k$ denote the (possibly complex) signaling points, assumed random for mathematical convenience, $\widetilde{\varepsilon}_p(t)$ is the optically filtered field of an isolated pulse, and $1/R$ denotes the bit duration. Note that this model leads to analytically tractable results, but has its formal shortcomings caused by pulse field overlap whenever the individual pulses $\widetilde{\varepsilon}_p(t)$ are longer than the bit duration. This can be the case for high duty cycle RZ signaling, for example, and, in particular, for NRZ modulation. Nevertheless, even in such limiting cases the sequence spectral envelope closely matches the isolated pulse spectrum, and therefore results in accurate approximations when used for the signal–MPI beat noise calculation.

As a next step, we calculate the sequence autocorrelation, which can generally be shown to be [42]

$$\Gamma_{\widetilde{\boldsymbol{\varepsilon}}_s}(\tau, \tau') = \langle \widetilde{\boldsymbol{\varepsilon}}_s^*(\tau)\widetilde{\boldsymbol{\varepsilon}}_s(\tau')\rangle = R \sum_{m=-\infty}^{\infty} \Gamma_{\boldsymbol{a}}(m)\Gamma_{\widetilde{\varepsilon}_p}(\tau - \tau' - m/R). \tag{A15.7}$$

This equation takes the form of a discrete-time convolution of the data autocorrelation

$$\Gamma_{\boldsymbol{a}}(m) = \langle \boldsymbol{a}_k \boldsymbol{a}_{k+m}^* \rangle \tag{A15.8}$$

and the pulse autocorrelation

$$\Gamma_{\widetilde{\varepsilon}_p}(\tau - \tau') = \int_{-\infty}^{\infty} \widetilde{\varepsilon}_p(\xi)\widetilde{\varepsilon}_p^*(\tau - \tau' + \xi)d\xi. \tag{A15.9}$$

To give some examples, the data autocorrelations for plain on-off keying and carrier-suppressed on-off keying (CS-OOK, used widely in the form of 67% duty cycle CS-RZ) can be shown to read

$$\Gamma_{\boldsymbol{a},\text{OOK}}(m) = \frac{1}{4}[\delta(m) + 1] \quad \text{and} \tag{A15.10}$$

$$\Gamma_{\boldsymbol{a},\text{CS-OOK}}(m) = \frac{1}{4}[\delta(m) + (-1)^m]. \tag{A15.11}$$

Assuming Gaussian pulses and filters (cf. Eqs. (15.90) and (15.91)), the pulse auto-correlation reads

$$\Gamma_{\tilde{\varepsilon}_p}(\tau - \tau') = \tilde{E}_p \exp[-\pi B_s^2/(1 + B_s^2/B_o^2)(\tau - \tau')^2], \qquad (A15.12)$$

where $\tilde{E}_p$ is the filtered pulse energy. The autocorrelations for the two optical sequences thus read

$$\Gamma_{\tilde{s},\text{OOK}}(\tau, \tau') = \frac{1}{4/R}\left\{2\Gamma_{\tilde{\varepsilon}}(\tau - \tau') + \sum_{m \neq 0}\Gamma_{\tilde{\varepsilon}_p}(\tau - \tau' - m/R)\right\} \quad \text{and}$$

$$\qquad (A15.13)$$

$$\Gamma_{\tilde{s},\text{CS-OOK}}(\tau, \tau') = \frac{1}{4/R}\left\{2\Gamma_{\tilde{\varepsilon}_p}(\tau - \tau') + \sum_{m \neq 0}(-1)^m\Gamma_{\tilde{\varepsilon}_p}(\tau - \tau' - m/R)\right\}.$$

$$\qquad (A15.14)$$

Translating these expressions to the frequency domain reveals the well-known facts that the power spectra are composed of

- a continuous part, corresponding to the $\delta(m)$ portion of the data autocorrelations (cf. Eqs. (A15.10) and (A15.11)), and
- a comb of tones spaced at integer multiples of R (OOK) and at integer multiples of R offset by $R/2$ (CS-OOK), corresponding to the sum of shifted pulse autocorrelations.

Equations (A15.13) and (A15.14) are visualized in Fig. A15.1, evaluated for $B_s = 2R$ and $B_o = 2.2R$. The figure shows that the shifted pulse autocorrelations (dotted for OOK and dashed for CS-OOK) may overlap, suggesting that it could be insufficient to consider only the center part of the optical signal autocorrelation (shaded). However, because the signal–MPI beat noise integral (15.64) only takes into account a range of the optical signal autocorrelation corresponding to $-T_h < \tau - \tau' < T_h$, where T_h stands for some effective duration of the detector impulse response (measured, e.g., by

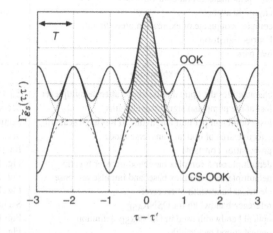

Fig. A15.1. Visualization of optical sequence autocorrelations for plain OOK and carrier-suppressed OOK. Gaussian pulses and filters are assumed, with $B_s = 2R$ and $B_o = 2.2R$.

a $1/e$ drop of $h(t)$), it can be justified within the frame of a reasonable approximation to neglect all shifted replicas of the pulse autocorrelation in the beat noise integral, and to keep only the main (centered) portion. The shorter the optically filtered pulses, and the larger the bandwidth of the receiver electronics, the better will this approximation match realistic receivers. In general, for plain OOK, the approximation will underestimate the signal–MPI beat noise, whereas for CS-OOK it will overestimate the beat noise, as evident from comparing the shaded portion of the autocorrelation in Fig. A15.1 with the full autocorrelation (solid lines).

Physically, the above approximation corresponds to neglecting the beat noise stemming from discrete tones in the MPI field spectrum. These tones fall on top of the tones in the signal spectrum, thus their main beating contributions will be at low electrical frequencies (determined by the laser linewidth). Because the integral effect of low frequencies is less pronounced for reasonably broadband receiver electronics, the error in neglecting these tones becomes small in this regime.

Taking note of all approximations made above, we can finally write the MPI field autocorrelation, for OOK and CS-OOK alike, as

$$\Gamma_{\widetilde{\boldsymbol{\mathcal{E}}}_s}(\tau, \tau') \approx P_{\mathrm{MPI}}^P \sqrt{1/(1 + B_s^2/B_o^2)}\, \exp[-\pi B_s^2/(1 + B_s^2/B_o^2)(\tau - \tau')^2], \quad (A15.15)$$

where the normalization to P_{MPI}^P, the copolarized optical MPI power measured *in front of* the WDM demultiplexer has been further introduced.

A15.C. List of Variables

General Notation	Explanation	Occurrence
$\widetilde{x}$	quantity x after optical demultiplexing filter	Fig. 15.16
$\widehat{x}$	peak of pulsed quantity x	Eqs. (15.87), (15.90)
$\widehat{\widetilde{x}}$	optically filtered peak of pulsed quantity x	Eq. (15.87)
$\overline{x}$	time average of quantity x	Eq. (15.73)
$\overline{\widetilde{x}}$	time average of optically filtered quantity x	Eq. (15.87)
$\boldsymbol{x}$	bold print indicates random variables	
$\langle \cdot \rangle$	ensemble average	
c.c.	complex conjugate of expression preceding it	
$\mathcal{F}\{\cdot\}$	Fourier transform	
$\mathrm{Re}[\cdot]$	real part	

Symbol	Explanation	
$\widehat{a}$	characterizes the optical pulse shape after demultiplexer	Eq. (15.79)
α_p, α_s	loss at the pump and signal wavelengths	Section 15.3.2
α_R	loss coefficient per unit length due to Rayleigh scattering	Section 15.3.1
A_{eff}	effective area of optical transverse mode	Section 15.3.1
β	propagation constant	Eq. (15.26)
$B(f)$	demultiplexer's complex baseband transfer function	Fig. 15.16
$b(t)$	demultiplexer's complex baseband impulse response	Fig. 15.16
B_e	electrical bandwidth of receiver	Fig. 15.19
B_{ref}	reference bandwidth for $\mathrm{OSNR_{ASE}}$	Section 15.4.1
B_o	optical bandwidth used in $\mathrm{OSNR_{ASE}}$ definition	Fig. 15.19
B_s	optical signal bandwidth	Fig. 15.19

Symbol	Explanation	Occurrence
C_R	Raman gain efficiency	Section 15.3.2
$C_{\widetilde{\boldsymbol{P}}}(t_1, t_2)$	autocovariance of optical power at detector	Eq. (15.61)
d_S, d_G	duty cycle of signal and gating pulses	Section 15.4.3
$\delta(t)$	Dirac's δ-functional	
Δf	linewidth of Lorentzian source spectrum full-width, half-maximum)	Section 15.2.1.2
$\Delta \boldsymbol{I}(t)$	time-dependent intensity fluctuations	Eq. (15.4)
$\Delta \text{OSNR}_{\text{ASE,dB}}$	OSNR penalty due to MPI	Section 15.5.2.3
ΔQ_{dB}	Q-factor penalty due to MPI	Section 15.5.2.2
Δz	distance between scatter sites at z_1 and z_2	Section 15.3.2
e	elementary charge	
$\text{erfc}(\cdot)$	complementary error function	Eq. (15.70)
$\boldsymbol{\varepsilon}$	amplitude of optical field at demultiplexing optical filter	Fig. 15.16
$\vec{\varepsilon}$	optical field amplitude vector	Eq. (15.1)
$\mathcal{E}_s(f)$	Fourier transform of $\varepsilon_s(t)$	Eq. (15.42)
f	frequency of optical or electrical quantities	
f_o	central optical frequency of source	Eq. (15.1)
f_s	peak signal attenuation factor due to filtering	Eq. (15.73)
$f_{\text{s-N}}^2$	attenuation factor for variance of beating between signal and ASE or MPI due to filtering	Eqs. (15.74), (15.75)
F_{OSNR}	OSNR-based MPI tolerance factor	Eqs. (15.83), (15.103)
F_Q	Q-based MPI tolerance factor	Eq. (15.84), Fig. 15.23
g	gain coefficient per unit length	Section 15.3.2
γ	nonlinear coefficient	Section 15.3.2
$\Gamma_{\boldsymbol{x}}(t, t+\tau)$	autocorrelation of the stochastic process $\boldsymbol{x}(t)$	Eq. (15.6)
G	net gain of an amplifier	Fig. 15.25
$\overrightarrow{G}\ [\overleftarrow{G}]$	net intensity gain in the forward [backward] directions	Section 15.2.1.1
$\overleftrightarrow{G}_f(\Delta z)$	integrated round-trip gain from all paths with the same Δz	Fig. 15.8
$\overrightarrow{G}_R\ [\overleftarrow{G}_R]$	on-off gain of a Raman amplifier in the forward [backward] propagating direction	Fig. 15.25
$H(f)$	transfer function of optoelectronic conversion chain	Fig. 15.16
$h(t)$	impulse response of optoelectronic conversion chain	Fig. 15.16
h	Planck's constant	
I	optical intensity	Section 15.2.1.1
$\mathbb{I}$	identity matrix	
$\mathbb{J}_{12}$	Jones matrix for propagation from z_1 to z_2	Eq. (15.26)
L	length of fiber amplifier	Section 15.3.2
λ_s, λ_p	wavelength of signal or pump light	Section 15.3.2
M	number of reflection sites	Section 15.2.1.3
$\mathbb{M}$	Müller matrix describing polarization evolution	Section 15.3.4
$\boldsymbol{m}_i$	elements of Müller transmission matrix $\mathbb{M}_t$	Section 15.3.4
N	number of interfering paths	Fig. 15.3
n	refractive index	
n_2	Kerr nonlinear coefficient	Section 15.6.1.2
$\vec{\boldsymbol{n}}(t)$	sum of all interferers' optical fields	Eq. (15.17)
N_{ASE}	ASE power spectral density per (polarization) mode in front of optical demultiplexing filter	Eqs. (15.49), (15.106)
NF	noise figure	Section 15.6.1.4
N_{Span}	number of spans in a transmission line	Fig. 15.31
$\boldsymbol{\phi}(t)$	instantaneous phase of optical field	Eq. (15.1)
$\phi_{\text{NL}}(z)$	accumulated nonlinear phase after transmission	Section 15.6.1.2

Symbol	Explanation	Occurrence
$\vec{p}_s, \vec{p}_r$	polarization vector for signal and doubly reflected fields	Section 15.2.1.1
P	optical power	
P'_s, P'_{DRB}	average signal and DRB power measured on gated OSA	Section 15.4.3
P_p	pump power	Section 15.3.2
P_f, P_b	forward and backward Raman pump power	Fig. 15.25
$\overline{P}_{in}, \overline{P}_{out}$	power in to and out of amplifier	Fig. 15.25
P_{in}, P_{out}	peak power of the 1-bit after filtering	Eq. (15.113)
$\overline{P}_{ref}$	reference average signal power for nonlinear effects	Section 15.6.1.2
Q	Personick's Q-factor	Eq. (15.71)
Q_{ref}	reference Q-factor in the absence of MPI	Eq. (15.80)
R_C	MPI inband crosstalk ratio	Eq. (15.50)
R_C^P	MPI inband crosstalk ratio, same polarization states	Eq. (15.51)
R_i	intensity reflectivity of discrete reflection at z_i	Fig. 15.4
$\boldsymbol{\rho}(z_i)$	differential Rayleigh backscatter coefficient at z_i	Eq. (15.28)
r	*inverse* extinction ratio of optical signal	Eq. (15.76)
$RIN(f)$	relative intensity noise	Eq. (15.11)
R	data rate	
R_D	overall optoelectronic conversion factor	Fig. 15.16
R_{NL}	nonlinear ratio for nonlinearity of class I	Section 15.6.1.2
R_{SNR}	ratio of SNRs at the output of a transmission line	Eq. (15.120)
S	dimensionless backscatter recapture fraction	Eq. (15.18)
$\boldsymbol{s}(t)$	electrical signal at decision gate ([A] or [V])	Fig. 15.16
$\langle \boldsymbol{s}_{0,1} \rangle$	sampled electrical signal for 1-bit and 0-bit	Eq. (15.77)
$\sigma^2(t)$	total variance of electrical signal at decision gate	Eq. (15.60)
$\sigma^2_{shot}(t)$	shot noise variance at decision gate	Eq. (15.100)
$\sigma^2_{s-N}(t)$	variance of electrical signal at decision gate due to beating between signal and ASE, MPI, or DRB	Eqs. (15.64), (15.68), (15.95)
σ^2_{N-N}	variance of electrical signal at decision gate due to beating of ASE, MPI, or DRB with themselves	Eqs. (15.65), (15.69)
$\sigma^2_{0,1}$	total variance of electrical signal sample for 0-bit and 1-bit at decision gate	Eq. (15.71)
$S_{\boldsymbol{x}}(f)$	power spectral density of random process $\boldsymbol{x}(t)$	Eq. (15.9)
$\vec{S} = (S_0, S_1, S_2, S_3)$	Stokes vector describing polarization of light	Section 15.3.4
SNR_{in}, SNR_{out}	SNR at the input and output of a span	Section 15.6.1.4
τ	correlation time variable	Eq. (15.6)
T_d	round-trip delay time	Fig. 15.4
$\overrightarrow{T}_F (\overleftrightarrow{T}_F)$	transmission of a passive fiber in the forward (backward) propagating direction	Section 15.6.1.1
T_L	transmission of a lossy element	Fig. 15.32
v_g	group velocity at signal wavelength	Section 15.3.3
z_i	coordinate at position i	

A15.C. List of Acronyms

Acronym	Explanation	Occurrence
AOM	acoustooptic modulator	Fig. 15.14
ASE	amplified spontaneous emission	Sections 15.4.1 and 15.6.1.1
BER	bit error ratio	Section 15.5.2
BW	backward	Fig. 15.32
ccG	circularly symmetric complex Gaussian	Section 15.2.2
CSRZ	carrier-suppressed return-to-zero	Fig. 15.10
CW	continuous wave	
DFA	distributed fiber amplifier	
DFB	distributed feedback laser	
DOP	degree of polarization	Section 15.3.4
DPSK	differential phase shift keying	Fig. 15.10
DRA	distributed Raman amplifier	
DRB	double-Rayleigh backscattering	
DUT	device under test	Fig. 15.11
ECL	external cavity laser	
EDFA	erbium-doped fiber amplifier	
ESA	electrical spectrum analyzer	Fig. 15.11
FEC	forward error correction	
FW	forward	Fig. 15.32
FWM	four-wave mixing	Section 15.6.1.2
IFWM	intrachannel four-wave mixing	Section 15.6.1.2
ISI	intersymbol interference	Section 15.5.2
IXPM	intrachannel cross-phase modulation	Section 15.6.1.2
MPI	multiple-path interference	
N	denotes either ASE or MPI	Section 15.5.1
NRZ	nonreturn-to-zero	
NZDF	nonzero dispersion fiber	
OOK	on-off keying	
OSA	optical spectrum analyzer	Fig. 15.11
OSNR	optical signal-to-noise ratio	Section 15.4
PDF	probability density function	Section 15.2.2
PC	polarization controller	
RZ	return-to-zero	Fig. 15.10
RF	radio frequency	
Rx	receiver	
SNR	signal-to-noise ratio	Section 15.6.1.4
SOP	state of polarization	Section 15.3.4
SPM	self-phase modulation	Section 15.6.1.2
SRB	single-Rayleigh backscattering	
SSMF	standard single-mode fiber	
Tx	transmitter	
VOA	variable optical attenuator	Fig. 15.11
WDM	wavelength-division-multiplexing	
WSS	wide-sense stationary	Eq. (15.6)
XPM	cross-phase modulation	Section 15.6.1.2

References

[1] H. Takahashi, K. Oda, and H. Toba, Impact of crosstalk in an arrayed-waveguide multi-plexer on N x N optical interconnection, *J. Lightwave Technol.*, 14:1097–1105, 1996.

[2] R. Khosravani, M.I. Hayee, B. Hoanca, and A.E. Willner, Reduction of coherent crosstalk in WDM add/drop multiplexing nodes by bit pattern misalignment, *IEEE Photon. Technol. Lett.*, 11:134–136, 1999.

[3] R. K. Staubli and P. Gysel, Crosstalk penalties due to coherent Rayleigh noise in bidirectional optical communication systems, *J. Lightwave Technol.*, 9:375–380, 1991.

[4] S. Radic and S. Chandrasekhar, Limitations in dense bidirectional transmission in absence of optical amplification, *IEEE Photon. Technol. Lett.*, 14:95–97, 2002.

[5] J.L. Gimlett, M.Z. Iqbal, L. Curtis, N.K. Cheung, A. Righetti, F. Fontana, and G. Grasso, Impact of multiple reflection noise in Gbit/s lightwave systems with optical fibre amplifiers, *Electron. Lett.*, 25:1393–1394, 1989.

[6] J.C. van der Plaats and F.W. Willems, RIN increase caused by amplified-signal redirected Rayleigh scattering in erbium-doped fibers. In *Proceedings of the Optical Fiber Communications Conference*, WM6, 1994.

[7] M. Nissov, K. Rottwitt, H.D. Kidorf, and M.X. Ma, Rayleigh crosstalk in long cascades of distributed unsaturated Raman amplifiers, *Electron. Lett.*, 35:997–998, 1999.

[8] J.L. Gimlett, M.Z. Iqbal, N.K. Cheung, A. Righetti, F. Fontana, and G. Grasso, Observation of equivalent Rayleigh scattering mirrors in lightwave systems with optical amplifiers, *IEEE Photon. Technol. Lett.*, 2:211–213, 1990.

[9] S. Ramachandran, B. Mikkelsen, L.C. Cowsar, M.F. Yan, G. Raybon, L. Bovin, M. Fishteyn, W.A. Reed, P. Wisk, D. Brownlow, R.G. Huff, and L. Gruner-Nielsen, All-fiber grating based higher order mode dispersion compensator for broad-band compensation and 1000-km transmission at 40 Gb/s, *IEEE Photon. Technol. Lett.*, 13:632–644, 2001.

[10] J.L. Gimlett, J. Young, R.E. Spicer, and N.K. Cheung, Degradations in Gbit/s DFB laser transmission systems due to phase-to-intensity noise conversion by multiple reflection points, *Electron. Lett.*, 24:406–407, 1988.

[11] A.F. Judy, Intensity noise from fiber Rayleigh backscatter and mechanical splices. In *Proceedings of the Fifteenth European Conference on Optical Communication*, 486–489, 1989.

[12] D.A. Fishman, D.G. Duff, and J.A. Nagel, Measurements and simulation of multipath interference for 1.7-Gb/s lightwave transmission systems using single and multifrequency lasers, *IEEE J. Lightwave Technol.*, 8:894–905, 1990.

[13] E.L. Goldstein and L. Eskildsen, Scaling limitations in transparent optical networks due to low-level crosstalk, *IEEE Photon. Technol. Lett.*, 7:93–94, 1995.

[14] S. Burtsev, W. Pelouch, and P. Gavrilovic, Multi-path interference noise in multi-span transmission links using lumped Raman amplifiers. In *Proceedings of the Optical Fiber Communications Conference*, TuR4, 2002.

[15] L. Eskildsen and E. Goldstein, High-performance amplified optical links without isolators or bandpass filters, *IEEE Photon. Technol. Lett.*, 4:55–58, 1992.

[16] S. Wu, A. Yariv, H. Blauveit, and N. Kwong, Theoretical and experimental investigation of conversion of phase noise to intensity noise by Rayleigh scattering in optical fibers, *Appl. Phys. Lett.*, 59:1156–1158, 1991.

[17] P. Wan and J. Conradi, Double Rayleigh backscattering in long-haul transmission systems employing distributed and lumped fibre amplifiers, *Electron. Lett.*, 31:383–384, 1995.

[18] K. Petermann and E. Weidel, Semiconductor laser noise in an interferometer system, *IEEE J. Quantum Electron.*, QE-17:1251–1256, 1981.

[19] J.A. Armstrong, Theory of interferometric analysis of laser phase noise, *J. Opt. Soc. Amer.*, 56:1024–1031, 1966.

[20] M.M. Choy, J.L. Gimlett, R. Welter, L.G. Kazovsky, and N.K. Cheung, Interferometric conversion of laser phase noise to intensity noise by single-mode fibre-optic components, *Electron. Lett.*, 23:1151–1152, 1987.

[21] J. W. Goodman, *Statistical Optics*, New York: Wiley, 2000.

[22] A.J. Stentz, T. Nielsen, S.G. Grubb, T.A. Strasser, and J.R. Pedrazzani, Raman ring amplifier at 1.3 μm with analog-grade noise performance and an output power of 23 dBm. In *Proceedings of the Optical Fiber Communications Conference*, PD16, 1996.

[23] A. Bononi and M. Papararo, Optimal placement of isolators in Raman amplified optical links. In *Optical Amplifiers and their Applications*, OTuA2, 2001.

[24] M. Born and E. Wolf, *Principles of Optics*, Oxford, UK: Pergamon, 1980.

[25] M. Nazarathy, W.V. Sorin, D.M. Baney, and S. A. Newton, Spectral analysis of optical mixing measurements, *J. Lightwave Technol.*, 7:1083–1096, 1989.

[26] D. Derickson, ed, *Fiber Optic Test and Measurement*, Upper Saddle River, NJ: Prentice-Hall, 1998.

[27] R.W. Tkach and A.R. Chraplyvy, Phase noise and linewidth in an InGaAsP DFB laser, *J. Lightwave Technol.*, LT-4:1711–1716, 1986.

[28] J.L. Gimlett and N.K. Cheung, Effects of phase-to-intensity noise conversion by multiple reflections on gigabit-per-second DFB laser transmission systems, *J. Lightwave Technol.*, 7:888–895, 1989.

[29] M. Tur and E.L. Goldstein, Probability distribution of phase-induced intensity noise generated by distributed-feedback lasers, *Optics Lett.*, 15:1–3, 1990.

[30] C.J. Rasmussen, F. Liu, R.J.S. Pedersen, and B.F. Jorgensen, Theoretical and experimental studies of the influence of the number of crosstalk signals on the penalty caused by incoherent optical crosstalk. In *Proceedings of the Optical Fiber Communications Conference*, TuR5, 1999.

[31] G. Einarsson, *Principles of Lightwave Communications*, New York: Wiley, 1996.

[32] D. Marcuse, Derivation of analytical expressions for the bit-error probability in lightwave systems with optical amplifiers, *IEEE J. Lightwave Technol.*, 8:1816–1823, 1990.

[33] M.E. Lines, W.A. Reed, D.J. DiGiovanni, and J.R. Hamblin, Explanation of analomalous loss in high delta singlemode fibres, *Electron. Lett.*, 35:1009–1010, 1999.

[34] E. Brinkmeyer, Analysis of the backscattering method for single-mode optical fibers, *J. Opt. Soc. Amer.*, 70:1010–1012, 1980.

[35] A.H. Hartog and M.P. Gold, On the theory of backscattering in single-mode optical fibers, *J. Lightwave Technol.*, LT-2:76–82, 1984.

[36] S. L. Hansen, K. Dybdal, and C. C. Larsen, Gain limit in erbium-doped fiber amplifiers due to internal Rayleigh backscattering, *IEEE Photon. Technol. Lett.*, 4:559–561, 1992.

[37] Y. Qian, J.H. Povlsen, S.N. Knudsen, and L. Grüner-Nielsen, Analysis and characterization of dispersion compensating fibers in fiber Raman amplification. In *Proceedings of Optical Amplifiers and Their Applications*, OMB6, 2000.

[38] P. Gysel and R.K. Staubli, Statistical properties of Rayleigh backscattering in single-mode fibers, *J. Lightwave Technol.*, 8:561–567, 1990.

[39] P. Gysel and R.K. Staubli, Spectral properties of Rayleigh backscattered light from single-mode fibers caused by a modulated probe signal, *J. Lightwave Technol.*, 8:1792–1798, 1990.

[40] L. Mandel and E. Wolf, *Optical Coherence and Quantum Optics*, New York: Cambridge University Press, 1995.

[41] S. Gray, M. Vasilyev, and K. Jepsen, Spectral broadening of double-Rayleigh backscattering in a distributed Raman amplifier, In *Proceedings of the Optical Fiber Communication Conference*, MA2, 2001.

[42] A. Papoulis, *Probability, Random Variables, and Stochastic Processes*, 3d edition. London: McGraw-Hill, 1991.

[43] C.R.S. Fludger and R.J. Mears, Electrical measurements of multipath interference in distributed Raman amplifiers, *J. Lightwave Technol.*, 19:536–545, 2001.

[44] M.O. van Deventer, Polarization properties of Rayleigh backscattering in single-mode fibers, *J. Lightwave Technol.*, 11:1895–1899, 1993.

[45] H. Kogelnik, R.M. Jopson, and L.E. Nelson, *Polarization-Mode Dispersion*. In Optical Fiber Telecommunications, Volume IVB. New York: Elsevier Science Chap. 15, 725–861, 2002.

[46] G.J. Foschini and C.D. Poole, Statistical theory of polarization dispersion in single mode fibers, *IEEE J. Lightwave Technol.*, 9:1439–1456, 1991.

[47] E. Brinkmeyer, Forward-backward transmission in birefringent single-mode fibers: Interpretation of polarization-sensitive measurements, *Optics Lett.*, 6:575–577, 1981.

[48] F.W. Willems, J.C. van der Plaats, and D.J. DiGiovanni, EDFA noise figure degradation caused by amplified signal double Rayleigh scattering in erbium doped fibres, *Electron. Lett.*, 30:645–646, 1994.

[49] J. Bromage, C.H. Kim, R.M. Jopson, K. Rottwitt, and A.J. Stentz, Dependence of double-Rayleigh backscatter noise in Raman amplifiers on gain and pump depletion. In *Proceedings of Optical Amplifiers and their Applications*, OTuA1, 2001.

[50] W.S. Wong, C.-J. Chen, H.K. Lee, G. Wilson, and M.-C. Ho, Multipath interference (MPI) measurements using a laser with a variable linewidth, In *Proceedings of Optical Amplifiers and Their Applications*, OME8, 2002.

[51] D.M. Baney and R.L. Jungerman, Optical noise standard for the electrical method of optical amplifier noise figure measurement. In *Proceedings of Optical Amplifiers and Their Applications*, MB3, 1997.

[52] S.A.E. Lewis, S.V. Chernikov, and J.R. Taylor, Characterization of double Rayleigh scatter noise in Raman amplifiers, *IEEE Photon. Technol. Lett.*, 12:528–530, 2000.

[53] V. Smokovdin, S.A.E. Lewis, and S.V. Chernikov, Direct comparison of electrical and optical measurements of double Rayleigh scatter noise. In *Proceedings of the European Conference on Optical Communications*, Symposium 3.0.5, 2002.

[54] C.-J. Chen and W.S. Wong, Transient effects in saturated Raman amplifiers, *Electron. Lett.*, 37:371–373, 2001.

[55] S.R. Chinn, Temporal observation and diagnostic use of double Rayleigh scattering in distributed Raman amplifiers, *IEEE Photon. Technol. Lett.*, 11:1632–1634, 1999.

[56] L. Gruner-Nielsen, Yujun Qian, B. Palsdottir, P.B. Gaarde, S. Dyrbol, T. Veng, Yifei Qian, R. Boncek, and R. Lingle, Module for simultaneous C + L-band dispersion compensation and Raman amplification. In *Proceedings of the Optical Fiber Communications Conference*, TuJ6, 2002.

[57] N.A. Olsson, Lightwave systems with optical amplifiers, *IEEE J. Lightwave Technol.*, 7:1071–1082, 1989.

[58] E. Desurvire, *Erbium-Doped Fiber Amplifiers*. New York: Wiley, 1994.

[59] R. G. Smith, Optical power handling capacity of low loss optical fibers as determined by stimulated Raman and Brillouin scattering, *Appl. Opt.*, 11:2489–2494, 1972.

[60] V. Curri and G. Rizzo, Statistical properties and system impact of multi-path interference in Raman amplifiers. In *Proceedings of the European Conference on Optical Communications*, 110–111, Tu.A.1.2, 2001.

[61] P.B. Hansen, L. Eskildsen, A.J. Stentz, T.A. Strasser, J. Judkins, J.J. DeMarco, R. Pedrazzani, and D.J. DiGiovanni, Rayleigh scattering limitations in distributed Raman pre-amplifiers, *IEEE Photon. Technol. Lett.*, 10:159–161, 1998.

[62] P. Wan and J. Conradi, Power penalties due to Rayleigh backscattering in long-haul transmission systems employing lumped and distributed EDFAs. In *Proceedings of Optical Amplifiers and Their Applications*, ThC6, 1994.

[63] P. Wan and J. Conradi, Impact of double Rayleigh backscatter noise on digital and analog fiber systems, *J. Lightwave Technol.*, 14:288–297, 1996.

[64] L. Eskildsen and P.B. Hansen, Interferometric noise in lightwave systems with optical amplifiers, *IEEE Photon. Technol. Lett.*, 9:1538–1540, 1997.

[65] F. Liu, C.J. Rasmussen, and R.J.S. Pedersen, Experimental verification of a new model describing the influence of incomplete signal extinction ratio on the sensitivity degradation due to multiple interferometric crosstalk, *IEEE Photon. Technol. Lett.*, 11:137–139, 1999.

[66] T. Zami, L. Noirie, F. Bruyére, and A. Jourdan, Crosstalk-induced degradation in an optical-noise-limited detection system. In *Proceedings of the Optical Fiber Communications Conference*, TuR4, 1999.

[67] H.K. Kim and S. Chandrasekhar, Dependence of in-band crosstalk penalty on the signal quality in optical network systems, *IEEE Photon. Technol. Lett.*, 12:1273–1274, 2000.

[68] M. H. Eiselt, L. D. Garrett, and V. Dominic, Optical SNR versus Q-factor improvement with distributed Raman amplification in long amplifier chains. In *Proceedings of the European Conference on Optical Communications*, 4.4.4, 2000.

[69] P. J. Winzer, R.-J. Essiambre, and J. Bromage, Combined impact of double-Rayleigh backscatter and amplified spontaneous emission on receiver noise. In *Proceedings of the Optical Fiber Communications Conference*, THGG87, 2002.

[70] L. Boivin and G. J. Pendock, Receiver sensitivity for optically amplified RZ signals with arbitrary duty cycle. In *Proceedings of Optical Amplifiers and Their Applications*, 106–109, ThB4, 1999.

[71] P. J. Winzer and A. Kalmar, Sensitivity enhancement of optical receivers by impulsive coding, *J. Lightwave Technol.*, 17(2):171–177, 1999.

[72] J. L. Rebola and A. V. T. Cartaxo, Power penalty assessment in optically preamplified receivers with arbitrary optical filtering and signal-dependent noise dominance, *J. Lightwave Technol.*, 20(3):401–408, 2002.

[73] P. J. Winzer, Optical transmitters, receivers, and noise. In John G. Proakis, ed., *Wiley's Encyclopedia on Telecommunications*, New York: Wiley, 2002.

[74] P. J. Winzer, Receiver noise modeling in the presence of optical amplification. In *Proceedings of Optical Amplifiers and Their Applications*, OTuE16, 2001.

[75] P.J. Winzer, S. Chandrasekhar, and H. Kim, Impact of filtering on DPSK reception, *IEEE Photon. Technol. Lett.*, 15(6):840–842, 2003.

[76] S. D. Personick, Receiver design for digital fiber optic communication systems, *Bell Syst. Tech. J.*, 52:843–874, 1973.

[77] P.J. Winzer and R.-J. Essiambre, Optical receiver design trade-offs. In *Proceedings of the Optical Fiber Communications Conference*, ThG1, 2003.

[78] J. Bromage, C.-H. Kim, P. J. Winzer, L. E. Nelson, R.-J. Essiambre, and R. M. Jopson, Relative impact of multiple-path interference and amplified spontaneous emission noise on optical receiver performance. In *Proceedings of the Optical Fiber Communication Conference*, TuR3, 2002.

[79] C. H. Kim, J. Bromage, and R. M. Jopson, Reflection-induced penalty in Raman amplified systems, *IEEE Photon. Technol. Lett.*, 14:573–575, 2002.

[80] B. E. A. Saleh and M. C. Teich, *Fundamentals of Photonics*. New York: Wiley, 1991.

[81] A. Yariv, H. Blauvelt, and S.-W. Wu, A reduction of interferometric phase-to-intensity conversion noise in fiber links by large index phase modulation of the optical beam, *J. Lightwave Technol.*, 10:978–981, 1992.

[82] P. K. Pepeljugoski and K. Y. Lau, Interferometric noise reduction in fiber-optic links by superposition of high frequency modulation, *J. Lightwave Technol.*, 10(7):957–963, 1992.

[83] S.L. Woodward, T.E. Darcie, and F. Liu, A method for reducing multipath interference noise, *IEEE Photon. Technol. Lett.*, 6:450–452, 1994.

[84] A. Yariv, H. Blauvelt, D. Huff, and H. Zarem, An experimental and theoretical study of the suppression of interferometric noise and distortion in AM optical links by phase dither, *J. Lightwave Technol.*, 15(3):437–443, 1997.

[85] P. V. Mamyshev, Rayleigh scattering impairments in systems with Raman amplification, Bell Labs internal technical memorandum, 2000 (unpublished).

[86] G. P. Agrawal, *Nonlinear Fiber Optics*, 3d ed., San Diego: Academic, 2000.

[87] C. Paré, A. Villeneuve, P.-A. Bélanger, and N. J. Doran, Compensating for dispersion and the nonlinear Kerr effect without phase conjugation, *Optics Lett.*, 21:459–461, 1996.

[88] A. Altuncu, L. Noel, W. A. Pender, A. S. Siddiqui, T. Widdowson, A. D. Ellis, M. A. Newhouse, A. J. Antos, G. Kar, and P. W. Chu, 40 Gbit/s error free transmission over a 68-km distributed erbium-doped fibre amplifier, *Electron. Lett.*, 32:233–234, 1996.

[89] L. F. Mollenauer, R. H. Stolen, and M. N. Islam, Experimental demonstration of soliton propagation in long fibers: Loss compensated by Raman gain, *Optics Lett.*, 10:229–231, 1985.

[90] J.-I. Kani, M. Jinno, T. Sakamoto, S. Aisawa, M. Fukui, K. Hattori, and K. Oguchi, Interwavelength-band nonlinear interactions and their suppression in multiwavelength-band WDM transmission systems, *J. Lightwave Technol.*, 17:2249–2260, 1999.

[91] Y. Aoki, Properties of fiber Raman amplifiers and their applicability to digital optical communication systems, *J. Lightwave Technol.*, 6:1225–1239, 1988.

[92] I. Kaminow and T. Li, *Optical Fiber Telecommunications IV*, San Diego: Academic, Vol. A, Chap. 5, 2002.

[93] I. Kaminow and T. Li, *Optical Fiber Telecommunications IV*, San Diego: Academic, Vol. B, Chap. 6, 232–304, 2002.

[94] R.-J. Essiambre, B. Mikkelsen, and G. Raybon, Intra-channel cross-phase modulation and four-wave mixing in high-speed TDM systems, *Electron. Lett.*, 35:1576–1578, 1999.

[95] I. Kaminow and T. Koch, *Optical Fiber Telecommunication IIIA*, chapter Solitons in High Bit-Rate, Long Distance Transmission, San Diego: Academic Press, 373–460, 1997.

[96] René-Jean Essiambre and G. P. Agrawal, *Soliton Communication Systems*, Vol. XXXVII of *Progress in Optics*, Amsterdam: Elsevier/North-Holland, Chap. Soliton Communication Systems, 185–256, 1997.

[97] M. O. Van Deventer, Polarization properties of Rayleigh backscattering in single-mode fibers, *J. Lightwave Technol.*, 11:1895–1899, 1993.

[98] P. C. Becker, N. A. Olsson, and J. R. Simpson, *Erbium-Doped Fiber Amplifiers Fundamentals and Technology*, San Diego: Academic Press, 1999.

[99] K. Shimizu, T. Horiguchi, and Y. Koyamada, Characteristics and reduction of coherent fading noise in Rayleigh backscattering measurement for optical fibers and components, *J. Lightwave Technol.*, 10:982–987, 1992.

[100] C. R. S. Fludger, Y. Zhu, V. Handevek, and R. J. Mears, Impact of MPI and modulation format on transmission systems employing distributed Raman amplification, *Electron. Lett.*, 37:970–971, 2001.

Chapter 16

Raman Impairments in WDM Systems

P. M. Krummrich

16.1. Introduction

In most chapters of this book, stimulated Raman scattering (SRS) is invoked intentionally. Pump radiation is coupled into the fiber carrying the signal radiation to generate Raman gain. The Raman gain can be used very advantageously, for example, to improve the optical signal-to-noise ratio (OSNR) budget by distributed amplification in the transmission fiber. However, SRS also occurs unintentionally in WDM transmission systems. Due to the large number of channels inside the Raman gain bandwidth, total power can add up to levels where considerable amounts of SRS are generated, with the signal channels acting as pumps. In contrast to the beneficial effects of intentional Raman pumping, the unintended generation of SRS usually degrades system performance.

This chapter addresses effects resulting from the unintended invocation of SRS and their impact on WDM signal transmission. Section 16.2 covers the generation of spontaneous emission and its effect on maximum channel launch powers. A number of system impairments result from the interaction between signal channels due to SRS. Effects with time scales well below the bit period affect the mean values of the individual channel powers. These slow interactions are discussed in Section 16.3, whereas Section 16.4 addresses fast interactions between individual bits. Such interactions change the variances of the respective channel powers and can be considered as noise. The last section (16.5) provides some selection criteria for transmission fibers with respect to Raman efficiency.

16.2. Raman Threshold

The Raman amplification process relies on stimulated emission of radiation. However, stimulated emission is inevitably linked to the generation of spontaneous emission [1, 2]. Consequently, launching signal radiation of WDM channels into an optical fiber leads to generation of new photons due to spontaneous Raman scattering. The peak intensity occurs at a frequency of one Stokes shift (approx. 13.2 Thz in silica fibers, corresponding to a wavelength shift of 106 nm for a signal at 1550 nm) below the frequency of the signal radiation.

The initial number of spontaneously generated photons is rather small and usually negligible. It can grow considerably due to amplification by stimulated emission with the signal radiation acting as a pump. The stimulated emission converts signal photons into amplified spontaneous emission (ASE) photons, thus contributing an additional loss mechanism for the signal radiation. The conversion rate is proportional to the intensity of the ASE. Consequently, the additional loss of the signal radiation due to SRS is negligible compared to other loss mechanisms in the fiber as long as the ASE intensity is sufficiently small. If the signal launch power increases and ASE starts to experience a very strong growth, the additional loss due to SRS will eventually exceed the other loss mechanisms in the fiber. The depletion of the signal power due to SRS grows so strongly that further increase of the launch power above the level of equal loss contributions results in a decrease of the signal power at the fiber output. This limitation of reasonable signal launch powers is called the Raman threshold [3].

There are several options to define the Raman threshold or the critical launch power. One would be the minimum launch power resulting in an excess loss due to SRS that is equal to other loss mechanisms at some point along the signal propagation. Another option is the signal launch power that achieves the maximum signal power at the fiber output. The most commonly used definition for the critical power seems to be the launch power that results in a total ASE power equal to the signal power at the fiber output [3, 4].

The critical signal launch power P_{crit} in W according to the definition with equal signal and ASE power at the fiber output can be estimated using the following equation [4].

$$P_{crit} = 8 \frac{A_{\text{eff}}}{g_R} \frac{\alpha_S}{1 - \exp(-\alpha_S L)}, \tag{16.1}$$

where A_{eff} denotes the effective mode field area in m^2, g_R the peak Raman coefficient in m/W for random polarizations, α_S the loss coefficient at the signal wavelength in 1/m, and L the length of the fiber in m. For a section of standard single-mode fiber (SSMF) with a length of 100 km, an effective mode field area of 80 μm^2, a peak Raman coefficient of 2.3×10^{-14} m/W, and a loss coefficient of 0.2 dB/km, this equation predicts a critical launch power of 1.3 W for a single signal with a center wavelength of 1550 nm.

According to Eq. (16.1), the critical launch power depends on the transmission fiber type and its characteristics. Experiments were carried out to determine the critical power for four different transmission fiber types [5]. Figure 16.1 shows the experimental setup. Eight continuous wave (CW) laser diode sources with different output wavelengths were employed to generate the signal radiation rather than a single one in order to reduce the impact of stimulated Brillouin scattering (SBS). The signal wavelengths were chosen on the ITU grid in the range from 1549.3 to 1560.6 nm with 200 GHz channel spacing and output powers of 2 dBm per channel at the input of the power combiner. SBS was further suppressed by applying a triangular amplitude modulation with a frequency of 15 kHz and a modulation index around 10% to

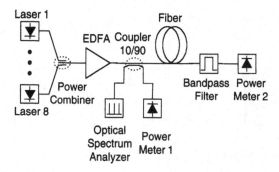

Fig. 16.1. Experimental setup that was used to measure the Raman threshold of different transmission fiber types.

Table 16.1. Characteristics of the Four Fibers Under Investigation

Fiber Type	Fiber Length km	Measured Loss Coeff. dB/km	Mode Field Area μm^2
SSMF	41.9	0.21	78
PSCF	100.0	0.17	75
LS	50.4	0.22	53
TW	53.2	0.22	53

all laser injection currents. The modulation resulted in a shift of the SBS threshold of each channel to launch powers above 27 dBm, corresponding to a total launch power of 36 dBm into the fiber.

The different channels were coupled into a single fiber by an 8:1 power combiner followed by a high-power EDFA with a maximum output power of 32 dBm. A tap coupler with a splitting ratio of 10% allowed monitoring of the EDFA output power as well as spectral components backscattered from the fiber input.

Table 16.1 summarizes the characteristics of the four transmission fiber samples that have been investigated. Each sample represented a different fiber type: standard single-mode fiber (SSMF), pure silica core fiber (PSCF), LS fiber, and TrueWave® classic (TW) fiber. The lengths of the fibers under study were chosen to be at least twice the effective Raman length avoiding artifacts arising from different interaction lengths.

The output of the transmission fiber was followed by an optical bandpass filter and an optical power meter. The purpose of the bandpass filter was to keep all spectral components that did not belong to the eight signal channels from reaching the detector.

Figure 16.2 shows a plot of spectral components measured at the fiber input that were propagating counterdirectionally to the signal channels. The narrow peaks around 1550 nm correspond to a small fraction of the eight signal channels being reflected by Rayleigh scattering and incompletely suppressed SBS. Most of the total power is contained in the ASE generated by SRS. The peak of the ASE spectrum is

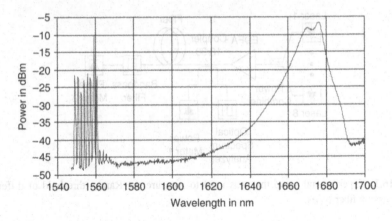

Fig. 16.2. Spectral components measured at the fiber input propagating counterdirectionally to the eight signal channels.

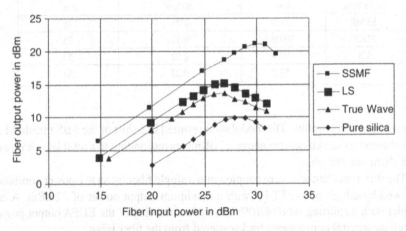

Fig. 16.3. Total power of the eight signal channels at the fiber output versus total launch power at the fiber input.

located approximately 110 nm above the center of the signal wavelengths. A similar ASE spectrum can be observed at the output of the transmission fiber.

The main results of the investigation are depicted in Fig. 16.3. The plot shows the total power of the signal channels at the fiber output versus launch power. For total launch powers into the fiber below 25 dBm, the output power increases linearly with input power. At a certain input power value depending on the fiber type, the output power saturates and even decreases for increasing launch powers.

The deviation from linear loss could be attributed to the conversion of signal radiation to ASE caused by SRS. Total ASE power at the fiber output was smaller than the total power in the eight channels for total launch power levels below 25 dBm. However, the total ASE power exceeded the remaining signal power at the fiber output as soon as the signal output power started to drop with increasing input power. This is a clear indication that the decrease of output power was caused by crossing the SRS threshold.

Figure 16.3 allows us to determine the Raman thresholds of the fibers under investigation. SSMF revealed a maximum signal output power for a total launch power of about 30.2 dBm, however, lower threshold values were obtained for PSCF (28.2 dBm) and the NZDSFs (about 26.6 dBm).

The relation of the Raman coefficients of the four transmission fiber types was determined experimentally by measuring the SRS-induced tilt, a configuration similar to the one described in [6]. As the method actually measures the slope of the Raman coefficient spectrum in a wavelength range of approximately 30 nm, it is quite difficult to determine the peak value from the experimental data very accurately. But very similar shapes of the spectra of the different fiber types enable a reliable prediction of the relation of the peak values. With the Raman coefficient of the SSMF as a reference, the Raman coefficients of the PSCF, LS, and TrueWave® fiber are higher by a factor of 1.13, 1.23, and 1.13, respectively.

Analyzing the relation of the Raman coefficients in combination with the fiber parameters given in Table 16.1 enables an interpretation of the results shown in Fig. 16.3. The lower Raman thresholds of the LS and TrueWave® fibers compared to SSMF can be explained by smaller effective mode areas and higher Raman coefficients, whereas the lower Raman threshold of PSCF results from a higher Raman coefficient and a smaller loss coefficient at the signal wavelengths.

Although the use of SSMF seems to be the best choice for applications requiring maximum launch power, a more detailed discussion is necessary in terms of span length within reach. It can be shown that bigger span lengths are achievable with PSCF instead of SSMF. This fact is due to a trade-off between the different usable launch power levels and the different loss coefficients.

Maximum launch power due to the Raman threshold is a critical parameter for repeaterless transmission systems. The main goal of these systems is to achieve maximum transmission length with a single span, avoiding the use of inline repeaters. Remotely pumped erbium-doped fiber amplifiers are a powerful technology to extend the system reach. Pump radiation at 1480 nm has to be carried from the receive terminal to these amplifiers either through the transmission fiber or a separate fiber. Launch power levels of the pump radiation up to 8 W could be achieved by using very large effective mode field area fibers [7].

16.3. Tilt of the Channel Power Distribution

The previous section addressed energy transfer due to SRS from signal channels to ASE. SRS can also cause considerable energy transfer between signal channels, if the WDM signal channels are distributed over a wide wavelength range and total launch power is high. Figure 16.4 tries to illustrate the concept. Many channels are launched into the transmission fiber with equal power at the fiber input. During propagation in the fiber, channels on the short-wavelength side act as Raman pumps for channels

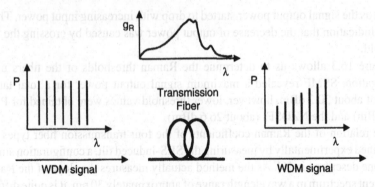

Fig. 16.4. Energy transfer between WDM signal channels caused by SRS in the transmission fiber.

on the long-wavelength side. The resulting energy transfer leads to amplification of the channels on the long-wavelength side whereas channels on the short-wavelength side experience additional loss. As a consequence, the channel power distribution at the fiber output looks tilted.

16.3.1. Channel Power Management

An expression for the amount of tilt at the output of a fiber section can be derived analytically under these assumptions [8]:

*   the spectrum of the Raman gain coefficient is approximated by a triangular profile (wavelength difference of band edge channels <110 nm);
*   the energy that is lost when a short-wavelength photon is converted into a long-wavelength photon is neglected (minor approximation for a maximum spacing <30 nm between the shortest and longest wavelength channels);
*   all channels experience the same attenuation due to other loss mechanisms; and
*   no energy transfer to ASE.

The difference of the power levels of the longest and shortest wavelength channels ΔP in dB can be calculated using the expression:

$$\Delta P = \frac{10}{\ln(10)} \frac{g_R}{A_{\text{eff}}} \frac{1}{\alpha_S} [1 - \exp(-\alpha_S L)] \frac{\Delta f_{ch}}{\Delta f_R} P_0, \qquad (16.2)$$

where g_R denotes the peak Raman coefficient in m/W for random polarizations, A_{eff} the effective mode field area in m², α_S the loss coefficient in 1/m, L the length of the fiber in m, Δf_{ch} the difference of the center frequencies of the shortest and longest wavelength channels in THz, Δf_R the frequency difference between the signal channel acting as a pump and the peak of the Raman gain spectrum, and P_0 the total launch power in W (sum of the time-averaged power of the channels).

For a fiber section with a length of 100 km, a maximum Raman coefficient of 2.3×10^{-14} m/W, an effective mode field area of 80 μm², a loss coefficient of

0.2 dB/km, and a peak Raman gain at a spacing of 13.2 Thz below the pump center frequency, Eq. (16.2) predicts a tilt of 0.8 dB for 40 channels with a spacing of 100 GHz and a total launch power of 20 dBm.

It is interesting to note that for a given fiber, the amount of tilt depends on the total launch power and the spacing between the shortest and longest wavelength channels only. The tilt does not depend on the distribution of power among channels at the fiber input, the number of channels, or the location of channels inside the wavelength band. The detailed distribution of channel powers at the fiber output does depend on these boundary conditions, of course.

According to Eq. (16.2), SRS introduces a linear tilt on a decibel/nanometer scale. It can be corrected by a single filter with a log-linear loss spectrum. The necessary slope of the filter depends on the fiber type, the wavelength separation between the shortest and longest wavelength channels, and the total launch power only.

Experimental investigations of the tilt induced by SRS for bidirectional and unidirectional propagation were presented at OFC 1999 [9, 10]. Current state-of-the-art systems (2002) feature unidirectional transmission of 80 channels in the C-band and the same number of channels in the L-band. With a channel spacing of 50 GHz, the separation between the shortest and the longest wavelength channels reaches 75 nm.

Figure 16.5 shows the comparison of measured loss spectra (difference between channel power levels at the fiber input and output) of a single span of SSMF with a length of 80 km for different launch powers. The 80 channels in the C-band (1528.7 to 1563.9 nm) and the 80 channels in the L-band (1570.4 to 1607.5 nm) had a total power of 23.4 and 22.4 dBm, respectively. One spectrum was measured with a fixed attenuation of 20 dB in front of the fiber input, the other without any attenuation (0 dB).

The loss spectra with attenuator correspond to the linear loss spectrum of the fiber, whereas the difference between the spectra with and without attenuator was caused by the amplification and attenuation of channels due to SRS. This difference is plotted

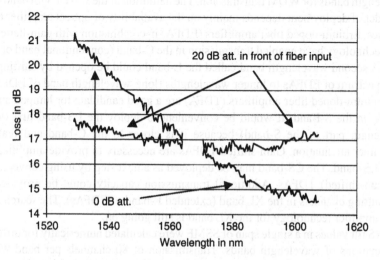

Fig. 16.5. Comparison of loss spectra for different launch power levels.

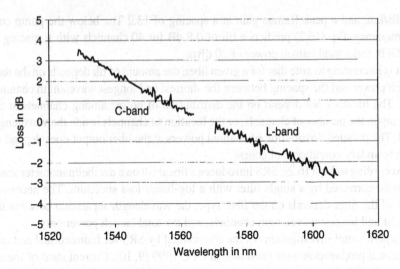

Fig. 16.6. Difference of the fiber loss spectra with and without attenuator in front of the fiber.

in Fig. 16.6. The tilt of the channel power distribution features a slight deviation from a log-linear shape and a total amount of approximately 6.1 dB. Equation (16.2) predicts a tilt of 7.1 dB for the same configuration, which is quite good considering the assumptions that were necessary for deriving the analytical expression.

The analytical expression is quite helpful in estimating the gain tilt of WDM systems operating with one or two wavelength bands, corresponding to a wavelength range of up to 80 nm. SRS tilt values for systems with wider wavelength ranges or more bands should be calculated numerically by solving the set of differential equations for all channel powers.

Figure 16.7 shows a plot of a typical fiber loss spectrum together with potential wavelength bands for WDM transmission. The definition of the wavelength bands and the order of deployment depends mainly on the availability of optical amplifier technologies. Erbium-doped fiber amplifiers (EDFA) in combination with gain-flattening filter technology have enabled transmission in the C-band (conventional band of ED-FAs). A second wavelength band called the L-band could be opened by shifting the gain spectrum of EDFAs to longer wavelengths (long-wavelength band of EDFAs).

Thulium-doped fiber amplifiers (TDFA) are a good candidate for lumped amplification in the S-Band. It would be convenient to deploy the LS-band first (long-wavelength part of the S-band) because it is closer to the C-band and features lower fiber attenuation. Gain-shifted TDFAs are necessary to provide amplification in the LS-band. The CS-band may be deployed as a next step by using conventional (nongain-shifted) TDFAs. Additional transmission capacity could be provided by transmitting channels in the XL band (extended L-band of EDFAs). The search for a good amplifier technology for the XL band is still going on.

SRS tilt values in a single span of SSMF were calculated numerically for different combinations of wavelength bands. Transmission of 80 channels per band with a launch power of +3 dBm per channel was assumed. The results are depicted in

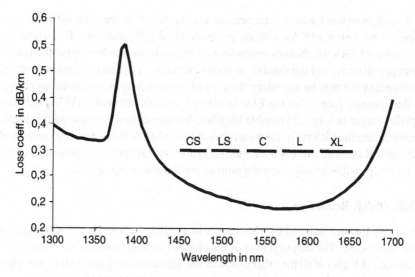

Fig. 16.7. Wavelength bands for WDM transmission.

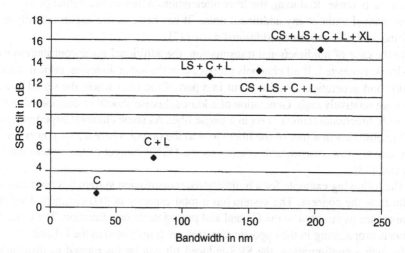

Fig. 16.8. SRS tilt values for combinations of different wavelength bands.

Fig. 16.8. According to the numerical results, the SRS tilt grows nearly exponentially until the width of the wavelength band used for transmission reaches the width of the Raman gain spectrum. Further growth of the wavelength band still increases the SRS tilt, but with a less steep slope.

The numerical investigation has shown that transmission of WDM channels in a wide wavelength range results in considerable tilt of the channel power distribution due to SRS. Tilt values up to several dB can be expected even for a single span. As the tilt accumulates from span to span [11], growth of channel power differences has to be limited in multispan systems by compensation of the SRS-induced tilt.

A straightforward way to compensate accumulation of the SRS tilt from span to span is to design EDFAs with an inversely tilted gain spectrum. Unfortunately, the amount of SRS tilt changes depending on the transmission fiber type, the launch power per channel, and the number of active channels. A dynamic adjustment of the tilt compensation may be necessary. Such an adjustment may be performed by changing the dynamic gain tilt of the EDFAs using a variable attenuator (VOA) between amplifier stages or a special tunable tilt filter. Another option is to use pump signals to compensate the SRS tilt in the transmission fiber [12]. If the system uses wideband distributed Raman amplifiers with multiple pump wavelengths, compensation of the tilt may be possible by adjusting the ratio of the Raman pump powers.

16.3.2. OSNR Reduction

The energy transfer induced by SRS results in an effectively reduced loss compared to the linear case. The output power of the channel with the longest wavelength can be several dB higher if all the other channels are activated compared to the case when it operates alone. However, the increased output power should be interpreted as a result of optical amplification rather than as a result of reduced loss. The fundamental difference is noise. Reducing the fiber attenuation increases the output power of a given channel without any additional noise. It has been shown experimentally that SRS-induced tilt contributes additional noise [13].

In the case of unidirectional transmission, the additional noise contribution can usually be neglected. If all channels propagate in the same direction, optical amplification and generation of noise occur in a part of the fiber where the signal power levels are relatively high. Generation of additional noise should be considered in the case of bidirectional transmission in a single fiber. As some channels experience optical amplifiction in a part of the fiber close to the output where signal power levels are relatively low, considerable impact on the OSNR budget of the system can be expected [13].

The following example for a bidirectional transmission system has been chosen to illustrate the concept. The system has a total capacity of 160 channels. Half the channels are transmitted in the C-band and propagate in one direction. The other 80 channels propagating in the opposite direction are transmitted in the L-band.

In such a configuration, the SRS-induced tilt can be interpreted as distributed amplification of the L-band channels with the C-band channels acting as counterdirectional Raman pumps. Figure 16.9 shows a plot of the gain spectrum in the L-band for a section of SSMF with a length of 80 km. The gain spectrum was calculated using a numerical simulation tool based on integration of a set of differential equations for all spectral components. The channels in the C- and L-bands were launched with a total power of 23.4 and 22.4 dBm, respectively. The tool predicts a Raman gain of nearly 3 dB for the L-band channel with the longest wavelength.

A convenient tool to analyze the noise contribution associated with distributed Raman gain is provided by the effective noise figure. It can be defined as

$$NF_{\text{eff}} = \frac{1}{G_{\text{on/off}}}(1 + \frac{2S_{\text{ASE}}}{h\nu}), \tag{16.3}$$

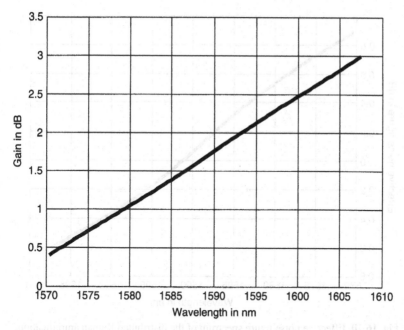

Fig. 16.9. Raman gain of the L-band channels pumped by the counterdirectionally propagating C-band channels.

where $G_{on/off}$ denotes the on/off Raman gain (ratio of the channel power at the fiber output with and without Raman pumping), S_{ASE} the spectral power density of the ASE in one polarization in W/Hz, h Planck's constant, and ν the center frequency of the channel in Hz.

A reduction of the span loss by reducing the fiber attenuation does not contribute additional ASE. In this case, the effective noise figure in dB is equal to (-1) times the decrease of the fiber loss in dB. For example, reducing the span loss by 3 dB corresponds to an effective noise figure of -3 dB.

Figure 16.10 shows a plot of the effective noise figure spectrum calculated for the L-band channels. The numerical tool predicts an effective noise figure of -0.6 dB for the channel with the longest wavelength. This number has to be compared with an effective noise figure of -3 dB which corresponds to an on/off Raman gain of 3 dB without additional ASE. The ASE generated during the amplification of the channel has increased the effective noise figure by 2.4 dB.

The impact of additional ASE generated by SRS on the OSNR budget of the system depends on the noise performance of the amplifier following the fiber section. In most cases, the lumped amplification will be provided by EDFAs. Assuming a noise figure of the lumped amplifier of 4 dB leads to a total noise figure of 2.1 dB for an on/off Raman gain of 3 dB and an effective noise figure of the distributed Raman amplifier of -0.6 dB. Amplification without additional ASE or reduction of the fiber loss by 3 dB results in a total noise figure of 1 dB. In this example, the additional

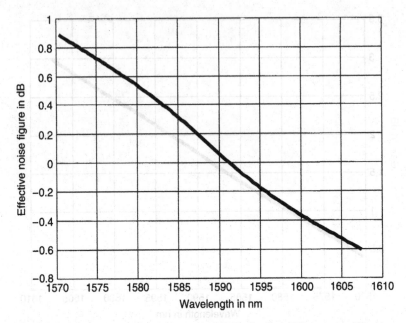

Fig. 16.10. Effective noise figure spectrum of the distributed Raman amplification.

ASE generated in combination with the SRS-induced tilt has deteriorated the OSNR budget of the system by 1.1 dB.

In most cases, it will not be a problem to tolerate the additional ASE generated by the SRS-induced tilt. The channels that are affected most are the channels on the long-wavelength side. The net effect in this wavelength range is still positive; that is, the OSNR improvement due to the increased power at the fiber output overcompensates the OSNR degradation due to additional ASE. Furthermore, broadband EDFAs generally feature a better noise performance on the long-wavelength side of their gain spectrum. As a consequence, the channels on the long-wavelength side of the transmission band will provide the best noise performance assuming equal launch power of all channels.

However, the additional ASE should be taken into account if the OSNR budget of bidirectional transmission systems has to be calculated very accurately. It has been mentioned in the previous section that the SRS-induced tilt has to be compensated to keep channel power differences below acceptable limits in wideband, multispan transmission systems. Without additional ASE, equal noise performance of all channels could be achieved by selecting launch power levels resulting in equal power levels of all channels at the fiber output. This approach relies on the additional assumption that the lumped amplifiers following the fiber sections provide a wavelength-independent noise figure spectrum. If the system OSNR budget has to take the wavelength-dependence of the noise figure spectrum of lumped amplifiers into account, it should also include the additional ASE generated by the SRS tilt.

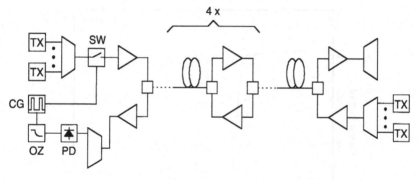

Fig. 16.11. Sketch of the experimental setup consisting of a bidirectional DWDM link and means to analyze channel add-drop scenarios: a clock generator (CG) controls an optical switch (SW) and triggers an oscilloscope (OZ) recording the signal of a photo diode (PD).

16.3.3. Channel Power Transients

The tilt of the channel power distribution due to SRS depends on the total launch power. Adding or dropping channels changes the total launch power. The resulting change of the SRS tilt introduces power transients for the surviving channels. The impact of such transients on a system featuring bidirectional transmission in a single fiber have been investigated experimentally [14].

Figure 16.11 shows a sketch of the experimental setup. It consists of a bidirectional DWDM link featuring transmission of sixteen 10 Gbps channels per direction over five spans of standard single-mode fiber. The channels were modulated with a $2^{31} - 1$ pseudorandom bit sequence. Each span had a length of 90 km and a loss of 29 dB with contributions from the fiber and a midspan attenuator. The total power launched into the fiber ends was 20 dBm. The channels were grouped into two counterdirectionally propagating wavelength bands. Each band was amplified by separate EDFAs. The channels of the blue band covered the wavelength range from 1530.3 to 1542.9 nm and the channels of the red band the wavelength range from 1547.7 to 1560.6 nm with a spacing of 100 GHz between channels.

An optomechanical switch (SW) was inserted between the multiplexer output and the input of the booster. It enabled us to activate or deactivate all channels propagating in one direction simultaneously. The switch was driven by a clock generator (CG). The power of a given channel propagating in the counterdirection was detected by a PIN photo diode (PD) at the system output. The electrical output signal of the PD was recorded by an oscilloscope (OZ), which was triggered by the clock generator.

The optomechanical switch had a transition time of approximately 0.4 ms. Figure 16.12 shows the time-resolved output signal of the switch with a CW input signal during a switching event. In a first series of experiments, the switch was used to deactivate all channels of the blue band. Figure 16.13 shows the time-resolved output signal of the channel with the longest wavelength in the red band (1560.6 nm) at the system output.

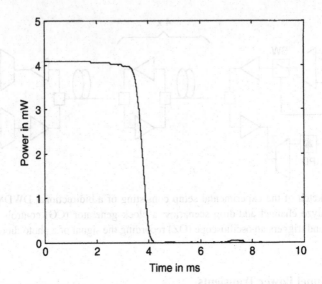

Fig. 16.12. Time-resolved output signal of the optomechanical switch.

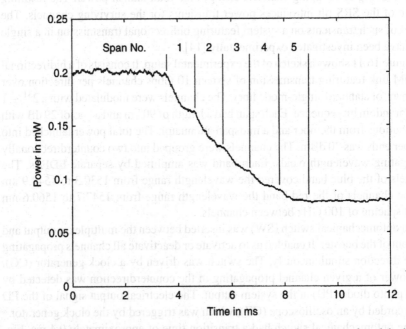

Fig. 16.13. Time evolution of the power of the channel with the longest wavelength in the red band (1560.6 nm) at the system output after activation of the switch.

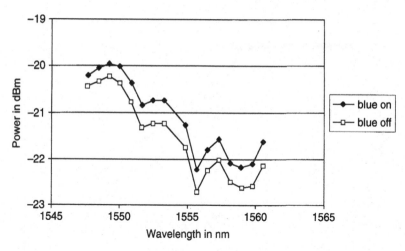

Fig. 16.14. Comparison of power levels of the channels in the red band at the output of the first span with and without channels in the blue band.

The deactivation of the channels in the blue band resulted in a decrease of the output power of the channels in the red band in five steps. This decrease cannot be caused by the erbium-doped optical amplifiers, because each counterdirectionally propagating band is processed by separate amplifiers.

Further experimental investigation provided convincing evidence that the decrease of the channel powers was caused by a reduction of SRS-induced tilt in the transmission fiber. Figure 16.14 shows a comparison of channel powers in the red band at the output of the first span with and without presence of the blue band channels. The channel with the longest wavelength in the red band experiences a change of the output power of approximately 0.7 dB caused by a change of the SRS-induced tilt.

Figure 16.15 shows a comparison of the power transients of three different channels in the red band after activation of the switch: the channel with the longest wavelength (1560.6 nm), a channel close to the center of the band (1553.3 nm), and the channel with the shortest wavelength (1547.7 nm). The amplitude of the power transients exhibits the same wavelength dependence as the SRS-induced tilt.

The structure in the time evolution of the power decrease for a given channel can be explained by the accumulation of the effect in each span. The EDFAs in the system were equipped with a fast gain control by pump power adjustment. A fast reduction of the input power level results in a proportional reduction of the output power. This means that the change of power levels in the red band generated in the first span superimpose to changes in the second span and so on. At the system output, the total power decrease accumulated to more than 4 dB in case of the red band channel with the longest wavelength.

The power transients observed in the red band can be explained as follows. After activation of the switch, a falling power slope of the blue band channels propagates through the system with the speed of light (approx. 2×10^8 m/s in silica fibers). The strongest energy transfer due to stimulated Raman scattering occurs at locations

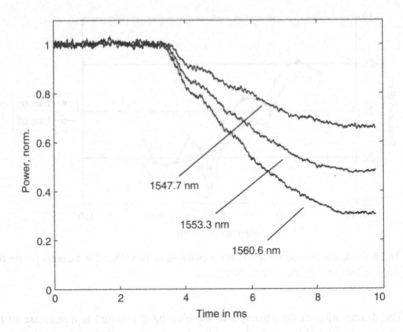

Fig. 16.15. Comparison of the power transients of three channels in the red band: the channel with the longest wavelength (1560.6 nm), the shortest wavelength (1547.7 nm), and a channel close to the center of the band (1553.3 nm).

along the fiber with the highest channel power levels in the blue band, that is, in the first kilometers following the fiber input. Consequently, the strongest changes of the red band power levels can be observed if the falling slope of the blue band enters a new fiber span. The rate of decay reduces as the slope propagates along the fiber span and reaches locations with lower channel powers in the blue band. The resulting pulsations of the power decrease can be observed in Fig. 16.13.

A similar behavior can be observed for the power levels of the channels in the blue band at the system output if the red band is switched off. However, in this case the channel output power increases after activation of the switch. This can be understood from the fact that the energy transfer from the blue band to the red band caused by stimulated Raman scattering stops if the channels in the red band are switched off.

BER measurements for the channel with the longest wavelength in the red band have revealed error bursts of several thousand bit errors in transmitted 10 Gbps signals immediately after the activation of the switch. The channel was transmitted error free (BER $< 10^{-13}$) before and a short time after the switching event. The number of errors clearly depends on the system configuration and the characteristics of the receiver. Severe error burst can be expected for systems with many spans and considerable SRS-induced tilt.

The stepwise time evolution of the power transients results from the counterdirectional propagation of the switched channels and the surviving channels. A different time evolution has to be expected for codirectionally propagating channels. In this case, the power slope of the switched channel propagates in the same direction as the induced transient of the surviving channel. As the response time of SRS is very short, the rise or fall time of the transient will basically correspond to the rise or fall time of the optical switch.

The accumulation of the contributions from the individual spans depends on the characteristics of the optical amplifiers. The amplitude of the transient in dB at the output of the system will be approximately the sum of the contributions from the individual spans in dB, if EDFAs are gain controlled or if the response time of the switching event is faster than the response time of the inversion (<1 ms). Compensation of the power transients by pump power adjustment is rather inefficient, because it mainly alters the sum of all channel powers and does not provide the necessary tilt. Better results can be expected from a tilt filter with a fast control loop.

An impression about the potential magnitude of SRS-induced transients in high-capacity systems can be obtained by considering the C+L-band system mentioned above. The unidirectional transmission of 80 channels in the C-band and the same number of channels in the L-band results in a SRS-induced tilt of 6.1 dB in a single span of SSMF. Deactivation of all channels in the L-band with a fall time of 1 ms increases the power of the channel with the shortest wavelength in the C-band (1528.8 nm) at the fiber output by 2.2 dB. After 10 spans, the amplitude of the transient will have accumulated to approximately 22 dB. Such a huge increase of the channel power within 1 ms will result in a severe error burst and can damage the receiver.

Fortunately, deactivation of all channels in a transmission band is an unlikely event under normal operating conditions. It may occur in the case of a component failure or if the service staff unplugs the wrong patchcord. Both events can result in a fall time of the channel power below 1 ms, especially if the connector of the patch cord is springloaded. The resulting transients of the surviving channels will have a severe impact on system performance.

SRS-induced transients may also deteriorate the performance of transparent photonic networks. Dynamic adding and dropping of channels has been proposed to adapt the network to a changing traffic pattern. In such networks, it is unlikely that many channels will be added or dropped at the same time. However, even the activation or deactivation of a few channels can result in considerable transients, if the channels are distributed over a wide wavelength range and transmitted over many spans.

16.4. Pattern Dependent Crosstalk

The effects mentioned in the previous sections result from the impact of SRS on the time-averaged channel power. SRS in silica fibers is a very fast effect with a response time on the order of several 10 femtoseconds. It is much faster than the bit length of currently installed transmission systems working with bit rates up to 10 Gbps, corresponding to a bit length of 100 ps. Consequently, SRS can cause power fluctuations within the timescale of a single bit.

Figures 16.16 and 16.17 try to illustrate the concept. In Fig. 16.16, three WDM channels with different wavelenths and bit patterns are launched into a section of

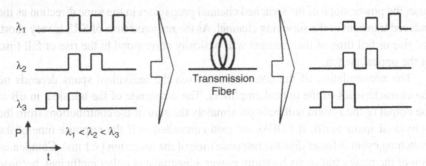

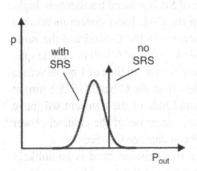

Fig. 16.16. Pattern-dependent energy transfer between WDM channels induced by SRS.

Fig. 16.17. Comparison of probability densities of the power of a given channel at the fiber output with and without SRS.

transmission fiber. SRS results in an energy transfer from shorter to longer wavelength channels. The amount of energy transfer in a given time slot depends on the number of marks transmitted in the individual channels in this time slot. For example, the first bit from the left side of the channel with the longest wavelength (λ_3) does not experience any energy transfer because the other channels are transmitting space bits in the same time slot. The second bit gains some energy which is transferred from the mark bits of the other channels due to SRS.

The pattern-dependence of the energy transfer due to SRS results in a very fast modulation of the power of a given channel. As the bit patterns transmitted in the channels of a WDM system are usually independent, the modulation can be interpreted as a noise contribution increasing the variance of the power of a given channel. Figure 16.17 depicts probability densities of the power level of a given channel at the output of the fiber. Without SRS and with negligible impact of other nonlinear effects and quantum noise, every mark bit arrives at the fiber output with the same power. SRS results in a dependence of the output power on the bits transmitted in the other channels. The shift of the mean of the probability distribution corresponds to the SRS-induced tilt discussed in the previous section. This section focuses on the broadening of the distribution.

The impact of pattern-dependent crosstalk induced by SRS (SRS-XT) on the performance of WDM transmission systems has been analyzed by many authors [15–34]. The initial work concentrated on a small number of channels: in many cases only two. Furthermore, it did not differentiate between the shift of the time-averaged channel power due to SRS and the additional variance. As the shift of the time-averaged power can be compensated, the initial studies usually overestimated the impact of SRS-XT on ultimate system capacity.

For deriving expressions describing the variance of the channel power, it is helpful to start by looking at the channel with the shortest wavelength. As SRS transfers energy from shorter to longer wavelengths, the shortest wavelength channel can only experience depletion. The depletion D in dB can be expressed as the power P_0 of a given channel at the fiber output with the other channels switched off divided by the output power of the same channel being transmitted together with the other channels:

$$D = 10 \log_{10} \frac{P_0}{P}. \tag{16.4}$$

The amount of depletion of the shortest wavelength channel depends on the bits transmitted in the other channels. It reaches its maximum value if all other channels transmit mark bits and is equal to 0 dB if all other channels transmit space bits. The depletion can be interpreted as a random variable, because the bits transmitted in the individual WDM channels are usually independent.

In the following discussion, we consider N equally spaced channels with a channel spacing of Δf. The channel with the shortest wavelength has the index $C = 1$. The channels are binary intensity modulated according to the nonreturn-to-zero (NRZ) format. Space and mark bits have equal probabilty: $p_0 = p_1 = \frac{1}{2}$. The Raman gain profile is approximated by a triangular shape. In the first step, the effect of pulse walk-off or dispersion will be neglected. Using these assumptions, the mean μ of the depletion of channel No. C can be calculated using the following expression.

$$\mu = \frac{(N - C)(N - C + 1) - (C - 1)C}{4} K. \tag{16.5}$$

The factor K in dB is defined as

$$K = 10 \log_{10} \frac{g_R \Delta f}{A_{\text{eff}} \Delta f_R} P \frac{1}{\alpha_S} [1 - \exp(-\alpha_S L)], \tag{16.6}$$

where g_R denotes the peak Raman coefficient in m/W for random polarizations, A_{eff} the effective mode field area in m^2, Δf the channel spacing THz, Δf_R the frequency difference between the signal channel acting as a pump and the peak of the Raman gain spectrum in THz, P the launch power per channel in W during a mark bit, α_S the loss coefficient in 1/m, and L the length of the fiber in m.

The mean of the depletion in dB corresponds to the difference of the channel power level at the fiber output in dBm for single-channel operation and the time-average output power level in dBm for transmission together with the other channels. For equal launch power levels of all channels, the depletion of the channel with the shortest wavelength ($C = 1$) equals half the SRS-induced tilt. Equation (16.5) can actually be converted into Eq. (16.2) by replacing $NP/2$ by the total time-averaged launch power P_0 and $(N - 1)\Delta f$ by the frequency separation of the shortest and the longest wavelength channel Δf_{ch}.

Figure 16.18 shows a plot of the mean of the depletion divided by K of a given channel C for a total number of $N = 80$ channels with equal launch powers. The

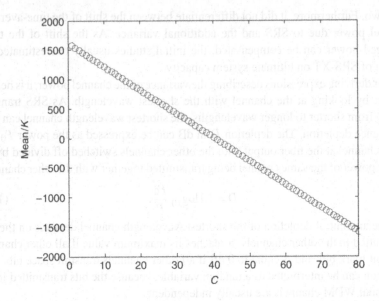

Fig. 16.18. Plot of the mean of the depletion divided by K of a given channel C for $N = 80$ channels.

channels at the band edges experience the strongest shift of the time- averaged channel power, whereas the channels in the center of the band arrive at the fiber output with the same time-averaged power as in the case of single-channel operation.

A different wavelength-dependence can be observed for the variance of the depletion. The variance of the SRS-induced depletion of channel C for a total number of N channels can be calculated using the following expression.

$$\sigma^2 = \frac{(N-C)(N-C+1)(2N-2C+1)+C(C-1)(2C-1)}{24}K^2. \qquad (16.7)$$

Figure 16.19 shows a plot of the variance of the depletion devided by K^2 of a given channel C for the same total number of channels ($N = 80$) as in the previous example. The maximum variance occurs for the channels at the band edges. Note that the variance does not drop to zero in the center of the band. This can be understood from the fact that the output power of the channel in the center of the band does still depend on the bits transmitted in the other channels. The output power reaches its maximum if all channels with shorter wavelength transmit a mark bit and all channels on the longer wavelength side transmit a space bit. Minimum output power can be observed if the mark and the space bit switch sides.

The nonvanishing variance of the depletion in the center of the band has an important impact on system performance. Pattern-dependent crosstalk introduced by SRS degrades the performance of all channels, even if a filter is used to compensate for the SRS-induced tilt. The strongest reduction of the signal quality will be observed for the channels at the band edges. However, some impairments also have to be expected in the center of the band.

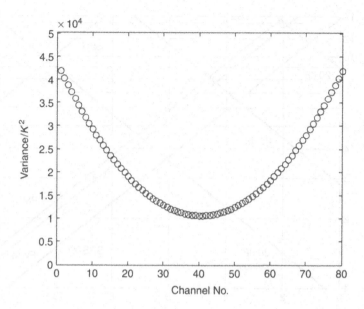

Fig. 16.19. Plot of the variance of the depletion divided by K^2 of a given channel C for $N = 80$ channels.

The observation that the channels at the band edges experience stronger performance degradations can be employed to differentiate between limiting nonlinear effects. Other multichannel nonlinear effects such as cross-phase modulation (XPM) and four-wave-mixing (FWM) grow stronger with the number of nearby interacting channels. Consequently, XPM and FWM result in a stronger performance degradation of channels in the center of the band. Stronger degradation of the channels at the band edges is a clear indication for dominating pattern-dependent crosstalk introduced by SRS.

In the previous discussion, the walk-off of bits in different channels due to dispersion was neglected. This assumption may be valid for a small number of channels located around the zero dispersion wavelength of a fiber with a small dispersion slope. In most fibers currently used for the transmission of WDM channels, dispersion will introduce a considerable walk-off between bits, especially if the channels are distributed over a wide wavelength range.

The walk-off of the bits increases the number of bits interacting with a mark bit of a given channel. The effect on the variance of the depletion is similar to an increase of the number of channels. However, fiber attenuation decreases the impact of bit interactions on the variance during propagation through the fiber.

The variance of the depletion of the channel with the longest wavelength for the case of fast walk-off can be estimated using the following expression [32].

$$\sigma^2 = \frac{N(N-1)\alpha_S T}{16|D\Delta\lambda|} K^2, \tag{16.8}$$

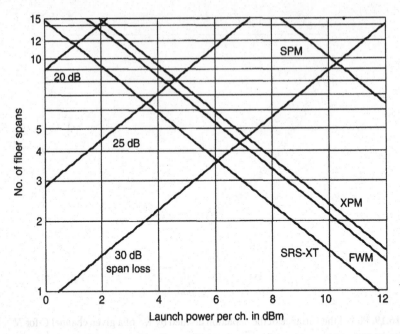

Fig. 16.20. Number of fiber spans as a function of launch power per channel for a WDM system transmitting 160 × 10 Gbps in the C- and L-bands over SSMF.

where α_S denotes the fiber attenuation in 1/km, T the bit length in ps, D the dispersion coefficient in ps/(nm km), and Δl the channel spacing in nm.

Due to its statistical nature, pattern-dependent crosstalk introduced by SRS can be treated as an additional noise contribution to the overall noise budget of a WDM transmission system. The system reach is limited by quantum noise from the optical amplifiers if launch powers per channel into the fiber sections are too low. The impact of quantum noise decreases with increasing launch power. According to Eqs. (16.7) and (16.8), the variance of the depletion increases with increasing launch power. As a consequence, SRS-XT results in an upper limit of the launch power similar to other nonlinear effects.

Figure 16.20 shows limitations imposed on the number of spans as a function of launch power per channel. The plot was calculated for transmission of 160 channels in the C- and L-bands with a channel spacing of 50 GHz over SSMF. The required OSNR at the system output was set to 22 dB to enable error-free detection of 10 Gbps signals without FEC. The dispersion-compensation scheme was optimized to minimize the impact of XPM and FWM. Gain-flattening filters compensated for the SRS-induced tilt and other wavelength-dependent effects.

The lines with the positive slope correspond to noise limitations for the span losses noted below the lines. The positive slope reflects the fact that increasing the launch power reduces the ASE noise at the system output. The lines with the negative slope correspond to OSNR penalties of 1 dB contributed by various nonlinear effects. The impact of these effects increases with increasing launch power.

It can be concluded from the plot that pattern-dependent crosstalk induced by SRS is the dominating nonlinear effect. The high dispersion coefficient of SSMF helps to reduce the impact of XPM and FWM, leaving SRS-XT as the limiting effect. A similar situation can be observed for various other configurations featuring transmission of many channels distributed over a wide wavelength range as well as for other fiber types. Consequently, SRS-XT has to be considered as an important limitation for the reach of state-of-the-art high-capacity WDM transmission systems.

16.5. Selecting Transmission Fibers

Many different transmission fiber types are currently commercially available. The loss spectra of these fibers are quite similar. The main difference can be found in the dispersion characteristics. The dispersion spectrum of a transmission fiber has a very important impact on system performance. For example, a high dispersion coefficient can help to suppress nonlinear effects. On the other hand, it generates a need for strong dispersion compensation, which in turn may be difficult to implement, expensive, or degrade the system performance.

The dispersion coefficient may be a very important parameter of transmission fibers, but it is not the only one that has an impact on system performance. Another parameter that has gained attention recently is the Raman efficiency. The Raman efficiency is defined as the Raman gain coefficient divided by the effective mode field area of the fiber. A high Raman efficiency can be achieved by designing a fiber with a small mode field area. Such a fiber will also exhibit strong nonlinear effects due to the Kerr effect, because the efficiency of these effects also grows inversely proportional to the effective mode field area. Another way to increase the Raman efficiency is to select a fiber material with a high Raman coefficient. Such a material does not necessarily come with a strong Kerr effect.

The previous discussion has shown that it is possible to select the Raman efficiency of transmission fibers within certain limits. But it did not answer the question of whether it is more desirable for transmission fibers to feature high or low Raman efficiencies. The answer seems to be clear, if the potential benefits of distributed Raman amplification are taken into account. This technology can be deployed very effectively to improve the OSNR budget of a system by virtually reducing the span loss. Increasing the Raman efficiency reduces the amount of pump power required for a given gain of the distributed amplifier. Thus higher Raman efficiency seems to be beneficial.

However, increasing the Raman efficiency also increases the strength of system impairments introduced by SRS. It has been shown in the previous section that pattern-dependent crosstalk caused by SRS is the dominating limitation of system reach for many high-capacity long-haul terrestrial system configurations. Means to compensate the signal degradation caused by SRS-XT are currently not available. On the other hand, it is quite easy to compensate smaller Raman gain of transmission fibers with less Raman efficiency by applying more pump power. Pump sources with sufficient output power to achieve the maximum useful gain even in fibers with a small Raman efficiency are commercially available. As a consequence, it seems to be more beneficial from a system design perspective to select transmission fibers with a small Raman efficiency if system impairments introduced by SRS contribute a potential limitation for the system reach.

References

[1] A. Einstein, Zur Quantentheorie der Strahlung, *Physikalische Zeitschrift*, 18:121–128, (March), 1917.

[2] A. Yariv, *Introduction to Quantum Electronics*. 3d ed., New York: Holt, Rinehart and Winston 129–132, 1985.

[3] R. G. Smith, Optical power handling capacity of low loss optical fibers as determined by stimulated Raman and Brillouin scattering, *Appl. Optics*, 11:11 (Nov.), 2489–2494, 1972.

[4] G. Agrawal, *Nonlinear Fiber Optics*, 2d ed., New York: Academic, 319–322, 1995.

[5] P. M. Krummrich, R. Neuhauser, and G. Fischer, Experimental comparison of Raman thresholds of different transmission fiber types. In *Proceedings of the 26th European Conference on Optical Communication (ECOC 2000)*, Sept. 3–7, Munich, Vol. 3, P1.5, 133–134, 2000.

[6] S. Bigo, S. Gauchard, A. Bertaina, and J.-P. Hamaide, Experimental investigation of stimulated Raman scattering limitation on WDM transmission over various types of fiber infrastructures, *IEEE Photon. Technol. Lett.*, 11:6 (June), 671–673, 1999.

[7] T. Miyakawa, I. Morita, and N. Edagawa, 40 Gbit/s × 25 WDM unrepeatered transmission over 362 km. In *Proceedings of the Conference on Optical Amplifiers and their Applications (OAA 2002)*, July 14–17, Vancouver, Canada, OTuA1, 1–3, 2002.

[8] M. Zirngibl, Analytical model of Raman gain effects in massive wavelength division multiplexed transmission systems, *Electron. Lett.*, 34:8, 789–790, 1998.

[9] P.M. Krummrich, E. Gottwald, A. Mayer, C.-J. Weiske, and G. Fischer, Influence of stimulated Raman scattering on the channel power balance in bidirectional WDM transmission. In *Proceedings of the Conference on Optical Fiber Communication (OFC 1999)*, San Diego, Feb. 21–26, Vol. 2, WJ6, 171–173, 1999.

[10] S. Bigo, S. Gauchard, A. Bertaina, and J. Hamaide, Investigation of stimulated Raman scattering limitation on WDM transmission over various types of fiber infrastructures. In *Proceedings of the Conference on Optical Fiber Communication (OFC 1999)*, San Diego, Feb. 21–26, Vol. 2, WJ7, 174–176, 1999.

[11] V. J. Mazurczuyk, G. Shaulov, and E. A. Golovchenko, Accumulation of gain tilt in WDM amplified systems due to Raman crosstalk, *IEEE Photon. Technol. Lett.*, 12:11, 1573–1575, 2000.

[12] M. Takeda, S. Kinoshita, Y. Sugaya, and T. Tanaka, ed. S. Kinoshita, J. C. Livas, and G. van den Hoven, Active gain-tilt equalization by preferentially 1.43 mu m- or 1.48 mu m-pumped Raman amplification. In *OSA Trends in Optics and Photonics Series* (TOPS 1999), Vol. 30, *Conference on Optical Amplifiers and their Applications*, Nara, Japan, June 9–11, 101–105, 1999.

[13] P.M. Krummrich, A. Mayer, and G. Fischer, Noise penalty caused by stimulated Raman scattering in bidirectional DWDM transmission. In *Proceedings of the Conference on Optical Amplifiers and their Applications (OAA 1999)*, Nara, Japan, ThB3, 102–105, 1999.

[14] P.M. Krummrich, E. Gottwald, A. Mayer, R. Neuhauser, and G. Fischer, Channel power transients in photonic networks caused by stimulated Raman scattering. In *Proceedings of the Conference on Optical Amplifiers and their Applications (OAA 2000)*, Quebec, July 9–12, OTuC6, 143–145, 2000.

[15] M. Ikeda, Stimulated Raman amplification characteristics in long span single mode silica fibers, Optics Commun. 39:148–152, 1981.

[16] A. Tomita, Cross talk caused by stimulated Raman scattering in single-mode wavelength-division multiplexed systems, *Optics Lett.*, 8:7 (July), 412–414, 1983.

[17] A. R. Chraplyvy and P. S. Henry, Performance degradation due to stimulated Raman scattering in wavelength-division-multiplexed optical fibre systems, *Electron. Lett.* 19:16 (Aug.), 641–643, 1983.

[18] A. R. Chraplyvy, Optical power limits in multi-channel wavelength-division multiplexed systems, *Electron. Lett.*, 20:2 (Jan.), 58–59, 1984.

[19] D. Cotter and A. M. Hill, Stimulated Raman crosstalk in optical transmission: Effects of group velocity dispersion, *Electron. Lett.*, 20:4 (Feb.), 185–187, 1984.

[20] R. H. Stolen and A. M. Johnson, The effect of pulse walk-off on stimulated Raman scattering in fibers, *IEEE J. Quantum Electron.*, QE-22:11 (Nov.), 2154–2160, 1986.

[21] D. N. Christodoulides and R. I. Joseph, Theory of stimulated Raman scattering in optical fibers in the pulse walk-off regime, *IEEE J. Quantum Electron.*, QE-25:273–279, 1989.

[22] M. S. Kao, ON/OFF ratio degradation of high density WDM systems due to Raman crosstalk, *Electron. Lett.*, 26:14 (July), 1034–1035, 1990.

[23] R. Comuzzi, C. de Angelis, and G. Gianello, Improved analysis of the effects of stimulated Raman scattering in a multi-channel WDM communication system, *European Trans. Telecommun. Related Technol.*, 3:3, 295–298, 1992.

[24] S. Tariq and J. C. Palais, A computer model of non-dispersion-limited stimulated Raman scattering in optical fiber multiple-channel communications, *IEEE J. Lightwave Technol.*, 11:12 (Dec.), 1914–1924, 1993.

[25] X. Zhang, B. F. Jorgensen, F. Ebskamp, and R. J. Pedersen, Input power limits and maximum capacity in long-haul WDM lightwave systems due to stimulated Raman scattering, *Optics Commun.* 107:5–6, 358–360, 1994.

[26] F. Forghieri, R. W. Tkach, and A. R. Chraplyvy, Effect of modulation statistics on Raman crosstalk in WDM systems, *IEEE Photon. Technol. Lett.* 7:1, 101–103, 1995.

[27] G. P. Agrawal, Raman-induced crosstalk. In *Nonlinear Fiber Optics*, New York: Academic, 334–336, 1995.

[28] S. Tariq and J. C. Palais, Stimulated Raman scattering in fiber optic systems, *Fiber Integ. Optics* 15:4 (April), 335–352, 1996.

[29] D. N. Christodoulides and R. B. Jander, Evolution of stimulated Raman crosstalk in wavelength division multiplexed systems, *IEEE Photon. Technol. Lett.*, 8:12 (Dec.), 1722–1724, 1996.

[30] J. Wang, X. Sun, and M. Zhang, Effect of group velocity dispersion on stimulated Raman crosstalk in multichannel transmission systems, *IEEE Photon. Technol. Lett.*, 10:4, 540–542, 1998.

[31] L. Rapp, Impact of stimulated Raman scattering in WDM systems using different types of fiber, *Int. J. Electron. Commun.*, 52:5, 302–309, 1998.

[32] K.-P. Ho, Statistical properties of stimulated Raman crosstalk in WDM systems, *IEEE J. Lightwave Technol.*, 18:7, 915–921, 2000.

[33] S. Norimatsu and T. Yamamoto, Waveform distortion due to stimulated Raman scattering in wide-band WDM transmission systems, *IEEE J. Lightwave Technol.*, 19:2 (Feb.), 159–171, 2001.

[34] A. G. Grandpierre, D. N. Christodoulides, W. E. Schiesser, C. M. McIntosh, and J. Toulouse, Stimulated Raman scattering crosstalk in massive WDM systems under the action of group velocity dispersion, *Optics Commun.*, 194:4–6, 319–323, 2001.

Chapter 17

Ultra-Long-Haul Submarine and Terrestrial Applications

Howard Kidorf, Morten Nissov, and Dmitri Foursa

17.1. Introduction

Optical communication is the science of transmitting information over a distance using light. The engineering difficulties vary greatly because the distances to be traversed differ significantly. For some, the task may be the need to interconnect electronic integrated circuits within a computer. For others, the distance to be covered may be between computers in a building or between the buildings in a city. The differences between these cases are the distances over which information must be carried. These differences dictate the nature of the technologies required to implement the connections.

Extreme situations are found in cases where communication is required between distant cities within a land mass or between the various landmasses of the globe. These communications links are often called ultra-long-haul (ULH) systems. As defined in this chapter, ULH systems span from 1500 to over 12,000 km.

A very reasonable question is: why would anyone want ULH transmission systems? To cross oceans, the case is clear. The lack of intermediate land upon which to build the equipment capable of optoelectronic regeneration creates a need for ULH systems. The demand to transmit high-capacity data signals 6000 km across the Atlantic and 9000 km across the Pacific Ocean drives this obvious need.

For terrestrial systems a good question is: why would anyone want ULH transmission systems when concatenating shorter systems can cover the same distances? For decades, terrestrial digital transmission systems were built from city to city. Traffic was reconfigured to redirect local and express traffic. Why might this network model change? There are two reasons.

The demand patterns for traffic have changed. Twenty years ago, a small fraction of total telecommunications traffic was data traffic. The vast majority was voice traffic. It has been shown that demand for voice traffic falls off very fast as the distance increases. A span distance of 600 km (as is common today) satisfies more than 60% of the connections for voice traffic [1]. In other words, most people communicate and do business close to where they are located. Their calling patterns reflect this.

The demand for the Internet has changed all of this. In the past few years, data traffic (i.e., mostly the Internet) has surpassed voice traffic in the major economic regions of the world (Europe, North America, and SE Asia). Interestingly, Internet traffic has a very different demand pattern than voice traffic. A span distance of 3000 km is required to satisfy 60% of the connections for Internet traffic [1]. Few people care or are even aware where the information that they are seeking resides when they follow a hyperlink. The result is that a higher ratio of traffic travels much farther from the user than ever before.

The other reason for a changing network model is the technical possibility and economic motivation for all-optical networking. With the commercial availability of all-optical cross-connects [2] (based on microelectromechanical systems, MEMS, technology among others), the opportunity for all-optical mesh networking has grown. The goal of reducing the cost of large networks has driven the market to look for technologies to allow transmission of a lightwave as far as possible so that as many optical–electrical–optical (O-E-O) regenerators as possible can be eliminated. The economic argument is that routing an optical signal with a simple mirror reflection is much less expensive (and takes less space and power) than an O-E-O regenerator. Further development of all-optical cross-connects, wavelength add-drop devices, and dynamic gain equalization technologies all promise to enhance the appeal of ULH transmission.

17.1.1. History of Ultra-Long-Haul

It is hard to imagine a time when rapid communications between distant points was not possible. Yet, for 5000 years of mankind's civilization, messages could not travel much faster than wind-propelled ships or the gallop of a horse. Kings, queens, and emperors, in addition to ordinary businessmen, frequently made history-changing decisions in the absence of timely or correct information.

The development of the telegraph would change all of this. In 1861 the Western Union Company completed the final leg of the first high-speed telegraph link between the east and west coasts of North America. This led to the rapid growth of a web of connectivity. Unfortunately, the web stopped at the shores of the continent. In 1866, after four previous unsuccessful attempts, entrepreneur Cyrus Field, numerous eminent scientists, and venture capitalists succeeded in spanning the Atlantic Ocean with the first telegraph cable [3]. This was the first introduction to a worldwide communications network.

In 1884, the cutting edge in long distance telephony was a telephone link between Boston and New York City [4]. However, without amplification, a transcontinental connection would have to wait until the summer of 1914 when a New York to San Francisco link could be established with the aid of Dr. DeForest's triode vacuum tube [4]. Voice connectivity between the continents would not be established until overseas radiotelephony was introduced in 1927 [5]. Despite technical improvements, high cost and unavoidable ionospheric disturbances kept usage low.

The world would have to wait until 1955 before the first of more than 20 trans-Atlantic telephone cables were installed [6]. This impressive 4000 km transmission feat was accomplished with amplified, analogue, frequency-division-multiplexed

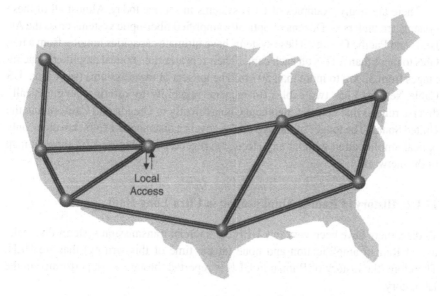

Fig. 17.1. National core network showing local access.

transmission over coaxial cables. Analogue ULH telephony cables would eventually be capable of carrying over 4200 voice channels on each coaxial cable. A web of ULH connectivity soon grew to connect the major cities of the world using these technologies (along with digital microwave transmission).

From April 1977 [1] through the mid1980s, three generations of optical fiber transmission technology allowed terrestrial fiber-optic transmission to spread and to gradually replace coaxial cables and digital microwave as the technology of choice for ULH connectivity [7]. Beginning in the early 1990s erbium-doped fiber amplifiers (EDFAs) allowed optically amplified transmission to greatly increase the capacity of optical transmission systems and to cement the role of optical transmission in inter-city transmission.

Throughout this evolution, terrestrial optical transmission systems used technologies that required signal regeneration at least every 600 to 1200 km (i.e., optical signals that are converted to electronic signals and regenerated after every fiber span). This regenerator spacing was selected partly because in Asia, Europe, and most of North America, major cities are spaced by these distances. At each of these sites, traffic is regroomed (i.e., local traffic added and dropped, express traffic rerouted) with electronic cross-connects. Therefore, electronic regeneration in metropolitan areas makes sense. This is illustrated in Fig. 17.1, which shows an example of a national core network. Only in submarine systems, where transoceanic transmission eliminates the need for local traffic access and the size of regenerative equipment makes it unappealing, do we see true ULH transmission systems deployed today.

---

[1] General Telephone and Electronics sent the first commercial telephone traffic though fiber optics in April 1977.

There are many examples of ULH systems in service today. Almost all of these systems are undersea. Dozens of optically amplified fiber-optic systems cross the Atlantic and Pacific Oceans. These systems have ultimate capacities ranging from a few Gb/s to more than 5 Tb/s on each cable. Their repeater (i.e., optical amplifier) spacing ranges from 33 km to more than 50 km. The longest of these systems (the China–US Cable Network) has two cables (to increase reliability by offering geographically diverse routes) that connect mainland China directly to Oregon and California in the United States. The longer of the two cables covers a distance of 12,460 km using only optical amplification without any electronic regeneration. Figure 17.2 shows a map of this network.

17.1.2. History of Raman Amplification in Ultra-Long-Haul

To date, there have been very few ULH commercial transmission systems that make use of Raman amplification and none (at the time of this writing) that are ULH. Therefore, the history of Raman in ULH is reported through work performed in the laboratory.

The first major experiment that used Raman amplification was, interestingly, not focused on Raman amplification, but on soliton transmission [8]. This work was the foundation of the first ULH demonstration, which traversed approximately 4000 km without regeneration [9]. This demonstration was made years before the EDFA was invented and entirely used Raman amplification to overcome the 9.2 dB loss of the 41.7 km spans. This experiment foretold work to be performed more than a decade in the future by demonstrating Raman amplification by pumping the transmission fiber at about 1497 nm to achieve optical amplification at a wavelength of about 1600 nm. This work also anticipated the need to avoid Brillouin backscattering of the intense pump light by spreading out the spectrum of the pump.

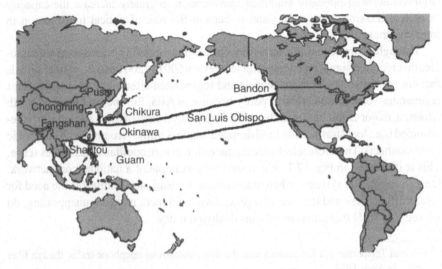

Fig. 17.2. The China–US cable network.

In 1997 and 1998, the availability of practical high-powered 1450 to 1500 nm pumps spurred interest in experiments that demonstrated the value of Raman amplifiers in ULH transmission systems. The pump technology that became available was the combination of a cladding-pumped fiber laser and a cascaded Raman resonator [11] (themselves an interesting application of Raman scattering) capable of producing an output power in excess of 1.2 W at 1480 nm.

In the first demonstration ever of ULH WDM transmission using only Raman amplification to overcome the loss of the transmission fiber, experimenters successfully demonstrated transmission over 7200 km (far enough to traverse the Atlantic Ocean or North America) by measuring error-free (bit error rate, BER, $< 10^{-9}$) performance for ten 10 Gb/s channels [12]. Furthermore, the optical signal-to-noise ratio (OSNR) of the transmission line was 1.5 dB better than equivalent EDFA chains with the same path-averaged signal power. This demonstrated for the first time the previously theorized advantages of distributed amplification as having superior noise performance to lumped amplification (as EDFAs are). It also demonstrated that some of the potentially large transmission impairments found in distributed amplification systems, such as multipath interference (MPI), could be managed effectively.

This experiment and many others are discussed in Section 17.4 (Applications of Raman Gain).

17.2. Ultra-Long-Haul Transmission Issues

Ultra-long-haul systems have fundamental design differences from their shorter length cousins. These differences are mostly (but not entirely) due to transmission distance and the fact that optically amplified fiber transmission behaves somewhat like analogue transmission where the impairments incurred in each span accumulate over the entire length of the transmission. That is, digital regeneration occurs only at the end of the transmission line that may be more than 10,000 km from the transmitter.

17.2.1. Differences Between Terrestrial and Undersea Systems

There are many differences between terrestrial and undersea ULH systems besides the opportunity for large regenerator sites along terrestrial routes. These differences have a great impact on the designs of these systems. They arise due to two primary reasons: in undersea systems, access to the amplifiers is limited to that which can be provided through the cable and all equipment in the transmission line must be able to fit into a housing that is capable of withstanding the pressure found on the bottom of the ocean.

All of the power needed for the inline amplifiers along an undersea transmission line comes from a pair of large power supplies on the shore (although sometimes only one is provided). This power is supplied through a power conductor in the cable to the repeaters where it is distributed to the amplifiers. The ocean serves as the second conductor between the two shores. Due to constraints on the dielectric material in the cable, the maximum voltage used in these power supplies is typically 10 to 12 kV.

The operating current is about one ampere supplying a cable that has a resistance of 0.7 to 1.6 Ω/km [13]. Over a 10,000 km system with repeater spacing of 50 km (200 repeaters), less than 100 W is available to power each repeater (which may contain up to eight pairs of amplifiers with today's technology). Whereas a terrestrial repeater can consume a much larger amount of power (subject to availability and cooling constraints), an undersea repeater has a significant limit on available power. The impact on Raman amplified systems is that, because undersea systems are constrained by available power, their system designers have reason to heavily favor use of the most power-efficient optical amplification. In general, this is delivered by EDFAs.

Another limitation imposed by the oceans is the need to house all optical, electrical, and mechanical equipment for powering, mechanical interconnection, amplification, and supervision and diagnostics inside a pressure housing that can be deployed from a ship, sink more than 7.5 km through the ocean, and rest under more than 730 atmospheres (74 MPa = 10,700 lbs/in^2) of pressure. Practical considerations call for restrictions on the size of the pressure housing to 0.065 m^3 (65 liters). See Figure 17.3 for an example. For comparison, the volume of a typical belowground terrestrial repeater hut is 16 to 100 m^3.

Unlike undersea systems, terrestrial systems have limitations caused by long repeater spacing and fiber selection. The first restriction arises from the fact that most terrestrial transmission systems are installed into already established rights-of-way. Therefore the repeater spacing is already predetermined. Also, the fiber type is selected once and deployed (often years in advance of the transmission system deployment) in a cable frequently with hundreds of fibers. The optical system designer must contend with the fibers that are already deployed.

Freedom to select the optimum fiber at the time of the system design is a great advantage in submarine systems. All-optical transmission systems require a match

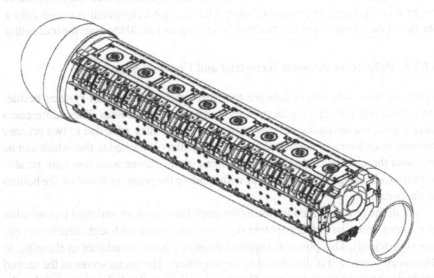

Fig. 17.3. View of repeater housing for undersea application. The housing shown contains up to eight pairs of optical amplifiers.

between the optical source and the transmission fiber. In high-capacity ULH systems that make use of optical amplifiers, this match is even more critical. Ever since the first ULH optically amplified systems, proper selection of the chromatic dispersion characteristics of the fiber has been critical. The advent of WDM systems brought focus on the inherently nonlinear characteristics of the transmission fibers and the desire to lower their impact through expanding the effective area [14]. Matching the desire for spans with large effective area and low dispersion slope with the chirped return-to-zero (CRZ) modulation format results in a hybrid span design (i.e., two types of fiber in one span) that is implemented primarily in undersea transmission systems.

Perhaps the greatest difference between terrestrial and submarine systems is their different approaches to reliability. Submarine systems are designed to be one of the most reliable systems on earth. Not only are the repeaters usually beneath 5 to 8 km of water, but they are also often in places where it takes up to a week to get a suitable ship into position for a repair. Unlike terrestrial equipment that can depend on service personnel being able to access and service the equipment (often within hours of a reported failure), submarine systems are created with a design target to suffer less than three failures in their lifetime of 25 years. This level of reliability requires extensive qualification testing of all components and subsystems that will be placed undersea. The requirement for reliability further underscores the need for simple, low-power undersea components.

Table 17.1 summarizes the design issues found in terrestrial and submarine ULH systems.

Table 17.1. Summary of ULH Design Issues

	Terrestrial	Submarine
Space for inline amplifiers	$>1.0 \text{ m}^3$	$<0.1 \text{ m}^3$
Powering	Less important	Limited due to remote supply
Transmission fiber	Legacy	Custom tailored
Reliability (assemblies)	~1000 FITs	≪100 FITs

The challenge of introducing Raman amplification into systems is to do it in a way that provides sufficient economic benefit for the owner/system supplier to warrant the substantial change of technology. For submarine systems, the introduction must be compatible with the limited physical space in the pressure vessels, limited power-feed capability, and ultra-high system reliability.

17.2.2. ULH Impairment Accumulation

All ULH transmission systems have at least one thing in common: all the impairments experienced by the signal along the transmission line accumulate from the transmitter to the receiver. Obtaining successful ULH transmission between the regenerative terminals is first a matter of engineering amplifiers and fiber spans so that the optical signal-to-noise ratio over the bandwidth of the system is acceptable at the receiver. This is achieved by carefully controlling amplifier and system gain shape.

Raman amplifiers in ULH systems add several challenges to obtaining spectral flatness. Broadband Raman amplifier designs need broadband components with a bandwidth sometimes even in excess of the signal bandwidth. One example is the optical component that couples the pump laser into the transmission fiber while at the same time letting the signals pass through. For a multiwavelength-pumped Raman amplifier with 100 nm of bandwidth it is not uncommon to require that this coupler must have a low-loss bandwidth of 200 nm. Achieving low loss and little spectral shape for both the signals and the pumps is important for good transmission performance and good pump efficiency. This can be challenging to component manufacturers whose main product lines until recently have been geared towards primarily C-band EDFAs. Another challenge affecting spectral flatness is how to manage the pump wavelengths over the life of the system. Unlike EDFAs, Raman gain is directly affected by deviations of pump wavelengths away from the designed values. This gives rise to two issues for an ULH system with many cascaded amplifiers: getting pump lasers at the correct wavelength when the system is being built, and dealing with pump laser aging/wavelength changes over the life of the system.

Some of the other significant impairments that require consideration in the design of any ULH system (including Raman amplified systems) are:

- polarization mode dispersion (PMD),
- polarization-dependent loss (PDL),
- chromatic dispersion,
- fiber nonlinearity, and
- multiple reflections from components and fiber (mostly Rayleigh scattering, MPI).

To control the impacts of these effects careful attention must be paid to the design of the system as well as the specifications for all the optical components used.

17.2.2.1. PMD and PDL

Polarization mode dispersion is introduced into the transmission path through the components in the optical amplifiers and the transmission fiber itself. This is a much bigger issue in ULH systems, inasmuch as there are more fiber and more components.

In single-mode fiber there exists a small modal birefringence due to core deformation and external stress. It has been shown that the PMD grows according to the square root of the length of the transmission line [15]. Fortunately, for ULH systems, this means that a 10,000 km system has only three times as much PMD as a 1000 km system. Still, standard fiber has about 0.1 ps/km$^{1/2}$ resulting in an average of 10 ps of differential group delay (DGD) over 10,000 km.

Within the optical amplifier, the short length of the devices and their inherent circular symmetry causes most components to exhibit low PMD. The largest contributor to PMD in the amplifier is the optical isolator/circulator. Typical isolators have less than 0.05 ps per device, thus contributing a few more of picoseconds differential group delay to the transmission line.

For the PMD to have a small impact on transmission, the total DGD in the transmission path must be a small fraction of the bit period. At a bit rate of 10 Gb/s (a bit period of 100 ps), a PMD of 10 to 15 ps due to both the fiber and amplifier components would likely cause a tolerable amount of transmission penalty. Transmission at higher bit rates would be difficult, however.

Another polarization effect associated with amplifier components is PDL. An optical component exhibiting PDL where the polarization state of the input signal slowly changes will cause the signal power to fluctuate. Due to the presence of amplifiers in the transmission line and the fact that ASE is mostly unpolarized, signal polarization fluctuations are converted by the amplifiers into changes in received OSNR [16]. Just as with PMD, the impairment caused by many components each having small values of PDL accumulates over the length of the transmission line.

17.2.2.2. Fiber Nonlinearity

This impairment is due to the index of refraction of silica fiber being dependent on the intensity of the light. This effect is given by

$$n = n_0 + n_2 \frac{P}{A_e},$$ (17.1)

where P is the power in the fiber, A_e is the effective area of the fiber, and n_2 is the nonlinear index of refraction. The accumulated phase change (where L_e is the effective length of each span, N is the number of spans in a system, and λ is the wavelength) is given by

$$\Phi_{nl} = \frac{2\pi n_2}{\lambda A_e} P N L_e.$$ (17.2)

Though the nonlinear index is small (about 2.6×10^{-20} m^2/W [18]), the resultant phase change for light propagating in the fiber becomes significant (about $\pi/2$ radians) when $P \cdot L_e$ reaches 1000 mW-km [19]. This value is easily reached in ULH systems.

The nonlinear index of refraction has a number of manifestations: self-phase modulation (SPM), cross-phase modulation, and four-wave-mixing. A complete treatment of these impairments [20] is beyond the scope of this book, but any ULH system with or without Raman amplifiers must account for their impact.

There are two ways to minimize the impact of optical nonlinearity (as can be seen in the equation above): reduce the power of the signal and increase the effective area. The impact of Raman amplification on the average power level of the signal is significant and is discussed later in this chapter. Increasing the effective area of the transmission fiber is an ongoing effort. The effective area of transmission fibers has been increased from 50 to 75 μm^2 in practice (dispersion-shifted fibers) and to over 120 μm^2 (non-dispersion-shifted fibers) in laboratories. This increase in effective area has clearly led to improvements in transmission performance [14]. However, it should be noted that Raman amplifiers often depend on pumping smaller effective area fibers to achieve reasonable efficiencies. Even though increasing the effective area of the transmission fibers improves performance it also makes Raman amplification less efficient and, subsequently, less attractive.

17.2.2.3. MPI

Reflections in transmission systems have long been known to cause several deleterious effects. Refractive index discontinuities due to components, connectors, and
splices were shown to cause large power penalties and bit error rate floors in even
simple, unamplified transmission systems [22]. The impact of the inherent Rayleigh
backscatter has also been shown to cause transmission penalty [23].

Rayleigh scattering (one of the dominant loss mechanisms in the 1.5 μm low-loss
transmission window in silica) results from random inhomogeneities in the fiber on
a scale small compared with the wavelength of light. Although somewhat reducible
through improved fabrication technique, the effect cannot be eliminated [24]. For
most single-mode fiber, the power reflected due to Rayleigh scattering results in a
return loss of about 32 dB.

If the signals were simply reflected by Rayleigh scattering, Rayleigh scattering
would have minimal effect on transmission. But because the reflected light itself is
subject to Rayleigh reflection and both of these reflections are subject to distributed
Raman amplification, the doubly reflected and amplified signal (often called multipath
interference) generates an inband noise source for the transmitted signal. In general,
analysis of this type of interference is complicated and has been ongoing since the
first evaluation of MPI in amplified systems in 1990 [25].

17.3. Raman Effects Applicable to ULH Transmission

In this section, the Raman effects that are most important for ULH systems are discussed. Several of the effects have already been discussed in greater detail in the
preceding chapters of this book. Discussion is included in this chapter relating their
importance to ULH systems. In this way, a better understanding of the use of Raman
amplification for such systems is provided. Most of the discussion in this section
is focused on single-channel Raman amplifiers. This is done to simplify the discussion and present the concepts more clearly. However, all the concepts are readily
extended to also cover multiwavelength-pumped, broadband Raman amplifiers, or
hybrid combinations of Raman amplifiers and EDFAs.

This section primarily discusses the use of distributed Raman amplification.
Lumped Raman amplification for ULH is not nearly as attractive as either distributed
Raman amplification or the use of conventional EDFAs. Lumped Raman amplification has the disadvantage of poorer noise performance than both distributed Raman
amplification and EDFAs. Furthermore, the total amount of optical fiber allocated
to the repeaters is significant compared to the spans. In submarine systems, the limited space in the repeater housings cannot easily accommodate the needed lengths of
fiber. So for ULH systems, lumped Raman amplifiers are most useful in the terminals
of systems operating at wavelengths where no other convenient lumped amplifier
technology is readily available. Another application for lumped Raman amplifiers is
in terrestrial systems where there is a strong desire and/or necessity to reuse previously installed fiber even though that fiber type might not be particularly suitable for
distributed Raman amplification.

Two fundamental concepts in ULH system design are the path-average power and effective noise figure [26]. These are discussed first. The path-average power provides a simple way of accounting for the significant changes to the channel power evolution that results from the use of Raman gain. Effective noise figure allows different lumped and/or distributed amplifier topologies to be compared with equivalent path-average power (which causes similar nonlinear impairments).[2] The main advantage of the effective noise figure concept is to allow the majority of the amplifier design trade-offs to be performed without having to resort to (time-consuming) split-step simulations.

The next two effects discussed are the impact of temperature on Raman amplification and noise and the additional noise effects generated by Rayleigh scattering. After all the basic concepts have been clarified, the benefits of lumped versus distributed amplification are analyzed. The section ends by briefly discussing additional penalties that are enhanced by Raman amplification that must be taken into account when Raman is used for ULH transmission.

17.3.1. Path-Average Power

Performance of ULH transmission systems is almost always limited by optical nonlinearity. To reduce accumulation of amplified spontaneous emission (ASE) noise, higher channel launch power is desired. However, higher launch power also increases nonlinear impairments. Channel launch power is therefore chosen as a trade-off between nonlinear distortions and noise accumulation. A comparison of lumped rare earth-doped fiber amplifier topologies is straightforward and can be performed based on noise figure, power efficiency, cost, and so on. However, comparing lumped and distributed amplifiers is not as easy. Noise figure is no longer an unambiguous metric. Any comparison between distributed and lumped amplifiers in such a case must take into account that the channel launch power of distributed amplification systems has to be reduced to make up for the increased power in the span.

Figure 17.4 shows the power evolution of a single channel in a transparent[3] 90 km Raman amplifier normalized to launch power. The power evolution is shown for various backward pump ratios (defined as the ratio between backward pump power and total pump power) as well as for an unpumped fiber where the signal only experiences loss. The figure shows that the power evolution is strongly affected by the presence of Raman gain and the pump direction. The signal power for the case of 100% backward pumping is the smallest everywhere in the span. As more and more pump power is injected into the forward direction the signal power increases. To avoid the system being degraded by additional nonlinear impairment, the launch power has to be reduced. This has an impact on the received OSNR.

---

[2] The performance of real system designs ultimately has to be optimized using nonlinear propagation models to accurately model the constituent amplifiers and transmission impairments, however.

[3] A transparent Raman amplifier is one with unity gain where the internal Raman gain exactly compensates the fiber loss.

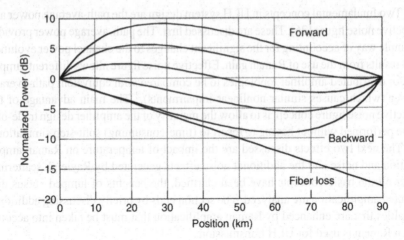

Fig. 17.4. Power evolution in a 90 km Raman amplifier for backward pump ratios of 0% (forward), 25%, 50% (bidirectional), 75%, and 100% (backward). For comparison a fiber span without Raman gain is also shown.

An important metric for characterizing the nonlinearity enhancement associated with the use of Raman gain is the path-average power. The path-average power $\overline{P}$ of a signal channel is the integrated power of the channel normalized by the span length.

$$\overline{P} = \frac{1}{L} \int_0^L P_s(z)\, dz, \tag{17.3}$$

where $P_s(z)$ is the signal power at position z and L is the span length. The normalized path-average power η is subsequently defined as the path-average power normalized by the launch power:

$$\eta = \frac{\overline{P}}{P_s(0)}. \tag{17.4}$$

All simple comparisons of distributed and lumped amplifiers should be performed for the same path-average power. This ensures similar nonlinear impairments. This is the basis for the definition of effective noise figure in the next section.

Figure 17.5 shows the normalized path-average power versus span length for forward, bidirectionally, and backward pumped transparent Raman amplifiers. The normalized path-average power of bidirectionally pumped transparent amplifiers is always very close to 0 dB. Forward pumped amplifiers have a normalized path-average power larger than and backward pumped amplifiers smaller than the bidirectional case. The figure also shows that the difference between the pump directions increases significantly with span length.

Figures 17.4 and 17.5 deal with transparent Raman amplifiers (unity gain). In all-Raman systems all amplification must be provided as Raman gain. Multiple optical components are needed in practical amplifiers such as the means for coupling pump power into the fiber, optical isolators, monitor couplers, and gain-equalizing

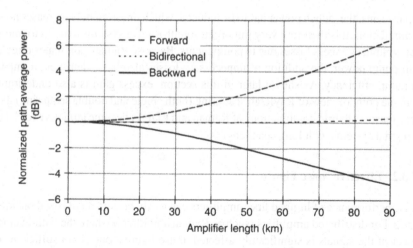

Fig. 17.5. Normalized path-average power versus span length for transparent Raman amplifiers. Curves are shown for forward, bidirectional, and backward pumping.

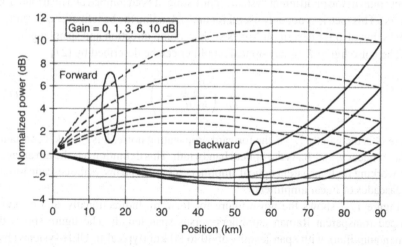

Fig. 17.6. Power evolution in 45 km forward and backward pumped Raman amplifiers for excess gain of 0 through 10 dB.

filters. The amplifier must therefore be able to provide excess gain to overcome these losses in the amount of multiple dBs. This is not the case for Raman-based systems. Figure 17.6 shows the power evolution for forward and backward pumped 45 km Raman amplifiers with an end-to-end gain of 0 through 10 dB. In conventional lumped EDFA-based systems, component losses have no effect on the power evolution of the signals.

The path-average power of forward pumped amplifiers is especially sensitive to component losses. But even for backward pumped amplifiers the extra gain needed to overcome the component losses increases path-average power. To avoid nonlinear impairments, the launch power has to be reduced; which unfortunately increases noise accumulation. It is therefore very important for all-Raman systems to use a topology that minimizes excess losses due to components. Because Raman gain scales directly with pump power, the addition of unnecessary losses also has a detrimental impact on pump efficiency. As shown later in this section, excess gain is also undesirable from the point of view of Rayleigh scattering (both single and double scattering). For these reasons, hybrid combinations of Raman and EDFAs become attractive for ULH terrestrial systems with long span lengths.

17.3.2. Effective Noise Figure

Noise performance of lumped fiber amplifiers can easily be compared based on noise figure. For distributed amplifiers (such as Raman amplifiers) where the path-average power of the signals is significantly affected, noise figure alone is not sufficient for comparison. For such amplifiers, it is important to perform the comparison with similar amounts of nonlinearity. This is achieved by performing the comparison for the same path-average power. An effective noise figure is defined in this section to allow direct comparison of different systems (consisting of both lumped and distributed amplifiers). This metric is not only useful for understanding the benefit over conventional EDFAs, but also for design trade-offs such as pump direction and span length.

The noise figure F of any optical amplifier can be described by [27]

$$F = \frac{1}{G} \cdot \left(\frac{P_{ASE}^+}{h v B_o} + 1 \right),$$ (17.5)

where G is the gain, P_{ASE}^+ is the forward propagating noise, h is Planck's constant, v is the frequency, and B_o is the optical bandwidth. This definition has the advantage of also working for transparent optical amplifiers ($G = 1$) and obeys the cascade formula for cascades of linear amplifiers.

Figure 17.7 shows the noise figure for forward, bidirectionally, and backward pumped transparent Raman amplifiers versus span length. The figure shows that Raman amplifiers with span lengths of 40 to 90 km (typical in ULH systems) have noise figures of 6 to 16 dB. These noise figures, however, include the loss of the span whereas for conventional EDFAs the quantum limited 3 dB noise figure only includes the lumped amplifier.

The figure shows that forward pumped amplifiers have the lowest noise figure and backward pumped amplifiers the highest. This could suggest that forward pumping would be the best choice. However, in the preceding discussion of path-average intensity, forward pumping also had the highest relative path-average power.

To take this into account the effective noise figure is therefore introduced. The effective noise figure F_{eff} is defined by comparison to a lumped amplifier system with the same received OSNR for the same path-average power after transmission through a long chain of amplifiers. This allows direct comparison of the noise figure of the lumped amplifier with the effective noise figure of the distributed amplifier. In other

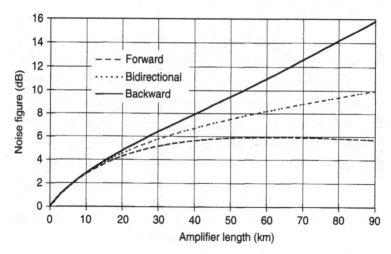

Fig. 17.7. Noise figure for transparent Raman amplifiers versus span length. Curves are shown for forward, bidirectional, and backward pumping.

words, the effective noise figure is the noise figure that a lumped amplifier must have to achieve the same noise performance as a distributed amplifier.

To develop the concept of effective noise figure for a distributed amplifier, we start with the received OSNR for a transmission span followed by a lumped amplifier with gain G (gain exactly compensates for the fiber losses). It can be expressed as

$$\mathrm{OSNR}_l = \frac{\overline{P}}{\eta_l \cdot (F_l G_l - 1) h v B_o}, \tag{17.6}$$

where the subscript l is used to denote values associated with the lumped amplifier. G_l and F_l are the gain and noise figure of a single lumped amplifier, respectively. The definition of normalized path-average power incorporates the launch power of the signal.

Similarly the received OSNR of a unity gain (amplifier gain exactly compensates fiber and component losses) distributed amplifier can be expressed as

$$\mathrm{OSNR}_d = \frac{\overline{P}}{\eta_d \cdot (F_d - 1) h v B_o}, \tag{17.7}$$

where the subscript d is used to denote values associated with the distributed amplifiers. The effective noise figure F_{eff} is now defined as the lumped amplifier noise figure that would result in the same received OSNR for the same path-average power as for the distributed case. An expression for the effective noise figure can now be written using Eqs. (17.6) and (17.7).

$$F_{\mathrm{eff}} = \frac{1}{G_l} \left[\frac{\eta_d}{\eta_l} (F_d - 1) + 1 \right]. \tag{17.8}$$

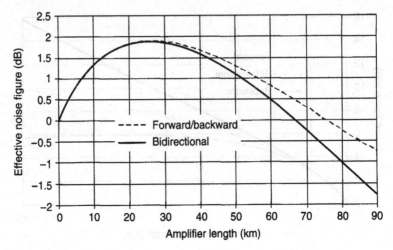

Fig. 17.8. Effective noise figure for forward, bidirectionally, and backward pumped transparent Raman amplifiers.

In the analysis, signals are assumed to experience only linear loss in the spans of the lumped system; signal–signal Raman interactions are ignored in the equivalent lumped system. The path-average power of the lumped system can therefore be simply expressed in closed form and inserted in Eq. (17.8), resulting in:

$$F_{\text{eff}} = \frac{\ln L}{L-1} \eta_d \cdot (F_d - 1) + \frac{1}{L}, \qquad (17.9)$$

where $L(> 1)$ is the loss of the span. This is the definition of the effective noise used in the remainder of this chapter.

Figure 17.8 shows the effective noise figure of forward, backward, and bidirectionally pumped transparent Raman amplifiers versus span length. Only considering power evolution and noise figure, the figure shows that pump direction has little impact on performance (<0.2 dB) for amplifiers shorter than 50 km. For span lengths in excess of 50 km bidirectional pumping performs best.

The additional component losses that are present in any practical Raman amplifier were discussed in Section 17.3.1 (Path-Average Power). It can be shown that the additional gain needed to make up for the losses actually reduces the noise figure, while at the same time increasing path-average power. The effective noise figures for 45 km Raman amplifiers versus gain are shown in Fig. 17.9. For 45 km Raman amplifiers backward pumping is optimal when the gain exceeds 2.5 dB. For less than 2.5 dB of excess gain bidirectional pumping works slightly better. Forward pumping is especially sensitive to excess gain. Regardless of pump direction the figure demonstrates that effective noise figure and therefore system performance degrades with increased excess gain.

Apart from performing design trade-offs, the effective noise figure also allows direct comparisons with other systems (including conventional EDFA-based systems),

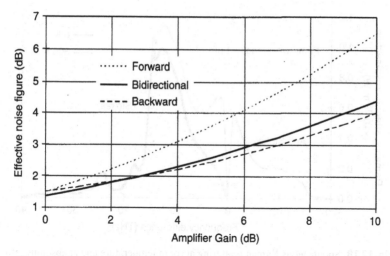

Fig. 17.9. Normalized noise figure versus gain for 45 km Raman amplifiers.

by simple subtraction of effective noise figures. The effective noise figure of an EDFA is the noise figure. This is used later in this chapter to quantify the performance improvement achieved by using distributed amplification.

17.3.3. Temperature-Dependence of Noise Figure

The noise figure of Raman amplifiers is temperature-dependent because at temperatures above absolute zero there is a nonzero probability of finding molecules thermally excited [28]. These thermally excited molecules can interact both with signal and pump photons and create spontaneous and stimulated emission (both Stokes and anti-Stokes). On average, anti-Stokes stimulated emission cancels out with the temperature-enhanced Stokes stimulated emission. The Raman gain is therefore temperature-independent. The spontaneous emission (both Stokes and anti-Stokes) is, however, temperature-dependent. Thus the noise figure of a Raman amplifier is temperature-dependent but the gain is not.

Figure 17.10 shows the Raman scattering cross-section both at absolute zero and at room temperature (300°K). It demonstrates that the anti-Stokes scattering as well as the enhanced noise generation disappears at absolute zero. The effect of the temperature-dependence is seen to be significant especially close to the pump wavelength; but even at the gain peak the effect is not insignificant. As an example: for a single-wavelength pumped Raman amplifier with close to unity gain the noise figure at the gain peak is increased by ~0.5 dB. This increase might not seem significant, but for shorter spans the inherent Raman benefit compared to EDFAs can be as small as 1 dB. Also for wideband multiwavelength pumped Raman amplifiers where the longest wavelength pumps are very close to the shortest wavelength signals significant excess noise can be added thereby increasing the noise figure well in excess of 0.5 dB.

Most of the simple models that have been published traditionally neglect this temperature-dependence [29, 30]. However, to realistically predict the performance

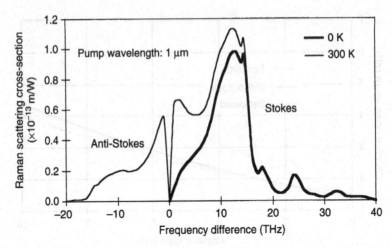

Fig. 17.10. Spontaneous Raman scattering at room temperature and at absolute zero.

advantage of Raman amplification over EDFAs or to compare alternate Raman amplifier topologies it is important to take this effect into account.

17.3.4. Noise Enhancement Through Rayleigh Scattering

Rayleigh scattering in EDFA systems mainly manifests itself as span loss. For Raman amplified systems, however, Rayleigh scattering is also responsible for contributing additional noise. The two main contributions are: double-Rayleigh scattering (DRS) of the signal leading to incoherent inband crosstalk and (although not as important as DRS) single reflections of backward traveling spontaneous noise into the forward direction. Both effects are also present in EDFA systems, but due to the fiber loss, they have very small impact. For distributed amplifiers with close to unity gain the effects are much enhanced.

DRS can significantly limit system performance in ULH systems. Because, both the single and double reflected signals are amplified, the DRS efficiency scales with the gain squared [26, 31]. This is another reason why amplifier topologies and components should be chosen to minimize excess gain. Figure 17.11 shows the crosstalk relative to the signal power at the output of a single 45 and 90 km Raman amplifier versus gain. Assuming that the crosstalk adds as power, the total crosstalk after 100 to 200 amplifiers is 20 to 23 dB higher than after a single amplifier. Virtually no penalty is seen when the crosstalk is 30 dB lower than the signal at the receiver [32].

The figure shows that the crosstalk increases rapidly with gain and span length. A system based on 90 km spans has half the amplifiers compared to a system with 45 km spans. Even taking this into account, the 90 km span system suffers ~7 dB more from double-Rayleigh scattering unless bidirectional pumping is used. This is another reason why bidirectional pumping is very advantageous for long Raman amplifiers. The figure also shows that even for unity gain and short repeater spacing it is very hard to keep the double-Rayleigh scattering for each amplifier below −50

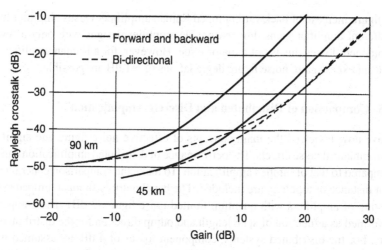

Fig. 17.11. Rayleigh crosstalk relative to signal power at output of 45 and 90 km Raman amplifiers versus net amplifier gain (assuming a fiber loss of 0.2 dB/km and no excess losses).

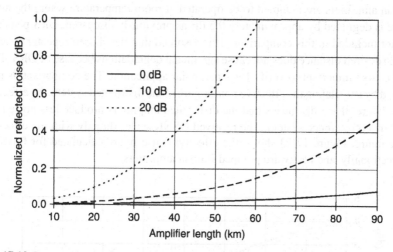

Fig. 17.12. Increased noise figure through Rayleigh reflection of backward ASE for backward pumped transparent Raman amplifiers.

dB so that hundreds of amplifiers can be cascaded with insignificant penalty. One way to limit the penalty is to combine the Raman amplifier with an EDFA to form a hybrid amplifier. In this way the Raman and EDFA gains can be chosen to optimize performance by trading off ASE noise and DRS. This has been used to obtain a 7000 km transmission distance with 100 km spans using terrestrial fibers [38].

In EDFA systems, the reflection of the backward traveling ASE is only an issue for high-gain amplifiers. In this case the extra noise can be avoided by the use of input and output optical isolators. This solution cannot be used for distributed amplifiers where both reflections and gain happen in the same fiber. Figure 17.12 shows the noise figure increase of backward pumped Raman amplifiers versus amplifier length for different amplifier gains. For reasonable amplifier designs with only a few dB of component losses this is not a major issue. However, for a Raman amplifier with significant excess gain, noise figure degradations of ~1 dB are possible.

17.3.5. Comparison of Distributed and Discrete Amplification

We have now reviewed the main concepts of effective noise figure and significant Raman enhanced noise effects. The performance of distributed amplification can now be compared to that of lumped amplification. To make the comparison general no details of distance or topology are included. The benchmark is an ideal lumped system consisting of amplifiers with quantum limited noise figure (3 dB). The comparison is performed as a function of span length and pump direction for the distributed amplifiers. For the distributed systems, component losses of 4 dB are assumed which include typical optical components such as input pump signal coupler, optical isolator(s), output tap, and gain-flattening filter. Because all distributed amplifiers are assumed to be transparent the excess Raman gain must therefore also equal 4 dB. The Raman amplifiers are assumed to be operated at room temperature where the noise figure is degraded by approximately 0.5 dB as previously discussed. DRS penalties are not included in this comparison. It is assumed that the chosen repeater spacing for the desired distance does not cause significant degradations because of DRS. This is of course important to verify for a particular application. The comparison is performed by calculating the effective noise figure of the distributed system. An effective noise figure of <3 dB means that the distributed amplifiers produce less noise than the lumped reference system. The received OSNR scales directly with the effective noise figure. Figure 17.13 shows the effective noise figure calculated for forward, bidirectionally, and backward pumped Raman amplifiers.

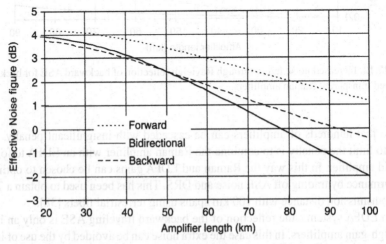

Fig. 17.13. Effective noise figures for forward, backward, and bidirectionally pumped Raman amplifiers with a temperature-dependent noise figure degradation of 0.5 dB.

The figure shows that for very short amplifier spacing, lumped amplifiers work as well as distributed amplifiers and there is therefore little reason for using distributed amplification. For intermediate length amplifier spacings (~50 km) backward pumped Raman amplifiers provide the best performance, but the benefit does not exceed 1 dB over lumped amplification until a span length of 55 km. For the longest span lengths, bidirectional pumping significantly improves performance (up to 5 dB compared to lumped amplification). It is important to emphasize that even though the smallest effective noise figure is seen for the longest spans, performance is still worse than for shorter spans. Having a lower effective noise figure just means that the distributed benefit over lumped amplification for the same span length is larger.

This comparison illustrates why especially for terrestrial systems, where the repeater hut spacing is fixed and typically on the order of 90 km, Raman amplification has been strongly embraced. On the other hand, for submarine applications where the repeater spacing can be chosen to optimize system performance and is typically on the order of 50 km, Raman amplification is not as competitive as EDFAs.

17.3.6. Other Degradation Mechanisms

There are several other issues and performance-degrading effects that are unique to or enhanced by the presence of Raman gain. Most of these are discussed in much greater detail in other parts of the book, but are briefly mentioned here for completeness.

Relative intensity noise (RIN) from the pump lasers can couple to the signals. This pump-RIN transfer is especially strong for forward pumping and virtually insignificant for backward pumping. For this reason all of the early experiments used only backward pumping. As discussed in connection with effective noise figure, in general there is no particular advantage of forward pumping as compared to backward pumping. However, for long Raman amplifiers, bidirectional pumping is very attractive. Because bidirectional pumping requires that part of the pump light propagate in the forward direction, pump lasers with low RIN are needed. Recently several pump laser designs that provide high power and low RIN have therefore been developed [33, 34].

Raman gain is polarization-dependent; maximum gain is obtained only when signal and pump are copolarized. To avoid unacceptable levels of polarization-dependent gain, the Raman pump has to be mostly unpolarized. This is especially important for forward pumping, but is also necessary for backward pumping. (That the backward pump and signal travel in opposite directions and follow independent evolutions of the polarization state helps to "scramble" the polarization-dependent interactions.) Early experiments used fiber lasers to provide the Raman pump. These sources are inherently unpolarized [12]. For practical use, semiconductor pump lasers are a more desirable choice. However, these sources are polarized and must therefore be depolarized for acceptable system performance. One popular way of ensuring depolarized Raman pumping is to combine two lasers with similar wavelengths using a polarization beam combiner. This also has the advantage of almost doubling available pump power at each wavelength. The downside is that more diodes are needed, which influences cost, space, etc.

17.4. Applications of Raman Gain

Due to its unique properties, Raman amplification has been studied extensively since the advent of low-loss optical fibers. Over the past few years, one of the last technological obstacles on the way to practical implementation of Raman amplifiers in telecommunication has been overcome with the development of compact, reliable, and powerful Bragg grating stabilized laser diodes (LDs).

The system designer can now use Raman amplification in a number of different ways. In some designs, Raman amplifiers allow the unregenerated reach of the system to be increased or the repeaters (i.e., amplifiers) to be spread farther apart. In other designs, the bandwidth of the system can be increased; or, systems can be operated at wavelengths where rare earth-doped fiber amplifiers are not available. To achieve these goals, Raman amplification can be used either in combination with EDFAs (as hybrid Raman/EDFA) or on their own in a transmission system that employs only Raman amplification (all-Raman) [35, 36]. Progress toward each of these goals is discussed in the following sections.

17.4.1. Extending Repeater Spacing Using Hybrid Raman Assisted EDFAs

The addition of Raman amplification to erbium-doped amplifiers is very much like adding a low-noise preamplifier to the EDFA. Combining Raman amplification with EDFAs has several benefits. Raman amplifiers are very good at providing low-noise preamplification for the signals, whereas EDFAs provide saturation and high output power with good pump efficiency. Saturation is desirable for long systems because the signal power is then self-limiting and active control of the amplifier can potentially be avoided. As shown in the previous section, it is very desirable to limit the excess gain that the Raman amplifier has to provide. Less excess gain has an impact on DRS, effective noise figure, pump efficiency, and so on. This way of using Raman amplification, however, limits the bandwidth of the system to that of the lumped amplifier.

The performance improvement obtained from adding the Raman assist to the EDFA can be used (among other things) to extend the unregenerated reach of the system or to extend the repeater spacing (distance between amplifiers) while still fulfilling the design targets (system length, capacity, etc.). For a given system length, the desire to reduce the system cost is a strong motivation to extend the amplifier spacing as much as possible while achieving the desired system capacity. This can be particularly attractive in ULH transmission systems where in excess of 200 amplifiers [44] are used in the transmission lines of the longest systems.

Many experiments have been conducted to explore the feasibility of adding margin to systems by adding Raman assist to the EDFAs. In the first attempt (1998) to demonstrate the use of distributed Raman amplification to increase the span length in ULH WDM systems, a record 240 km between repeaters (defined as optoelectronic units requiring power and supervision) was obtained [37]. This was achieved with distributed Raman amplifiers and remotely pumped EDFAs. The configuration of the 240 km transmission spans is shown in Fig. 17.14(a). Remote pumping by a fiber laser of EDF1 through a low-loss pure silica-core fiber forms the first amplifier. The dedicated pump fiber is used to avoid interactions between the noisy copropagating pump and the signal. A second fiber laser backward pumped the remaining three amplifiers. After passing through EDF3, the pump power was then launched into the transmission fiber creating a distributed Raman amplifier with approximately 14 dB of internal gain. The pump power remaining after passing through the Raman amplifier was used to pump EDF2. Isolators were used along the transmission line to reduce MPI. In total, two pump sources, each providing 1.2 W (at 1480 nm) of optical power were required for each span. The signal power evolution between the repeaters, depicted in Fig. 17.14(b), was reduced to less than 20 dB of total excursion. Without the remote EDFAs and the Raman amplifier a signal excursion in excess of 48 dB would be expected. The impact of the remotely pumped EDFAs is similar to that of distributed Raman amplification, that is, making the power distribution along the propagation path more even.

Hybrid Raman/EDFA amplifiers have also been shown to be useful in creating ULH transmission lines using fibers of a type that is already deployed in much of the world. Raman amplification in combination with an EDFA has been used to achieve transcontinental (7500 km) transmission with 90 km spans built from LEAF™ fiber

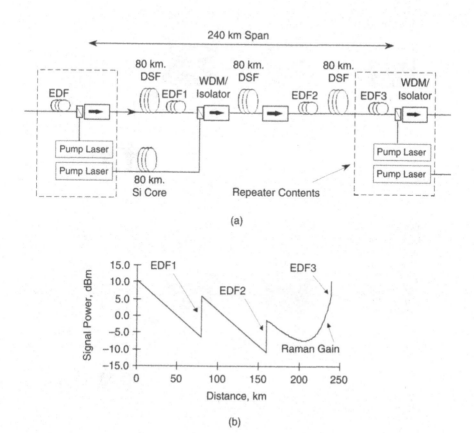

(a)

(b)

Fig. 17.14. (a) Configuration of the 240 km repeater span; (b) power evolution in the span.

618   H. Kidorf et al.

[38]. Two unpolarized pump sources with wavelengths of 1430 and 1457 nm backward pumped the transmission fiber to provide 10 dB of internal Raman gain. The remaining 10 dB of gain needed to overcome the loss of the transmission fiber was provided by the EDFA. In this way, both amplifier technologies were used: distributed Raman preamplification ensured good noise performance and the EDFA provided the output power with good pump efficiency.

In the section on effective noise figure it was shown that bidirectional pumping, which achieves a more distributed gain, improves noise performance for longer spans (in excess of 55 km). This has also been demonstrated in several experiments [39, 40].

Another approach to improve the distribution of Raman gain is achieved by changing the order of the fibers in the span. In most system designs, most of the Raman amplification comes from a small effective area (and hence better Raman pump efficiency) fiber at the end of the span; see Fig. 17.15(a). Improvements were demonstrated by placing the fiber with smaller effective area in the middle portion of the transmission span (Fig. 17.15(b)) [41]. This fiber arrangement improved the system

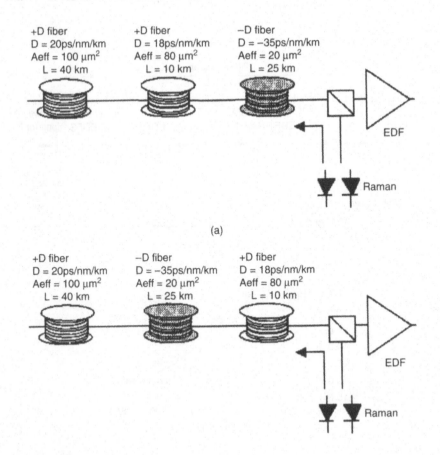

Fig. 17.15. (a) Conventional map $(+D/-D)$, (b) A_{eff} managed map $(+D/-D/+D)$.

OSNR by 0.9 dB. By shifting the Raman gain into the middle section, a more distributed amplification was achieved. However, doing this also significantly reduces pump efficiency.

Yet another approach to distribute the Raman gain more uniformly is the use of higher-order pumping, where the pump signals are Raman amplified by the addition of strong shorter wavelength pumps. This concept has the disadvantage of needing even more pump wavelengths, which increases complexity and also suffers from reduced efficiency. Higher-order pumping is described in detail in Chapter 10.

Following the first demonstration in 1998, numerous experiments have since confirmed the benefits of ULH transmission using Raman-assisted EDFAs supporting a very high transmission capacity [42]. Most recent ULH transmission experiments performed using terrestrial span lengths (~100 km) depend on Raman amplification to achieve their goals.

17.4.2. Extending Bandwidth Using Raman/EDFAs

One of the main advantages that Raman amplification has over rare earth-doped amplifiers is that Raman amplification is not limited to wavelength bands linked to atomic transitions. Raman amplification can be realized across the whole transparency spectrum of silica fibers. It can therefore be used to expand the transmission bandwidth of ULH WDM systems beyond the EDFA's C-band (and even L-band).

The first use of Raman amplification to extend the bandwidth of ULH systems was the demonstration of a hybrid combination of Raman amplification and an erbium-doped fluoride fiber (EDFFA) amplifier with overlapping gain spectra [43]; see Fig. 17.16(a). To construct the hybrid amplifier, the experimenters followed a Raman amplifier (with a gain peak at 1610 nm) with a two-stage EDFFA. The resultant gain bandwidth was 75 nm. This type of amplifier has the advantage that its gain is continuous. That is, there is no break in the gain spectrum as there is in most C+L-band EDFA amplifiers.

A similar hybrid approach resulted in the realization of a continuous bandwidth of 80 nm in the demonstration of transmission over 11,000 km of 256 channels modulated at 10 Gb/s [44]. The 80 nm Raman/EDFA hybrid was constructed by using a single-wavelength, unpolarized, backward pumped Raman amplifier in front of a single-stage EDFA. The gain spectra of the Raman and EDFA sections complemented each other and provided a wide continuous bandwidth with minimum gain ripple (Fig. 17.17). A pump wavelength of 1497 nm was chosen to best complement the gain shape of the highly inverted EDFA. The pump radiation from two polarization-multiplexed, grating stabilized lasers was coupled into the span using a circulator to achieve simultaneous low-loss and isolation for both pump and signals. The transmission spans used a dispersion-matched combination of large effective area ($110 \ \mu m^2$, +20 ps/nm/km) and small effective area ($30 \ \mu m^2$, −40 ps/nm/km) fibers in a 2:1 ratio. The pump was coupled into the fiber with the small effective area to achieve good efficiency. A gain-flattening filter at the output was designed to equalize the gain shape of the combined Raman/EDFA. The amplifier had a total output power of 18.6 dBm. The OSNR as measured after 11,000 km of transmission corresponded to an effective NF of a single Raman/EDFA ranging from 2.7 dB in the L-band to

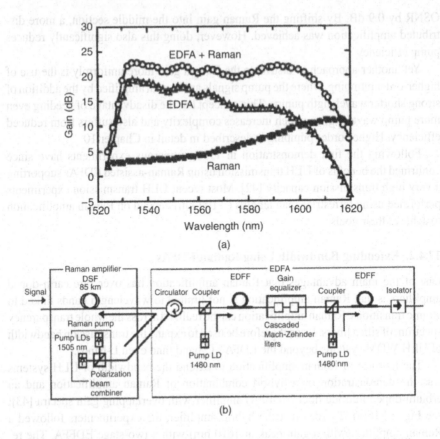

Fig. 17.16. (a) Overlapping gain spectra of EDFF and Raman Amplifiers; (b) block diagram of hybrid EDFF and Raman amplifier.

4.8 dB at the shortest wavelength. The improvement of the OSNR towards longer wavelengths is attributed to the increased contribution from distributed Raman gain (Fig. 17.17(b)).

This ability to significantly increase the bandwidth of a system by adding only a few extra components compares very favorably to the C+L-band EDFA alternative. Compared to C+L-band amplifiers, wider bandwidth, better noise performance, and potentially better electrical efficiency can be achieved. The repeater spacing, however, is limited to what is possible using just EDFAs unless Raman assist is added to preamplify the channels that are predominantly amplified by the EDFAs.

17.4.3. All-Raman Amplification

Despite the lure of easily constructed and pump-efficient EDFAs, many researchers have successfully demonstrated ULH amplifier chains based entirely on Raman amplification. The first WDM demonstration of all-Raman amplification in an ULH transmission line was made in 1997 with an extremely simple Raman amplifier architecture [12]. A wavelength selective coupler (to inject the pump while passing the

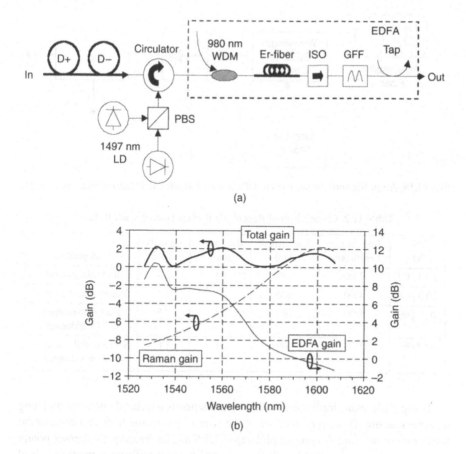

Fig. 17.17. (a) Hybrid Raman/EDFA; (b) Raman, EDFA, and total gain.

signal), isolator, and tap were used in each repeater (see Fig. 17.18). Despite the simplicity of the amplifier, a chain of these amplifiers was successfully used to transmit
ten 10 Gb/s channels over 7200 km. The bandwidth limitation was due, in part, to
the narrow bandwidth caused by single-wavelength pumping and the lack of gain
equalization in the amplifier.

All-Raman amplifier chains have since matured significantly. Most notably, multiwavelength pumping has been employed and careful fiber span design has been
employed to permit wideband operation. Some recent results are shown in Table 17.2.

Apart from the unique ability of Raman amplifiers to provide broadband gain
through the use of multiple wavelength pumping, these amplifiers also offer new ways
of dealing with gain equalization. For example, 100 nm bandwidth Raman amplifiers
have been realized using a 12 wavelength pump unit comprising high-power laser
diode modules. Gain-flatness of better than ±0.5 dB was achieved without the use of
any gain-equalization filters [47]. This level of gain-flatness is, however, not sufficient
for most ULH systems and in general gain-flattening filters will be necessary even
for multiwavelength pumped Raman amplifiers.

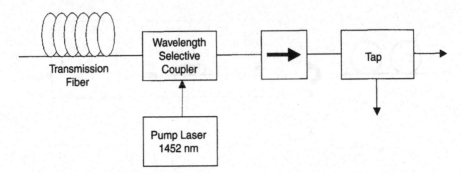

Fig. 17.18. Amplifier configuration used with first all-Raman ULH transmission experiment.

Table 17.2. Comparison of Recent All-Raman Transmission Results

Year [Ref.]	Transmission Length (km)	Span Length (km)	Bandwidth (nm)	Capacity (Tb/s)	Configuration
2001 [45]	7400	40	74	2.4	Backward pumped
2001 [40]	4000	100	53	1.3	Bidirectionally pumped
2002 [46]	4000	100	53	2.5	Backward pumped 42.7 Gb/s channels
2002 [39]	3600	100	31	1.6	Bidirectionally pumped 42.7 Gb/s channels

Using all-Raman amplification allows simultaneous wideband operation and long repeater spacing. However, since all of the gain to overcome fiber and component losses must come from Raman amplification, DRS and increased path-average power can affect performance. Whether all-Raman amplification performs better than hybrid Raman/EDFAs is strongly dependent on the details of the system design. It is clear from Table 17.2 that multi-Tb/s fiber capacity is available over ULH fiber spans with all-Raman amplification. Early concerns about insurmountable obstacles caused by pump power demand, MPI, channel–channel crosstalk, and the like, have all been overcome (but still need to be taken into account) to produce a technology that can potentially compete with the EDFA.

17.5. Conclusion

Ultra-long-haul optically amplified transmission systems are some of the most technically challenging systems designed today. Raman amplifiers have proven their usefulness in certain categories of these systems, that is, terrestrial systems. In terrestrial systems, the marriage of Raman amplification technology and erbium-doped fiber amplifiers has demonstrated great benefit by expanding the bandwidth of amplifiers, extending the distance between amplifiers, and allowing longer distances to be spanned. Because of the long repeater spacing in typical terrestrial networks the benefit of adding Raman is substantial. Many manufacturers of the terrestrial systems, therefore, already offer a Raman product.

For submarine systems where the systems are designed to achieve a desired capacity over often the longest transmission distances (6000 to 11,000 km), shorter span length (than for terrestrial systems) often has to be chosen. For such shorter spans (~50 km) the benefits of Raman amplification are not nearly as substantial. Furthermore, submarine systems have much more stringent requirements for power efficiency, volume, and reliability. These requirements put Raman amplification at a further disadvantage. In addition, ongoing transmission fiber development has continuously increased the effective area of the transmission fibers to lower optical nonlinearity and increase capacity and span length. This has a negative impact on Raman amplifier designs, because the pump efficiency is further reduced. Even faced with all these challenges most submarine suppliers have ongoing research efforts to find applications for Raman in undersea systems. Presently, the most promising candidate use of Raman amplification in submarine systems is the wideband hybrid Raman/EDFA. For systems that require a very wide bandwidth this seems like an attractive way to more than double the transmission bandwidth without doubling the component count.

References

[1] R. E. Wagner, L. Nederlof, and S. De Maesschalck, The potential of optical layer networks, *Technical Digest of Optical Fiber Communications 2001*, TuT3-1, 2001.

[2] Lucent's new all-optical router uses Bell Labs microscopic mirrors, Lucent Press Release, November 10, 1999. http://www.bell-labs.com/news/1999/november/10/1.html.

[3] B. Dibner, *The Atlantic Cable*, Norwalk, CT: Burndy Library, Inc., 1959.

[4] J. Brooks, *Telephone: The First Hundred Years*, New York: Harper & Row, 90, 1975.

[5] E. H. Ehrbar, Undersea cables for telephony. In *Undersea Lightwave Communications*, ed. P. K. Runge and P. Trischitta, New York: IEEE Press, 1986.

[6] A. Clarke, *Voice Across The Sea*, New York: Harper & Row, 1959.

[7] J. Hecht, *City of Light: The Story of Fiber Optics*, New York: Oxford University Press, 1999.

[8] L. F. Mollenauer, R. H. Stolen, and M. N. Islam, Experimental demonstration of soliton propagation in long fibers: Loss compensated by Raman gain, *Optics Lett.*, 10:5, (May), 229–231, 1985.

[9] L. F. Mollenauer, R. H. Stolen, and M. N. Islam, Demonstration of soliton transmission over more than 4000 km in fiber with loss periodically compensated by Raman gain, *Optics Lett.*, 13:8, (August), 1988.

[10] R. H. Stolen and E. P. Ippen, Raman gain in glass optical waveguides, *Appl. Phys. Lett.*, 22:6, (March), 276–278, 1973.

[11] S. G. Grubb, T. Strasser, W. Y, Cheung, W. A. Reed, V. Mizrahi, T. Erdogan, P. J. Lemaire, A. M. Vengsarkar and D. J. DiGiovanni, High power 1.48 μm cascaded Raman laser in germanosilicate fibers. In *Proceedings of Optical Amplifiers and Their Applications*, Davos, Switzerland, (June), 197–199, 1995.

[12] M. Nissov, C. R. Davidson, K. Rottwitt, R. Menges, P. C. Corbett, D. Innis, and N. S. Bergano, 100 Gb/s (10×10Gb/s) WDM transmission over 7200 km using distributed Raman amplification. In *Proceedings of ECOC'97*, 1997.

[13] P. Lancaster, P. Mejasson, A. Cordier, C. Little, T. Shirley, P. Dupire, and T. Farrar, Efficient powering of long haul and high capacity submarine networks. In *Proceedings of SubOptic 2001 International Convention*, Kyoto, Japan, May 20–24, T4.5.2, 2001.

[14] Suzuki, H. Kidorf, et al., 170 Gb/s transmission over 10,850 km using large core transmission fiber. In *Postdeadline Papers of OFC*, 1998.

[15] F. Corti, B. Daino, G. de Marchis, and F. Matera, Statistical treatment of the evolution of the principal states of polarization in single-mode fiber, *J. Lightwave Technol.*, 8:8, (August), 1162–1166, 1990.

[16] C. D. Poole and J. Nagel, Polarization effects in lightwave systems. In *Optical Fiber Telecommunications IIIA*, ed. I. Kaminow and T. Koch, 153, San Diego: Academic Press, 1997.

[17] E. A. Golovchenko, A. N. Pilipetskii, N. S. Bergano, C.R Davidson, F. I. Khatri, R. M. Kimball, and V. J. Mazurczyk, Modeling of transoceanic fiber-optic WDM communication systems, *IEEE J. Select. Topics Quantum Electron.*, 6:2, (March/April), 337–347, 2000.

[18] K. S. Kim, W. A. Reed, R. H. Stolen, and K. W. Quoi, Measurement of the non-linear index of silica core and dispersion shifted fibers, *IEE Electron. Lett.*, 32:570.

[19] F. Forghieri, R. W. Tkach, and A. R. Chraplyvy, Fiber nonlinearities and their impact on transmission systems. In *Optical Fiber Telecommunications IIIA*, ed. I. Kaminow and T. Koch, San Diego: Academic Press, 1997.

[20] G. P. Agrawal, *Non-Linear Fiber Optics*, 2d ed., San Diego: Academic, 1995.

[21] F. Forghieri, R. W. Tkach, and A. R. Chraplyvy, Fiber nonlinearities and their impact on transmission systems. In *Optical Fiber Telecommunications IIIA*, ed. I. Kaminow and T. Koch, San Diego: Academic Press, 1997.

[22] J. L. Gimlett and N. K. Cheung, Effects of phase-to-intensity noise conversion by multiple reflections on gigabit-per-second DFB laser transmission systems, *J. Lightwave Technol.*, 7:6, (June), 888–895, 1989.

[23] A. F. Judy, The generation of intensity noise from fiber Rayleigh backscatter and discrete reflections. In *Proceedings of the European Conference on Optical Communications*, TuP11, 1989.

[24] J. M. Senior, *Optical Fiber Communications*, Englewood Cliffs, NJ: Prentice-Hall, 69, 1985.

[25] J. L. Gimlett, M. Z. Iqbal, N. K. Cheung, A. Righetti, F. Fontana, and G. Grasso, Observations of equivalent Rayleigh scattering mirrors in lightwave systems with optical amplifiers, *IEEE Photon. Technol. Lett.*, 2:3, (March), 211–213, 1990.

[26] M. Nissov, Long-haul optical transmission using distributed Raman amplification, PhD thesis, Department of Electromagnetic Systems, Technical University of Denmark, December 1997.

[27] B. Pedersen, A. Bjarklev, J. H. Povlsen, K. Dybdal, and C. C. Larsen, The design of erbium doped fiber amplifiers, *J. Lightwave Technol.*, 9:9, (Sept.), 1105–1112, 1991.

[28] R. Stolen, Stimulated Raman scattering (SRS). In *Optical Fiber Telecommunications*, ed. S. E. Miller and A. G. Chynoweth, Chapter 5.2, 127–133, New York: Academic, 1979.

[29] G. P. Agrawal, *Non Linear Fiber Optics*, 2d ed. Chapter 8, San Diego: Academic, 1995.

[30] R. G. Smith, Optical power handling capability of low loss optical fibers as determined by stimulated Raman and Brillouin scattering, *Appl. Optics*, 11:11, 2489–2494, 1972.

[31] M. Nissov, K. Rottwitt, H. D. Kidorf, and M. X. Ma, Rayleigh crosstalk in long cascades of distributed unsaturated Raman amplifiers, *IEE Electron. Lett.*, 35:12, (June), 997–998, 1999.

[32] H. Takahashi et al., Impact of crosstalk in an arrayed waveguide multiplexer on n×n optical interconnection, *J. Lightwave Technol.*, 14:6, (June), 1097–1105, 1996.

[33] N. Tsukiji et al., Advantage of inner-grating-multimode laser (iGM laser). In *Proceedings of OMB4, Optical Amplifiers and Their Applications*, Vancouver Canada, July 14–19, 2002.

[34] L. L. Wang et al., Linewidth limitations of low noise wavelength stabilized Raman pumps. In *Proceedings of OMB5, Optical Amplifiers and Their Applications*, Vancouver Canada, July 14–19, 2002.

[35] A. Evans, Raman amplification in broadband WDM systems. In *Technical Digest of OFC'01*, Anaheim, CA, TuF4: 2001.

[36] M. N. Islam, Raman amplifiers for telecommunications, *IEEE J. Select. Topics Quantum Electron.*, 8:3, 548–559, 2002.

[37] M. X. Ma, H. D. Kidorf, K. Rottwitt, F. W. Kerfoot, and C. R. Davidson, 240-km repeater spacing in a 5280-km WDM system experiment using 8 2.5 Gb/s NRZ transmission, *IEEE Photon. Technol. Lett.*, 10:6, (June), 1998.

[38] C. B. Clausen, S. Ten, B. Cavrak, C. R. Davidson, A. N. Pilipetskii, M. Nissov, E. A. Golovchenko, R. Ragbir, and K. Adams, Modeling and experiments of Raman assisted ultra long-haul terrestrial transmission over 7500 km, *Technical Digest of ECOC'01*, Amsterdam, The Netherlands, 315-1, 2001.

[39] F. Liu, J. Bennike, S. Dey, C. Rasmussen, B. Mikkelsen, and P. Mamyshev, 1.6 Tbit/s (40×42.7 Gb/s) transmission over 3600 km UltraWave™ fiber with all-Raman amplified 100 km terrestrial spans using ETDM transmitter and receiver. In *Postdeadline Papers of OFC'02*, Anaheim, CA, FC7, 2002.

[40] D. F. Grosz, A. Küng, D. N. Maywar, L. Altman, M. Movassaghi, H. C. Lin, D. A. Fishman, and T. H. Wood, Demonstration of all-Raman ultra-wide-band transmission of 1.28 Tb/s (128×10 Gb/s) over 4000 km of NZ-DSF with large BER margins. *Postdeadline Papers of ECOC'01*, Amsterdam, The Netherlands, 72–73, 2001.

[41] K. Shimizu, K. Ishida, K. Kinjo, T. Kobayashi, S. Kajiya, T. Tokura, T. Kogure, K. Motoshima, and T. Mizuochi, 65 × 22.8 Gb/s WDM transmission over 8,398 km employing symmetrically collided transmission with A_{eff} managed fiber. In *Technical Digest of OFC'02*, Anaheim, CA, WX4, 2002.

[42] T. Tanaka, N. Shimojoh, T. Naito, H. Nakamoto, I. Yokota, T. Ueki, A. Suguyama, and M. Suyama, 2.1-Tbit/s WDM transmission over 7,221 km with 80-km repeater spacing. *Postdeadline Papers of ECOC'00*, Munich, PD1.8, 2000.

[43] H. Masuda, S. Kawai, K.-I. Suzuki, and K. Aida, 75-nm 3-dB gain-band optical amplification with erbium-doped fluoride fibre amplifiers and distributed Raman amplifiers in 9 × 2.5 Gb/s WDM transmission experiment. In *Proceedings of the European Conference on Optical Communication (ECOC'97)*, 73–75, 1997.

[44] D. G. Foursa, C. R. Davidson, M. Nissov, M. A. Mills, L. Xu, J. X. Cai, A. N. Pilipetskii, Y. Cai, C. Breverman, R. R. Cordell, T. J. Carvelli, P. C. Corbett, H. D. Kidorf, and N. S. Bergano, 2.56 Tb/s (256 × 10 Gb/s) transmission over 11,000 km using hybrid Raman/EDFAs with 80 nm of continuous bandwidth. In *Postdeadline Papers of OFC'02*, Anaheim, CA, FC8, 2002.

[45] N. Shimojoh, T. Naito, T. Tanaka, H. Nakamoto, T. Ueki, A. Sugiyama, K. Torii, and M. Suyama, 2.4-Tbit/s WDM transmission over 7400 km using all Raman amplifier repeaters with 74-nm continuous single band. In *Postdeadline Papers of ECOC'01*, Amsterdam, The Netherlands, 8–9, 2001.

[46] A H. Gnauck, G. Raybon, S. Chandrasekhar, J. Leuthold, C. Doerr, L. Stulz, A. Agarwal, S. Banerjee, D. Grosz, S. Hunsche, A. Kung, A. Marhelyuk, D. Maywar, M. Movassagbi, X. Liu, C. Xu, X. Wei, and D. M. Gill, 2.5 Tb/s (64×42.7 Gb/s) transmission over 40 × 100 km NZDSF using RZ-DPSK format and all-Raman-amplified spans. In *Postdeadline Papers of OFC'02*, Anaheim, CA, FC2, 2002.

[47] Y. Emori, K. Tanaka, and S. Namiki, 100nm bandwidth flat-gain Raman amplifiers pumped and gain-equalised by 12-wavelength-channel WDM laser diode unit, *IEE Electron. Lett.*, 35:16, 1355–1356, 1999.

Chapter 18

Ultra-Long-Haul, Dense WDM Using Dispersion-Managed Solitons in an All-Raman System

Linn F. Mollenauer

18.1. All-Optical Transmission

In the late 1990s, the telecommunications industry began to see dense WDM as the way to provide for the seemingly explosive growth in demand for transmission capacity. The usual industry practice of using electronic regeneration at every node point (typically, once every 400 to 600 km; see Fig. 18.1), however, promised to use far too much capital equipment and office space, especially if the net transmission rates were to be at terabit levels. For example, a system carrying one terabit/s in each direction, at the practical and increasingly popular per-channel rate of 10 Gbit/s, would require no less than 200 (expensive and bulky) regenerators, or OT units per node (one for each direction and channel). In the meantime, it was already well known, principally from undersea practice, that such dense WDM could be successfully carried out, without regeneration, over transoceanic distances, at least under the special conditions of the undersea environment. Thus the idea of developing an all-optical terrestrial system (which had in fact been advanced many years ago [1–3]) began to take root and to undergo engineering development by many firms.

The principal requirements for such ultra-long-haul, all-optical systems, are that the growth of spontaneous emission noise be held to the minimum possible, that the transmission mode be chosen to yield the fewest nonlinear penalties, and that the pulse behavior be periodic. The third of these requirements stems from the needs of optical networking, where the data must be instantly readable, and where standard pulses can be introduced, at least at all node points along the path. With the additional requirement to work with existing terrestrial fiber spans of typically 80 to 100 km in length between amplifier huts, the only way to meet the first requirement is to take advantage of the greatly reduced noise figure provided by distributed gain from the Raman effect. The second requirement is best, and the third uniquely met with the use of dispersion-managed solitons (henceforth abbreviated as DMS).

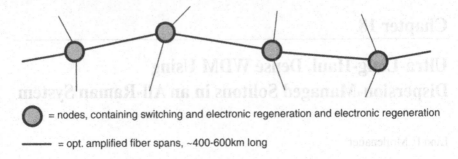

= nodes, containing switching and electronic regeneration and electronic regeneration

─────── = opt. amplified fiber spans, ~400-600km long

Fig. 18.1. Schematic of a typical terrestrial transmission system.

18.2. Dispersion-Managed Solitons

18.2.1. Introduction

With dispersion management, the transmission line consists of segments of fiber whose individual dispersion parameters (D_{local}) are of alternating algebraic sign and with absolute values typically at least 3 or 4 ps/nm km (Fig. 18.2). Furthermore, this arrangement, or *dispersion map*, is ideally periodic (although, in practice, it need not be exactly so). For each map period, the accumulated dispersions of the two segments nearly cancel, so that the path-average dispersion parameter of the map $\bar{D}$ is small, typically no greater than about 0.2 ps/nm km. To support solitons, $\bar{D}$ is also positive (anomalous dispersion).

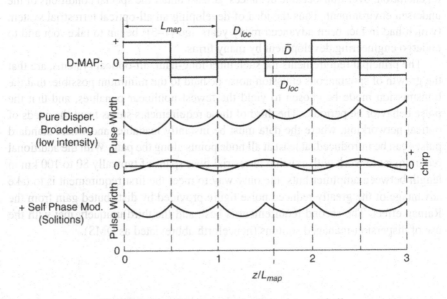

Fig. 18.2. Dispersion-managed solitons in a nutshell.

It is instructive first to consider pulse behavior at very low intensities, when only the dispersive term of the nonlinear Schrödinger (NLS) equation is important. In response to the relatively large, alternating D_{local} values, the pulse width tends to undergo a significantly large fractional change, periodic with the map. This pulse "breathing" is accompanied by a similarly periodic variation in the chirp parameter, with the chirp passing through zero at or near the center of each fiber segment, and alternating between large positive and negative peaks. But on a distance scale typically many times greater than the map period, there is also a gradual net broadening of the pulses, in response to the effect of $\bar{D}$ (again, see Fig. 18.2).

Now, to obtain dispersion-managed solitons, we merely need to increase the pulse intensity until self-phase modulation (SPM, from the nonlinear term in the NLS equation) produces a phase shift across the pulse that just cancels out, at the end of each map period, the net phase shift produced by the dispersive term. This cancellation in turn eliminates the net pulse broadening from $\bar{D}$, so that the pulse behavior now becomes truly periodic (the bottom curve of Fig. 18.2). The cancellation of phase shifts is similar to that obtaining with ordinary solitons (see Fig. 18.3).

Although the pulse field-envelope shape function of the ordinary soliton is $\text{sech}(t)$, that of the dispersion-managed soliton is essentially Gaussian, because in each segment, the large dispersive term (which scales with D_{local}) dominates the much smaller nonlinear term (which scales with $\bar{D}$), and the solution to the NLS equation, in the case of pure dispersion, is Gaussian.

The scheme of Fig. 18.2 is very general so that, in principle, one can have any relation one likes between the map period and the spacing between amplifier huts; also, so far, the exact fiber types have not been specified. In practice, however, for terrestrial systems, the typically 80 to 100 km long spans between amplifier huts are already filled with one or another type of nonzero dispersion-shifted fiber, such as OFS-Fitel Raman Reduced Slope TrueWave® or Corning LEAF. In that case, the negative D fiber is a coil of DCF, or dispersion-compensating fiber, with $D \approx -100$

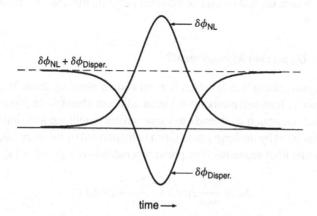

time →

Fig. 18.3. The nonlinear and dispersive phase shifts of an ordinary soliton; note that they sum to a constant [4]. A similar cancellation of the nonconstant parts occurs at the end of each map period with dispersion-managed solitons.

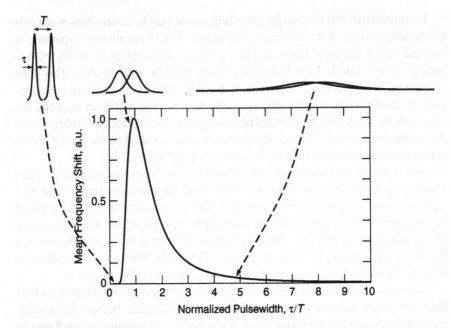

Fig. 18.4. Adjacent pulse interaction (a frequency shift resulting from cross-phase modulation, or XPM) as a function of the pulse width τ, expressed as a fraction of the bit period T. The path-average dispersion $\bar{D}$ then converts these frequency shifts into temporal displacements (timing jitter) [5]. Note that as long as the maximum pulse width is less than approximately $T/2$, the adjacent pulse interaction is negligible.

ps/nm km, conveniently located in each amplifier hut. Note that this scheme makes the dispersion map period essentially the same as the amplifier hut spacing. This shortest possible map period (for a given amplifier hut spacing) is in general desirable because it produces the smallest possible extent of the pulse breathing. Limited pulse breathing is in turn needed to control adjacent pulse interaction [5–7]; see Fig. 18.4 and its caption.

18.2.2. Why Dispersion Management?

Dispersion management was invented to meet certain needs of dense WDM [8–11]. In the first place, four-wave-mixing between adjacent channels (a potentially most harmful effect, inasmuch as it tends to cause severe amplitude and timing jitter) is efficiently repressed by the large phase mismatch provided by the large $|D_{\text{local}}|$ values of the individual fiber segments. The phase mismatch is computed as [4]

$$\Delta k = \frac{\partial^2 k}{\partial \omega^2} \Delta \omega^2 = -\frac{2\pi \lambda^2}{c} D (\Delta f)^2,$$

where Δf is the frequency separation between the adjacent channels. For example, consider a system with the parameters $D_{\text{local}} \approx 6$ ps/nm km and $\Delta f = 50$ GHz (typical for a 10 Gbit/s per channel system). The above equation then yields $\Delta k = 0.76$/km.

TYPE	D (ps/nm-km) (@ 1560 nm)	S = ∂D/∂λ (ps/nm²-km)	D/S	$A_{eff}(\mu m)^2$
low-slope T.W.	+5.5	+0.037	148	55
ultra-slope D.C.F.	−115	−0.78	147	18
Standard	+17.0	+0.056	304	80
IDF	−17.7	−0.057	310	35

Fig. 18.5. Parameters for two examples of $\pm D$ fiber pairs yielding nearly constant $\bar{D}$ because they have nearly identical ratios of D/S (dispersion to dispersion slope parameters) in the middle of the WDM band. The first is a combination of an experimental version of low-slope TrueWave® fiber with a matching DCF; the second is standard SMF with a matching inverse dispersion fiber (IDF). (Note that the second combination would tend to be suitable only for new installations, however, because the IDF segment would have to be nearly as long as the standard SMF segment it is compensating.)

Because of this large phase mismatch, the E field of the four-wave-mixing product spirals rapidly in tight circles in the complex plane, and hence cannot grow to significant size. (Note that, in this example, the circle is completed (and hence nearly closes on itself, especially in Raman-amplified systems) once every $2\pi/0.76 = 8.2$ km.) Second, fibers having the rather small D values needed for ordinary solitons, constant over the wide wavelength bands required for dense WDM, simply do not exist, and probably never will. For dispersion management, however, it is possible to use *combinations* of fiber for which the path-average dispersion is nearly constant. Two examples of fiber pairs permitting this extra degree of freedom in map design are shown in Fig. 18.5.

Figure 18.6 shows the variation in the path-average dispersion parameter typically obtained from such fiber combinations. Note that the variation in $\bar{D}$ for a 50 nm wide band (sufficient for 125 channels at 50 GHz/channel), centered about the peak in the curve, is only about ±27% of the median value for that band.

18.2.3. Why Dispersion-Managed Solitons?

Note that, thus far, the cited needs of dense WDM to be met by dispersion management do not necessarily require dispersion-managed *solitons*. But when the goal is to provide the backbone of an all-optical network, then the periodicity of the solitons' behavior, unique to them, becomes vital. The many important issues surrounding the periodic nature of soliton transmission can be best discussed in terms of Fig. 18.7. Although the choice is arbitrary, it is convenient to let the map periods begin and end at the unchirped pulse positions in each coil of DCF, as in Fig. 18.7. Note that the accumulated linear dispersion values shown there are the discrete values obtaining at the end of each period, so that they correspond to the product of $\bar{D}$ and the particular transmission distance. Note further that the precompensation (pre-comp.)

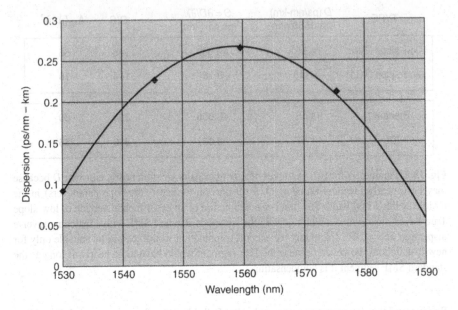

Fig. 18.6. Measured path-average dispersion parameter ($\bar{D}$) for a 100 km span low-slope TrueWave® fiber, compensated with an ≈4.5 km long coil of matching DCF (see Fig. 18.5), as a function of wavelength. The data make an excellent fit to a shallow parabola whose peak is at the center of the intended WDM band.

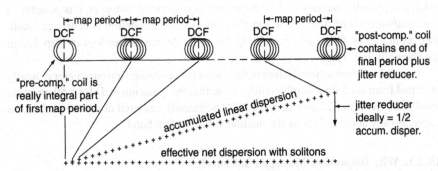

Fig. 18.7. Dispersion compensation in a DMS system. In this view, the map periods begin and end at the unchirped pulse positions in each coil of DCF; the accumulated linear dispersion values shown here are those discrete values obtaining at the end of each such period. Note that the pre- and postcompensation coils are really just integral parts of the map periods (see text).

coil is really an integral part of the first map period, and that the postcompensation (post-comp.) coil, save for an additional jitter-reducer, is likewise an integral part of the final map period. The jitter-reducer ideally represents a dispersion equal to $-1/2$ of the accumulated linear dispersion, but the exact value is not at all critical and thus, in practice, it can be set to the best value for the longest distance to be encountered in the system. Finally, note that the effective net dispersion for solitons is always zero at the end of each map period. This scheme of dispersion compensation has the following important consequences.

1. The pulse parameters (temporal width, bandwidth, energy, chirp, etc.), are identical at the end of each map period, and the pulses are always well resolved from each other in time. This periodic behavior in turn means that:

   (a) The data can be read instantly anywhere, or at least at the end of any map period;
   (b) Standard soliton pulses can be injected anywhere (i.e., at the beginning of any map period).

2. The pre- and postcompensation dispersion values are independent of distance.

These properties are exactly as required for the creation of an all-optical network and for efficient, inexpensive system monitoring. Their compatibility with the use of standard parts (pre- and postcompensation coils) is also very important for the reduction of system cost and for the ease of system assembly. Once again, these properties are uniquely supplied by dispersion-managed solitons.

It is instructive to look at the dispersion compensation scheme most often used in non-DMS systems. There it is common to use much greater precompensation dispersion, so that the accumulated linear dispersion tends to pass through zero somewhere near the halfway point of the net transmission distance (see Fig. 18.8). This scheme represents an attempt to reduce cross-phase modulation (XPM) from interchannel collisions by greatly broadening the pulses and thus making their peak intensities lower, over at least most of the path. Unfortunately, however, that action simultaneously greatly increases nonlinear penalties from certain intrachannel effects, such as the adjacent-pulse interaction of Fig. 18.4, and intrachannel four-wave mixing. (The four-wave-mixing tends to produce ghost pulses in the positions of zeros (bit slots where there are no pulses) by transferring energy from adjacent ones [5].) Furthermore, in strong contrast to solitons, most of the accumulated linear dispersion is not compensated by self-phase modulation. In consequence, one has the following.

1. Over much of the path, the pulses are strongly overlapped, so that the data are not immediately readable.
2. The pre- and postcompensation dispersion values must be carefully tuned for each distance. (The total of pre- and postdispersion compensation required is roughly proportional to the total distance.)
3. Even for a fixed distance, dispersion tends to make it impossible to properly compensate all wavelengths of a wide WDM band with just one set of pre- and postcompensation coils.

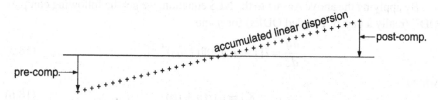

Fig. 18.8. Dispersion compensation in a non-DMS system; see text.

These facts mitigate strongly against the creation of an all-optical network and efficient, inexpensive system monitoring.

18.2.4. A Shortcut for Computing DMS Behavior

Thus far the discussion of dispersion-managed solitons has been largely qualitative. For real system design, however, we must compute exact pulse behavior, often for many different possible dispersion maps, amplifier span gain profiles, and initial pulse parameters. To do all this computation by exact numerical solution of the NLS equation is tedious and time consuming. One can create an efficient shortcut, however, by taking advantage of the fact that, as already stated, in a DMS system, the pulse shape is Gaussian to a very good approximation. That is, by applying that assumed pulse shape, or Ansatz, to the NLS equation, one can create an equivalent set of ordinary differential equations, or ODEs, that are much easier and faster to solve. Although several other ODE (largely variational [12–14]) approaches have been used, the special ODE method (nonvariational) invented and used with great success in our laboratory [15] is briefly outlined here.

We write the (Gaussian) signal pulse in the general form:

$$u(t) = \sqrt{W}(\eta/\pi)^{1/4} \exp[-\frac{1}{2}(\eta + i\beta)t^2], \qquad (18.1)$$

where $1/\sqrt{\eta}$ is a measure of the pulse width, and β is the chirp parameter. Let η_0 refer to the unchirped pulse; that is, $\eta = \eta_0$ when $\beta = 0$. Clearly, if we know the complex number $\eta + i\beta$, and the pulse energy W, we then know all the pulse properties. In particular, for the pulse width in time, we have:

$$1/\sqrt{\eta} = \tau/\sqrt{4\ln 2} = \tau/1.6651\ldots, \qquad (18.2)$$

where τ is the intensity FWHM. The phase and frequency shifts across the pulse are, respectively:

$$\phi(t) = -\frac{1}{2}\beta t^2 \qquad (18.3a)$$

and

$$\delta\omega(t) = -\beta t. \qquad (18.3b)$$

Finally, the spectrum of the pulse (the Fourier transform $\tilde{u}(\omega)$ of Eq. (18.1)), yields the spectral intensity

$$|\tilde{u}|^2 \propto \exp[-\eta\omega^2/(\eta^2 + \beta^2)], \qquad (18.4)$$

which has a FWHM of $\Delta f = (1.6651../2\pi)\sqrt{(\eta^2 + \beta^2)/\eta}$. Thus the time-bandwidth product is $\tau\Delta f = 0.441\sqrt{1 + (\beta/\eta)^2}$.

By applying the above Ansatz to the NLS equation, we get the following complex ODE (really a pair of coupled ODEs) for η and β.

$$\frac{dq}{dz} = i[1 - Kq^2(\Re(1/q))^{3/2}], \qquad (18.5)$$

where

$$q(z, K) = \eta_0/(\eta + i\beta). \qquad (18.6)$$

The distance z is always measured in units of the characteristic dispersion length, which, for a Gaussian pulse, is:

$$z_c = \frac{1}{4\ln 2} \frac{2\pi c}{\lambda^2} \frac{\tau_0^2}{D}. \tag{18.7}$$

The nonlinear coefficient K is calculated as

$$K = \frac{(2\pi)^2}{\sqrt{2}\,4\ln 2} \frac{n_2 c}{\lambda^3 A_{\text{eff}}} \frac{\tau_0^2}{D} P = \frac{P}{P_c}, \tag{18.8}$$

where A_{eff} is the fiber core area, τ_0 refers to the unchirped pulse, P is the peak pulse power, and P_c is the peak power of an ordinary soliton of pulsewidth τ_0 in fiber of the (local) dispersion parameter D. Note that $K = 1$ corresponds to ordinary solitons (although the Gaussian pulse shape is not quite right in that case), and that dispersion-managed solitons tend to correspond to $K \ll 1$. Note also that both z_c and K are negative when D is negative. Although this convention and, in particular, the concept of a negative dispersion length, may seem strange at first, it is self-consistent, and avoids a certain awkwardness that would occur without it.

If we let $z = 0$ correspond to the unchirped pulse ($\eta = \eta_0$, $\beta = 0$), then $q(0, K) = 1$. The solution to Eq. (18.5) then has the general form:

$$q(z, K) = 1 + iz + K\, f(z, K). \tag{18.9}$$

Note that the linear solution $q = 1 + iz$ is the well-known solution for a Gaussian pulse subject to pure dispersion.

We have developed an efficient computer program to obtain solutions to Eq. (18.5), based on the Maple mathematics package. From input data consisting of details of the dispersion map, Raman pumping conditions, and input pulse parameters, our program first calculates the "gain profile" (relative signal pulse energy versus z), and from that, $K(z)$. It then uses the Maple program "dsolve" to obtain solutions of Eq. (18.1), and finally, it graphs the various pulse parameters as functions of z. The program is very efficient, so that on a reasonably fast PC, one can obtain a full picture of pulse behavior in a given map in just a matter of minutes. Thus our program has proven to be a very useful engineering tool, enabling the exploration of system performance over a wide range of map designs and pulse parameters in a relatively short time. It has also engendered a deeper understanding of the fundamentals of dispersion-managed soliton transmission itself. Much of the pulse behavior presented in this chapter was computed with this program.

18.2.5. Pulse Behavior in Lossless Fiber

Before going on to study pulse behavior in "real" dispersion maps, where loss and gain tend to complicate matters, it is instructive to survey behavior in lossless fiber. Accordingly, Figs. 18.9 to 18.11 plot the most important pulse parameters (pulse width, bandwidth, and chirp) as functions of distance (normalized to $|z_c|$), with the

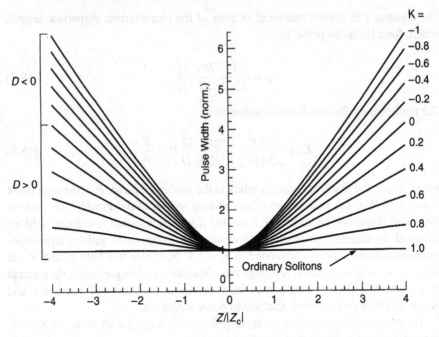

Fig. 18.9. Pulse width (normalized to that of the unchirped pulse) in lossless fiber as a function of distance, for various values of the nonlinear parameter K.

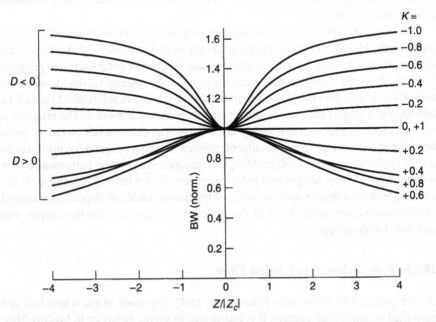

Fig. 18.10. Pulse bandwidth (normalized to that of the unchirped pulse) in lossless fiber as a function of distance, for various values of the nonlinear parameter K.

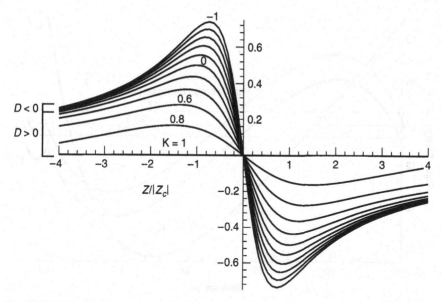

Fig. 18.11. Chirp parameter (normalized to η_0) in lossless fiber as a function of distance, for various values of the nonlinear parameter K. Note that beginning with the unchirped pulse, the chirp at first increases almost linearly with distance (as the frequency components of the pulse just begin to separate), then peaks and declines as the separation becomes complete (whence the range of frequencies is spread out over ever greater time).

nonlinear coefficient K as parameter. Note that for all three parameters, the purely dispersive effect (i.e., that at $K = 0$) is always enhanced for $D < 0$, but tends to be decreased for $D > 0$. And, of course, for the ordinary soliton ($K = 1$), the dispersive effects disappear altogether.

18.2.6. Pulse Behavior When Loss and Gain Are Included

Even with Raman gain, pulse energies in the 80 to 100 km long spans typical of terrestrial systems tend to vary by at least several dB along the span; see Fig. 18.12 for graphs of typical behavior.

Figures 18.13 to 18.15 show pulse width, chirp parameter, and pulse band-width, respectively, as functions of z for the dispersion map and the several pumping conditions of Fig. 18.12. (The color code of Fig. 18.12 is preserved.) The three figures are worthy of close study. First, note that in each case, the minima of pulse widths and the extrema (maxima for $D > 0$, minima for $D < 0$) of pulse bandwidths correspond exactly to the positions of zero chirp in all cases. Second, note the asymmetry in the pulse breathing for the first two cases (red and green curves), and the corresponding displacements of the zero-chirp positions toward the high-intensity end of the main span. These asymmetries are brought about by the higher intensity, hence higher non-linear coefficient K in the first part of each span. (Recall from Fig. 18.9 that higher K causes shallower breathing in $D > 0$ fiber.)

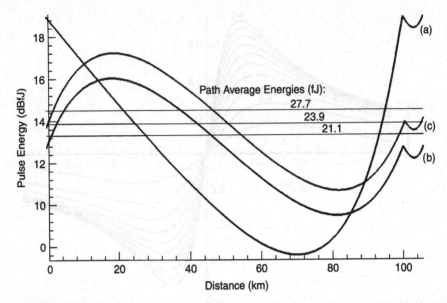

Fig. 18.12. Pulse energies as a function of distance in one period of a map consisting of 100 km of TrueWave® Reduced Slope fiber, followed by a matching coil of DCF; in all cases the unchirped pulse width is ≈30 ps. (a) Main span backward pumped only. (b) Main span backward and forward pumped in a 50:50 ratio. DCF (the range $z > 100$ km) backward pumped for all three cases. (c) Same as curve b, but with mirror reflectivity $R = 4\%$, free spectral range $FSR = 50$ GHz, Fabry–Perot guiding filter at the beginning of each map period.

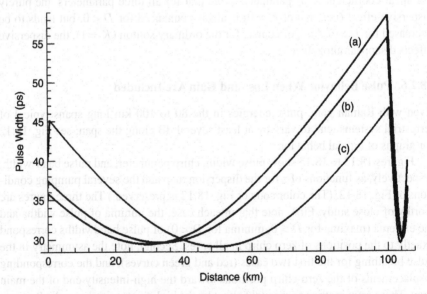

Fig. 18.13. Pulse widths as a function of distance for the dispersion map of Fig. 18.12. The labels a, b, and c refer to the same pumping conditions as in Fig. 18.12.

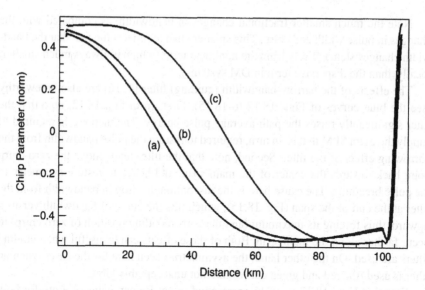

Fig. 18.14. Chirp as a function of distance for the dispersion map of Fig. 18.12. The labels a, b, and c refer to the same pumping conditions as in Fig. 18.12.

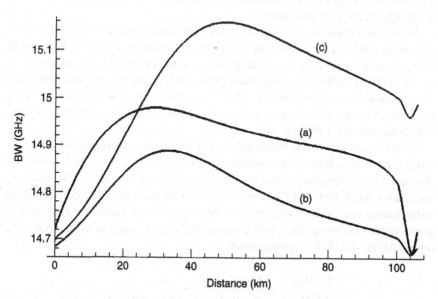

Fig. 18.15. Pulse bandwidths as a function of distance for the dispersion map of Fig. 18.12. The labels a, b, and c refer to the same pumping conditions as in Fig. 18.12.

Note the much smaller fractional changes in bandwidth, as compared with the changes in pulse width and chirp. This smaller change reflects the fact that the bandwidth changes stem entirely from the nonlinear term (which is always much smaller, locally, than the dispersive term in DM systems).

The effects of the narrow-bandwidth (guiding) filter [4, 16] are also noteworthy (see the blue curves of Figs. 18.12 to 18.15). First, from Fig. 18.12, note that the filter significantly raises the path-average pulse energy. The increase is required to supply the extra SPM that is, in turn, required to restore the pulse bandwidth from the narrowing effect of the filter. Second, note that the filter tends move the zero-chirp point back towards the center of the main span, and hence to restore symmetry to the pulse breathing. The cause here is the discontinuous drop in bandwidth from the filter at the end of the span (Fig. 18.15), which tilts the curve of bandwidth versus z upward, thus forcing its maximum (and the corresponding position of zero chirp) to occur farther along the main span. Both of these effects can be useful when guiding filters are used. On the other hand, the asymmetries seen here for the cases when no filter is used (the red and green curves) are not unacceptably large.

Note that Figs. 18.12 to 18.15 correspond to net Raman gains of unity for both the main and DCF spans. As detailed later, however, to reduce noise from double-Rayleigh backscattering of the signals themselves, the usual practice is to provide a few dB less than unity gain in the main span, and to make up for that deficit with extra Raman gain in the DCF. Nevertheless, the pulse behaviors just shown here are not greatly affected by that change.

From time to time, someone worries that nonlinear effects in DCF will be excessive, presumably on account of the relatively small core area of DCF. Fortunately, the worry is unfounded. As far as the nonlinear effects of WDM are concerned, the very high $|D|$ of DCF renders four-wave-mixing truly negligible, because of the huge phase mismatch it creates (see Sec. 18.2.2), and the relative velocities of colliding pulses are so large that the frequency shifts from XPM are almost always much smaller than those from the $+D$ fiber.

SPM is necessary for the maintenance of dispersion-managed solitons and hence, in that context, should never be considered as a harmful effect. Nevertheless, the DCF of a map tends to contribute only a small fraction of the total required SPM. The dominant factor here is that the DCF is relatively short, and that the SPM, everything else being equal, grows in direct proportion to the propagation distance. If we temporarily ignore signal power variation with distance, the relative fraction of the total SPM contributed by the DCF is approximately:

$$\frac{SPM_{DCF}}{SPM_{D+}} = \frac{L_{DCF}}{L_{D+}} \frac{A_{D+}}{A_{DCF}} = \frac{D^+}{D_{DCF}} \frac{A_{D+}}{A_{DCF}}.$$

For the 100 km span maps referred to here in Figs. 18.12 to 18.15, for example, this fraction is small, no more than about 15%; see Fig. 18.16, where the quantity $q - 1 - iz$, or nonlinear residue (see Eq. (18.9)), is displayed in the complex plane. Note that the real part of that quantity corresponds to the change in inverse squared

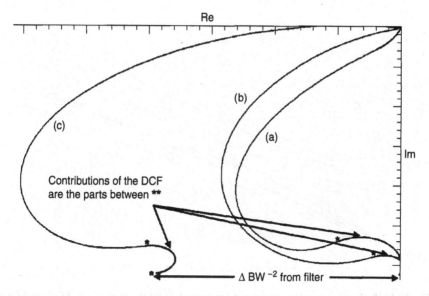

Fig. 18.16. Nonlinear residue, the quantity $q - 1 - iz$, for the dispersion map of Fig.18.12. The labels a, b, and c refer to the same pumping conditions as in Fig. 18.12.

bandwidth, whereas the imaginary part is that which cancels the residual dispersion [15, 16].

The final but very important issue is that of the path-average pulse energies of the dispersion-managed solitons. To anticipate a bit from the next section on noise growth, error-free transmission over transcontinental distances, in a well-designed all-Raman system, tends to require minimum pulse energies in the 20 to 30 fJ range. Figure 18.17 shows the energies for the systems of Fig. 18.12, as a function of $\bar{D}$, both with and without the use of guiding filters. Note that the energy of the dispersion-managed soliton, for a given combination of pulse width and $\bar{D}$, is several times greater than that of the ordinary soliton. This energy-enhancement effect ($\rho \equiv W_{DMS}/W_{OS}$) is a well-known [17–20] and important advantage of dispersion-managed solitons. Note that for the map in question here, the DMS with the pulse widths most desirable for 10 Gbit/s ($\approx$30 to 32 ps) have adequate energies for $\bar{D} \approx 0.15$ to 0.2 ps/nm km, whereas the energies of ordinary solitons, for the same parameters, are several times too small.

The energy enhancement is related to the map strength parameter S, defined as $S = (|L_+| + |L_-|)/2) \cong L_+$, where the quantities L are the lengths of the map segments as measured in units of z_c (see Eq. (18.7)). For the case where no guiding filters are used, $\rho \approx S$, as shown by the results of many numerical simulations and as borne out in Fig. 18.17, where $S \cong 2.3$ for the map it refers to and for a pulse width of 30 ps. The energy enhancement results from the small but significant change in pulse bandwidth across the map (see Fig. 18.15), and from the fact that the phase shift across the pulse caused by the dispersive term scales as the product of $D_{local} \times BW^2$. Thus the net dispersive phase shift across the map is larger than would exist without the change in bandwidth. In turn, then, the compensating nonlinear phase shift and

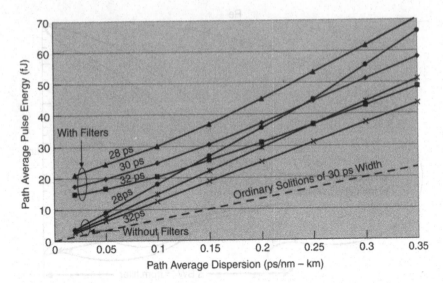

Fig. 18.17. Path-average pulse energy as a function of $\bar{D}$ for the map of Fig. 18.12, and for a range of unchirped pulse widths. In this graph, the main span is backward/forward pumped in a 70:30 ratio. The energies of ordinary solitons refer to a map where $D = \bar{D}$ everywhere.

the pulse energy required to create it must also be larger. When guiding filters are used, the change in bandwidth across the map (again, see Fig. 18.15) and, hence, the resulting phase shift from the dispersive term, are even greater.

18.3. Amplifier Spontaneous Emission: Growth and Effects

18.3.1. Calculating the Noise Growth

In the final analysis, the performance of any ultra-long-haul system is limited by the growth of spontaneous emission noise. Without noise, after all, signal intensities could be reduced to arbitrarily low levels, where all nonlinear penalties would disappear. This section briefly outlines a unique and very clear approach, initiated many years ago [21], to the analysis of noise growth and the problem of selection of signal intensities that minimize the combined effects of noise and nonlinear penalties. As that problem is still all too often surrounded by confusion, the tutorial summary given here is an integral and important part of this chapter on ultra-long-haul transmission.

The basic facts of Raman gain, on the other hand, have been treated many times over in this volume, so only a brief recap is needed here. The Raman effect in silica-glass fibers begins with a pump-induced transition to a virtual state, followed by emission from it, where the emission terminates on an excited state of the lattice; emission of an optical phonon (which typically takes place within a few femtoseconds) then completes the return to the ground state. Because of the extremely fast relaxation,

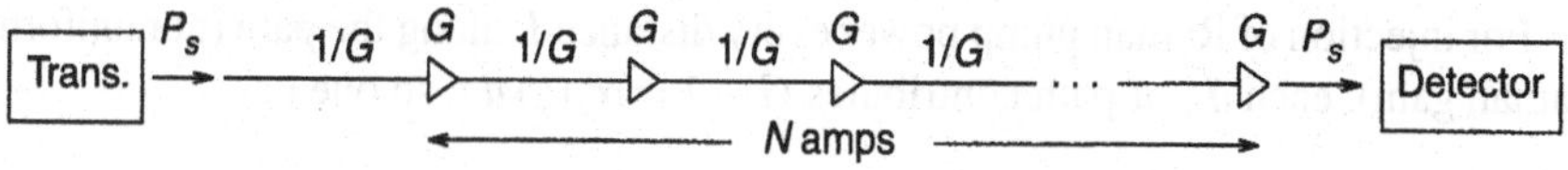

Fig. 18.18. Prototypical ultra-long-haul system, with N amplifiers of power gain G preceded by N fiber spans of loss factor $1/G$.

the population of the terminal state of the optical emission tends to be determined by equilibrium with the surrounding phonon bath, and hence is almost independent of the rates of optical pumping and emission. Thus, in contrast to erbium amplifiers, both the shape of the Raman gain band and the excess spontaneous emission factor are essentially independent of pump and signal levels. For gain in the neighborhood of the very broad peak of the Raman gain band, and when the fiber is at or near room temperature, the excess spontaneous emission factor $n_{sp} \cong 1.1$, or about 0.5 dB.

Nevertheless, Raman gain is prized, more than anything else, for the great noise reduction enabled by its distributed gain. In order to fully appreciate this matter, we must begin with a general model of a long-haul system that includes the possibility of lumped amplification. Figure 18.18 shows the prototypical system.

In the system of Fig. 18.18, each amplifier contributes a noise power per unit bandwidth, or equipartition energy, of $W_{eq} = (G - 1)n_{sp}h\nu$, where $h\nu$ is the photon energy. (Note that the power/unit bandwidth has units of energy. Note further that W_{eq} is the expected value of the noise energy in each mode of the radiation field and hence corresponds to just one polarization state at a time. Finally, because the unit time and unit bandwidth can be taken as the bit period and its reciprocal, respectively, it is often useful to think of W_{eq} as the expected value of the noise energy in each bit period and in the corresponding bandwidth.)

The gain from the output of each amplifier to the system output $(z = Z)$ is unity, therefore the noise at the detector is just

$$w_{eq} = N(G - 1)n_{sp} = \frac{\alpha Z}{\ln G}(G - 1)n_{sp}, \qquad (18.10)$$

where α is the ln of the fiber loss rate and where, to simplify appearances, we have substituted $w_{eq} = W_{eq}/h\nu$.

For perfectly uniform Raman gain, we can let $G \to 1 + \epsilon$, where $\epsilon << 1$, let N become very large, and set $n_{sp} = 1.1$, so Eq. (18.10) becomes

$$w_{eq} = \alpha Z \times 1.1. \qquad (18.11)$$

From Eq. (18.11), we can immediately see that uniform Raman gain provides the lowest possible noise at the system output. Furthermore, from a comparison with Eq. (18.10), it is clear that the noise with high-gain lumped amplifiers is much higher. Consider, for example, amplifiers of 20 dB gain (as would be required for spans of length approaching 100 km); the noise at system output is then nearly 22 times, or 13.4 dB greater than with uniform Raman gain!

For injection of Raman pump power every distance L along the path (nonuniform Raman gain), each dz of path contributes $G - 1 = \alpha_g(z)\,dz$, so one has

$$w_{eq} = \alpha Z \times 1.1 \times \frac{1}{\alpha L} \int_0^L \alpha_g(z)\, g(z)\, dz, \tag{18.12}$$

where $g(z)$ is the net gain from z to L. Although the noise here is intermediate between that of Eqs. (18.10) and (18.11), as we show, for fiber spans of 100 km or less, the result is much closer to that of Eq. (18.11).

The range of acceptable signal levels is bounded on the low side by the onset of significant errors from the inadequate signal-to-noise ratio (SNR), and on the high side by the onset of significant errors from nonlinear effects. Because the most important nonlinear effects (primarily SPM and XPM) tend to scale with the path-average signal power, to facilitate comparison, we should calculate the corresponding path-average value of the noise. Thus we must multiply the span input noise powers just calculated by the appropriate ratios of path-average to span input power. For the case of lumped amplifiers, that path-average factor is:

$$\frac{\bar{P}}{P_0} = \frac{1}{L} \int_0^L \exp(-\alpha z)\, dz = \frac{G - 1}{G \ln G}.$$

Multiplying Eq. (18.10) by the above factor, we get the path-average noise for the case of lumped amplifiers:

$$\bar{w}_{eq} = \alpha Z \times \left\{ n_{sp} \frac{1}{G} \left[\frac{G - 1}{\ln G} \right]^2 \right\}. \tag{18.13}$$

Note that the path-average noise for lumped amplifiers, although still considerable for high gains, is nevertheless substantially smaller than the noise at amplifier output (Eq. (18.10)) for the same gain.

For the case of Raman gain, where the pump power is injected every distance L, a similar calculation yields:

$$\bar{w}_{eq} = \alpha Z \times \left\{ 1.1 \times \frac{1}{\alpha L} \int_0^L \alpha_g(z)\, g(z)\, dz \times \int_0^L sig.(z)/sig.(0)\, dz \right\}. \tag{18.14}$$

In both Eqs. (18.13) and (18.14), the quantities in large {} are the penalty factors, which can be interpreted equally well as

1. The factor by which the path-average noise increases (over αZ) for constant path-average signal power, or
2. The factor of increase in path-average signal power (hence, increase in nonlinear penalties) required to maintain a given SNR.

Although the above expressions for these penalty factors may not be immediately transparent, they have been evaluated numerically and are plotted in Fig. 18.19 for lumped amplifiers and for various situations of Raman pumping. Note that although

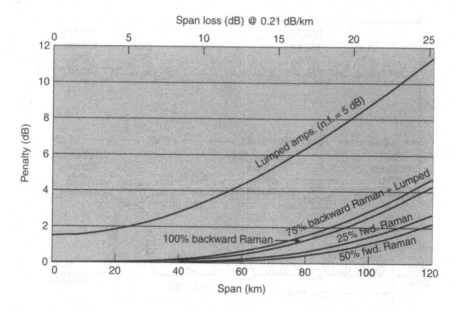

Fig. 18.19. Noise penalty factors from Eqs. (18.13) and (18.14), plotted as a function of span loss. The assumed noise figure for the lumped (erbium) amplifiers is assumed to be 5 dB, a fairly typical value. All penalty factors have been normalized to the $n_{sp} = 1.1$ of Raman gain, so that the Raman penalty curves will begin at 0 dB.

the difference in penalty between the lumped and Raman amplifier curves begins with the modest difference in their n_{sp} values, and does not change much for span lengths of just a first few tens of km, eventually it becomes substantial, ≈ 6 to 8 dB in the neighborhood (≈ 100 km) of typical terrestrial amplifier hut spacings. To the extent that the limits of error-free transmission are governed by the growth of spontaneous emission noise, this 6 to 8 dB difference represents the factor (four to six times) by which the maximum transmission distance is increased when all-Raman amplification is substituted for lumped erbium amplifiers. It was this great increase in reach that first attracted systems developers to the all-Raman approach. Note also that although the penalty with purely backward Raman pumping is nearly 3 dB at 100 km, the addition of a mere 25% of forward pumping cuts that penalty to about a half. Finally, note that if the ≈ 100 km spans could be backward pumped at midspan as well as at their far ends, the noise penalty is reduced to <1 dB, making it almost negligible. Such midspan pumping would also have the practical advantage of reducing the powers required of the individual pump lasers by a factor of two [22].

18.3.2. Experimental Test

Recently, we have been able to make an accurate experimental test of the predictions of Eq. (18.14) by using the recirculating loop shown schematically in Fig. 18.20. As shown there, the loop consists of six 100 km long spans of TrueWave® Extra Reduced

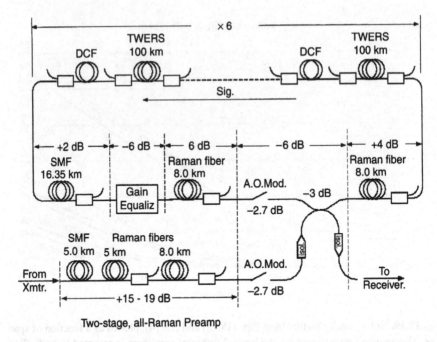

Two-stage, all-Raman Preamp

Fig. 18.20. All-Raman amplified recirculating loop used in test of spontaneous emission noise growth. The net dispersion of the loop-closing amplifiers is essentially zero. The small rectangular boxes are WDM couplers for the introduction of Raman pump light.

Slope fiber, each span properly compensated by a coil of DCF, with the DCF backward pumped, and provision made for forward as well as backward pumping of the 100 km spans. (For the experimental results reported here, the forward pumping WDM couplers were moved to allow the normally forward pumps to be used as midspan backward pumps.) When the loss of the DCF coils, of the WDM couplers, and of all the items (acoustooptic modulator, gain equalizer, etc.) used to close the loop on itself is factored in, the effective loss per 100 km of transmission fiber is about 30 dB. (The length of fiber in the DCF and loop-closing amplifiers is not included in the reported transmission distance.)

For the noise measurements, we begin with determination, on a polarization insensitive OSA (optical spectrum analyzer), of the ratio of the spectral intensity of a 10 Gbit/s data stream with the usual half-occupancy of bit periods, to the spectral intensity of the noise in an adjacent empty channel, and with the OSA's spectral resolution wide enough to completely take in the entire spectrum of the pulse stream. That raw SNR is then corrected by adding 6 dB to correct to just one polarization mode and for the unoccupied bit periods, and is further corrected to reflect the noise in the bandwidth (10 GHz in this case) corresponding to the bit period. (It should be noted that others often report the raw SNR, but without the listed corrections, and as measured with somewhat arbitrarily chosen spectral resolution.) Our corrected measurement then yields the fundamental quantity SNR $= W_{sol}/W_{eq}$. An independent determination of the signal pulse energy then lets us compute the noise itself; the

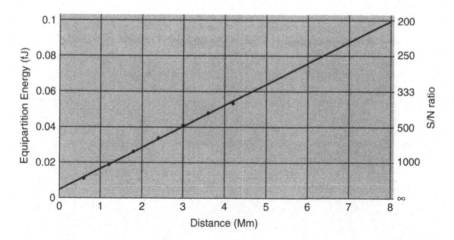

Fig. 18.21. Experimentally measured noise as a function of distance for the loop of Fig. 18.20. The SNR indicated by the vertical scale on the right is based on an assumed signal pulse energy of 20 fJ.

result (reported as equipartition energy) is plotted in Fig. 18.21 as a function of the total transmission distance. The slope of the best-fit straight line makes an almost perfect fit to the prediction of Eq. (18.14), based on the effective loss per 100 km just cited, and upon the ≈1 dB penalty factor from the midspan pumping. The slight offset at the origin represents the noise contributed by the transmitter, and from the preamplifier shown in Fig. 18.20. In the absence of nonlinear penalties, the minimum SNR required for a BER $< 1 \times 10^{-9}$ is ≈100. Thus the SNR shown in Fig. 18.21 for 8000 km, even if degraded a dB or so for double-Rayleigh backscattering of the signal (see next section) or modest nonlinear penalties, is more than adequate for error-free transmission.

18.3.3. Double-Rayleigh Backscattering

One disadvantage with Raman pumping of long spans is that double-Rayleigh backscattering of the signal can add significantly to the spontaneous emission noise. The problem arises because large Raman gain near the end of a backward pumped span amplifies the backscattered signal just as much as it amplifies the signal itself; thus the double backscattered signal experiences the Raman gain twice. The backscattered signal appears only in active channels, therefore it is not included in the noise measurement of Fig. 18.21, and so must be accounted for separately. Figure 18.22 shows the calculated per-span NSR (noise-to-signal ratio) from the doubly backscattered signal only, calculated for various Raman gain configurations, and plotted as a function of span length. The relative importance of the double-Rayleigh backscattered signal is, of course, in direct proportion to the absolute signal level itself. Consider, for example, the W_{eq} and the 20 fJ signal pulses of Fig. 18.21. From the NSRs of Fig. 18.22, we can calculate that for the case of 25/75% forward/backward pumping

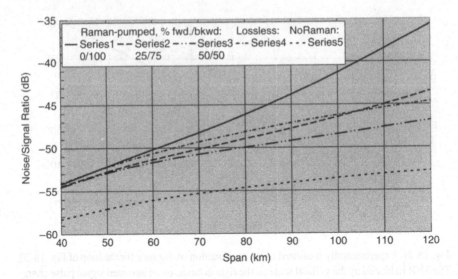

Fig. 18.22. Per-span NSR from a double-Rayleigh backscattered signal pulse train, for the Raman gain configurations shown, as a function of span length. The assumed fiber core cross-sectional area is 50 μm^2 (appropriate for TrueWave® fiber) and, for the cases with Raman gain, the spans have been pumped to unity net gain. A 3 dB correction has been made for the half-occupancy of the bit periods in the original signal.

of 100 km spans, at any distance, the double-Rayleigh backscattering increases the net noise by about 1/3 (so that it reduces the net SNR to about 3/4 (−1.25 dB) of the values shown in Fig. 18.21, already a significant reduction. For the case of 100% backward pumping (again of 100 km spans), on the other hand, the decrease in net SNR is by a (probably intolerable) −3 dB. From this example, we can see why the Raman pumping of long spans to complete transparency by back pumping alone is not generally done. Rather, one either uses a combination of forward/backward pumping, or excess gain in the following coil of DCF. Finally, note that midspan pumping (of the same 100 km spans), and insertion of an isolator there as well, tends to make the double-Rayleigh backscattered signal almost insignificantly small.

18.3.4. The Gordon–Haus effect

The Gordon–Haus effect is a jitter in pulse arrival times, caused by noise-induced frequency shifts, which are then translated into timing shifts by $\bar{D}$. For dispersion-managed as well as for ordinary solitons, the standard deviation σ of the Gordon–Haus jitter can be computed as

$$\sigma(\text{ps}) = 1.005..(\text{ps nm}) \sqrt{\frac{W_{eq}}{W_{sol}}} \sqrt{\frac{Z}{L}} \frac{\bar{D}(\text{ps/nm km})}{\tau(\text{ps})} Z^2(\text{km}^2), \qquad (18.15)$$

where W_{eq} is the equipartition energy per span, L is the span length, and τ is the unchirped pulse width. Equation (18.15) is the original formula of Gordon and Haus

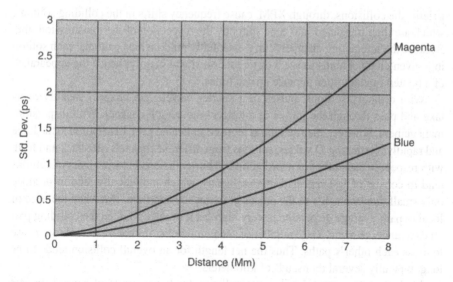

Fig. 18.23. Standard deviation of the Gordon–Haus jitter versus distance for a DMS system like that shown in Fig. 18.20. Magenta curve: Without postdispersion compensation (the full result from Eq. (18.15)). Blue curve: With optimum postdispersion compensation (half the result of Eq. (18.15)).

[23] rewritten to display the quantities W_{sol} and $\bar{D}/\tau$ explicitly, as those two quantities are not rigidly coupled for dispersion-managed as they are for ordinary solitons. (For ordinary solitons, $W_{sol} \propto \bar{D}/\tau$.) Now, due to the energy enhancement effect discussed earlier, for the same W_{sol}, the value of $\bar{D}/\tau$ is required to be several times larger for ordinary than for dispersion-managed solitons. Then σ (by virtue of Eq. (18.15)) is also larger for ordinary solitons by the same factor. Thus, although the Gordon–Haus effect tends to impose a serious penalty in ultra-long-haul transmission with ordinary solitons at 10 Gbit/s, it poses much less of a threat for the same with dispersion-managed solitons, at least when it is the sole source of timing jitter. That is, note from the data of Fig. 18.23, that the total spread in arrival times out to the 10^{-9} probability level ($\approx 13\sigma$) is a small fraction of the bit period at 10 Gbit/s, especially when the optimal postdispersion compensation is used. (On the other hand, note that the same spread is comparable to the bit period at 40 Gbit/s.)

18.4. Dense WDM

18.4.1. Soliton–Soliton Collisions

We now turn our attention to the most serious nonlinear penalty in dense WDM with OOK (on-off shift keying), viz., the jitter in pulse arrival times created by the collisions between pulses of different channels. (As noted earlier, with dispersion-managed solitons, those collisions are essentially the *only* significant source of nonlinear penalty.)

Briefly, the collisions, through XPM, cause frequency shifts in the colliding solitons, which are then translated into time shifts by the dispersion of the transmission line. Because the interacting channels carry essentially random data patterns, each soliton in a given channel tends to see a different collisional history; hence the appearance of a related random jitter in pulse arrival times.

With ordinary solitons, pulses in a shorter wavelength channel steadily overtake and pass through the pulses of a longer wavelength channel. With dispersion-management, however, the situation is more complicated [24]. In response to the large and rapidly alternating D values, solitons from different channels race back and forth with respect to each other in retarded time. Thus collisions between pairs of solitons tend to consist of fast repeated minicollisions, which individually tend to produce only small displacements of the pulses in frequency and time. But when the ratio of local to path-average dispersion is very high (as it usually is), then the colliding pair tends to undergo a very large number of such minicollisions before the solitons cease to cross each other's paths. Thus the net length for an overall collision tends to be long, typically several thousands of kilometers.

Looked at in greater detail, a net collision tends to consist of several distinct phases. First, it begins where the pulses tend to achieve maximum overlap at the junction between the $+$ and $-D$ fibers; note that each of these half-collisions tends to produce a net frequency shift that is approximately twice as great (when $|D_+| \approx |D_-|$) as the peak shift of a collision completed in just one kind of fiber. Thus the net effect of these half-collisions is to produce a steep wall of rise (or fall) in frequency shift. The middle part consists of complete collisions that tend to produce only small net effects, especially when they take place in a region of small intensity gradient. Finally, at the end, once again we have half-collisions, but this time at the other junction of the $+$ and $-D$ fibers; these produce a steep decline (or increase) of frequency, back to zero net shift.

Figures 18.24 and 18.25 illustrate a typical case of the relative motion of a pair of colliding, dispersion-managed solitons in adjacent channels, and the resultant frequency shift of one of them, respectively.

If we define L_{coll} as beginning and ending when the pulses completely overlap during the half-collisions, a study of Fig. 18.24 reveals that

$$L_{coll} \cong \frac{D_+ L_+}{\bar{D}} \cong \frac{\tau_{eff}}{\bar{D} \, \Delta\lambda}, \tag{18.16}$$

where $\tau_{eff} \equiv L_+ D_+ \Delta\lambda$ is an effective pulse width for the colliding soliton, $\Delta\lambda$ is the channel separation, and where, of course, D_+ and L_+ refer to the $+D$ segment of the dispersion map. Note that when the specific parameters of the map of Fig. 18.24 are entered into it, Eq. (18.16) yields $L_{coll} = 3200$ km, in excellent agreement with Fig. 18.25. Also, note that $\tau_{eff} = 192$ ps, in accord with the vertical span of the blue lines in Fig. 18.24.

Although the absolute states of polarization of the colliding solitons tend to evolve rapidly along the fiber (with major changes typically occurring every few meters), in low PMD fiber, their relative states of polarization tend to remain fixed over long distances (many thousands of km); this is especially true for pulses in adjacent chan-

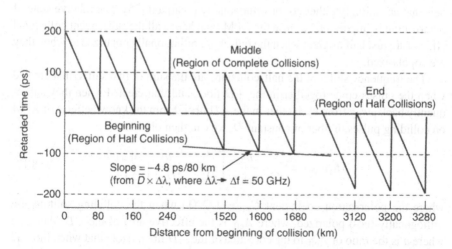

Fig. 18.24. Relative motion in retarded time of a pair of soliton pulses from adjacent channels separated by 50 GHz; for convenience, the lower-frequency pulse (red line) is fixed in retarded time; thus the higher-frequency pulse displays the entire relative motion (blue line). The dispersion map consists of 80 km spans of $D = 6$ ps/nm km fiber, compensated by a coil of DCF to yield $\bar{D} = 0.15$ ps/nm-km; the channel spacing is 50 GHz. Note the two breaks in the distance scale, necessitated by the extremely great length of the overall collision.

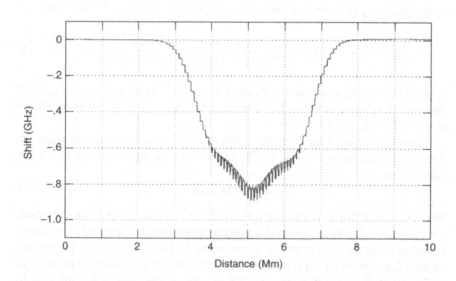

Fig. 18.25. Frequency shift, as a function of distance, of the lower-frequency pulse of the colliding pair of dispersion-managed solitons of Fig. 18.24, as determined by exact numerical solution of the NLS equation. (An equal but opposite frequency shift is induced into the higher-frequency pulse.) The colliding pulses are orthogonally polarized. Midspan Raman pumping minimizes signal intensity variation along the main span, in order to make the frequency shift curve symmetric.

nels that are initially either co- or orthogonally polarized [25]. The relative state of polarization is important, because the XPM (and hence all the subsequent collisional effects) are just half as great when the pulses are orthogonally polarized as when they are copolarized.

The frequency shifts of the half-collisions are the sum of two shifts: that created when the pulses come together in the $+D$ fiber, and that created when they subsequently back away from each other in the $-D$ fiber. From the known effects of XPM on colliding pulses in fiber of constant D, we can then obtain:

$$\delta f_{\text{half-coll}} = C_{pol}\, \frac{n_2}{A_{\text{eff}}\,\lambda} \left[\frac{1}{D_+} + \frac{a}{|D_-|} \right] \frac{W_{sol}}{\Delta\lambda\,\tau}, \tag{18.17}$$

where the polarization coefficient $C_{pol} = 1/2$ (1) when the colliding solitons are orthogonally (co-) polarized, where A_{eff} is the effective area of the $+D$ fiber, and where a is the ratio of A_{eff} to the core area of the $-D$ fiber. (Note that when the $-D$ fiber is DCF, the major contribution comes from the $+D$ fiber.) The net frequency shift of the overall collision can then be obtained by multiplying Eq. (18.17) by the effective number of half-collisions:

$$\delta f = C_{pol}\, \frac{n_2}{A_{\text{eff}}\,\lambda} \left[\frac{1}{D_+} + \frac{a}{|D_-|} \right] \frac{W_{sol}}{\Delta\lambda\,\tau} \times \frac{2\tau}{L_{\text{map}}\,\bar{D}\,\Delta\lambda}. \tag{18.18a}$$

For $A_{\text{eff}} = 50\,\mu\text{m}^2$, $\lambda = 1550$ nm, and expressing W_{sol} in fJ and the other quantities in the usual units of ps/nm km, nm, and ps, respectively, Eq. (18.18a) becomes

$$\delta f\,(\text{GHz}) = \pm 0.335\, C_{pol} \left[\frac{1}{D_+} + \frac{a}{|D_-|} \right] \frac{2W_{sol}}{L_{\text{map}}\,\bar{D}\,(\Delta\lambda)^2}. \tag{18.18b}$$

If we put the particular parameters (including $C_{pol} = 0.5$) for the collision of Fig. 18.25 into Eq. (18.18b), we get $\delta f = 0.74$ GHz, in good agreement with the exact solution as shown by Fig. 18.25.

The time shift associated with the collision scales as the area under the frequency shift curve, such as that of Fig. 18.25. Accordingly, the time shift can be roughly estimated as

$$\delta t \approx \bar{D}\,\delta\lambda\, L_{\text{coll}}, \tag{18.19}$$

where $\delta\lambda = -(\lambda^2/c)\,\delta f$ is the wavelength shift corresponding to δf. (Strictly speaking, the area of the sharp "needles" representing the complete minicollisions should be multiplied by the appropriate D_{local}, and not by $\bar{D}$. The error caused by the simplification in Eq. (18.19) is usually not great, however.) For the parameters of the collision of Fig. 18.25, Eq. (18.19) yields $\delta t \approx 2.5$ ps. (Again, the result is for orthogonally polarized colliding pulses; δt is twice as great when the pulses are copolarized.)

It is extremely important, as we discuss shortly, that the XPM- induced frequency shifts scale as the inverse *square* of the channel separation ($\Delta\lambda$). This scaling is nicely borne out in exact numerical simulation. For example, Fig. 18.26 shows the same collision as Fig. 18.25, but for a three times greater channel separation. Note

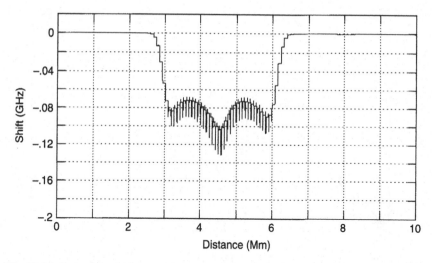

Fig. 18.26. Frequency shift, as a function of distance, for a collision like that of Fig. 18.25, except that now the channel spacing is 150 GHz.

that the peak frequency shift is indeed nine times smaller. Note further, however, that except for even steeper walls, the collision has about the same shape and essentially the same width as with the smaller channel spacing.

It may happen that, at or near the transmitter, the colliding solitons are on top of each other in the middle of a span, so that the overall collision begins near the middle of a complete collision (see Fig. 18.27). Such half-collisions can produce even greater time shifts, inasmuch as they produce a large residual frequency shift that does not disappear. In that case, the length factor in Eq. (18.19) is not L_{coll}, but nearly the entire distance of the transmission. Also note that the algebraic sign of the frequency shift (and hence that of the corresponding time shift), is opposite to those produced by the complete collision.

Finally, we note that the more usual distributions of signal intensity with distance over the span (see Fig. 18.12) tend to introduce a certain asymmetry into the curves of the collision-induced frequency shift (see Fig. 18.28). We note, however, that the area under the curve does not seem to change significantly, as we have verified through many numerical simulations.

18.4.2. Calculating the Jitter

Thus far we have discussed the collision interaction of just two solitons. In the course of a long-haul transmission, however, each pulse of a given channel suffers many collisions with the pulse trains of many other channels. Thus, at first, calculating the net timing jitter from collisions would seem to be a daunting task. Nevertheless, there are a number of factors that make the influence of all but the very closest channels of very rapidly diminishing importance. The first of these is the fact that the maximum possible frequency shift from simultaneous interaction with many channels is not

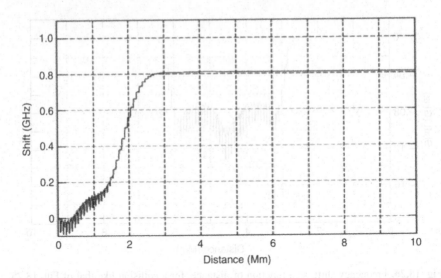

Fig. 18.27. Frequency shift, as a function of distance, of the half-collision version of the collision of Fig. 18.25. Note that the time shift is now proportional to the area underneath the tail of the curve.

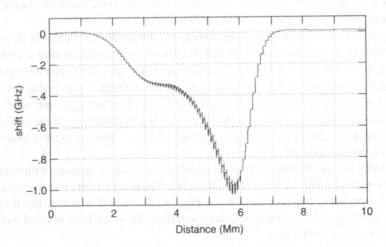

Fig. 18.28. Frequency shift, as a function of distance, of the collision like that of Fig. 18.25, except that in this case, the lack of midspan pumping produces a considerably greater variation of signal intensity across the main span.

much greater than that from interaction with just the nearest channel. This is because, as already noted, the frequency shifts from collisions fall off as $(\Delta\lambda)^{-2}$, and the fact that $\sum_{n=1}^{\infty} n^{-2} = \pi^2/6 = 1.64\ldots$. Thus, if we consider simultaneous collision with just one bit from each channel, the maximum frequency shift will be no more than about 64% greater than that from collision with just the adjacent channel. The second is that collisions with channels of lower frequency produce effects (frequency and

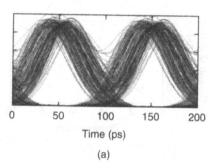 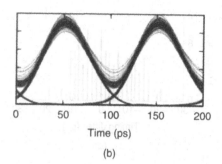

(a)   (b)

Fig. 18.29. Numerically simulated electrical eye diagrams after 8000 km transmission through a system like that of Fig. 18.20, for the worst of eight channels. ASE is not included; the only effect is that of the collisions. A different, 2^7 bit random pattern was used for each channel. (a) The unaltered eye; (b) the eye with timing jitter artificially removed.

time shifts) of opposite sign from those of higher frequency. Therefore, the effects of the lower and upper frequency sets of channels tend to cancel each other out.

Nevertheless, the minimum distance between subsequent collisions with pulses from a given channel, l_{coll}, is:

$$l_{coll} \cong \frac{T}{\bar{D}\,\Delta\lambda}, \tag{18.20}$$

where T is the bit period. Because Eq. (18.20) implies that the number of collisions increases in direct proportion to the channel spacing, one might at first think that the net effect of collisions with the pulses of many channels would then fall off as $\sum_{n=1}^{N} n^{-1} \approx \ln N$, where N is the number of interacting channels. But as $\Delta\lambda$ becomes large, with high probability, the number of colliding pulses from the interfering channel will remain close to half the maximum possible number. To make a long story short, statistics saves the day, so the net effect of collisions with many channels falls off much faster than $\ln N$ [26]. Therefore numerical simulations involving a modest number of channels (typically, about eight) tend to yield a rather accurate picture of the collision-induced timing jitter. Figure 18.29 shows some results of such a simulation, the eye diagram of the worst-behaved (one in the middle) of eight channels after 8000 km transmission through a system like that of the loop of Fig. 18.20. Note from Fig. 18.29 that artificial removal of the timing jitter almost completely opens the eye (there remains only a very small residual closure from modest amplitude jitter). Thus we can conclude that the rather severely penalizing closure of the unaltered eye does indeed stem almost entirely from the collision-induced timing jitter. There is one more important fact about this timing jitter, namely, that its statistical distribution tends to be strongly bounded (see Fig. 18.30). There are two reasons for the bounded nature of this distribution: as already discussed, the major damage is done by the immediately adjacent channels, and the maximum possible number of collisions with those adjacent channels is small. (For an adjacent channel spacing of 50 GHz and a path-average dispersion parameter of 0.15 ps/nm km, the minimum spacing between collisions (l_{coll}, Eq. (18.20)) is ≈ 1700 km.)

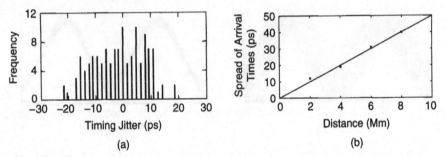

Fig. 18.30. Statistics of timing jitter in a numerically simulated version of the dense WDM system of Fig. 18.20; no ASE; adjacent channels orthogonally polarized. (a) Statistical distribution of pulse arrival times from the eye diagram of Fig. 18.29; (b) maximum spread in pulse arrival times versus distance.

18.4.3. The Temporal Lens as Jitter Killer

We have just seen that the jitter in pulse arrival times tends to be bounded. Of course, in a real system with ASE noise, Gordon–Haus jitter (which has a Gaussian distribution) adds to the nonlinear jitter. Fortunately, in dispersion-managed soliton systems, as discussed earlier, the Gordon–Haus jitter tends to be small. By combining the results shown in Figs. 18.23 and 18.30, one can infer that the probability that the net jitter will be greater than ± 1-3 of a bit period is less than 1×10^{-9} up to distances of at least 8000 km. This fact enables one to overcome the penalty from timing jitter by using a simple device [27–29] just prior to the receiver to remove the jitter itself. As can be seen from Fig. 18.31, the device itself is very straightforward, consisting of nothing more than a phase modulator, appropriately driven in synchronism with the locally recovered clock, and followed by an essentially linear dispersive element (a coil of fiber or a fiber Bragg grating). The phase modulator gives each incoming pulse a frequency shift proportional to its jitter displacement, and the dispersive element then serves to translate, or "focus" each incoming pulse on to the mean arrival time (modulo the bit period); hence we call the device a "temporal lens." To the extent that it can accomplish such translation for all incoming pulses, the temporal lens completely removes the eye closure, and hence there is no longer a nonlinear penalty. (It should be noted that in this basic principle of operation, at least, the temporal lens of Fig. 18.31 is very similar to a device described earlier [30]. We note further, however, that in [30], the device was apparently intended solely for use as a compensator of higher-order PMD, and not as a jitter-killer.) It is also important to note that the clock recovery must come before the phase modulator. (Thus, in a real system, the jitter-killer must have its own clock, independent of the one located at the detector, to avoid problems of varying delay with temperature in the dispersive element.)

In the ideal mode of operation, the phase shift $\phi(t)$ produced by the modulator is a series of truncated parabolas, centered about the middle of each bit period, such that the corresponding frequency shift (the time derivative of $\phi(t)$) is directly proportional to time as measured from the center of each period (see Fig. 18.32). Nevertheless,

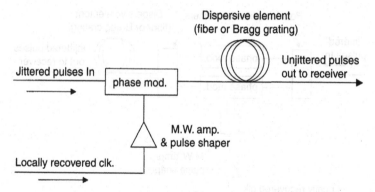

Fig. 18.31. Basic scheme of the temporal lens. The phase modulator, driven by the locally re-covered clock, shifts the frequencies of the incoming pulses in proportion to their temporal dis-placements; the dispersive element then converts these frequency shifts into jitter-compensating time shifts (see text).

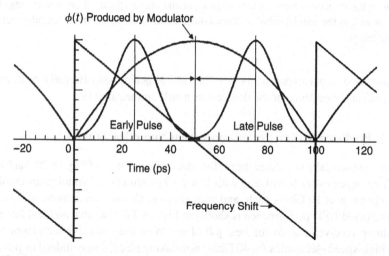

Fig. 18.32. Temporal details of the ideal mode of operation, showing: parabolic phase shift. (saw-toothed) curve: resultant frequency shift induced on pulses, and pulses that have arrived 25 ps early and late, and which are shifted to the center of the period by the temporal lens.

even a simple sinusoidal drive of the modulator works well, at least over the middle half of the bit period. In the following experiments, to further increase the lens's capture range, we have used a sinusoidal drive, but as modified by the addition of an appropriate amount of second harmonic.

Any device used at the receiver needs to have an essentially polarization-independent response. Unfortunately, however, most modulators, and in particular those made of $LiNbO_3$, are very strongly polarization-dependent. The most straight-forward way around this problem is to use a polarization diversity scheme. The one

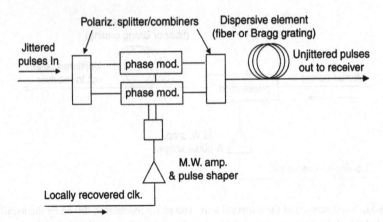

Fig. 18.33. Polarization diversity scheme for the temporal lens. The first polarization splitter separates the input of arbitrary polarization into its two, orthogonal, linearly polarized components. Each of those components is then sent through its own phase modulator. The outputs of the two phase modulators are then combined by a second polarization combiner before being sent on to the dispersive fiber. Note that the input and output (polarization-maintaining) fiber leads, as well as the coaxial cables to the modulators, must constitute three carefully matched pairs in length.

we have used in our experiments (see Fig. 18.33) performed extremely well, as the polarization-dependence of the device was nearly immeasurable.

18.4.4. Predicted System Performance from Numerical Studies

We have numerically simulated the performance of the system of Fig. 18.20, backward Raman pumped every 50 km, and with $\bar{D} = 0.15$ ps/nm km. All simulations involved eight channels at 10 Gbit/s each, and with adjacent channel separations of 50 GHz. The predicted BER performance is shown in Fig. 18.34. Note that an ideal integrate and dump receiver can do the best job of all. With the extensive development of ultra-high-speed electronics for 40 Gbit/s now taking place, it may indeed be possible to make a viable integrate and dump detector for 10 Gbit/s; if so, that would be the simplest and cheapest solution.

18.4.5. Experimental Tests of Dense WDM Using DMS

To date, we have made a great many tests of ultra-long-haul, dense WDM at 10 Gbit/s per channel using DMS. For all tests, the transmitter contained two sets of up to 80 DFB lasers, the lasers of each set on a grid of 100 GHz spacing, with their combined outputs fed through a common, LiNbO$_3$, MZ modulator-based pulse carver, and then through a second similar modulator used to impose data. The pulse carvers were driven sinusoidally at 5 GHz, and biased to yield two pulses per cycle in a very good approximation to Gaussian pulses of 33 ps FWHM. Each data modulator was driven with its own independent, typically 2^{15} bit long, pseudorandom pattern. Finally, the

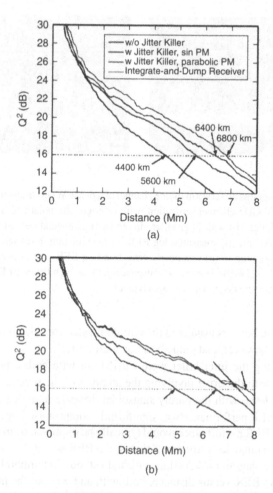

Fig. 18.34. Predicted BER performance of dense WDM system with and without jitter-killer, and with ideal integrate and dump detector: (a) all channels copolarized; (b) adjacent channels orthogonally polarized. ($Q^2 = 16$ should yield BER $= 1 \times 10^{-10}$.)

two sets of channels, with their frequencies offset from each other by 50 GHz, were brought together with a polarization combiner to yield an array of as many as 160 channels, with adjacent channels of orthogonal polarization.

The latest round of tests have involved the 600 km long recirculating loop shown in Fig. 18.20, both with and without the aid of the temporal lens jitter killer. Pre- and postdispersion compensation was carried out as discussed in Section 18.2.3; for details of the precompensation, see Fig. 18.20. About half of the final DCF coil, plus the SMF and 8 km of Raman fiber ($D = -18$ ps/nm km) shown in Fig. 18.20 form the major part of the postcompensation; the remainder was in one or another Raman pumped coil of DCF just ahead of the receiver (and ahead of the temporal lens jitter killer, when used). Thus the "jitter-reducer" part of the postdispersion compensation

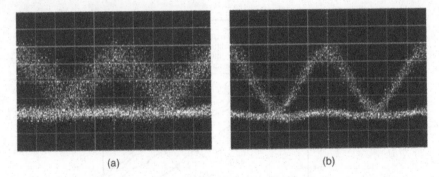

Fig. 18.35. Eye diagrams at 7200 km of one channel of a dense WDM transmission at 10 Gbit/s per channel and 50 GHz channel separation, with orthogonally polarized adjacent channels: (a) without jitter killer; (b) with jitter killer, driven by fundamental second harmonic combination. Although the directly measured log BER for the situation in (a) was in the −4 to −5 range, that corresponding to (b) was nearly error-free (log BER −9). A histogram of the eye in (b) implies a $Q^2 \cong 19$ dB, however, corresponding to an even better BER by several more orders of magnitude; the discrepancy is also typical.

could be adjusted, but was usually at the optimum value for a distance of about 5000 km; once again, however, that value was not at all critical.

Tests involving the jitter killer were carried out with both a purely sinusoidal drive, and later with the drive involving the addition of the second harmonic. Even in the very first tests, with the purely sinusoidal drive, at long distances ($Z > 5000$ km), where the BER performance was significantly compromised by eye closure from timing jitter, the BER performance would typically be improved from values in the log BER $= -4$ to −6 range to "error-free," that is, log BER $\ll -9$. Figure 18.35 shows a typical result, eye diagrams at 7200 km, with and without the temporal lens jitter killer; Fig. 18.36 shows BER versus distance, both with and without the jitter killer. From evidence such as that shown in Figs. 18.35 and 18.36, it is clear that the jitter killer consistently makes a substantial improvement in the BER performance of our dense WDM system at large distances. Nevertheless, several modest discrepancies remain. In the first place, note that the actual performance is still short of that predicted by the numerical simulations (compare the results of Fig. 18.36, where the distance, with jitter killer, for log BER $= -10$, is about 6800 km, with the graph of Fig. 18.34(b), where the corresponding distance is almost 9000 km). The discrepancy most probably has to do with an inadequately wide capture range of the temporal lens. That is, with probability much too small to be seen in an eye diagram, yet large enough to be significant relative to the 10^{-9} level, a few pulses are jittered outside the capture range.

It should be noted that FEC (forward error correction) was never used in the experiments. On the other hand, FEC is used by virtually all systems builders, because it can, at very low additional cost, render error-free transmission that would otherwise have error rates as great as $\approx 10^{-3}$. Nevertheless, once the uncorrected error rate passes a certain threshold, the FEC tends to completely lose control. Thus FEC can mask the

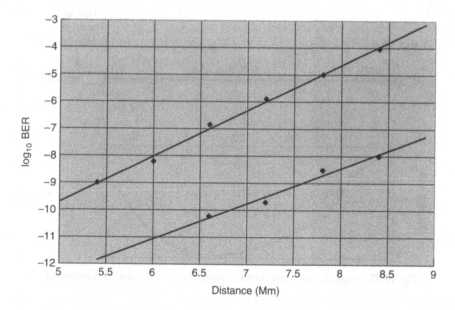

Fig. 18.36. Measured BER versus distance for: magenta points, with optimum postcompensation coil only; blue points, with jitter killer as in Fig. 18.35; no FEC was used for either case.

existence of dangerously thin margins, and in so doing, has no place in fundamental investigations, such as those reported here, of the intrinsic system quality. However, by the standard often employed today, where the performance of a system is defined by what can be achieved with FEC, the system we have just described would have a reach of ≈9 mm without the jitter killer, and ≈12 mm with it!

18.4.6. Test of a DMS System with Guiding Filters

Several years ago, we built a dispersion-managed recirculating loop containing guiding filters [31]. Although it was originally built as a model for a transoceanic system, it is now mostly of scientific interest as a test of the performance of an ultra-long-haul DMS system using guiding filters. Guiding filters tend to reduce frequency jitter and, consequently, they reduce timing jitter. They have the disadvantage, however, that although they act to directly reduce amplitude jitter, they also act to translate frequency jitter into more amplitude jitter. With ordinary solitons, because of the direct proportionality between pulse energy and pulse bandwidth, the guiding filters reduce both the frequency and amplitude jitter with great efficiency, so there is a net reduction of amplitude jitter as well. With DMS the pulse energy is typically several times less dependent on the pulse bandwidth. Then while the guiding filters are reducing frequency and timing jitter, they may actually cause a net *increase* in amplitude jitter. Nevertheless, they can still provide a net benefit, as our experiments showed.

   Figure 18.37 is a schematic of the dispersion map; as can be seen there, the amplification was a hybrid system, using backward Raman pumping to overcome about

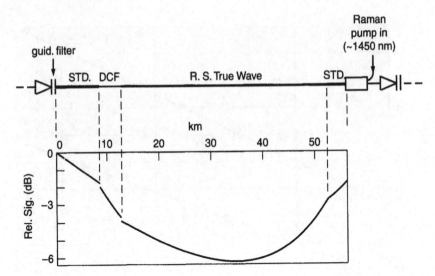

Fig. 18.37. Dispersion map and intensity profile for the transoceanic experiment.

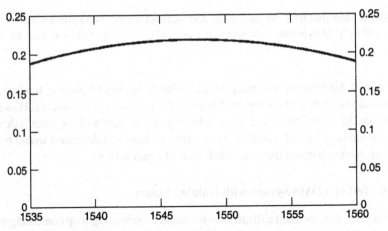

Fig. 18.38. Measured $\bar{D}$ for the dispersion map of Fig. 18.37.

75% of the span loss, and a low gain, C-band erbium fiber amplifier to compensate the rest. The path-average dispersion of the map was nearly constant over the wavelength span of the C-band (see Fig. 18.38).

Sliding the guiding filter frequencies at the modest rate of −2.7 GHz/Mm produced the best compromise between reduction of noise growth and reduction of frequency and timing jitter. Figure 18.39 shows a typical eye diagram at 9 Mm, and Fig. 18.40 plots the measured BER at 9 Mm as a function of channel wavelength for every other one out of the 27 channel wavelengths. The adjacent channel spacing, determined by the free spectral range of the guiding filters, was 75 GHz (≈0.6 nm), so the 27 channels spanned a good fraction of the entire C-band.

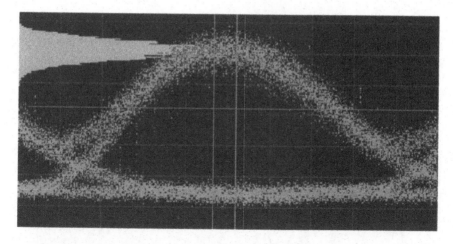

Fig. 18.39. Typical eye diagram at 9072 km for one channel in a transmission involving the setup of Fig. 18.37. The measured BER in this case is 3.7×10^{-10}.

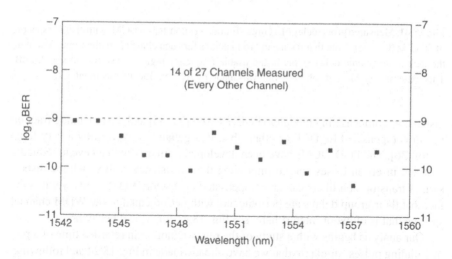

Fig. 18.40. Measured BER versus wavelength at 9072 km using the setup of Fig. 18.37.

18.4.7. Novel Technique Using Periodic-Group-Delay-Complemented Dispersion Compensation for Major Improvement of Dense WDM

We have just seen, both through theoretical argument and experimental verification, that collision-induced timing jitter is the principal limiting nonlinear penalty in dense WDM with dispersion-managed solitons, *at least when the dispersion compensation involves only fiber.* Very recently, however, in our laboratory, we have come to realize that when a certain modest fraction of the transmission span's dispersion is compensated by one or another of the recently developed periodic-group-delay

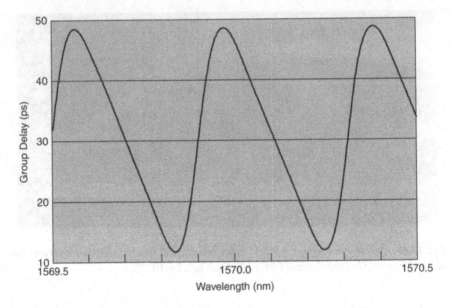

Fig. 18.41. Measured group delay of a Gires–Tournois etalon-based DCM, with channel spacing of 50 GHz (0.4 nm). Note that the mean group delays for each channel are the same. Also note the lack of detectable delay ripple in the usable frequency regions. Insertion loss = 2.6 dB. This extremely stable and robust device was made by Avanex, Inc. of Freemont, CA.

dispersion-compensating modules (PGD-DCMs) [32–35] and the remaining fraction is still compensated by DCF or other fiber, that penalty is reduced by a very large factor [36]. The PGD-DCMs have been developed primarily with an eye to reducing cost and insertion losses, and to improving the consistency with which the dispersion of transmission fibers can be compensated over wide WDM bands, yet it is the fact that their group delays are periodic and with period equal to the WDM channel spacing that is of major interest here; see Fig. 18.41 for typical behavior.

Our analysis begins with a study of the relative motion in retarded time of a pair of colliding pulses, similar to that we have already made in Fig. 18.24 and following. Now, however, we must consider the more general case, where a fraction f of the span dispersion is compensated by a PGD device, and the remainder is compensated by fiber; see Fig. 18.42.

Note from Fig. 18.42 that the mean inverse group velocity with which the pulses move with respect to each other (the slope of the sawtooth) ranges from the very small value $\bar{D}\Delta\lambda$ at $f = 0$ to the many times greater value $D^{+}\Delta\lambda$ at $f = 1$, where $\bar{D}$ is the path-average dispersion, D^{+} refers to the transmission span, and $\Delta\lambda$ is the channel spacing. Thus L_{coll} decreases very rapidly as f increases.

Our formula for the collision length (Eq. (18.16)) must now be modified. From a study of Fig. 18.42, one can easily deduce that

$$L_{\text{coll}} = \frac{2\tau + \tau_{\text{eff}}}{\bar{D}_{inter}\Delta\lambda},$$

(18.21)

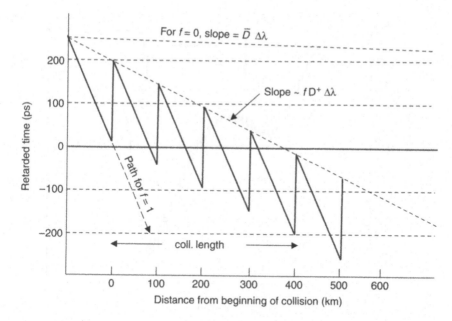

Fig. 18.42. Relative motion in retarded time of a pair of colliding pulses from neighboring channels separated by 50 GHz, where 100 km spans of $D = 6$ ps/nm km fiber, save for a small $\bar{D}$, are fractionally compensated f by a PGD device, and $(1 - f)$ by a DCF module. (For the particular behavior shown here, $f = 0.2$.)

where the effective pulse width

$$\tau_{eff} = (1-f)(D^+ - \bar{D})L^+\Delta\lambda \tag{18.22}$$

still equals the (now reduced) recovery time in the DCF, and where $\bar{D}_{inter} = f(D^+ - \bar{D}) + \bar{D}$ is the inter-channel $\bar{D}$ associated with the slope of the sawtooth of Fig. 18.42. Our formula for the minimum possible spacing between successive collisions with pulses of a given channel (Eq. (18.20)), must also be modified by substituting $\bar{D}_{inter}$ for $\bar{D}$ in the denominator; thus we now have

$$l_{coll} = \frac{T}{\bar{D}_{inter}\Delta\lambda}. \tag{18.23}$$

The rapid decline in L_{coll} as f is increased from zero Eq. (18.21) has profound effects, as can be clearly seen from the curves of collision-induced frequency shifts shown in Fig. 18.43. First, note from Fig. 18.43 that for $f = 0$ and $f = 0.2$, the frequency shifts ultimately return to zero. The most striking feature of these "frequency-conserving" collisions is the great shrinkage in all measures of the collision size as f goes from 0 to 0.2: first, in accord with the prediction of Eq. (18.21), L_{coll} is reduced by a factor of nearly 10 times, and the peak frequency shift is reduced by a similar factor. From the same numerical simulations, we find that the corresponding time displacements are 4.62 and 0.07 ps, respectively, for a 66-fold reduction. More

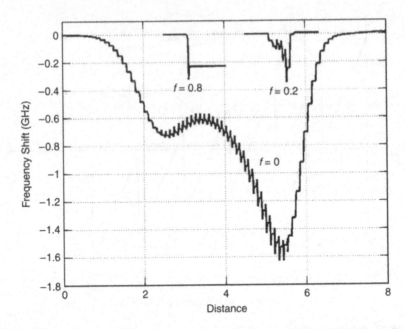

Fig. 18.43. Frequency shifts of the lower frequency of two colliding pulses from adjacent channels in the system of Fig. 18.2, for the indicated values of f; in all cases the channels are copolarized, the effective core area $= 50\ \mu m^2$, $\bar{D} = 0.15$ ps/nm km, and the spans are backward Raman pumped only.

generally, we find that the time shifts scale approximately as $L_{coll}^{1.85}$ as f increases (and L_{coll} decreases) throughout the region of frequency-conserving collisions.

We also find that as f becomes significantly greater than about 0.2 (so that L_{coll} ceases to be at least several span lengths long), the collisions are in general no longer frequency-conserving, as illustrated in Fig. 18.43 for the particular case $f = 0.8$. The situation is similar to that discovered many years ago for dense WDM with ordinary solitons, and is principally associated with the tendency of intensity gradients to destroy the symmetry of the collisions [37]. The effect is to be avoided, because otherwise the residual frequency shifts, when compounded with the dispersion remaining at the end of the transmission, can once again produce large time shifts.

With respect to the timing jitter itself, we are primarily concerned here with the way it scales down as f increases from zero (or as L_{coll} decreases), until the collisions become no longer frequency-conserving. As noted earlier, the spread in arrival times scales as the product of the time-shift per collision, times the number of collisions in the transmission. the latter scaling as $1/l_{coll}$. From Eqs. (18.23) and (18.21), respectively, we can see that for small f, both l_{coll} and L_{coll} scale in nearly the same way with f. Combining that fact with our above-given estimate for the scaling of the per-collision time shifts, we conclude that the timing jitter should scale down (for a given channel spacing) as about $L_{coll}^{0.85}$. Thus inasmuch as we have already seen that L_{coll} can be reduced by a factor of ≈ 10 times within the region of frequency-

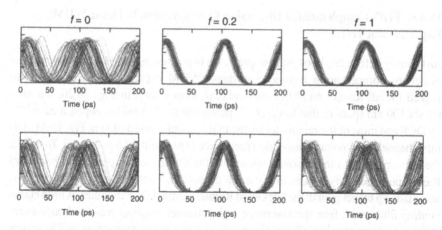

Fig. 18.44. Optical eye diagrams at 8000 km in dense WDM with 10 Gbit/s channels spaced 50 GHz apart, with all channels copolarized, and with no ASE, for the values of f shown. Top row: lossless fiber; bottom row: with 25/75% forward/backward Raman pumping. All other parameters are the same as those of the system of Figs. 18.2 and 18.3.

conserving collisions, we anticipate a similar reduction (≈ 7 times) in the timing jitter itself. This prediction is well borne out both by numerical simulations and real-world experiment.

The simulations are for 100 km spans of fiber with $D = +6$ ps/nm km, dispersion-compensated with various combinations of DCF and PGD-DCMs. Other details were essentially as earlier in this chapter (unchirped pulse width $\tau = 33$ ps, channel spacing 50 GHz, etc). Figure 18.44 shows a representative set of results from those many simulations, the eye diagrams as seen after 8000 km. As is immediately obvious from the figure, the quality of the eyes changes dramatically as f is varied. Note in particular that although the eyes are more or less uniformly bad at $f = 0$ (100% compensation by DCF), independently of the Raman pumping conditions, at $f = 0.2$ they are excellent for the case of uniform intensity and still very good for 25/75% forward/backward Raman pumping. (Also please note that here the adjacent channels are copolarized, yielding a factor of 2 times greater jitter than the orthogonally polarized case of Fig. 18.29.) The improvement here over the behavior for $f = 0$ corresponds rather well to the predictions of the simple scaling theory given earlier. Finally, note that in the $f = 1$ column (100% compensation by the PGD-DCM), the behavior is also as expected from the different intensity gradients represented there.

From simulations we have done with many other values of f, we observe that the timing jitter tends to remain very small over a rather wide range of values of f, and that the position of the absolute minimum depends somewhat on the exact profile of intensity over the spans. The tolerance for considerable variation in f is important, as it provides room to use the PGD-DCMs for complete correction of both slope and curvature of the span dispersion. (The etalon-based PGD-DCMs are particularly amenable to making such sophisticated corrections of the dispersion.)

18.4.8. PGD-Complemented Dispersion Compensation in Dense WDM: Experimental Test

Just shortly before this book went to press, we began testing the PGD-complemented dispersion-compensation technique in dense WDM experimentally. Our tests involved the all-Raman amplified recirculating loop shown in Fig. 18.20, but with the six 100 km spans of that loop ($D = 7$ ps/nm km at 1575 nm), compensated $\approx 79\%$ by DCF and most of the remainder by the PGD modules referred to in Fig. 18.41. The intrachannel path-average dispersion parameter $\bar{D}$ was still 0.15 ps/nm km. To reduce the noise penalty to the absolute minimum, the transmission spans were backward Raman pumped every 50 km, to just a few dB less than net zero gain; the excess loss was made up by net gain in the backward Raman pumped DCF modules. A mild etalon guiding filter (with free spectral range = the channel spacing) was used once every 300 km to overcome the effects of a weak adjacent pulse interaction, and to reduce the Gordon–Haus jitter (both intrachannel effects). The pre- and post-compensations were just as before for the experiments of Section 18.4.5, as was the transmitter. We used the temporal lens just ahead of the detector to remove any residual timing jitter from the collisions and from the Gordon–Haus effect. (It should be noted, however, that for the error-free distances achieved here, the net timing jitter would have been far outside the capture range of the temporal lens without the PGD-complemented dispersion compensation.) Finally, we inserted a polarization scrambler, operating at about 1 MHz, into the recirculating loop, to more realistically simulate the random PMD and PDL effects of a real system.

Figure 18.45 shows the measured BER versus distance for a typical WDM channel. First, note that the new PGD-complemented dispersion compensation technique

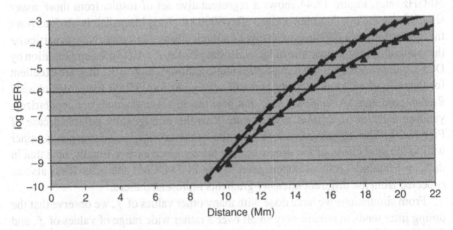

Fig. 18.45. Measured BER vs. distance for: diamonds, a WDM channel near the middle of a group of 37; triangles: the same, but with all other channels turned off (single channel performance). The distances reported here are just multiples of 600 km, and thus do not include the DCF and lumped-Raman amplifier fiber lengths. While the polarization scrambler was turned on for both sets of measurements, it tended to make negligible increase in BER except at the shorter distances.

has allowed us to greatly increase the transmission distances achievable with the more conventional compensation using DCF alone. (Compare with the results shown in Fig. 18.36.) Second, note that the BER performance in dense WDM is now very close to that for a single isolated channel. Taken together, these two facts offer clear and dramatic evidence that our new PGD-complemented dispersion compensation technique has succeeded in reducing the nonlinear penalty from interchannel collisions to relative insignificance. Indeed, the performance with the new technique, whether WDM or single channel, is now primarily limited by the combined effects of the various single-channel penalties, viz., spontaneous emission noise, PMD, PDL, Gordon–Haus jitter, and so on. Finally, note that by any measure, whether the 10^{-8} BER distance, or the "super-FEC distance" (where, with a mere 7% overhead, an uncorrected BER of $<10^{-3}$ can be corrected electronically to $<10^{-15}$), the distances here ($>20,000$ km with the use of super-FEC) are a factor of $\approx 2\times$ greater than those ever before achieved by any dense WDM transmission mode.

18.5. Summary

We have examined dispersion-managed solitons, their special periodic pulse behavior, their advantages over other transmission modes, the conditions required to create and maintain them, and we have examined closely the one serious nonlinear penalty they suffer, viz., the timing jitter from collisions with solitons of neighboring channels. Nevertheless, we have also seen how even that one nonlinear penalty can be made nearly insignificant by the use of the new PGD-complemented dispersion compensation. Thus we have seen how such dispersion-managed solitons, in an all-Raman dense WDM system at 10 G per channel, make a natural and comfortable fit with existing terrestrial fiber spans, and can provide for transmission to distances of 20,000 km or more. We have also seen how such DMS transmission is uniquely suited to provide the backbone of an all-optical network.

On the other hand, dense WDM at 40 G per channel has not been discussed, simply because the kind of DMS transmission described here is not possible over the typical 80 to 100 km fiber spans that occupy the US terrestrial network. That is, for the unchirped pulse widths required for 40 Gbit/s, the pulse breathing in such spans extends far into the quasilinear regime, where each pulse overlaps many neighboring pulses. Rather, DMS transmission at 40 G requires the use of so-called dispersion-managed cable, transmission fiber whose dispersion changes sign frequently, at least once every 10 km or so, and fiber which, unfortunately, does not yet exist commercially. (Note from Eq. (18.7) that when the unchirped pulse width is scaled down by a factor of 4 times, the scale length for dispersive broadening, z_c, scales down by a factor of 16 times.) In addition, transmission at 40 G is much more susceptible to the ravages of PMD and timing jitter, such that transmission over just 3000 or 4000 km has thus far required application of the most powerful FEC technologies. Thus, whatever merits transmission at 40 G may have, it does not fit into the purview of this chapter, which is about the foundation of an all-optical, ultra-long-haul network.

Perhaps a few words about the relatively new technology of DPSK (differential phase-shift keying) would be appropriate here. DPSK may well play an important role in quasilinear transmission at 40 G, however, it should be noted that, in general, DPSK is seriously handicapped by the fact that the nonlinear term in the NLS equation efficiently turns amplitude jitter into phase jitter (the so-called Gordon–Mollenauer effect) [38]. Thus, for dense WDM at 10 G with the usual 50 GHz channel spacing, the OOK with DMS presented here outperforms any mode that may include DPSK, and the margin becomes exceptionally wide with use of the PGD-complemented dispersion compensation.

Acknowledgments

A great many colleagues have made major contributions to the work that I have tried to distill and present here; in that regard, the contributions of Jim Gordon are of the most pervasive fundamental import; the input of my former colleague Pavel Mamyshev has been vital as well on a number of issues. Also the extensive numerical simulations of Nadja Mamysheva and, more recently, of Chongjin Xie have been invaluable, and it has been highly stimulating to discuss ideas with Chris Xu and Xiang Liu. It is especially important to note that the idea for the highly successful PGD-complemented dispersion compensation originated with Xing Wei and Xiang Liu. The experimental work has had the skillful input of Jay Cloonan, Andrew Grant, and Inuk Kang. Finally, but by no means least, I have had the most patient help with computer programming and administration from Jürgen Gripp, Andrew Grant, and Laura Luo.

References

[1] A. Hasegawa, Numerical Study of optical soliton transmission amplified periodically by the stimulated Raman process, *Appl. Optics*, 23: Oct., 3302–3305, 1984.

[2] L. F. Mollenauer, J. P. Gordon, and M. N. Islam, Soliton propagation in long fibers with periodically compensated loss, *IEEE J. Quantum Electron.*, QE-22, Jan., 157–173, 1986.

[3] L. F. Mollenauer and K. Smith, Demonstration of soliton transmission over more than 4000 km in fiber with loss periodically compensated by Raman gain, *Optics Lett.*, 13, Aug. 675–677, 1988.

[4] L. F. Mollenauer, J. P. Gordon, and P. V. Mamyshev, Solitons in high bit-rate, long-distance transmission. In *Optical Fiber Telecommunications IIIA*, I. P. Kaminow and T. L. Koch, eds., Vol. IIIA, Chap. 12, San Diego: Academic, 373–460, 1997.

[5] P. V. Mamyshev and N. A. Mamysheva, Pulse-overlapped dispersion-managed data transmission and intrachannel four-wave mixing, *Optics Lett.*, 24, Nov., 1454–1456, 1999.

[6] M. Matsumoto and H. A. Haus, Stretched-pulse optical fiber communications, *Optics Lett.*, 9, June, 785–787, 1997.

[7] T. Yu, E. A. Golovchenko, and A. N. Pilipetskii and C. R. Menyuk, Dispersion-managed soliton interactions in optical fibers, *Optics Lett.*, 22, June, 793–795, 1997.

[8] A. R. Chraplyvy, A. H. Gnauck, T. W. Tkach, and R. M. Derosier, 8 × 10 Gb/s transmission through 280 km of dispersion-managed fiber, *IEEE Photon. Technol. Lett.*, 5, Oct., 1233–1235, 1993.

[9] A. R. Chraplyvy, A. H. Gnauck, R. W. Tkach, and R. M. Derosier, One-third terabit/s transmission through 150 km of dispersion-managed fiber, *IEEE Photon. Technol. Lett.*, 7, Jan., 98–100, 1995.

[10] M. Suzuki, I. Morita, N. Edagawa, S. Yamamoto, H. Taga, S. Akiba, Reduction of Gordon–Haus timing jitter by periodic dispersion compensation in soliton transmission, *Electron. Lett.*, 31:2027–2028, 1995.

[11] N. S. Bergano and C. R. Davidson, Wavelength division multiplexing in long-haul transmission systems, *J. Lightwave Technol.*, 14, June, 1299–1308, 1996.

[12] D. Anderson, Variational approach to nonlinear pulse propagation in optical fibers, *Phys. Rev. A*, 27:3135, 1983.

[13] S. K. Turitsyn and E. G. Shapiro, Variational approach to the design of optical communication systems with dispersion management, *Opt. Fiber Technol.*, 4:151–188, 1998.

[14] J. N. Kutz, P. Holmes, S. G. Evangelides, and J. P. Gordon, Hamiltonian dynamics of dispersion-managed breathers, *J. Opt. Soc. Am. B*, 15, Jan., 87, 1998.

[15] J. P. Gordon and L. F. Mollenauer, Scheme for the characterization of dispersion-managed solitons, *Optics Lett.*, 24, Feb., 323–325, 1999.

[16] L. F. Mollenauer, P. V. Mamyshev, and J. P. Gordon, Effect of guiding filters on the behavior of dispersion-managed solitons, *Optics Lett.*, 24, Feb., 220–222, 1999.

[17] N. J. Smith, F. M. Knox, N. J. Doran, K. J. Blow, and I. Bernnion, Enhanced power solitons in optical fibres with periodic dispersion mangement, *Electron. Lett.*, 32, Jan., 54–55, 1996.

[18] N. J. Smith, N. J. Doran, F. M. Knox, and W. Forysiak, Energy-scaling characteristics of solitons in strongly dispersion-managed fibers, *Optics Lett.*, 21, Dec., 1981–1983, 1996.

[19] V. S. Grigoryan, T. Yu, E. A. Golovchenko, C. R. Menyuk, and A. N. Pilipetskii, Dispersion-managed soliton dynamics, *Optics Lett.*, 22, Nov., 1609–1611, 1997.

[20] T.-S. Yang and W. L. Kath, Analysis of enhanced-power solitons in dispersion-managed optical fibers, *Optics Lett.*, 22, July, 985–987, 1997.

[21] J. P. Gordon and L. F. Mollenauer, Effects of fiber nonlinearities and amplifier spacing on ultra-long distance transmission, *J. Lightwave Technol.*, 9, Feb., 170–173, 1991.

[22] L. F. Mollenauer, R. Bonney, J. P. Gordon, and P. V. Mamyshev, Dispersion-managed solitons for terrestrial transmission, *Optics Lett.*, 24, March, 285–287, 1999.

[23] J. P. Gordon and H. A. Haus, Random walk of coherently amplified solitons in optical fiber transmission, *Optics Lett.*, 11, Oct., 665–667, 1986.

[24] P. V. Mamyshev and L. F. Mollenauer, Soliton collisions in wavelength-division-multiplexed dispersion-managed systems, *Optics Lett.*, 24, April, 448–450, 1999.

[25] L. F. Mollenauer, J. P. Gordon, and F. Heismann, Polarization scattering by soliton–soliton collisions, *Optics Lett.*, 20, 2060–2062, 1995.

[26] C. Xu, and C. Xie, and L. F. Mollenauer, Analysis of soliton collisions in a wavelength-division-multiplexed dispersion-managed soliton transmission system, *Optics Lett.*, 27, Aug., 1303–1305, 2002.

[27] L. F. Mollenauer and C. Xu, Time-lens timing-jitter compensator in ultra-long haul DWDM dispersion-managed soliton transmission. In *Proceedings of the Conference on Lasers and Electro-Optics*, post deadline papers, CPDB1-1, Optical Society of America, May 19–24, 2002.

[28] L. F. Mollenauer, C. Xu, C. Xie. and I. Kang, The temporal lens as jitter-killer, In *Technical Digest of the Nonlinear Optics Conference*, Optical Society of America, July 29–Aug 2, 66–68, 2002.

[29] L. A. Jiang, M. E. Grein, B. S. Robinson, E. P. Ippen, and H. A. Haus, Experimental demonstration of a timing jitter eater. In *Technical Digest of the Conference on Lasers and Electro-Optics*, Optical Society of America, May 19–24, CTuF7, 2002.

[30] M. Romagnoli, P. Franco, R. Corsini, A. Schiffini, and M. Midrio, Time-domain Fourier optics for polarization-mode dispersion compensation, *Optics Lett.*, 24, Sept., 1197–1199, 1999.

[31] L. F. Mollenauer, P. V. Mamyshev, J. Gripp, M. J. Neubelt, and N. Mamysheva, Demonstration of massive wavelength-division multiplexing over transoceanic distances by use of dispersion-managed solitons, *Optics Lett.*, 25, May, 704–706, 2000.

[32] M. Shirasaki, Chromatic-dispersion compensator using virtually imaged phased array, *IEEE Photon. Technol. Lett.*, 9:1598–1600, 1997.

[33] C. R. Doerr, L. W. Stulz, S. Chandrasekhar, and L. Buhl, Multichannel integrated tunable dispersion compensator employing a thermooptic lens. In *Technical Digest of the Optical Fiber Communication Conference OFC 2002*, Optical Society of America and IEEE, March 17–22, PD FA6-2, 2002.

[34] C. K. Madsen and G. Lenz, Optical all-pass filters for phase response design with applications for dispersion compensation, *IEEE Photon. Technol. Lett.*, 10:994–996, 1998.

[35] D. J. Moss, S. McLaughlin, G. Randall, M. Lamont, M. Ardekani, P. Colbourne, S. Kiran, and C. A. Hulse, Multichannel tunable dispersion compensation using all-pass multicavity etalons, In *Technical Digest of the Optical Fiber Communication Conference OFC 2002*, Optical Society of America and IEEE, March 17–22, 132–133, 2002.

[36] X. Wei, X. Liu, C. Xie, and L. F. Mollenauer, Reduction of collision-induced timing jitter in dense WDM by the use of periodic group delay dispersion compensators, *Optics Lett.*, 28:xxx, 2003.

[37] L. F. Mollenauer, S. G. Evangelides, and J. P. Gordon, Wavelength division multiplexing with solitons in ultra-long distance transmission using lumped amplifiers, *J. Lightwave Technol.*, 9, March, 362–367, 1991.

[38] J. P. Gordon and L. F. Mollenauer, Phase noise in photonic communications systems using linear amplifiers, *Optics Lett.*, 15, Dec., 1351–1353, 1990.

Chapter 19

40 Gb/s Raman-Amplified Transmission

L. Nelson and B. Zhu

19.1. Introduction

High-capacity terrestrial optical transmission systems are now being deployed, offering aggregate capacities of 1 Tb/s or higher over distances of more than 1000 km. These systems use dense wavelength-division-multiplexing (DWDM) operating with more than 100 channels at a 10 Gb/s line rate and channel spacing of 50 GHz in both the C- and L-bands. In order to offer scalable solutions for future traffic growth in the backbone network, the 40 Gb/s line rate appears to be the natural successor to 10 Gb/s. However, with the downturn in the telecommunications industry in 2001 and 2002, natural questions to ask are: why are 40 Gb/s line rates needed, and why are researchers pursuing 40 Gb/s technologies and transmission? The answers are twofold: 40 Gb/s line rates will help to meet increased bandwidth demands and 40 Gb/s line rates reduce cost.

Although investments in new technologies and new network builds are low at the present time, there has been no substantial slowdown in the growth rate of Internet traffic [1]. According to Coffman and Odlyzko [2], Internet traffic has been approximately doubling each year since 1990 and is expected to continue to grow at this rate for the rest of this decade. Although Coffman and Odlyzko believe that supply and demand will grow at comparable rates, and that there will be neither a bandwidth glut nor a bandwidth shortage, it has always been difficult to predict which communications services people will accept and how these services will be used. Prices will also be important in determining the evolution of traffic. As prices fall, usage may increase. Coffman and Odlyzko predict that capacity will most likely continue to grow at rates somewhat faster than traffic, perhaps reflecting the fact that data traffic needs more "head-room" than voice traffic. Work on 40 Gb/s technologies today prepares us for the anticipated future demand.

Perhaps more importantly, 40 Gb/s line rates will eventually offer carriers the opportunity for lowering the cost-per-transmitted-bit [3]. A standard rule of thumb is that when technologies are mature, a factor of four times the capacity will be possible at about 2.5 times the cost. To date, many of the 40 Gb/s transmission demonstrations have used time-division-multiplexed (TDM) systems based on noncarrier-class prototype electronics or optical time-division-multiplexed (OTDM) systems based on impractical expensive architectures. However, due to advances in high-speed electronics [4–8] and optical components, cost-effective 40 Gb/s terminal equipment will soon become available. 40 Gb/s line rates lower the system cost for a number of reasons. Implementing one 40 Gb/s channel versus four 10 Gb/s channels uses fewer components, both optical and electrical, and takes up less physical space. Power consumption is also reduced. In addition, with fewer components, the network has fewer potential failure points, thus improving its reliability. It can also be argued that higher line rates are the better path to enabling higher spectral efficiencies (or, more precisely, information spectral density), allowing more capacity in a narrower spectral band. Extending the spectral bandwidth of the system or, for example, adding the L-band can increase cost. Finally, 40 Gb/s channels versus 10 Gb/s channels ease the network management, maintenance, and installation load due to the reduction in the number of WDM channels. For the same capacity there are fewer wavelength channels to manage and route. Turn-up of a single 40 Gb/s channel instead of four 10 Gb/s channels is faster and less expensive. Additionally, there are cost savings in sparing, as one-quarter the number of 40 Gb/s line cards must be available compared to the number of line cards required for 10 Gb/s. In summary, system houses are pursuing 40 Gb/s because they see significant cost advantages in the future.

If long-haul and ultra-long-haul terrestrial and submarine networks are to take advantage of the eventual lower cost per bit, 40 Gb/s line rates must be capable of bridging several thousand kilometers without electrical regeneration. The challenges of DWDM transmission at 40 Gb/s are addresssed in Section 19.2, along with the technologies enabling 40 Gb/s terrestrial transmission. Section 19.3 discusses advanced experiments and demonstrations at 40 Gb/s using Raman amplification, and Section 19.4 provides some concluding remarks. Appendix A19 provides a list of acronyms used in the chapter.

19.2. 40 Gb/s WDM Systems

Due to the promise of higher spectral efficiency and lower per Gb/s cost, 40 Gb/s technologies are attractive for the construction of high-capacity WDM backbone networks. However, when migrating from 10 to 40 Gb/s line rates, system designers face a number of technical challenges. In addition to requiring a higher optical signal-to-noise ratio than 10 Gb/s signals, 40 Gb/s signals have lower tolerance to group velocity dispersion and polarization mode dispersion. These technical issues must be addressed before cost-effective 40 Gb/s WDM systems can be installed. Recent laboratory experiments have demonstrated that broadband distributed and discrete Raman amplification, advanced modulation formats, optimized transmission fibers and dispersion maps, and forward error correction are key technologies for 40 Gb/s DWDM terrestrial transmission. In this section, we discuss the challenges and enabling technologies for developing multiterabit 40 Gb/s WDM systems.

19.2.1. Challenges of 40 Gb/s Transmission

19.2.1.1. Optical Signal-to-Noise Ratio

One of the fundamental limitations in an optically amplified transmission system is the signal–spontaneous beat noise at the receiver caused by accumulated amplified spontaneous emission (ASE). The signal–spontaneous noise impairment can be characterized in terms of the optical signal-to-noise ratio (OSNR), defined as the ratio of the optical signal power to the power of the ASE in a specified optical bandwidth, often referred to as the resolution bandwidth (RBW), which is usually 0.1 nm. In practical optical fiber transmission systems, the OSNR must be sufficiently high to achieve the required system performance, which, for example, is a bit error rate (BER) below 10^{-15} for modern commercial systems. The required OSNR should also have sufficient margin to include any impairments arising from chromatic dispersion, polarization mode dispersion, fiber nonlinearities, and the distortions introduced by the transmitter and receiver. When migrating systems from a 10 Gb/s to 40 Gb/s line rate, the required OSNR must theoretically increase by 6 dB in order to compensate for the four times wider receiver bandwidth. The actual increase in required OSNR with the channel bit rate may be greater than the theoretical value because it is more difficult to achieve comparable transmitter and receiver performance at higher bit rates, especially in 40 Gb/s systems, where there are great challenges for the development of high-speed integrated electronic circuits.

In an optically amplified transmission system, each optical amplifier contributes ASE, and these contributions are added cumulatively along the amplifier chain. As the length of the system increases, the OSNR at the end of the system decreases. The maximum unregenerated reach of an optically amplified system is the distance where the OSNR equals the target OSNR for the required system performance. This maximum unregenerated reach is also determined by the effective management of transmission impairments that can generate signal distortion. For a system containing N fiber spans, where each span is optically amplified, the OSNR of a 1550 nm signal channel at the end of the system is approximately [9]:

$$\text{OSNR[in dB / 0.1 nm RBW]} = 58 + P_{ch} - L_{sp} - NF - 10\log_{10}(N), \qquad (19.1)$$

where L_{sp} is the span loss (in dB), NF is the noise figure of the optical amplifier (in dB), and P_{ch} is the per-channel power (in dBm) launched into the span. Note that Eq. (19.1) assumes that the signals at the input to the first span have infinite OSNR. In reality, due to amplifiers in the transmitter, channels typically have OSNRs between 35 and 45 dB (0.1 nm RBW). This initial OSNR degradation also limits the transmission distance. It can be seen from Eq. (19.1) that the OSNR can be increased by one dB if P_{ch} is increased by one dB, if the noise figure is decreased by one dB, or if the span loss is reduced by one dB. If the OSNR is increased by three dB, the length of the system can be doubled, assuming the optically amplified system has fixed amplifier spacing and operates in the linear regime. Maintaining sufficiently high OSNR is a challenge for 40 Gb/s long-haul WDM transmission.

There are a number of possible ways to improve the OSNR in long-haul WDM systems. The OSNR can be increased by using shorter spans to reduce span loss, and this is done in commercial undersea systems where span lengths are usually 50 km or less. However, for terrestrial systems, which usually have shorter overall reach than undersea systems, shortening the span length is not cost effective, and the amplifier sites are more or less fixed by pre-existing equipment huts spaced about 80 km or more apart. In both types of systems, reducing the loss of the transmission fiber and components will increase the OSNR. Increasing the launch power per span is another possible way to increase OSNR while maintaining terrestrial span lengths; however, increased launch power can strongly increase nonlinear impairments associated with self-phase modulation (SPM), cross-phase modulation (CPM), four-wave-mixing (FWM), and stimulated Raman scattering (SRS) effects. Distributed Raman amplification combined with erbium-doped fiber amplifiers (EDFAs) [10], and all-Raman amplification have been demonstrated as powerful techniques for improving the OSNR for 40 Gb/s with terrestrial span losses [11, 12]. The advantages of distributed Raman technologies are discussed in Section 19.2.2.1. In addition, forward-error-correction coding can be used to decrease the required OSNR at the receiver, as discussed in Section 19.2.2.4. Most recent long-haul 40 Gb/s transmission experiments have used a combination of Raman amplification and forward error correction.

19.2.1.2. Chromatic Dispersion

Chromatic dispersion is one of the important sources of distortion for 40 Gb/s signals. It originates from the frequency dependence of the refractive index of the optical fiber (material dispersion) and the frequency dependence of the mode distribution (waveguide dispersion). As a result of chromatic dispersion, different frequencies of light travel at different speeds. In on-off keyed data transmission, where 1s and 0s are represented by the presence and absence of light, respectively, the 1 pulses contain a range of different frequencies, and chromatic dispersion causes the pulses to spread when they propagate along the fiber. The signal pulses can spread into the time slots for adjacent bits, leading to distortion.

The effect of chromatic dispersion is cumulative and increases linearly with transmission distance. More important, it increases quadratically with the data rate because of the combination of a wider spectrum and shorter pulse width when doubling the data rate. The dispersion-limited length, corresponding to the distance after which a pulse has broadened by one bit interval, is inversely proportional to the chromatic dispersion of the fiber, the bit rate, and the spectral width of signal pulses [13]. For high bit rate long-haul transmission, external modulation of continuous-wave diode lasers is usually preferred to direct modulation of the lasers because of the narrower spectrum that results. For signals produced by external modulation, the spectral width approximates the bit rate B. Assuming linear transmission using external nonreturn-to-zero (NRZ) modulation, the maximum allowable accumulated dispersion for a 1 dB eye-closure penalty is given by

$$DL[\text{ps/nm}] < \frac{104{,}000}{B^2},$$

(19.2)

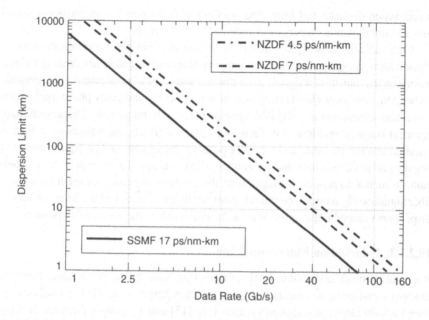

Fig. 19.1. Linear NRZ dispersion-limited transmission distance for optical fibers having chromatic dispersions of 4.5, 7, and 17 ps/nm-km.

where B is the bit rate in Gb/s, D is the chromatic dispersion in ps/nm/km, and L is the distance in km. It can be found from Eq. (19.2) that the dispersion limit is 1040 ps/nm for 10 Gb/s, and the dispersion limit is only 65 ps/nm for 40 Gb/s. Figure 19.1 shows a plot of the dispersion-limited transmission distance versus data rate when using common transmission fibers having chromatic dispersions of 4.5, 7, and 17 ps/nm-km. The precise limit depends on the details of the modulation format and design of the receiver circuitry. It can also be seen from Eq. (19.2) that a fourfold increase in the bit rate causes a decrease in the dispersion limit by a factor of 16. Precise dispersion compensation at 40 Gb/s is therefore critical, and for broadband WDM systems, dispersion slope compensation is also necessary. Dispersion-compensating fibers (DCF) with negative dispersion and negative dispersion slope are the most widely used compensation method for transmission fibers with positive chromatic dispersion, as discussed in Chapter 6. Today, slope-matching DCFs are commercially available for all common transmission fibers, including standard single-mode fiber (SSMF) and several nonzero dispersion fibers (NZDF).

In the field environment, the temperature-dependence of the chromatic dispersion can cause impairments by increasing the accumulated dispersion beyond the dispersion limit. The temperature coefficient of dispersion is inversely proportional to the dispersion slope, and coefficients can vary from -0.003 ps/nm/km/°C for high-slope NZDF to -0.0012 ps/nm/km/°C for low-slope NZDF [14]. For a 2000 km system and a 25°C temperature change, the dispersion for high-slope NZDF could vary by 150 ps/nm, clearly outside the 40 Gb/s dispersion window. Hence, depending on the transmission distance and fiber type, 40 Gb/s ultra-long-haul transmission systems may require dynamic dispersion compensation at the receiver.

Chromatic dispersion plays a dual role in modern fiber-optic communication systems: it hurts system performance by temporally spreading the pulses, leading to limitations on the transmitted signals as mentioned above, but, in the absence of chromatic dispersion, nonlinear effects such as four-wave-mixing and cross-phase modulation make high-channel-count DWDM systems practically inoperable. Dispersion management is the reconciliation of these two opposite effects; the transmission fiber is required to have nonzero, and typically positive, dispersion, and an equal amount of negative dispersion is introduced to obtain a link average of zero or near-zero dispersion. In such a dispersion-managed link, the nonzero dispersion of the transmission fiber significantly reduces the interchannel nonlinear effects, and the linear effects of dispersion-induced pulse spreading can be corrected by the negative dispersion.

19.2.1.3. Polarization Mode Dispersion

Polarization mode dispersion (PMD), originating from optical birefringence and the random variation of its orientation along the fiber length, is another critical issue in long-haul 40 Gb/s transmission systems (see [15] and references therein). To first-order, PMD causes a differential group delay (DGD) between the two orthogonal principal states of polarization (PSP), and when the DGD is a significant fraction of the bit period, pulse distortion and system penalties occur. Environmental changes including temperature and stress cause the fiber PMD to vary stochastically in time, making PMD particularly difficult to manage. In addition, although amplifiers or other components such as add-drop multiplexers in an optical system may have constant birefringence, variable polarization rotations between them due to the environment cause these components to add randomly to the PMD of the total system.

The DGD is a random variable having a Maxwellian probability density function [16]. Such a probability function can take on extremely high DGD values that exist in the tail of the distribution and correspond to system outages, when the power penalty exceeds a certain allowed value. On average, these outages may happen for only a few seconds or minutes per year and will depend on the mean DGD value and environment changes. In reality, the outages are likely to occur less often but with longer duration than the average. The PMD of a fiber span is typically specified in terms of a mean DGD, an average of the DGD over time or wavelength. Because of the statistical nature of PMD, for fiber lengths of several km or more, the mean DGD increases with the square root of the fiber length. The mean DGD of a fiber is often specified using a "PMD coefficient" having units of $ps/km^{1/2}$.

A commonly used rule of thumb is that systems can tolerate a mean DGD, $\overline{\Delta\tau}$, of approximately 10% of the bit period T. This rule arises from an outage probability calculation assuming that the power penalty contributions from PMD are less than N_p dB for all but a specified cumulative duration per year. The pulse shape (i.e., modulation format) [17–19] and details of the receiver [20] are other important factors in this calculation. An estimate for the power penalty due to first-order PMD can be obtained from [21]

$$\varepsilon(\text{dB}) = A(\Delta\tau/2T)^2 \sin^2\Theta, \tag{19.3}$$

where A is a dimensionless parameter depending on the pulse shape and receiver, $\Delta\tau$ is the instantaneous DGD, and Θ is the angle between the Stokes vectors of the launch polarization and input principal state of polarization of the fiber system. Knowing the probability density of the launch penalty factor, $\sin^2\Theta$, and the Maxwellian probability density for $\Delta\tau$, the limit for the mean DGD of the fiber can be shown to be

$$\overline{\Delta\tau}/T \leq 4\sqrt{N_p}/\sqrt{\pi A \ln(1/P_{\text{out}})}, \qquad (19.4)$$

where the outage probability P_{out} is the fraction of time per year that the PMD penalty is greater than N_p dB [15]. From 10 Gb/s PMD penalty measurements with an optically preamplified receiver, NRZ signals had $A = 70$, and 50% duty-cycle return-to-zero (RZ) signals had $A = 30$ [18]. Evaluation of Eq. (19.4) for these A values and allowing $N_p = 2$ dB for not more than 3 seconds per year ($P_{\text{out}} = 10^{-7}$) shows that $\overline{\Delta\tau} \leq 0.095T$ for NRZ, and $\overline{\Delta\tau} \leq 0.145T$ for RZ. It should be emphasized again that the A values are strongly dependent on pulse shape and receiver characteristics and must be determined for each specific system.

Fiber manufacturers currently use the link design value (LDV) to specify the PMD of cabled fiber. The LDV defines the maximum value for the PMD coefficient in terms of the probability Q_{PMD} for links with at least n concatenated sections, where typically $Q_{\text{PMD}} = 10^{-4}$ and $n \geq 20$ (IEC SC 86A/WG1). The LDV thus serves as a statistical upper bound for the PMD coefficient of the concatenated fibers comprising an optical cable link. This specification allows for small variations of the PMD coefficient from section to section. Although legacy fibers installed in the 1980s may exhibit high PMD coefficients, a recent study [22] showed that more than 80% of post-1994 installed fibers measured have sufficiently low PMD to transmit 40 Gb/s over at least 500 km of fiber. In addition, PMD measurements of 432/864 fiber count ribbon cable in the deployed configuration indicate sufficiently low PMD for 40 Gb/s [23]. Manufacturers are now routinely specifying LDVs as low as 0.04 ps/(km)$^{1/2}$.

The total system PMD is determined by all the components through which the signal travels, including the transmission fiber as well as the amplifiers, dispersion-compensating modules (DCM), gain equalizers, and multiplexers/demultiplexers for add-drop. Figure 19.2 shows a plot of the PMD-limited transmission distance for 40 Gb/s NRZ and RZ signals versus the transmission fiber's LDV. The calculations use Eq. (19.4) and assume A parameters of 30 and 70 for RZ and NRZ, respectively, and a penalty of 2 dB for not more than three seconds per year on average. The solid curves show the distance limits from transmission fiber PMD only, and the dashed curves show the limits considering both transmission fiber and component PMD. If the component DGD associated with each fiber span is 0.4 ps, then for 100 km fiber spans with LDV of 0.04 ps/(km)$^{1/2}$, the mean fiber DGD and mean component DGD will be the same. Thus the total distance over which a 40 Gb/s signal could be transmitted before being limited by PMD is reduced by a factor of two. The component DGD has the potential to vary significantly from one system to another, depending on the specifications for the various amplifier components and the amplifier design as well as the length of the DCF. Transmission fibers with lower dispersion require shorter DCF lengths and thus tend to have lower system PMD.

Note that recent measurements of installed cables have shown that DGD spectra are remarkably similar when the ambient temperature is the same [24, 25]. This

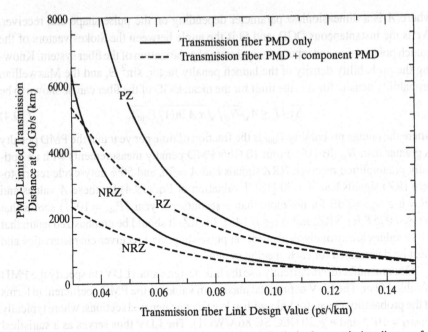

Fig. 19.2. Effect of fiber and component PMD on the PMD-limited transmission distance for 40 Gb/s systems, assuming 0.4 ps of component PMD per 100 km span ($A = 30$ for RZ, $A = 70$ for NRZ, and an allowed penalty of 2 dB for not more than 3 seconds per year on average).

behavior has been ascribed to the plant structure, where the buried sections of cable act as time-stable DGD elements and the short exposed cable sections experience the temperature variations and act as polarization rotators between two buried cable sections. Further work is necessary, but these measurements suggest that calculation of outage statistics may require a new nonstochastic model of DGD behavior to reflect its dependence on temperature.

19.2.2. Enabling Technologies for 40 Gb/s WDM Transmission

19.2.2.1. Raman Amplification

As discussed in several of the preceding chapters, Raman amplification can be used in telecommunications for both improving noise performance and expanding the WDM signal bandwidth. Distributed Raman amplification can dramatically improve the effective noise figure of an amplifier and break the 3 dB barrier [9]. This noise figure improvement increases system OSNR and satisfies the high OSNR requirement for 40 Gb/s transmission systems. Distributed Raman gain can be induced along the transmission fiber by backward Raman pumping, typically with a laser at a wavelength approximately 100 nm shorter than that of the signals. This distributed Raman amplification overcomes the attenuation in the latter part of the span, and the minimum signal power in the span is increased. This improvement in noise performance is typically represented by an equivalent (or effective) noise figure, which can be evaluated by representing the distributed Raman amplification by a discrete amplifier located at the end of the span that would produce the same gain and the same contribution to the accumulated ASE [26]. The theoretical (or quantum) limit of the noise figure for a discrete optical amplifier located at the end of the span is 3 dB, and the noise figures of commercial EDFAs are typically ~4 to 8 dB, depending on the design. The equivalent noise figure from a distributed Raman amplifier was measured to be as low as −3.7 dB for silica-core fiber [26]. Thus an improvement of 7 dB or more over a discrete EDFA can be achieved. Although this equivalent noise figure is referenced to the end of the span, the distributed Raman amplification is not actually located at the end of the span, but provides distributed amplification over an appreciable part of the preceding fiber span.

The performance of Raman amplification, such as the Raman gain efficiency and effective noise figure, depends on the properties of the transmission fibers used. The Raman gain efficiency, determining how much Raman gain can be obtained from a given amount of pump power, depends on a number of factors, including the Raman effective area (i.e., the overlap of the pump and signal modefields) [27], the composition of the fiber [28], and the pump and signal wavelengths [29]. A comparison of Raman gain efficiencies for several fiber types is shown in Table 19.1. The properties of a number of these fiber types are listed in Table 6.2 of Chapter 6. The effective area (A_{eff}) of the fiber should be optimized so that the best Raman gain efficiency can be achieved while maintaining sufficiently low nonlinear effects and a low Rayleigh backscattering coefficient, which is inversely proportional to the fiber's effective area [30]. Double-Rayleigh backscattering can limit the noise improvement from distributed Raman amplification due to Raman-enhanced multiple path interference (MPI) [26], as discussed in detail in Chapter 15. The optimum Raman gain efficiency of the transmission fiber may also depend upon the system architecture. The

Table 19.1. Comparison of Raman Gain Efficiencies for Various Fiber Types[a]

Fiber Type	Peak Raman Gain Efficiency (1/W km)	Ref.	Pump Wavelength (nm)	Required Pump Power (mW) for 20 dB Raman Gain
SSMF	0.39	[122]	1455	680
ELEAF[b]	0.46	[122]	1455	580
TrueWave® REACH	0.63	[121]	1450	440
TrueWave® RS	0.71	[121]	1454	380
AllWave®	0.39	[121]	1451	680
UltraWave™ SLA	0.29	[109]	1450	920
UltraWave™ IDF	1.2	[109]	1450	n.a.
DCF	3.3	[121]	1450	n.a.

[a] Pump powers assume a 100 km span, 0.25 dB/km loss at pump wavelength, and no pump depletion.
[b] LEAF is a registered trademark of Corning Incorporated.

performance of very broadband WDM systems could face a limitation from stimulated Raman scattering crosstalk from the short to long wavelength channels [31, 32].

Fibers used for distributed Raman should have low loss at the pump wavelengths, so that less pump power is required to obtain a certain level of Raman gain. Lower pump loss also means that the pump power penetrates farther into the span, reducing the span noise figure [33]. In addition, fiber properties such as dispersion and dispersion slope determine whether Raman noise associated with pump–pump and/or pump–signal nonlinear interactions occurs. Low dispersion slope fibers move the zero dispersion wavelength away from the pump wavelength range, preferably below 1400 nm, to prevent signal OSNR degradation from pump–pump four-wave-mixing [34] and from pump–signal four-wave-mixing for copumped Raman amplification [35]. Given these considerations for Raman pumping, new nonzero dispersion fibers have been developed with $\lambda_0 < 1400$ nm, loss at 1450 nm <0.26 dB/km, and Raman gain efficiency of $\sim$0.6 /W km [36].

The performance of Raman amplifiers also depends on the system architecture. There are two possible schemes for Raman amplifiers: copumped and counterpumped. Due to the "instantaneous" response of the Raman effect (<1 ps), any pump fluctuation slower than 1 ps can cause gain fluctuations, and hence signal fluctuation through the Raman gain. In the counterpumped distributed Raman amplification scheme, the long interaction between signal and pumps has an averaging effect, which reduces the intensity noise induced by the pump laser's fluctuation. In copumped schemes, where the signal and pump propagate through the fiber together, the averaging effect occurs only through the dispersive delay caused by walk-off between pump and signal. Therefore, low relative intensity noise pump lasers are required in order to avoid intensity noise on the signal. In addition, the Raman gain highly depends on the relative orientation of the pump and signal polarizations for copumping architectures and has been observed to have some degree of polarization-dependence for counterpumping configurations as well [37, 38]. Depolarized pump beams are thus required to minimize polarization-dependent gain.

Distributed Raman amplification is a powerful technique to ease noise performance and to increase available bandwidth. However, there are many issues to be considered in the system design. For example, the ratio of Raman gain in the transmission fiber to the gain in the EDFA or discrete Raman amplifier must be considered as well as gain ripple, ASE noise, MPI, and nonlinearities. Higher Raman gain, smaller effective area, longer fiber length, and lower data rates will result in a higher MPI-induced penalty from double-Rayleigh scattering. For the same launched signal power, fiber nonlinearities are also increased due to the higher path-average powers that depend on the Raman gain and pumping scheme for distributed Raman amplification [39]. Therefore, complex trade-offs must be met for optimization of the system performance.

As discussed in Section 19.3, distributed Raman amplification combined with erbium-doped fiber amplifiers has been demonstrated as a powerful technique for improving the OSNR for 40 Gb/s systems with terrestrial span losses [10, 40–42]. All-Raman amplified spans have been demonstrated for 40 Gb/s transmission [11, 12], where both co- and counterpumping are sometimes used in order to achieve both flat Raman gain and flat OSNR across the entire operating band [43].

19.2.2.2. New Optical Fibers

The optical fiber is a critical medium to be optimized for high capacity 40 Gb/s (ultra-) long-haul transmission. New fiber technologies should support broadband dispersion and dispersion slope compensation, have low PMD, enhance Raman amplification, reduce the total system penalties due to nonlinear impairments, and simplify optical networking.

In terrestrial systems, chromatic dispersion compensation is achieved in the amplifier stations with lumped dispersion-compensating fibers. For such systems, the transmission fiber must meet a complex trade-off: the chromatic dispersion should be small enough to avoid SPM-induced penalty and to ease dispersion compensation, but it should be large enough to avoid interchannel nonlinear effects [44]. Low dispersion slope of the transmission fiber is also highly desired, enabling an easier match of the transmission fiber to a commercially available and cost-effective DCF. When the dispersion and dispersion slope are compensated, all WDM channels experience nearly the same dispersion map, resulting in optimum performance of all channels. In the system design, this requirement can be met when the transmission fiber and DCF have almost the same relative dispersion slope (RDS = ratio of dispersion slope to dispersion at a specified wavelength). Tables 6.2 and 6.3 of Chapter 6 list the properties of common commercial transmission fibers from several companies and the DCFs available from one company.

Fibers with moderate chromatic dispersion around +7 to +8 ps/nm/km and low RDS have been developed recently and used in transmission experiments [45, 12]. These fibers have the advantage of sufficient dispersion for reduction of interchannel nonlinearities (a problem particularly at 10 Gb/s with tight channel spacing), and yet a lower dispersion than standard single-mode fibers (D = 17 ps/nm/km at 1550 nm) to minimize costs associated with the DCF. Dispersion compensation for standard single-mode fiber is more costly both in terms of the dispersion-compensating fiber lengths required as well as the higher noise figure and increased design complexity of the EDFA when up to 10 dB of interstitial loss must be compensated. In addition, moderate fiber dispersion has been shown to be optimal for 40 Gb/s, nonreturn-to-zero signals [46, 47]. A new NZDF fiber with dispersion of 7 ps/nm/km and a low RDS of 0.0058/nm at 1550 nm has been used for a large-scale long-haul transmission experiment [12], where the dispersion of each 100 km TrueWave® REACH fiber span was compensated by a single DCF module covering the entire C+L-bands. The variation in the span path-average dispersion was less than ±0.043 ps/nm/km over the 75 nm band, as shown in Fig. 19.3. Furthermore, these DCF modules can function as lumped Raman gain modules, enabling all-Raman amplified spans without bandsplitting.

Broadband dispersion and dispersion slope compensation can also be achieved using so-called hybrid or dispersion-managed spans, where large effective area positive dispersion fiber and negative dispersion fiber having a matching RDS are employed in the transmission line. This technique has been used routinely in submarine links to achieve transoceanic transmission lengths [48–50]. Use of dispersion-slope managed fiber pairs in terrestrial ultra-long-haul systems is a natural extension of their use in

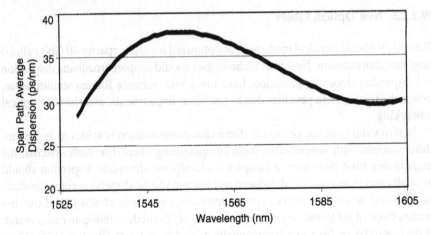

Fig. 19.3. Residual dispersion of a 100 km span of TrueWave® *REACH* fiber compensated by a single DCF module covering both C- and L-bands for the 40 Gb/s, WDM experiment described in [12]. (Reprinted with permission from the Optical Society of America.)

undersea systems, and several recent experiments have been performed at 10 and 40 Gb/s [51–55], discussed in more detail in Section 19.3. The latter four experiments employed super-large area fiber (SLA) and inverse dispersion fiber (IDF), which currently have the properties listed in Table 19.2. Figure 19.4(b) shows the residual dispersion obtained for a 100 km span consisting of 42.5 km of SLA, 34.5 km of IDF, and 23 km of SLA. This span configuration (see Fig. 19.4(a)) is often referred to as an "ABA" span, where "A" refers to the SLA fiber and "B" refers to the IDF. In earlier submarine systems, the span configuration was typically AB due to the shorter span lengths and EDFA-only amplification [48, 49]. For terrestrial Raman-pumped transmission, placing the smaller effective area IDF in the center of the span (as in the ABA span) can effectively reduce the nonlinear effect and improve the noise performance for the system, as discussed in Section 19.3.2.2.

In addition to providing slope compensation across the C- and L-bands, dispersion-managed spans eliminate several disadvantages of the lumped DCMs used in single-fiber systems. The DCMs contribute loss, PMD, and cost to the system without in-

Table 19.2. Current Properties of UltraWave™ Fibers [56] for the Dispersion-Managed Spans Used in Several Experiments[a]

	UltraWave™ SLA	UltraWave™ IDF
Dispersion @ 1550 nm (ps/nm km)	20	−44
Dispersion Slope @ 1550 nm (ps/nm² km)	0.062	−0.13
RDS (1/nm)	0.0031	0.0031
Attenuation (dB/km)	0.187	0.234
Effective Area (μm²)	107	31

[a] [52–55].

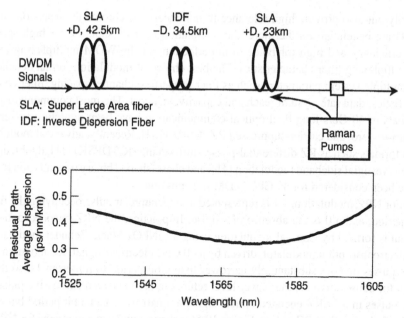

Fig. 19.4. Residual dispersion for the 100 km SLA + IDF + SLA spans used in [57].

creasing total system length. Systems based on dispersion-managed spans can be expected to have a lower amplifier noise figure and lower system PMD. There are a number of issues in the design of dispersion-managed spans, including the optimal span configuration [58] and how many times to change the sign of the dispersion within the span. Recent submarine experiments have used double (ABAB) and quadruple (ABABABAB) hybrid spans [59, 50] in an effort to mitigate the waveform distortion induced by the fiber nonlinearity. An inverse double-hybrid span configuration (BABA) has also been proposed, where the A fiber is a pure silica-core fiber with 195 $\mu m^2 A_{eff}$ [60]. Another issue is the exact dispersion values for the positive and negative dispersion fibers [61, 62]. Medium dispersion fibers have been proposed to lower the accumulated dispersion within each span; however, there is still a debate about how much accumulated dispersion a system can tolerate. Most likely, this value depends both on the modulation format and exact details of the system. To date, a problem with these medium dispersion fiber pairs has been that the effective areas of both the positive and negative dispersion fibers are smaller than those of the higher dispersion fiber pairs. Research work is now focusing on these issues to assess whether terrestrial systems using dispersion-managed spans offer sufficient performance advantage to offset the increased complexity in the installation and management of the outside plant.

19.2.2.3. Optimized Modulation Formats

The modulation format must be chosen so that both linear and nonlinear performance are optimized at 40 Gb/s. Modulation formats that not only have better receiver sensitivity but also provide high tolerance to nonlinearities, chromatic dispersion, and PMD are beneficial. Furthermore, the modulation formats should offer high spectral efficiency and high tolerance to imperfections of the WDM multiplexing and demultiplexing filter characteristics. The best choice of modulation format depends on multiple system parameters such as fiber properties, channel spacing, filter characteristics, data rate, system reach and capacity, and cost. The most commonly used formats in 40 Gb/s long-haul optical communication have been nonreturn-to-zero, return-to-zero, and carrier-suppressed RZ (CSRZ) [63]. Recently, advanced modulation formats such as RZ differential-phase-shift-keying (RZ-DPSK) [11], duobinary, single/vestigial sideband modulation [41], and phase-shaped binary transmission [64] have been considered for 40 Gb/s WDM transmission.

For NRZ modulation, a 1 is represented as a rectangular pulse occupying the full bit period, and a 0 is the absence of a pulse. In practice, the NRZ-encoded optical signal is formed by external modulation using a LiNbO$_3$ Mach–Zehnder modulator or electroabsorption modulator, driven by a 40 Gb/s electrical signal. RZ modulation uses pulses that are substantially narrower than a bit period to represent 1s, so that even for consecutive 1s, the power level returns to zero between successive pulses. The pulses in a CSRZ-encoded optical signal are narrower than a bit period but are typically longer than RZ pulses. Figure 19.5(a) shows optical eye diagrams for NRZ, RZ, and CSRZ. Usually, RZ and CSRZ pulses are generated using two modulation stages, the first to encode data, and the second to form pulses. The second modulation stage has conventionally occurred in a second dual-electrode LiNbO$_3$ Mach–Zehnder modulator, driven with a pair of half bit rate clock signals and biasing either at the transmission peak (for RZ) or the null point (for CSRZ). The carrier component of the CSRZ signal spectrum is suppressed, and the spectral bandwidth of the CSRZ signal is smaller than that of the conventional RZ signal, as shown by the optical spectra

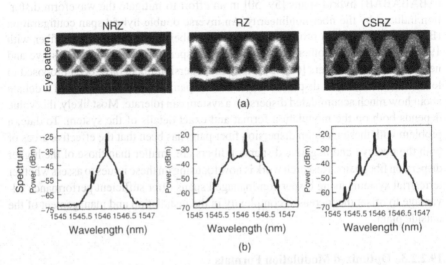

Fig. 19.5. Eye diagrams and spectra at 40 Gb/s line rate for NRZ, RZ, and CSRZ modulation formats.

in Fig. 19.5(b). For practical receiver designs, RZ and CSRZ modulation result in receiver performance superior to that for NRZ modulation by 1 to 2 dB depending on the duty cycle. Various RZ and CSRZ formats have narrower pulse widths and hence wider optical spectra and thus are inherently more resilient to nonlinearity and first-order PMD [19, 63]. NRZ has a narrower optical bandwidth and is more tolerant to dispersion slope, but less resilient to nonlinearity. CSRZ has a narrower optical bandwidth than that of RZ, and thus it has the potential for higher spectral efficiency.

For DPSK modulation, the data information is encoded in the relative phases of the optical pulses. Adjacent bits of the electrical NRZ data signal are compared, and the phase of the optical signal is modulated depending on the outcome of the comparison. When the data signal has two adjacent 1s or two adjacent 0s, the optical signal receives a zero relative phase shift; when the data signal changes from a 0 to 1 or 1 to 0, the optical signal receives a π-phase shift. The intensity of the optical signal is therefore constant or, for RZ-DPSK, is a constant pulse train. A high-speed phase modulator or dual-drive intensity modulator is usually employed to generate the PSK, and an integrated Mach–Zehnder delayed interferometer demodulator is used at the receiver to convert the differential phase modulation back to amplitude modulation. This allows a balanced receiver to be used to detect the signal [11, 65]. In principle, compared to amplitude shift-keyed (ASK) modulation, DPSK provides superior receiver sensitivity when a balanced receiver is used [66]. In addition, DPSK reduces nonlinear effects such as CPM due to its constant channel power; however, it may suffer from nonlinear phase noise caused by amplitude-to-phase-noise conversion from signal–ASE noise beating through SPM [67]. About 3 dB reduction in the OSNR required to achieve a given BER has been demonstrated using balanced detection, hence a larger increase of system reach can be achieved. For example, 2.5 Tb/s (64×42.7 Gb/s) transmission over 4000 km of TrueWave® RS fiber [11] and 3.2 Tb/s (80×42.7 Gb/s) over 5200 km of fiber with dispersion-managed spans [65] were demonstrated recently.

In order to increase spectral efficiency in WDM systems, it is essential to reduce the optical spectrum width without losing information. Vestigial sideband (VSB) filtering of unequally spaced NRZ signals [41] and optical duobinary coding or phase-shaped binary transmission (PSBT) [68, 69] have been considered recently for 40 Gb/s WDM transmission. For VSB filtering, one of the two sidebands of an NRZ optical spectrum is filtered out, as these two sidebands generally contain redundant information. However, VSB is difficult to implement at the transmitter because the suppressed sidebands can rapidly reconstruct through fiber nonlinearities [41]. The VSB filtering can be employed at the receiver to increase spectral efficiency, and an experiment has demonstrated a spectral efficiency of 0.64 bit/s/Hz with 40 Gb/s channels alternately spaced by 50 and 75 GHz using VSB filtering [41]. Duobinary, also referred to as PSBT, is basically a three-level coding scheme ($-1/0/+1$) that is optimized to reduce the channel spectral width. High spectral density thus can be achieved; however, duobinary modulation has poor receiver sensitivity due to its poor extinction ratio of the signal. Enhanced PSBT (EPSBT) was proposed to overcome the low tolerance to noise by superimposing NRZ modulation on the PSBT modulation by cascading two modulators in series [70]. The low extinction ratio also can be improved by passing the PSBT signal through a bandwidth-limited optical filter located at the transmitter or at the receiver [71]. A transmitter composed of a 10 Gb/s low-pass filtered duobinary modulator followed by an optical filter having a 3 dB bandwidth of 7 GHz was shown to enhance the spectral efficiency by ~20% and improve the receiver sensitivity by more than 1 dB. This bandwidth-limited PSBT (BL-PSBT) also has been reported to reduce single-channel nonlinear effects [64] and has high dispersion tolerance. Such advanced modulation formats have been considered to increase overall capacity by increasing the spectral efficiency. However, this increase in spectral efficiency is often associated with a sacrifice in receiver sensitivity and, hence, a considerable reduction in the overall system reach.

19.2.2.4. Forward Error Correction

In order to relax the constraints in terms of OSNR and signal distortion and to meet the target reach for a system, forward error correction (FEC) is virtually a prerequisite for 40 Gb/s systems. With FEC, extra bits are appended to the data by the FEC encoder at the transmitter. The FEC decoder at the receiver then uses these extra bits to detect and correct erroneously received bits. As a result, FEC enables the system to operate at a required OSNR far below that required without FEC, while maintaining an acceptable BER. Thus FEC enables a very large system margin that can be used to increase the transmission distance and system capacity.

The performance of the FEC code in correcting errors is characterized in terms of the coding gain, defined as the difference in the Q-factor (as $10 \log Q^2$) between the uncorrected and corrected BER. The coding gain is usually defined at the system target Q or BER (e.g., 10^{-15}). The line rate increases when extra bits are added due to the FEC, and systems have additional penalties associated with impairments due to increased bandwidths. Thus the coding gain is often quoted as net effective coding gain (NECG) [72], defined as the coding gain corrected for the penalties due to the higher line rate when the extra overhead is added to the system. Typically, for a given type of error-correcting code, the stronger the FEC coding gain and the higher the overhead, the better the FEC code will be at correcting a severely corrupted signal. In practice, the best FEC code is the one offering the greatest NECG, but requiring the least line rate increase. In addition to the coding gain and added overhead, the complexity of the encoding algorithm and the feasibility of implementing it on a high-speed integrated circuit (IC) are critical considerations in developing an FEC code.

Several types of FEC schemes have been employed and considered in optical communication systems [72]. A commonly used FEC code is the (255,239) Reed–Solomon (RS) code, which is the ITU G.975 standard (ITU 1999). The RS (255,239) code increases the bit rate by 7%, from 9.953 to 10.664 Gb/s, providing a coding gain of about 6 dB at 10^{-15} BER. This code was first adopted for commercial undersea systems, and its use has now been extended to terrestrial systems. A straightforward improvement of coding gain would be to use a stronger RS code, and a coding gain increase of 1.2 dB has been reported by increasing the line rate from 7% for the G.975 RS code to 14% for the RS(255,223) [73]. To further improve the coding gain concatenated FEC codes can be used, where two FEC codes, an inner code and outer code, are employed sequentially. At the transmitter the data are sequentially encoded with the outer code and then the inner code, whereas at the receiver the data are sequentially decoded with the inner code followed by the outer code. An interleaver and iterative coding process are often used in concatenated FEC codes. A concatenation of the RS (255,223) with the RS (255, 239) code was proposed for 10 Gb/s systems, providing an additional 2 dB of coding gain relative to the RS (255,239) with an increased line rate of about 25% [74]. An optimum FEC overhead of 10% with additional NECG increase of 1.99 dB using concatenated turbo codes and low-density parity check codes was also reported for 20 Gb/s ultra-long-haul transmission [75].

For 40 Gb/s transmission FEC is even more critical because of the high OSNR required at the receiver. Because the penalties from transmission and from the transmitter/receiver increase dramatically for line rates near 40 Gb/s, higher coding gain with less overhead redundancy would be very attractive. One way is to use interleaving and iteration of several shorter codes to increase the FEC coding gain without significantly increasing the overall redundancy. A trade-off has to be made between the coding gain (and overhead) and the transmission impairments due to a higher ratio of redundant bits. The optimum choice will be determined by the progress of high-speed electronic ICs [6] and error-correction coding technologies.

19.3. Advanced Experiments and Demonstrations

The previous section has reviewed the issues and challenges of 40 Gb/s transmission as well as several of the key enabling technologies. The reach and capacity of 40 Gb/s transmission experiments have increased in large part due to advancements in Raman amplification, particularly for terrestrial-length spans (i.e., >75 km). The lower span noise figure afforded by Raman can be utilized for a number of different system upgrades, including increased span loss, higher bit rates, increased number of spans, and lower launch power per channel to decrease fiber nonlinearities. As also discussed in previous chapters, Raman amplification opens new transmission bands and allows seamless broadband systems. This section discusses a number of 40 Gb/s terrestrial transmission experiments illustrating how Raman amplification has enabled multi-Tb/s capacities over long-haul and ultra-long-haul distances. Table 19.3 at the end of Section 19.3 summarizes the various transmission experiments discussed. Note that this section does not attempt to cover the specific nonlinear impairments influencing transmission at 40 Gb/s. These are primarily single-channel effects (i.e., intrachannel cross-phase modulation and intrachannel four-wave-mixing), and are explained in detail by Essiambre et al. [76] and Bayvel and Killey [77].

19.3.1. Typical 40 Gb/s Transmission Experiment Setup

As many of the transmission experiments reported in the literature have used similar configurations, we show a schematic diagram of a typical 40 Gb/s transmission experiment in Fig. 19.6. The WDM transmitters consist of distributed feedback (DFB) lasers with channel spacing on the ITU grid (i.e., 200, 100, or 50 GHz). The odd

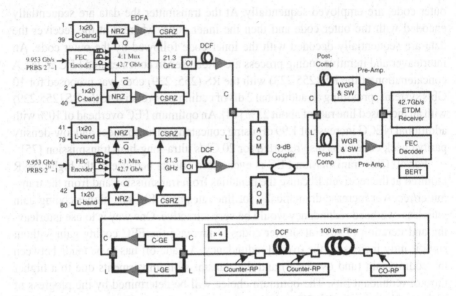

Fig. 19.6. Schematic diagram of a general 40 Gb/s, C+L-band recirculating loop experiment: OI: optical interleaver, AOM: acousto-optic modulator, RP: Raman pump(s), C-GE: C-band dynamic gain equalizer, L-GE: L-band dynamic gain equalizer, WGR: waveguide grating router, SW: optical switch.

and even channels are multiplexed separately by waveguide grating routers (WGR) and modulated independently. The optical signals pass first through LiNbO₃ Mach–Zehnder modulators (MZM) driven by 42.7 Gb/s electrical data streams. Electrical time-division-multiplexing (ETDM) is employed to generate the 42.7-Gb/s signals from four 10.664 Gb/s data streams, each consisting of $2^{31} - 1$ pseudorandom bit sequence (PRBS), 9.953 Gb/s data encoded with 255,239 Reed–Solomon FEC. Ideally, these 10.664 Gb/s data streams should have relative delays of a quarter of the PRBS length, so that the resulting 42.7 Gb/s signal is also a PRBS. The PRBS is chosen as the test data signal for experiments because it closely resembles the genuinely random signals found in real data [78]. Pseudorandom and genuinely random sequences have nearly identical autocorrelation properties and spectral characteristics. Use of long PRBS lengths (i.e., $2^{23} - 1$ or $2^{31} - 1$) is generally recommended for several reasons. Because the lowest frequency content of a PRBS extends down to B/N (where B is the bit rate and N is the PRBS length), a long PRBS length will test the bandwidth of the receiver electronics. In addition, long PRBS lengths are the more stringent test for transmission impairments because most penalties arise from isolated 0s within long strings of 1s. Note that some experiments use a simple 10.664 Gb/s PRBS ($2^{31} - 1$) to emulate the 7% overhead FEC, instead of encoding 9.953 Gb/s data with 255,239 Reed–Solomon FEC code.

The second modulator in each transmitter is used for pulse carving to generate a 42.7 GHz RZ or CSRZ pulse train by driving it with a 21.329 GHz clock. An amplifier (possibly polarization-maintaining) can be inserted between the two modulators to maintain a high OSNR at the transmission line input. The odd and even channels in each band are then combined by an optical interleaver (OI). Such a scheme provides effective decorrelation of the data carried by neighboring WDM channels, assuring good emulation of real transmission where there are random data patterns on all channels. For precompensation of dispersion, the WDM channels of each band can be transmitted through DCF modules, prior to being amplified, combined, and finally launched into the transmission line.

Figure 19.6 also shows the recirculating transmission loop, where the optical signals are switched into the loop and allowed to circulate a number of times through the fiber spans before being switched out for error analysis. Until recently, recirculating loops were used primarily for submarine system testing [79], as they allow testing of long-haul and ultra-long-haul transmission distances with a modest number of components and fiber spans. Loop experiments offer a significant cost advantage when compared to straightline experiments, but these loop experiments are more complex and must be done with care to avoid misleading results. The loop should be at least as long as one period of the dispersion map, and preferably longer, as longer loops also ensure that the signals pass through the loop components (acousto-optic modulators (AOM) and coupler) as few times as possible for the target transmission distance. The AOMs often have polarization-dependent loss (PDL), and an extra EDFA is usually required to compensate for the loss of the AOM and output coupler. Passing through these components less frequently minimizes their contributions to the system impairments.

As an aside, it should be noted that recirculating loops do not necessarily correctly reproduce the polarization behavior of real systems. Recent studies have shown that the loop's PDL can unrealistically improve the performance when a polarization controller is used to optimize the polarization evolution in the loop [80, 81], particularly when the loop is short and the transmission distance is very long. With polarization adjustment the PDL can extinguish the ASE orthogonal to the signal, thus improving the degree of polarization and OSNR [82]. The PMD in loop transmission experiments is also a concern, because recirculating loops tend to exhibit some measure of deterministic behavior that can produce unrealistic PMD distributions [83]. Polarization adjustment of the signals can minimize (or maximize) the effects of PMD. Using a polarization scrambler for the signals before the loop as well as a fast polarization scrambler within the loop (operating synchronously, on the time scale of one round-trip) has recently been proposed to better emulate the results that would be obtained for straightline transmission [83, 81, 82]. However, these polarization scramblers add additional loss and thus lower the OSNR, a critical impairment to 40 Gb/s transmission. Due to this OSNR impairment, to date, most 40 Gb/s transmission experiments have not used polarization scrambling and have relied on long loop lengths and relatively short transmission distances to minimize the unrealistic loop polarization effects. But as 40 Gb/s transmission experiments increase to thousands of km, at least some amount of polarization scrambling of the signals undoubtedly will be required.

The recirculating loop of Fig. 19.6 contains several spans of fiber with amplifiers. In the 40 Gb/s experiments discussed in this section, these amplifiers are comprised

of an EDFA with Raman counterpumps or the amplifiers are all-Raman, where co-
and counterpumps at multiple wavelengths provide all the gain. For single-fiber type
spans (i.e., SSMF or NZDF), DCF is also required in the amplifiers. This DCF can be
located between the two stages of the EDFA or at the input to the EDFA. In addition,
dynamic gain equalizers (GE) (see, e.g., [84] or [85]) are usually required in the loop
to achieve loop gain equalization. Figure 19.6 shows splitting of the C- and L-bands
and a separate GE for each band, followed by additional C- and L-band EDFAs to
compensate for the loss of the GE and loop components. In the following, descriptions
of the specific experiments stress the unique components or amplification techniques
used and the results.

After circulating in the loop for the targeted transmission distance, the optical
signals are switched out and sent to the receiver for error analysis. The signal bands
can be split, as shown in Fig. 19.6, before individual channels are selected by a WGR
and optical switch (SW) and sent to the preamplifier. Postdispersion compensation is
often required. This can be a fixed amount of dispersion compensation (i.e., a spool of
DCF) for all channels, if the dispersion slope of the transmission fiber and inline DCFs
match sufficiently well. Or, the postcompensation can be optimized on a channel-by-
channel basis using a tunable dispersion compensator [86–88] or different lengths of
DCF. After preamplification and filtering to suppress the ASE, the signal is sent to a
40 Gb/s (42.7 Gb/s) receiver comprised of a clock and data recovery, as well as a 1:4
demultiplexer to enable bit error rate measurements on the 10 Gb/s tributaries. 40 Gb/s
clock recovery is usually accomplished using a phase-locked loop or by selecting out
the 40 Gb/s component using a high-Q filter. In some experiments, the demultiplexing
is done optically, where an electroabsorption modulator (or lithium niobate modulator)
is used as an optical gate to choose a 10 Gb/s tributary [75] that is then sent to a 10
Gb/s receiver. More recently, commercial electronic demultiplexers (40 to 10 Gb/s)
have become available and are now frequently used in laboratory experiments. When
FEC encoding is done on the 9.953 Gb/s signals at the transmitter, the 10.664 Gb/s
electrical signals are sent to the FEC decoder before BER measurement. The BER
of each channel can be measured twice, once with the FEC decoder turned off (to
measure the "line BER" or "uncorrected BER"), and a second time with the FEC
decoder turned on (to measure the "corrected BER"). For proper analysis of the
system performance, the BERs of all four 10 Gb/s tributaries should be measured.
However, in loop experiments, during each gating period the clock recovery must
regain phase lock and thus it randomly selects a tributary. The average performance
of all four tributaries is then ascertained when the measurement time is set sufficiently
long to include a large number of gating periods.

In properly designed systems, the longest transmission distance is achieved when
the signal power is chosen to balance the OSNR impairment with the nonlinear impairments. In single-channel systems, the optimum launch power for the longest reach
usually can be found fairly easily. An illustrative example is shown in Fig. 19.7 from
a single-channel 40 Gb/s transmission experiment using hybrid Raman/EDFAs and
100 km spans of NZDF [89]. The plot shows the range of allowed launch powers
for Q-factors greater than 15.6 dB (i.e., BER $> 10^{-9}$) versus transmission distance.
The Raman gain in the spans was kept constant as the launch power and distance

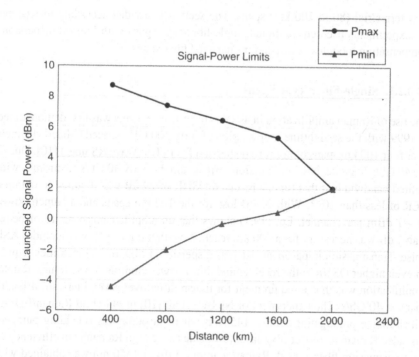

Fig. 19.7. The range of allowed launch powers for Q-factors greater than 15.6 dB versus transmission distance for the single-channel, 40 Gb/s experiment described in [89]. (Adapted and reprinted with permission from the Optical Society of America.)

were varied. The lower curve indicates the launch power at which the performance is OSNR-limited, and the upper curve indicates the launch power at which performance is limited by nonlinearities. In this experiment, a launch power of +2 dBm resulted in the longest transmission. In multichannel systems, however, the situation is more complicated due to gain ripple from the Raman amplification and EDFAs. As the gain ripple accumulates after transmission over multiple spans, the high-power channels can be limited by nonlinearities, whereas the low-power channels are limited by OSNR. Both the nonlinearity and the span noise figure are determined by factors such as the proportion of Raman gain in the transmission fiber versus EDFA gain or the ratio of copumped gain to counterpumped gain in all-Raman systems. In WDM experiments, therefore, a complicated optimization must be done involving adjustment of the launch powers of the signals as well as Raman gain from the various pumps to achieve a balance between the OSNR and nonlinearities on average, across all the WDM channels.

19.3.2. 40 Gb/s Terrestrial Transmission Using Hybrid Raman/EDFAs

The subject of hybrid Raman/EDFAs, their design and performance, has been covered in detail in Chapter 13. This section concentrates on high-capacity 40 Gb/s experiments that have employed hybrid Raman/EDFAs for C- and L-band transmission over terrestrial (80 to 100 km) spans. The section is divided according to whether the experiments used conventional single-fiber-type spans with lumped dispersion-compensating modules or dispersion-managed fiber spans.

19.3.2.1. Single-Fiber-Type Spans

The use of Raman amplification in 40 Gb/s terrestrial systems was first demonstrated in 1999 with the straightline transmission of forty 100 GHz spaced C-band channels over four 100 km spans of nonzero dispersion fiber (TrueWave RS fiber) [10]. Due to the 40 Gb/s line rate, NRZ modulation format, and the early 40 Gb/s electronics with limited bandwidth at that time, a received OSNR of 29 dB was required to obtain a BER of less than 10^{-9}. With 21 dB loss for the 100 km spans and a launch power of -1 dBm per channel, Eq. (19.1) shows that an amplifier noise figure of better than 1 dB was necessary for a 400 km reach. Note that Eq. (19.1) only includes ASE noise. Transmission impairments such as dispersion, PMD, or nonlinearities require an even higher OSNR at the receiver and thus a lower amplifier noise figure. Raman amplification was thus a requirement for transmitting over several terrestrial-length spans at 40 Gb/s. The experiment of Nielsen et al. [10] used hybrid Raman/EDFAs with a counterpropagating pump at 1450 nm, where the pump was a cladding-pumped, cascaded Raman resonator fiber laser [90]. Due to the high Raman gain efficiency of the transmission fiber, a peak Raman gain of 23 dB at 1550 nm was obtained with 520 mW of pump power. Figure 19.8 shows a schematic of the hybrid Raman/EDFA along with the measured gain flatness and effective noise figure [26]. The hybrid amplifiers consisted of distributed Raman amplification in the transmission fiber for the first stage, followed by two erbium-doped fiber stages. A dispersion-compensating fiber module was inserted between the Raman stage and the first EDFA stage, and a gain-flattening filter (GFF) was placed before the second EDFA stage. The composite gain ripple of less than 1 dB and the effective noise figure of -0.3 to -1.5 dB resulted in OSNRs greater than 30 dB (0.1 nm RBW) and BERs of less than 1×10^{-10} for all forty channels after 400 km transmission.

The first experiment to combine dual C+L-band transmission, distributed Raman amplification, and 40 Gb/s line rates was the transmission of 3.28 Tb/s total capacity over 3×100 km of nonzero dispersion fiber [40]. Dual-band hybrid Raman/EDFAs were used to amplify the 40 C-band and 42 L-band NRZ channels, as shown in Fig. 19.9. These amplifiers consisted of counterpumps at 1447 and 1485 nm for distributed amplification in the transmission fiber, followed by separate two-stage gain-flattened EDFAs for the C- and L-band channels. Pump powers of 600 mW at 1447 nm and 220 mW at 1485 nm provided peak Raman gains of 25 and 24 dB for the C- and L-band channels, respectively. Also shown in Fig. 19.9 are the gain and effective noise figure of the hybrid amplifiers. The better effective noise figure of the L-band channels is a result of the C-band pump providing gain for the L-band pump, thus enabling the L-band pump to penetrate farther into the transmission span and provide gain for the L-band channels closer to the span input. In addition, the L-band channels experience less loss through the fiber when the C-band channels are present due to stimulated Raman scattering [44]. After transmission over the three

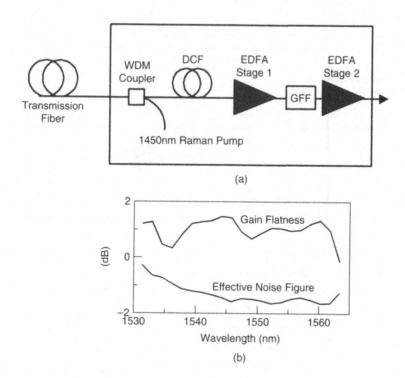

(a)

(b)

Fig. 19.8. (a) Schematic diagram of the C-band hybrid Raman/EDFA used in the experiment demonstrating transmission of 1.6 Tb/s over 400 km of TrueWave® RS fiber [10]. (b) Gain flatness of hybrid Raman/EDFA at output of EDFA stage 2 and effective noise figure referenced to the end of the transmission fiber.

spans the OSNRs were greater than 29 dB for all 82 channels, and the maximum BER was 8×10^{-11} (without FEC). The authors attributed the limitation to be OSNR in the C-band and cross-phase modulation in the L-band. In this experiment, the transmission fiber was a prototype TrueWave fiber with dispersion of 5.7 ps/nm/km and dispersion slope of 0.037 ps/nm²/km, both at 1550 nm. The low RDS of the transmission fiber enabled good slope match with the high-slope DCF (HS-DCF) resulting in dispersion variations of 60 and 140 ps/nm over the C- and L-bands, respectively, after transmission. The 60 ps/nm variation in accumulated dispersion in the C-band was within the 1 dB dispersion tolerance for 40 Gb/s NRZ signals, and thus per-channel postdispersion trimming was not required. Further improvement of the RDS of DCF has enabled 500 km, 40 Gb/s transmission over TrueWave RS fiber with no per-channel postcompensation [91].

The first multi-Tb/s transmission of 40 Gb/s channels over more than 1000 km of fiber used similar dual-band, hybrid Raman/EDFAs [42]. The demonstration of 3.08 Tb/s capacity (77 channels each at 42.7 Gb/s) was done with a recirculating loop setup close to that shown in Fig. 19.6. FEC (7% overhead Reed–Solomon) was implemented to achieve error-free operation of all 77 NRZ channels after twelve 100 km spans of prototype TrueWave fiber. Figure 19.10 shows the measured on-off

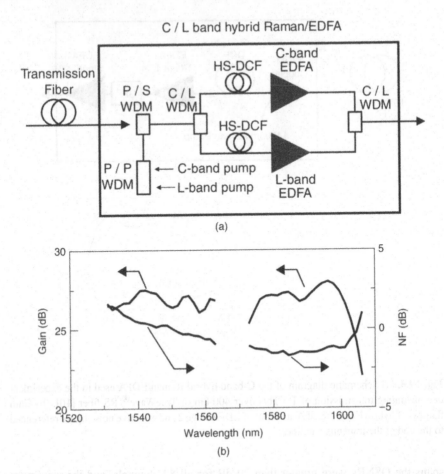

Fig. 19.9. (a) Diagram of the C+L-band hybrid Raman/EDFA used in the experiment demonstrating transmission of 3.28 Tb/s over 3 × 100 km of NZDF [40]. (P/S WDM: pump–signal combiner, P/P WDM: pump–pump combiner, C/L WDM: C–L band combiner). (b) Gain flatness of the hybrid Raman/EDFA at amplifier output and effective noise figure referenced to the end of the transmission fiber. (Reprinted with permission from the Optical Society of America.)

Raman gains with only the 1447 pump on, only the 1485 nm pump on, and both pumps on. The large increase (>10 dB) in the L-band channel powers when both pumps are on shows the effect of the C-band pump providing gain for the L-band pump. In this experiment, dynamic gain equalizers placed after the three spans in the loop were important elements in minimizing the accumulated gain ripple. Improvements in the dispersion slope match between the transmission fiber and the high-slope DCF modules resulted in dispersion variations of 47 ps/nm across the C-band and 61 ps/nm across the L-band after 1200 km, and no per-channel postcompensation in the C-band was required. At the time, this experiment probed the limit for the capacity × distance product at 40 Gb/s possible with simple NRZ modulation format, hybrid Raman/EDFAs with counterpumping only, and standard 7% overhead FEC.

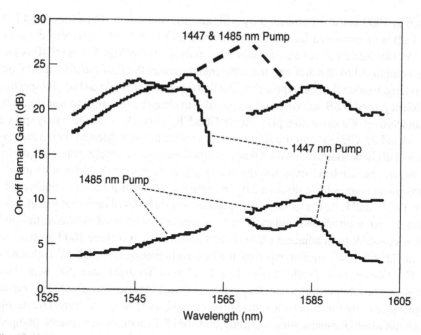

Fig. 19.10. The measured on-off Raman gains with only the 1447 pump on, only the 1485 nm pump on, and both pumps on for the experiment reported in [42]. (Adapted and reprinted with permission from the Optical Society of America.)

Several other hybrid Raman/EDFA experiments have demonstrated the use of Raman amplification to increase the amplified span lengths beyond 100 km. At 10 Gb/s, Terahara et al. [92] showed transmission of 128 channels over six 140 km spans of standard single-mode fiber (30.4 dB loss) with dual-band distributed Raman amplification and average Raman gains of 5.8 and 5.4 dB in the C- and L-bands, respectively. Srivastava et al. [93] transmitted one hundred, 25 GHz spaced, 10 Gb/s channels in the L-band over three spans of 125, 132, and 140 km of TrueWave fiber. The hybrid amplifiers used a single pump wavelength at 1480 nm, achieving a peak gain of 10 dB at 1585 nm, and the launch power per channel was −3 dBm. In addition, transmission of thirty-two 10 Gb/s, C-band channels over 750 km of fiber with 250 km spans was also demonstrated [94]. This experiment was targeted at a medium-haul submarine system with land-based optical amplifiers or coastal festoons. One-watt Raman fiber lasers at 1455 nm were used to achieve Raman preamplification in the pure silica-core fiber spans having 44.5 dB loss.

Transmission over multiple long spans has also been achieved at 40 Gb/s using Raman amplification. An experiment demonstrating a 1000 km reach for 32 C-band channels with five spans of 160 km and one span of 200 km was reported by Zhu et al. [95]. Hybrid Raman/EDFAs with pumps at 1427 and 1455 nm were used, and the Raman gain profile was tailored to minimize the accumulated gain tilt from the ED-FAs. The CSRZ modulation format allowed a launch power of +5 dBm/channel into the NDSF (nondispersion-shifted fiber) spans, and analysis showed that the system reach at 1000 km was dominated by ASE accumulation. Raw Q-factors of 10.1 to 13.1 dB were measured for the 40 Gb/s channels; FEC was not implemented.

A combination of advanced modulation formats and hybrid Raman/EDFAs has been employed to transmit spectral efficiencies greater than 0.4 bit/s/Hz at 40 Gb/s over terrestrial single-fiber type spans. The first demonstration of 0.64 bit/s/Hz spectral efficiency used VSB filtering and a special wavelength allocation to transmit 128 channels over 300 km of fiber [41]. The 40 Gb/s NRZ channels were alternately spaced by 50 and 75 GHz, and in the receiver a given channel was selected with a narrow filter (3 dB bandwidth = 30 GHz) that was tuned from the channel's center frequency to isolate the sideband experiencing the smallest overlap with adjacent channels. Subsequent studies have discussed optimization of the VSB filtering technique [96, 97]. The C- and L-band channels were polarization-interleaved (adjacent wavelength channels have orthogonal polarizations) to reduce interchannel nonlinearities and linear crosstalk from adjacent channels and transmitted over three 100 km spans of Teralight[1] fiber. In a similar experiment with a recirculating loop, 125 NRZ channels at 42.7 Gb/s were transmitted over 12×100 km of Teralight fiber [98] with VSB filtering. The erbium-doped fiber length in the hybrid amplifier was adjusted to deliver higher gain to the lower-wavelength regions of the C- and L-bands to correct for the tilt from stimulated Raman scattering. In addition, the DCF modules were counterpumped with two wavelengths per band to mask the DCF loss and improve the noise figure of the EDFAs. After transmission, OSNRs of better than 22.1 dB (0.1 nm RBW) and BERs better than 1.4×10^{-4} were measured. The performance of the worst channels was attributed to optical fiber nonlinearities.

Spectral efficiencies of 0.8 bit/s/Hz for transmission over terrestrial single-fiber spans have been achieved using PSBT format or bandwidth-limited modulation formats. In one experiment, eighty 40 Gb/s channels were placed on the 50 GHz grid in the C-band by using polarization-interleaving (PI) and PSBT format, where the bandwidth-limiting occurs in the electrical domain [69]. Although PSBT has disadvantages including a reduced extinction ratio and increased sensitivity to OSNR, transmission over three 100 km spans of Teralight fiber was achieved. Distributed Raman amplification providing 14 dB gain in the transmission fiber and 6 dB gain in the DCF was a key to maintaining OSNRs greater than 29.5 dB (0.1 nm RBW). Polarization-demultiplexing was not used at the receiver; instead an interleaver separated the odd and even channels, followed by a tunable filter with 0.24 nm full-width–half-maximum (FWHM). The 10 Gb/s tributaries ($2^{11} - 1$ PRBS) had BERs less than 9×10^{-9}, attesting to the difficulties of implementing a duobinary format and the tight channel spacing. Another demonstration of 0.8 bit/s/Hz spectral efficiency employed vestigial sideband NRZ format with optimized filtering [99]. Eighty 42.7 Gb/s, C-band channels at 50 GHz spacing were transmitted over 900 km of fiber by using narrow periodic filters to clip the NRZ-modulated spectra and provide VSB filtering at the transmitter side, as well as at the receiver side. An extended transmission distance of 21×100 km with 0.8 bit/s/Hz spectral efficiency was achieved using BL-PSBT [64]. As shown in [71], a narrow optical filter after the low-pass filtered duobinary mod-

---

[1] Teralight is a trademark of Alcatel S.A.

ulator improves the tolerance to noise and also allows denser channel spacing. By employing narrow filters in the transmitter and receiver, 159 C- and L-band channels at 42.7 Gb/s were polarization interleaved and spaced by 50 GHz for 2100 km transmission [64]. The authors reported that the neighboring 50 GHz spaced, BL-PSBT channels caused less than a 0.3 dB penalty due to linear and nonlinear crosstalk, as compared to their previous experiment with bandwidth-limited NRZ [99]. In both experiments first- and second-order Raman amplification [100–102] were used in the transmission fiber with four first-order counterpumps and a 1346 nm Raman fiber laser counterpump acting as a secondary pump to the lower-wavelength 1427 and 1439 nm pumps. About 1 dB OSNR improvement was attributed to the second-order pump. After the 2100 km transmission the OSNRs of the 159 WDM channels were at least 19.4 dB (0.1 nm RBW), and the BERs were better than 7×10^{-4}. It was assumed that this represented a 1 dB Q-factor margin above the concatenated FEC [247, 239] + [255,247] threshold for BERs less than 10^{-13}.

Spectral efficiencies of 1.28 bit/s/Hz have been demonstrated using polarization-division-multiplexing (PDM) in combination with VSB filtering. A record capacity of 10.2 Tb/s (256 channels at 42.7 Gb/s) was transmitted over 100 km [103] and then over 300 km [104] of Teralight fiber. The 256 C- and L-band channels were composed of 128 wavelengths (alternately spaced by 50 and 75 GHz) with two orthogonally polarized channels per wavelength. PDM with a polarization-demultiplexer at the receiver suffers from low tolerance to PMD [105–107], and thus the transmission reach is limited. PMD causes the output polarization to vary with optical frequency, resulting in PMD-induced coherent crosstalk at the receiver [107]. Raman amplification was necessary even for the 100 km transmission distance to maintain sufficiently high OSNR to overcome the linear and PMD-induced crosstalk penalties. Counterpumped Raman amplification in the transmission fiber, as well as in the DCF modules for pre- and postdispersion compensation, resulted in OSNRs greater than 28.4 dB (0.1 nm RBW) for all channels and BERs less than 1×10^{-4} for all channels that were measured. To achieve the 300 km transmission distance, several improvements were made. Analysis of the wavelength allocation revealed that a shift of 40 GHz between the carriers of the orthogonally polarized channels would minimize the polarization crosstalk. With this shift, the channels were, in effect, polarization-interleaved, and the carriers of the adjacent orthogonally polarized channels were located as far as possible from the filter center. On the transmitter side, tight filtering was implemented after the modulators, where a narrow periodic filter was inserted to limit the channel bandwidth and reduce linear crosstalk, and VSB filtering was done at the receiver. In addition, second-order Raman amplification was used, along with the four first-order Raman pumps. Adding the 1346 nm counterpropagating pump for each of the three transmission spans resulted in a reported 1 dB improvement in the received OSNR. After 300 km transmission, the OSNRs were greater than 27 dB (0.1 nm RBW), and the 42.7 Gb/s uncorrected BERs were better than 1×10^{-4}. Note that PDM with polarization-demultiplexing will require fast polarization tracking in the receiver for installed systems, creating challenges both in terms of implementation and system costs.

19.3.2.2. Dispersion-Managed Fiber Spans

Section 19.2.2.2 and Fig. 19.4 introduced the concept of dispersion-managed spans, where both positive and negative dispersion fibers are incorporated into the transmission path. In addition to the advantages of excellent dispersion slope match across a broad bandwidth and elimination of lumped DCMs, dispersion-managed spans in the ABA configuration have important advantages for terrestrial transmission using hybrid Raman/EDFAs or all-Raman amplification. In this subsection we focus on hybrid amplification for 10 Gb/s and 40 Gb/s systems.

The first large-scale WDM experiment using terrestrial-length dispersion-managed spans was the transmission of 211 C- and L-band, 10.7 Gb/s channels over 7221 km of fiber [51]. The 80 km AB spans consisted of 57 km of positive dispersion fiber and 23 km of negative dispersion fiber (effective areas were not specified). For the same launch power per channel, 80 km spans reduce the path-average power and the number of repeaters when compared to 40 km repeater spacing, thus reducing the effect of fiber nonlinearities. However, Raman gain is necessary to maintain the OSNR. In the experiment, the hybrid Raman/EDFAs had a similar configuration to those shown in Fig. 19.9(a), except DCF was not necessary. Four counterpropagating pumps with wavelengths between 1430 and 1505 nm were coupled into the $-D$ fiber, providing Raman gain for the 64.4 nm signal band. Due to the high gain in the small A_{eff}, $-D$ fiber, a total of only 180 mW was required to obtain an average Raman gain of 7 dB, resulting in an OSNR improvement of about 3.5 dB. The 37.5 GHz spaced, RZ channels were polarization-interleaved, and the residual dispersion of the spans was -1.5 ps/nm/km to reduce CPM-induced waveform distortion, a particularly difficult impairment for 10 Gb/s. After 7221 km transmission, uncorrected BERs for all channels were between 3×10^{-4} and 8×10^{-6}, and the authors identified OSNR, and not nonlinearities, as the transmission limitation.

Transmission of 10 Gb/s channels over 100 km ABA spans has been demonstrated using only Raman amplification to compensate for the fiber loss [52] and also with hybrid Raman/EDFAs [53]. The latter experiment showed transmission of eighty 10.7 Gb/s, 50 GHz spaced, C-band channels over 5200 km using the same hybrid Raman/EDFAs as in Fig. 19.8, but without the DCF. The ABA spans consisted of 38 km of UltraWave[TM] SLA, 35 km of UltraWave[TM] IDF, and 27 km of SLA. This scheme distributes the gain more evenly in the span compared to an AB fiber map, yielding better noise performance of the system. Due to the slope match of the SLA and IDF fiber, the accumulated residual dispersion difference across the C-band channels was less than 260 ps/nm after transmission over 5200 km. No pre- and no postdispersion compensation were used, and no per-channel dispersion trimming was required at the receiver. All eighty channels had uncorrected BERs less than 1×10^{-4}, and with the FEC enabled, no errors were detected when counting more than 10^{11} bits.

Transmission simulations have predicted that due to their lower span noise figure (defined as the combined noise figure of the entire span consisting of transmission fiber, discrete and/or distributed amplification, and passive components), ABA dispersion-managed spans offer a 2 to 3 dB increase in transmission reach over single-fiber type spans. This prediction has been roughly verified with 10 and 40 Gb/s experiments, where similar hybrid Raman/EDFAs were used in all cases. The experiment reported in [108] where 80×10.66 Gb/s NRZ channels were transmitted over 3200 km using 100 km NZDF spans was repeated using the same setup and hybrid amplifiers, but with the NZDF spans replaced by ABA dispersion-managed fiber spans. With the NRZ modulation format, the transmission distance was increased to 4800 km, and for 50% duty-cycle RZ, a distance of 5200 km was achieved [53], showing about 2 dB reach increase over NZDF.

Transmission of forty 42.7 Gb/s C-band channels over 2000 km of fiber [54] and, later, 2400 km of fiber [57] was demonstrated using 100 km dispersion-managed spans, consisting of the fiber lengths shown in Fig. 19.4(a), and hybrid Raman/EDFAs. The calculated signal power evolution for Raman gain of 6 dB below transparency is shown in Fig. 19.11(a), where it is apparent that the signal power is lowest in the small A_{eff} IDF fiber, thus minimizing nonlinearities [109]. Figure 19.11(b) shows the calculated span noise figure as a function of the starting position of the IDF fiber in the ABA map. A more than 2 dB improvement in the span noise figure is produced by placing the IDF near the middle of the span. The minimum noise figure appears to occur over more than a 10 km range for the IDF starting position, thus ensuring that the exact placement of the IDF is not critical. In the experiment of [57], the forty C-band channels were modulated with the CSRZ format, the average on-off Raman gain was 15.5 dB, and the launched power per channel was -1 dBm. BERs for all 42.7 Gb/s channels were better than 1×10^{-4}, which would correspond to a BER performance well below 10^{-12} with standard 7% overhead Reed–Solomon FEC. This experiment provides some verification of the 2 to 3 dB advantage of dispersion-managed spans at 40 Gb/s if its 2400 km reach (for the C-band) is compared to the 1200 km reach for the C- and L-bands over NZDF [42].

19.3.3. 40 Gb/s Terrestrial Transmission Using All-Raman Amplification

Before the early 1990s when EDFAs emerged as practical optical amplifiers for fiber communication systems, distributed Raman amplification was investigated for soliton transmission in simulations [110] and in experiments [111]. As interest in Raman amplification has grown, due to its improved noise performance, for 40 Gb/s terrestrial and ultra-long-haul transmission, a number of experiments have been reported demonstrating the use of Raman amplification only, or all-Raman amplification, to compensate for the transmission fiber and DCF loss. Many of these experiments have used both co- and counterpumping. Critical issues of copumping including pump–signal crosstalk, signal–pump–signal crosstalk, polarization-dependence of the gain, and pump laser requirements have been covered in Chapters 5, 8, and 9. In addition, design of pumped DCFs for discrete Raman amplifiers has been considered in Chapter 6. This section describes several 40 Gb/s WDM transmission experiments using all-Raman amplification. The focus is primarily on terrestrial span lengths for both single-fiber-type spans as well as dispersion-managed spans.

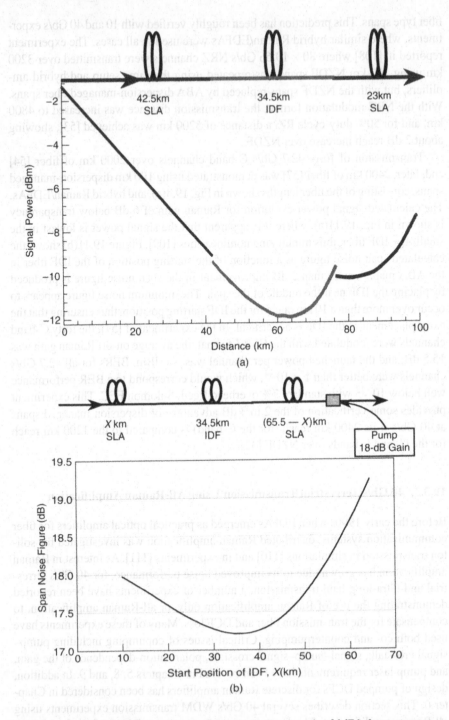

Fig. 19.11. (a) The calculated signal power evolution for Raman gain of 6 dB below transparency for the experiment reported in [54]. (b) The calculated span noise figure as a function of the starting position of the IDF fiber in the ABA map for the same experiment. (Adapted and reprinted from [109] with permission from the Optical Society of America.)

19.3.3.1. Single-Fiber-Type Spans

Several 10 Gb/s WDM experiments using all-Raman amplified spans consisting of a single fiber type were reported prior to work at 40 Gb/s. For submarine applications, Nissov et al. [112] demonstrated the first transoceanic-length transmission based on Raman gain with 100 Gb/s (10 × 10 Gb/s) capacity over 7200 km of fiber. The 45 km dispersion-shifted fiber spans were counterpumped to transparency. Raman amplification provided a 2 dB noise performance improvement compared to 980 nm pumped EDFAs, assuming equal path-averaged powers. In this experiment, as in most of the experiments described below, EDFAs were included at the end of the loop to compensate for the loss of the gain equalizers and loop-specific elements. Copumping was first reported in an experiment demonstrating 1 Tb/s transmission over four 80 km spans of dispersion-shifted fiber [113]. With 4 dB of forward gain and 18 dB of backward gain at 1550 nm, the launch power for the 100, 25 GHz spaced channels could be kept low enough (−18 dBm/ch) that four-wave-mixing was not observed, even in the vicinity of the average zero dispersion wavelength.

All-Raman amplification of a single, 53 nm extended L-band was shown first for transmission of 128 × 10 Gb/s over 4000 km of TrueWave RS fiber [114]. Figure 19.12 shows a typical pumping configuration. The 100 km spans were co- and counterpumped for a total of 20 dB on-off gain, and the slope-matched DCF module following each span was counterpumped for 12 dB on-off gain. The 10 Gb/s dispersion-managed solitons had BERs better than 10^{-6} after transmission with no per-channel postcompensation.

At 40 Gb/s, transmission of 3.2 Tb/s capacity over twenty 100 km spans of True-Wave *REACH* fiber was demonstrated using distributed Raman amplification only to compensate for the transmission fiber and DCF loss [12]. The location of the transmission fiber's zero dispersion wavelength at <1400 nm enabled co- and counterpumping for flat Raman gain and noise figure across the entire C+L-bands by eliminating pump–pump four-wave-mixing [34] and pump–signal four-wave-mixing [35]. The low dispersion slope of the TrueWave *REACH* fiber also enabled C+L-band dispersion and dispersion slope compensation of each span with a single Raman-pumped DCF module, as shown in Fig. 19.3. The on-off Raman gains from the three copump wavelengths, four counterpump wavelengths, and five DCF pump wavelengths are

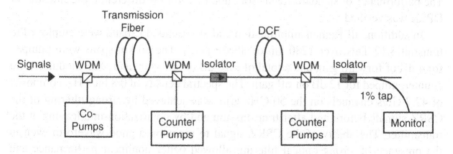

Fig. 19.12. Typical pumping configuration for all-Raman amplification of a single-fiber type transmission span with a DCF module. WDM: pump–signal combiner or circulator.

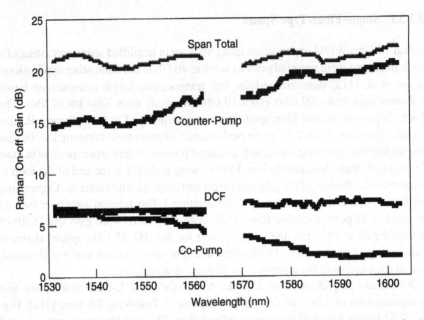

Fig. 19.13. The on-off gains for a Raman-pumped 100 km NZDF span and DCF module for the experiment reporting 3.2 Tb/s transmission over twenty 100 km NZDF spans [12]. (Reprinted with permission from the Optical Society of America.)

shown in Fig. 19.13. Although in the experiment the transmitters, receivers, and dynamic gain equalizers required splitting into separate C- and L-wavelength bands, the continuous Raman gain and single DCF module showed the potential for transmitting a wide, continuous single band, and eliminating the cost and loss of band splitters.

Using a similar extended L-band as in Grosz et al. [114], a record-breaking 2.56 Tb/s (64 channels at 42.7 Gb/s) was transmitted over 4000 km of TrueWave RS fiber [11]. This experiment was the first to demonstrate a 40 Gb/s RZ-DPSK modulation format in transmission and used a balanced receiver, resulting in a ~3 dB improvement in receiver sensitivity. With only counterpumping of the 100 km spans and DCF modules, after transmission the OSNRs of the 64 channels were 15.2 dB or higher, and with the balanced receiver the uncorrected BERs were 1.2×10^{-4} or better. The performance of standard Reed–Solomon FEC for the different noise statistics of DPSK was verified.

In addition, all-Raman amplification and an extended L-band were employed to transmit 5.12 Tb/s over 1280 km of SSMF [115]. The 80 km spans were pumped for 4 dB of forward gain and a total of ~16 dB on-off gain, and DCF modules were counterpumped for 12 dB on-off gain. The spectral density of 0.8 bit/s/Hz, composed of 42.7 Gb/s channels on the 50 GHz grid, was achieved by strong filtering of the CSRZ signals before and after transmission and by polarization-interleaving at the transmitter. The ability of the CSRZ signal to maintain a good waveform even in the presence of strong optical filtering allowed better nonlinear performance and

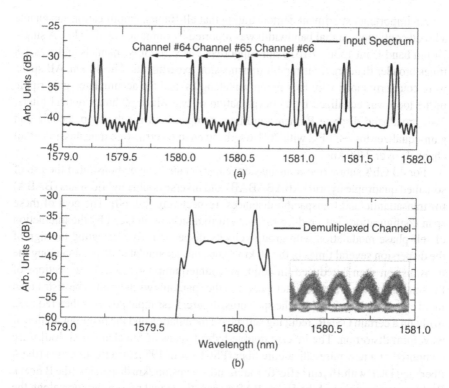

Fig. 19.14. (a) A small section of the input spectrum and (b) the spectrum of a demultiplexed channel (in back-to-back) along with its eye diagram for the experiment demonstrating 40 Gb/s CSRZ channels at 50 GHz spacing [115]. (Courtesy of Diego Grosz, Lucent Technologies.)

received sensitivity than other modulation formats when placed on the 50 GHz grid. Figure 19.14 shows part of the input spectrum and the spectrum of a demultiplexed channel along with its eye diagram.

19.3.3.2. Dispersion-managed spans

The longest WDM transmission distances at 10 and 40 Gb/s have been achieved using a combination of dispersion-managed spans and all-Raman amplification. Knudsen et al. [52] first investigated the possibility of extending the dispersion-managed span length to 100 km with the transmission of forty-two 10 Gb/s channels over 4000 km. The ABA spans were counterpumped to transparency by a single-wavelength Raman pump, and BERs less than 10^{-9} were achieved without FEC and without per-channel pre- or postcompensation. The gain flatness achievable with a single pump wavelength, and not the dispersion slope match of the SLA and IDF fibers, limited the bandwidth of the system to the forty-two channels. All-Raman amplification over 80 km, ABA dispersion-managed spans has also been reported as part of a 10 Gb/s network demonstration with optical add-drop multiplexers [116].

An important experiment demonstrating that all-Raman amplification can enable a broad continuous signal bandwidth was reported by Shimojoh et al. [117]. A single 74 nm band from 1536 to 1610 nm supported 240 × 10 Gb/s channels and was 1.8 times broader than previous terabit transmission experiments. The 40 km, AB spans were counterpumped with four pump wavelengths, and the accumulated gain ripple in the loop was equalized by a concatenation of two Mach–Zehnder optical filters. Two broadband discrete Raman amplifiers compensated for the loop-specific and gain-equalizer losses. FEC with 20% overhead enabled error-free transmission of all channels over 7400 km.

For 40 Gb/s submarine transmission, experiments have demonstrated the use of so-called quadruple hybrid (ABABABAB) and inverse double hybrid spans (BABA) for transatlantic and transpacific distances, respectively [50, 60]. The goal of these span configurations is to reduce the waveform distortion induced by the interaction of self-phase modulation with group-velocity dispersion. By alternating the sign of the dispersion several times in the 50 km spans, the accumulated dispersion along the transmission span is reduced. In addition, counterpumping with several sections of the small effective area fiber interspersed in the spans allows the gain to be distributed more evenly and can improve the span noise figure. Less input power is then required to obtain a certain OSNR, reducing self-phase modulation and the resulting nonlinear waveform distortion. The inverse double-hybrid spans of Sugahara et al. [60] were composed of a new pure-silica-core fiber (PSCF) with 195 μm^2 effective area (the A fiber) and DCF with 19 μm^2 effective area and −63 ps/nm/km dispersion (the B fiber). With counterpumping of the 50 km BABA span, the signal power variation along the span was only 3.5 dB, compared to 5 dB in the quadruple hybrid span with the new PSCF. Simulations showed that the eye-opening penalty after 9000 km transmission was smallest for the inverse double hybrid span, despite the fact that the signals were launched into the DCF and the accumulated dispersion was higher in the BABA spans. As mentioned previously, it is not yet clear how much accumulated dispersion a 40 Gb/s system can tolerate. This value could depend on the modulation format and other details of the system. In fact, transmission over transoceanic distances (6200 km) has been achieved using 51 km AB spans [118], where the accumulated dispersion within each span was two to four times higher than that in the inverse double or quadruple hybrid spans.

All-Raman amplification and dispersion-managed spans have been combined in several experiments to meet the stringent requirements of ultra-long-haul 40 Gb/s transmission with terrestrial span lengths. Forty C-band CSRZ channels were transmitted over 3600 km using 100 km UltraWave fiber spans with the properties listed in Table 19.2 [55]. Each symmetrical ABA span was co- and counterpumped to 3.5 dB above transparency to accommodate the WDM coupler/circulator and fixed gain-flattening filter inserted after the span. A semiperiodical dispersion map was used, where each span was undercompensated and an extra spool of IDF was inserted after four spans, resulting in −100 ps/nm accumulated dispersion per loop round-trip. A later experiment by the same authors extended the transmission distance to 5200 km using the same 100 km dispersion-managed spans [119]. Enhanced FEC increased the

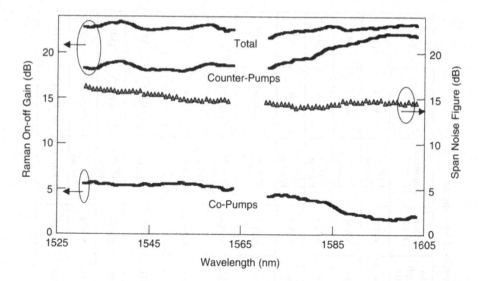

Fig. 19.15. The on-off Raman gains from the copumps and counterpumps, as well as the total on-off gain and span noise figure for the experiment demonstrating 3.2 Tb/s transmission over 5200 km of dispersion-managed fiber [65].

coding gain to 7.8 dB at 10^{-12} BER with the same 7% overhead, assuming Gaussian noise statistics. Back-to-back measurements showed that EFEC reduced the required OSNR by more than 10 dB at 10^{-12} BER.

Ultra-long-haul 40 Gb/s terrestrial transmission over 5200 km has also been achieved with RZ-DPSK modulation format and dispersion-managed spans [65]. The 3.2 Tb/s total capacity consisting of 80 C- and L-band channels was transmitted over symmetrical 100 km UltraWave fiber spans, using co- and counterpumping. Figure 19.15 shows the Raman on-off gains from the copumps and counterpumps, as well as the total on-off gain and span noise figure. The higher copumped gain in the lower C-band helped to partially offset the stimulated Raman scattering-induced power transfer to the L-band. This experiment showed the excellent slope-match achievable with these dispersion-managed spans, as the residual dispersion variation was only ±30 ps/nm from 1530 to 1603 nm after the 400 km loop. The RZ-DPSK format and balanced receiver enabled the 42.7 Gb/s channels to have BERs better than 1×10^{-4} after 5200 km transmission. It is clear that by applying stronger FEC and improving the transmitter waveforms and receiver sensitivities, longer 40 Gb/s transmission distances will be possible.

Table 19.3 summarizes the transmission experiments discussed in Section 19.3, both for hybrid Raman and all-Raman amplification, along with several other important 40 Gb/s long-haul experiments.

Table 19.3. Selected WDM 40 Gb/s Raman-Amplified Transmission Experiments

Date	Number of Ch.	Capacity (Tb/s)	Distance (km)	Capacity-Distance (Pb.km/s)	Ch. Spacing (GHz)	Density (bit/s/Hz)	Format	Span Length (km)	Fiber Type	Amplifier type	BER	Ref
09/99	40	1.6	400	0.64	100	0.4	NRZ	100	NZDF	EDFA/Raman	9e-11	[10]
03/00	82	3.28	300	0.98	100	0.4	NRZ	100	NZDF	EDFA/Raman	8e-11	[40]
09/00	128	5.12	300	1.54	50/75	0.64	NRZ/VSB	100	NZDF	EDFA/Raman	1e-9	[41]
09/00	32	1.28	1000	1.28	100	0.4	CSRZ	160	SSMF	EDFA/Raman	7e-4	[95]
03/01	256	10.24	100	1.02	50/75	1.28	NRZ/VSB/PDM	100	NZDF	EDFA/Raman	1e-4(FEC)	[103]
03/01	77	3.08	1200	3.7	100	0.4	NRZ	100	NZDF	EDFA/Raman	1e-4(FEC)	[42]
03/01	273	10.92	117	1.28	50	0.8	NRZ/PI	58	DMFS	TDFAs/Raman + EDFAs	1e-9	[120]
10/01	125	5.0	1200	6.0	50/75	0.64	NRZ/VSB/PI	100	NZDF	EDFA/Raman	1e-4(FEC)	[98]
10/01	40	1.6	2000	3.2	100	0.4	CSRZ	100	DMFS	EDFA/Raman	1e-4(FEC)	[54]
10/01	80	3.2	300	0.96	50	0.8	PSBT/PI	100	NZDF	EDFA/Raman	9e-9	[69]
03/02	64	2.5	4000	10	100	0.4	RZ-DPSK	100	NZDF	All Raman	1e-4(FEC)	[11]

Table 19.3. Continued

Date	Number of Ch.	Capacity (Tb/s)	Distance (km)	Capacity-Distance (Pb.km/s)	Ch. Spacing (GHz)	Density (bit/s/Hz)	Format	Span Length (km)	Fiber Type	Amplifier type	BER	Ref
03/02	256	10.2	300	3.06	50/75	1.28	NRZ/VSB/PDM	100	NZDF	EDFA/Raman	1e-4(FEC)	[104]
03/02	32	1.28	6050	7.74	100	0.4	CSRZ/PI	52	DMFS	All Raman	1e-4(FEC)	[50]
03/02	40	1.6	3600	5.76	100	0.4	CSRZ	100	DMFS	All Raman	1e-4(FEC)	[55]
03/02	80	3.2	2000	6.4	100	0.4	CSRZ	100	NZDF	All Raman	1e-4(FEC)	[12]
06/02	40	1.6	2400	3.84	100	0.4	CSRZ	100	DMFS	EDFA/Raman	1e-4(FEC)	[57]
07/02	32	1.28	9000	11.52	100	0.4	CSRZ/PI	50	DMFS	All Raman	4e-4	[60]
07/02	80	3.2	900	2.88	50	0.8	PSBT/PDM	100	NZDF	EDFA/Raman	1e-4(FEC)	[99]
09/02	159	6.36	2100	13.36	50	0.8	BL-PSBT/PI	100	NZDF	EDFA/Raman	7e-4(FEC)	[64]
09/02	80	3.2	5200	16.64	100	0.4	RZ-DPSK	100	DMFS	All-Raman	1e-4(FEC)	[65]
09/02	40	1.6	5200	8.32	100	0.4	CSRZ	100	DMFS	All Raman	1.5e-3 (FEC)	[119]
09/02	128	5.12	1280	6.55	50	0.8	CSRZ/PI	80	SSMF	All Raman	1e-4(FEC)	[115]

VSB: vestigial sideband; RZ-DPSK: return-to-zero differential-phase- shift-keyed; PSBT: Phased-shaped binary transmission; PDM: polarization division multiplexing; PI: polarization interleaving; NZDF: non-zero dispersion fiber; DMFS: dispersion-managed fiber span; SSMF: standard single-mode fiber

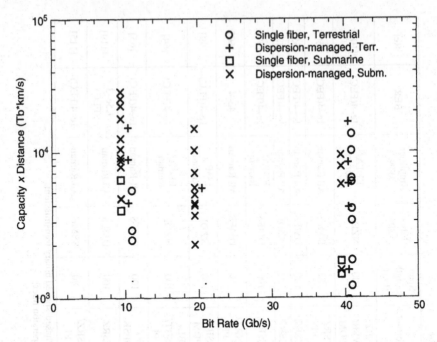

Fig. 19.16. Summary of recent high-capacity ETDM transmission experiments (1999–2002) plotted as the product of total capacity and distance versus the bit rate. Results are divided into terrestrial span lengths (>70 km) and submarine span lengths (<70 km). Also indicated is whether the transmission spans were of a single-fiber type with lumped dispersion-compensating modules at the optical amplifiers or were dispersion-managed spans.

19.4. Conclusion

Figure 19.16 shows a plot of recent laboratory transmission experiments at 10, 20, and 40 Gb/s using the product of total capacity and transmission distance as the performance metric. The results are divided into terrestrial span lengths (>70 km) and submarine span lengths (<70 km). The results are also grouped by whether the transmission spans are of a single-fiber type with lumped dispersion-compensating modules at the optical amplifiers or are dispersion-managed spans. It is clear from the plot that submarine experiments have concentrated primarily on dispersion-managed spans, whereas recent experiments have begun to assess the benefit of dispersion-managed spans for terrestrial systems. The fact that submarine experiments at 10 Gb/s lead those at 40 Gb/s may, in part, reflect the fact that work at 40 Gb/s for transoceanic transmission has begun more recently than work at 40 Gb/s for terrestrial transmission. The record of 16.6 petabit km/s for 40 Gb/s terrestrial dispersion-managed spans used the RZ-DPSK format and a balanced receiver. It can be expected that this record will be broken, perhaps in the very near future, by utilizing new modulation formats or fibers and/or improvements in FEC.

In summary, high-capacity 40 Gb/s transmission systems offer scalable solutions for future traffic growth in the core network. Raman amplification is likely to be a key driver to ease the noise performance and increase the available bandwidth for 40 Gb/s DWDM systems. New fiber technologies provide high system performance and enable a simple and cost-effective dispersion compensation scheme. More system margin can also be expected from high-coding-gain FEC. Optimized modulation formats and high-speed optoelectronics will make practical deployment of 40 Gb/s DWDM systems possible, facilitating multiple terabit transmission over Mm distance at low cost per bit per kilometer.

Appendix A19. List of Acronyms

Acronym	Definition	Section
A_{eff}	effective area	19.2.2.1
AOM	acousto-optic modulator	19.3.1
ASE	amplified spontaneous emission	19.2.1.1
ASK	amplitude shift key(-ed) (-ing)	19.2.2.3
BER	bit error rate	19.2.1.1
BERT	bit error rate test set	19.3.1
BL-PSBT	band-limited phase-shaped binary transmission	19.2.2.3
C-band	conventional band of EDFA (~1530 to 1565 nm)	19.2.2.2
CPM	cross-phase modulation	19.2.1.1
CSRZ	carrier-suppressed return-to-zero	19.2.2.3
D	group-velocity dispersion	19.2.2.2
DCF	dispersion-compensating fiber	19.2.1.2
DCM	dispersion-compensating module	19.2.1.3
DFB	distributed feedback laser	19.3.1
DGD	differential group delay	19.2.1.3
DPSK	differential phase-shift-key(-ed) (-ing)	19.2.2.3
DWDM	dense wavelength-division-multiplex(-ed) (-ing)	19.1.
EDFA	erbium-doped fiber amplifier	19.2.1.1
EPSBT	enhanced phase-shaped binary transmission	19.2.2.3
ETDM	electrical time-division-multiplex (-ed) (-ing)	19.3.1
FEC	forward error correction	19.2.2.4
FWHM	full-width at half-maximum	19.3.2.1
FWM	four-wave-mixing	19.2.1.1
GE	dynamic gain equalizer	19.3.1
GFF	gain-flattening filter	19.3.2.1
HS-DCF	high-slope dispersion-compensating fiber	19.3.2.1
IC	integrated circuit	19.2.2.4
IDF	inverse dispersion fiber	19.2.2.2
ITU	International Telecommunications Union	19.2.2.4
L-band	long-wavelength band of EDFA (~1570 to 1605 nm)	19.1.
LDV	link design value	19.2.1.3
MPI	multiple-path interference	19.2.2.1
MZM	Mach–Zehnder modulator	19.3.1
NDSF	nondispersion shifted fiber	19.3.2.1
NECG	net effective coding gain	19.2.2.4
NRZ	nonreturn-to-zero	19.2.1.2
NZDF	nonzero dispersion fiber	19.2.1.2
OI	optical interleaver	19.3.1
OSNR	optical signal-to-noise ratio	19.2.1.1

Acronym	Definition	Section
OTDM	optical time-division-multiplex(-ed) (-ing)	19.1.
PDL	polarization-dependent loss	19.3.1
PDM	polarization-division-multiplex(-ed) (-ing)	19.3.2.1
PI	polarization-interleav(-ed) (-ing)	19.3.2.1
PMD	polarization mode dispersion	19.2.1.3
PRBS	pseudorandom bit sequence	19.3.1
PSBT	phase-shaped binary transmission	19.2.2.3
PSCF	pure silica-core fiber	19.3.3.2
PSK	phase-shift-key(-ed) (-ing)	19.2.2.3
PSP	principal state of polarization	19.2.1.3
Q	Q-factor	19.2.2.4
RBW	resolution bandwidth	19.2.1.1
RDS	relative dispersion slope	19.2.2.2
RS	Reed–Solomon	19.2.2.4
RZ	return-to-zero	19.2.2.3
RZ-DPSK	return-to-zero differential-phase-shift-key(-ed) (-ing)	19.2.2.3
SLA	super large area fiber	19.2.2.2
SPM	self-phase modulation	19.2.1.1
SRS	stimulated Raman scattering	19.2.1.1
SSMF	standard single-mode fiber	19.2.1.2
SW	optical switch	19.3.1
TDM	time-division-multiplex(-ed) (-ing)	19.1.
VSB	vestigial sideband	19.2.2.3
WDM	wavelength-division-multiplex(-ed) (-ing)	19.2.2.3.
WGR	waveguide grating router	19.3.1

References

[1] RHK Telecommunications Industry Analysis, Internet traffic soars, but revenues glide, May 2002.

[2] K. Coffman and A. Odlyzko, Growth of the Internet. In *Optical Fiber Telecommunications IVB*. I. P. Kaminow and T. Li, eds., San Diego: Academic, 17–56, 2002.

[3] C. Videcrantz, S. Danielsen, B. Mikkelsen, and P. Mamyshev, Advances in 40G long haul WDM. In *Proceedings of the National Fiber Optics Engineers Conference 2002*, Dallas, TX, 325–332, 2002.

[4] A. Umbach, T. Engel, H. G. Bach, S. V. Waasen, E. Droge, A. Strittmatter, W. Ebert, W. Passenberg, R. Steingruber, W. Schlaak, G. G. Mekonnen, G. Unterborsch, and D. Bimberg, Technology of InP-based 1.55μm ultrafast OEMMICs: 40Gbit/s broadband and 38/60GHz narrow-band photoreceivers, *IEEE J. Quantum Electron.*, 35:7, 1024–1031, 1999.

[5] M. Yoneyama, Y. Miyamoto, T. Otsuji, H. Toba, Y. Yamane, T. Ishibashi, and H. Miyazawa, Fully electrical 40-Gb/s TDM system prototype based on InP HEMT digital IC technologies, *IEEE J. Lightwave Technol.*, 18:1, 34–42, 2000.

[6] Y. K. Chen, Y. Baeyens, C.-T. Liu, R. Kopf, C. Chen, Y. Yang, J. Frackoviak, A. Tate, A. Leven, P. Paschke, M. Berger, J. Weiner, K. Tu, G. Georgiou, P. Roux, V. Houstma, and U. Koc, High speed electronics for lightwave communications. In *Proceedings of the Optical Fiber Communications Conference 2002*, Anaheim, CA, WN2, 2002.

[7] M. Heins, J. M. Carroll, M.-Y. Kao, C. F. Steinbeiser, T. R. Landon, and C. F. Campbell, An ultra-wideband GaAs pHEMT driver amplifier for fiber optic communications at 40 Gb/s and beyond. In *Proceedings of the Optical Fiber Communications Conference 2002*, Anaheim, CA, WN3, 2002.

[8] B. L. Kasper, O. Mizuhara, and Y.-K. Chen, High bit-rate receivers, transmitters, and electronics. In *Optical Fiber Telecommunications IVA*. I. P. Kaminow and T. Li, eds., San Diego: Academic, 784–851, 2002.

[9] J. L. Zyskind, J. A. Nagel, and H. D. Kidorf, Erbium-doped fiber amplifiers for optical communications. In *Optical Fiber Telecommunications IIIB*. I. P. Kaminow and T. L. Koch, eds., San Diego: Academic, 13–63, 1997.

[10] T. Nielsen, A. J. Stentz, P. B. Hansen, Z. J. Chen, D. S. Vengsarkar, T. A. Strasser, K. Rottwitt, J. H. Park, S. Stulz, S. Cabot, K. S. Feder, P. S. Westbrook, S. G. Kosinski, 1.6 Tb/s (40 × 40 Gb/s) transmission over 4 × 100km non-zero dispersion fiber using hybrid Raman/erbium-doped amplifiers. In *Proceedings of the European Conference on Optical Communication 1999*, Nice, France, PD2-2, 1999.

[11] A. H. Gnauck, G. Raybon, S. Chandrasekhar, J. Leuthold, C. Doerr, L. Stulz, A. Agarwal, S. Banerjee, D. Grosz, S. Hunsche, A. Kung, A. Marhelyuk, D. Maywar, M. Movassaghi, X. Liu, C. Xu, X. Wei, and D. M. Gill, 2.5 Tb/s (64 × 42.7 Gb/s) transmission over 40 × 100 km NZDSF using RZ-DPSK format and all-Raman-amplified spans. In *Proceedings of the Optical Fiber Communications Conference 2002*, Anaheim, CA, postdeadline paper FC2, 2002.

[12] B. Zhu, L. Leng, L. E. Nelson, L. Gruner-Nielsen, Y. Qian, J. Bromage, S. Stulz, S. Kado, Y. Emori, S. Namiki, P. Gaarde, A. Judy, B. Palsdottir, and R. L. Lingle, Jr., 3.2 Tb/s (80 × 42.7Gb/s) transmission over 20 × 100 km of non-zero dispersion fiber with simultaneous C+L-band dispersion compensation. In *Proceedings of the Optical Fiber Communications Conference 2002*, Anaheim, CA, postdeadline paper FC8, 2002.

[13] A. H. Gnauck and R. M. Jopson, Dispersion compensation for optical fiber systems. In *Optical Fiber Telecommunications IIIA*. I. P. Kaminow and T. L. Koch, eds., San Diego: Academic, 162–190, 1997.

[14] T. Kato, Y. Koyano, and M. Nishimura, Temperature dependence of chromatic dispersion in various types of optical fibers. In *Proceedings of the Optical Fiber Communications Conference 2000*, Baltimore, MD, TuG7, 2000.

[15] H. Kogelnik, L. E. Nelson, and R. M. Jopson, Polarization mode dispersion. In *Optical Fiber Telecommunications IVB*. I. P. Kaminow and T. Li, eds., San Diego: Academic, 725–861, 2002.

[16] F. Curti, B. Daino, G. De Marchis, and F. Matera, Statistical treatment of the evolution of the principal states of polarization in single-mode fibers, *IEEE J. Lightwave Technol.*, LT-8:1162–1166, 1990.

[17] H. Taga, M. Suzuki, and Y. Namihira, Polarization mode dispersion tolerance of 10 Gbit/s NRZ and RZ optical signal, *Electron. Lett.*, 34:2098–2100, 1998.

[18] R. M. Jopson, L. E. Nelson, G. J. Pendock, and A. H. Gnauck, Polarization-mode dispersion impairment in return-to-zero and non-return-to-zero systems. In *Proceedings of the Optical Fiber Communication Conference 1999*, San Diego, WE3, 1999.

[19] H. Sunnerud, M. Karlsson, and P. A. Andrekson, A comparison between NRZ and RZ data formats with respect to PMD-induced system degradation. In *Proceedings of the Optical Fiber Communications Conference 2001*, Anaheim, CA, WT3, 2001.

[20] P. J. Winzer, H. Kogelnik, C. H. Kim, H. Kim, R. M. Jopson, and L. E. Nelson, Effect of receiver design on PMD outage for RZ and NRZ. In *Proceedings of the Optical Fiber Communication Conference 2002*, Anaheim, CA, TuI1, 2002.

[21] C. D. Poole, R. W. Tkach, A. R. Chraplyvy, and D. A. Fishman, Fading in lightwave systems due to polarization-mode dispersion, *IEEE Photon. Technol. Lett.*, 3:68–70, 1991.

[22] P. Noutsios and S. Poirier, PMD assessment of installed fiber plant for 40 Gbit/s transmission. In *Proceedings of the National Fiber Optics Engineers Conference 2001*, Baltimore, MD, 1342–1347, 2001.

[23] K. W. Jackson, A. Bhat, V. Chandraiah, R. E. Fangmann, L. R. Dunn, A. F. Judy, J. Kim, H. Ly, and S. C. Mettler, Polarization mode dispersion in high fiber count outside plant cable. In *Proceedings of the National Fiber Optics Engineers Conference 2001*, Baltimore, MD: 1333–1341, 2001.

[24] R. Caponi, B. Riposati, A. Rossaro, and M. Schiano, WDM design issues with highly correlated PMD spectra of buried optical cables. In *Proceedings of the Optical Fiber Communications Conference 2002*, Anaheim, CA, ThI5, 2002.

[25] M. Brodsky, P. D. Magill, and N. J. Frigo, Evidence of parametric dependence of PMD on temperature in installed 0.05 ps/km$^{1/2}$ fiber. In *Proceedings of the European Conference on Optical Communication 2002*, Copenhagen, 9.3.2, 2002.

[26] P. B. Hansen, L. Eskildsen, A. J. Stentz, T. A. Strasser, J. Judkins, J. J. DeMarco, R. Pedrazzani, and D. J. Digiovanni, Rayleigh scattering limitation in distributed Raman pre-amplifiers, *IEEE Photon. Technol. Lett.*, 10:1, 159–161, 1998.

[27] W. P. Urquhart and P. J. Laybourn, Effective core area for stimulated Raman scattering in single-mode optical fibers, *Proc. Inst. Elect. Eng.*, 132:201–204, 1985.

[28] J. Bromage, K. Rottwitt, and M. E. Lines, A method to predict the Raman gain spectra of germanosilicate fibers with arbitrary index profiles, *IEEE Photon. Technol. Lett.*, 14:24–26, 2002.

[29] K. Rottwitt, J. Bromage, and L. Leng, Scaling the Raman gain coefficient of optical fibers. In *Proceedings of the European Conference on Optical Communication 2002*, Copenhagen, symposium paper 3.03, 2002.

[30] A. H. Hartog and M. P. Gold, On the theory of backscattering in signal-mode optical fibers, *J. Lightwave Technol.*, LT-2:2, 76–82, 1984.

[31] A. Chraplyvy, Optical power limits in multichannel wavelength-division-multiplexed systems due to stimulated Raman scattering, *Electron. Lett.*, 20:58, 1984.

[32] F. Forghieri, R. W. Tkach, and A. R. Chraplyvy, Effect of modulation statistics on Raman crosstalk, *IEEE Photon. Technol. Lett.*, 7:1, 101–103, 1995.

[33] J. Bromage, H. Thiele, and L. E. Nelson, Raman amplification in the S-band. In *Proceedings of the Optical Fiber Communications Conference 2002*, Anaheim, CA, ThB3, 2002.

[34] R. E. Neuhauser, P. M. Krummrich, H. Bock, and C. Glingener, Impact of nonlinear pump interactions on broadband distributed Raman amplification. In *Proceedings of the Optical Fiber Communications Conference 2001*, Anaheim, CA, MA4, 2001.

[35] J. Bromage, P. J. Winzer, L. E. Nelson, and C. J. McKinstrie, Raman-enhanced pump-signal four-wave mixing in bidirectionally pumped Raman amplifiers. In *Proceedings of Optical Amplifiers and their Applications 2002*, Vancouver, OWA5, 2002.

[36] L. Grüner-Nielsen, Y. Qian, B. Palsdottir, P. B. Gaarde, S. Dyrbol, T. Veng, Yifei Qian, R. Boncek, and R. Lingle, Module for simultananeous C+L-band dispersion compensation and Raman amplification. In *Proceedings of the Optical Fiber Communications Conference 2002*, Anaheim, CA, TuJ6, 2002.

[37] J. Zhang, V. Dominic, M. Missey, S. Sanders, and D. Mehuys, Dependence of Raman polarization dependent gain on pump degree of polarization at high gain levels. In *Proceedings of Optical Amplifiers and their Applications 2000*, Quebec City, OMB4, 2000.

[38] H. H. Kee, C. R. S. Fludger, and V. Handerek, Statistical properties of polarization dependent gain in fibre Raman amplifiers. In *Proceedings of the Optical Fiber Communications Conference 2002*, Anaheim, CA, WB2, 2002.

[39] R.-J. Essiambre, Effects of Raman noise and double Rayleigh backscattering on bidirectionally Raman-pumped systems at constant fiber nonlinearity. In *Proceedings of the European Conference on Optical Communication 2001*, Amsterdam, Tu.A.1.1, 2001.

[40] T. Nielsen, A. Stentz, K. Rottwitt, D. Vengsarkar, Z. Chen, P. Hansen, J. Park, K Feder, T. Strasser, S. Cabot, S. Stulz, D. Peckham, L. Hsu, C. Kan, A. Judy, J. Sulhoff, S. Park, L. Nelson, and L. Gruner-Nielsen, 3.28 Tb/s (82 × 40 Gb/s) transmission over 3 × 100 km nonzero-dispersion fiber using dual C- and L-band hybrid Raman-erbium-doped inline amplifiers. In *Proceedings of the Optical Fiber Communications Conference 2000*, Baltimore, MD, PD23, 2000.

[41] S. Bigo, A. Bertaina, Y. Frignac, S. Borne, L. Lorcy, D. Hamoir, D. Bayart, J.-P. Hamaide, W. Idler, E. Lach, B. Franz, G. Veith, P. Sillard, L. Fleury, P. Guenot, and P. Nouchi, 5.12 Tb/s (128 × 40 Gbit/s WDM) Transmission over 3 × 100 km of Teralight fiber. In *Proceedings of the European Conference on Optical Communication 2000*, Munich, PD.1.2, 2000.

[42] B. Zhu, L. Leng, L. E. Nelson, Y. Qian, L. Cowsar, S. Stulz, C. Doerr, S. Chandrasekar, S. Radic, D. Vengsarkar, Z. Chen, H. Thiele, J. Bromage, L. Gruner-Nielsen and S. Knudsen, 3.08 Tb/s (77 × 42.7Gb/s) transmission over 1200 km of non-zero dispersion-shifted fiber with 100-km spans using C- and L-band distributed Raman amplification. In *Proceedings of the Optical Fiber Communications Conference 2001*, Anaheim, CA, PD23, 2001.

[43] S. Kado, Y. Emori, S. Namiki, N. Tsukiji, J. Yoshida, and T. Kimura, Broadband flat-noise Raman amplifier using low-noise bi-directionally pumping sources. In *Proceedings of the European Conference on Optical Communication 2001*, Amsterdam, PD.F.1.8, 2001.

[44] A. R. Chraplyvy, Limitations on lightwave communication imposed by optical fiber non-linearities, *J. Lightwave Technol.*, 8:10, 1548, 1990.

[45] S. Bigo, S. Gauchard, S. Borne, P. Bousselet, P. Poignant, L. Lorcy, A. Bertaina, J.-P. Thiery, S. Lanne, L. Pierre, D. Bayart, L.-A. de Montmorillon, R. Sauvageon, J.-F. Chariot, P. Nouchi, J.-P. Hamaide, and J.-L. Beylat, 1.5 Terabit/s WDM transmission of 150 channels at 10 Gbit/s over 4 × 100km of TeraLight fiber. In *Proceedings of the European Conference on Optical Communication 1999*, Nice, France, PD2-9, 1999.

[46] A. F. Judy, Optimum fiber dispersion for multi-wavelength 40 Gbit/s NRZ and RZ transmission. In *Proceedings of the European Conference on Optical Communication 1999*, Vol. II, Nice, France, 280–281, 1999.

[47] B. Konrad and K. Petermann, Optimum fiber dispersion in high-speed TDM systems, *IEEE Photon. Technol. Lett.*, 13:4, 299–301, 2001.

[48] T. Tsuritani, N. Takeda, K. Imai, K. Tanaka, A. Agata, I. Morita, H. Yamauchi, N. Deagawa, and M. Suzuki, 1 Tbit/s (100 × 10.7 Gb/s) transoceanic transmission using 30 nm wide broadband optical repeaters with Aeff-enlarged positive dispersion fiber and slope-compensation DCF. In *Proceedings of the European Conference on Optical Communication 1999*, Nice, France, PD2-10, 1999.

[49] J.-X. Cai, M. I. Hayee, M. Nissov, M. A. Mills, A. N. Pilipetskii, S. G. Evangelides Jr., N. Ramanujam, C. R. Davidson, R. Menges, P. C. Corbett, D. Sutton, G. Lenner, C. Rivers, and N. S. Bergano, 1.12 Tb/s transmission over trans-Atlantic distance (6200 km) using fifty-six 20 Gb/s channels. In *Proceedings of the European Conference on Optical Communication 2000*, Munich, postdeadline paper 1.6, 2000.

[50] H. Sugahara, K. Fukuchi, A. Tanaka, Y. Inada, and T. Ono, 6050km transmission of 32 × 42.7 Gb/s DWDM signals using Raman-amplified quadruple-hybrid span configuration.

In *Proceedings of the Optical Fiber Communications Conference 2002*, Anaheim, CA, postdeadline paper FC6, 2002.

[51] T. Tanaka, N. Shimojoh, T. Naito, H. Nakamoto, I. Yokota, T. Ueki, A. Sugiyama, and M. Suyama, 2.1-Tbit/s WDM transmission over 7,221 km with 80-km repeater spacing. In *Proceedings of the European Conference on Optical Communication 2000*, Munich, post-deadline paper 1.8, 2000.

[52] S. Knudsen, B. Zhu, L. E. Nelson, M. O. Pedersen, D. W. Peckham, and S. Stulz, 420 Gbit/s (42 × 10Gbit/s) WDM transmission over 4000km of UltraWave fiber with 100 km dispersion-managed spans and distributed Raman amplification, *Electron. Lett.*, 37:15, 965–967, 2001.

[53] B. Zhu, S. N. Knudsen, L. E. Nelson, D. W. Peckham, M. O. Pedersen, and S. Stulz, 800 Gb/s (80 × 10.664Gb/s) WDM transmission over 5200 km of fibre employing 100 km dispersion-managed spans, *Electron. Lett.*, 37:24, 1467–1469, 2001.

[54] B. Zhu, L. Leng, L. E. Nelson, S. Knudsen, J. Bromage, D. Peckham, S. Stulz, K. Brar, C. Horn, K. Feder, H. Thiele, and T. Veng, 1.6 Tb/s (40 × 42.7 Gb/s) transmission over 2000 km of fiber with 100-km dispersion-managed spans. In *Proceedings of the European Conference on Optical Communication 2001*, Amsterdam, PD.M.1.8, 2001.

[55] F. Liu, J. Bennike, S. Dey, C. Rasmussen, B. Mikkelsen, P. Mamyshev, D. Gapontsev, and V. Ivshin, 1.6 Tbit/s (40 × 42.7 Gb/s) transmission over 3600 km UltraWave fiber with all-Raman amplified 100 km terrestrial spans using ETDM transmitter and receiver. In *Proceedings of the Optical Fiber Communications Conference 2002*, Anaheim, CA, postdeadline paper FC7, 2002.

[56] M. O. Pedersen, S. N. Knudsen, T. Geisler, T. Veng, and L. Gruner-Nielsen, New low-loss inverse dispersion fiber for dispersion matched fiber sets. In *Proceedings of the European Conference on Optical Communication 2002*, Copenhagen, 5.1.3, 2002.

[57] B. Zhu, L. E. Nelson, L. Leng, S. Stulz, S. Knudsen, and D. Peckham, 1.6 Tb/s (40 × 42.7 Gb/s) WDM transmission over 2400 km of fiber with 100 km dispersion-managed spans, *Electron. Lett.*, 38:13, 647–648, 2002.

[58] R. Hainberger, J. Kumasako, K. Nakamura, T. Terahara, H. Onaka, and T. Hoshida, Optimum span configuration of Raman-amplified dispersion-managed fibers. In *Proceedings of the Optical Fiber Communications Conference 2001*, Anaheim, CA, MI5, 2001.

[59] Y. Inada, K. Fukuchi, T. Ono, T. Ogata, and H. Okamura, 2400-km transmission of 100-GHz-spaced 40-Gb/s WDM signals using a "double-hybrid" fiber configuration. In *Proceedings of the European Conference on Optical Communication 2001*, Amsterdam, We.F.1.6, 2001.

[60] H. Sugahara, M. Morisaki, T. Ito, K. Fukuchi, and T. Ono, 9000-km transmission of 32 × 42.7 Gb/s dense-DWDM signals using 195-μm^2-Aeff fiber and inverse double-hybrid span configuration. In *Proceedings of the Optical Amplifiers and their Applications Topical Meeting 2002*, Vancouver, postdeadline paper PD3, 2002.

[61] K. Schuh, M. Schmidt, E. Lach, B. Junginger, A. Klekamp, G. Veith, and P. Sillard, 4 × 160 Gbit/s DWDM/OTDM transmission over 3 × 80 km Teralight – Reverse Teralight fiber. In *Proceedings of the European Conference on Optical Communication 2002*, Copenhagen, 2.1.2, 2002.

[62] K. Mukasa, H. Moridaira, T. Yagi, and K. Kokura, New type of dispersion management transmission line with MDF for long-haul 40 Gb/s transmission. In *Proceedings of the Optical Fiber Communications Conference 2002*, Anaheim, CA, ThGG2, 2002.

[63] Y. Miyamoto, A. Hirano, K. Yonenaga, A. Sano, H. Toba, K. Murata, and O. Mitomi, 320 Gb/s (8 × 40 Gb/s) WDM transmission over 367 km with 120 km repeater spacing using carrier-suppressed return-to-zero format, *Electron. Lett.*, 35:23, 2041, 1999.

[64] G. Charlet, J.-C. Antona, S. Lanne, P. Tran, W. Idler, M. Gorlier, S. Borne, A. Kelkamp, C. Simonneau, L. Pierre, Y. Frignac, M. Molina, F. Beaumont, J.-P. Hamaide, and S. Bigo, 6.4 Tb/s (159 × 42.7Gb/s) capacity over 21 × 100 km using bandwidth-limited phase-shaped binary transmission. In *Proceedings of the European Conference on Optical Communication 2002*, Copenhagen, PD4.1, 2002.

[65] B. Zhu, L. Leng, A. H. Gnauck, M. O. Pedersen, D. Peckham, L. E. Nelson, S. Stulz, S. Kado, L. Gruner-Nielsen, R. L. Lingle, Jr., S. Knudsen, J. Leuthold, C. Doerr, S. Chandrasekhar, G. Baynham, P. Gaarde, Y. Emori, and S. Namiki, Transmission of 3.2 Tb/s (80 × 42.7 Gb/s) over 52 × 100 km of UltraWave fiber with 100-km dispersion-managed spans using RZ-DPSK format. In *Proceedings of the European Conference on Optical Communication 2002*, Copenhagen, PD2.4, 2002.

[66] S. R. Chinn, D.M.Boroson, and J.C. Livas, Sensitivity of optically pre-amplified DPSK receivers with Fabry-Perot filters, *IEEE J. Lightwave Technol.*, 14:370–376, 1996.

[67] J. P. Gordon and L. F. Mollenauer, Phase noise in photonic communications systems using linear amplifiers, *Optics Lett.*, 15:23, 1351–1353, 1990.

[68] K. Yonenaga, Y. Miyamoto, H. Toba, K. Murata, M. Yoneyama, Y. Yamane, and H. Miyazawa, 320 Gb/s, 100-km WDM repeaterless transmission using fully encoded 40Gbit/s optical duobinary channels with dispersion tolerance of 380 ps/nm. In *Proceedings of the European Conference on Optical Communication 2000*, Munich, Vol. 1, 75–79, 2000.

[69] H. Bissessur, G. Charlet, C. Simonneau, S. Borne, L. Pierre, C. De Barros, P. Tran, W. Idler, and R. Dischler, 3.2 Tb/s (80 × 40 Gb/s) C-band transmission over 3 × 100km with 0.8 bit/s/Hz efficiency. In *Proceedings of the European Conference on Optical Communication 2001*, Amsterdam, PD.M.1.11, 2001.

[70] H. Bissessur, L. Pierre, D. Penninckx, J.-P. Thiery, and J.-P. Hamaide, Enhanced phased-shaped binary transmission for dense WDM systems, *Electron. Lett.*, 37:1, 45–46, 2001.

[71] H. Kim and C. X. Yu, Optical duobinary transmission system featuring improved receiver sensitivity and reduced optical bandwidth, *IEEE Photon. Technol. Lett.*, 7:8, 1205–1207, 2002.

[72] F. Kerfoot, Forward error correction for optical transmission systems. In *Proceedings of the Optical Fiber Communications Conference 2002*, Anaheim, CA, tutorial talk WL, 2002.

[73] H. Kidorf, N. Ramanujam, I. Hayee, M. Nissov, J.-X. Cai, B. Pedersen, A. Puc, and C. Rivers, Performance improvement in high capacity, ultra-long distance, WDM systems using forward error correction codes. In *Proceedings of the Optical Fiber Communications Conference 2000*, Baltimore, MD, ThS3, 2000.

[74] S. Ait, FEC techniques in submarine transmission systems. In *Proceedings of the Optical Fiber Communications Conference 2001*, Anaheim, CA, TuF1, 2001.

[75] J. X. Cai, M. Nissov, A. N. Pilipetskii, C. R. Davidson, R. M. Mu, M. A. Mills, L. Xu, D. Foursa, R. Menges, P. C. Corbett, D. Sutton, and N. S. Bergano, 1.28 Tb/s (32 × 40 Gb/s) transmission over 4,500 km. In *Proceedings of the European Conference on Optical Communication 2001*, Amsterdam, PD.M.1.2, 2001.

[76] R.-J. Essiambre, G. Raybon, and B. Mikkelsen, Pseudo-linear transmission of high-speed TDM signals: 40 and 160 Gb/s. In *Optical Fiber Telecommunications IVB*. I. P. Kaminow and T. Li, eds., San Diego: Academic, 232–304, 2002.

[77] P. Bayvel and R. Killey, Nonlinear optical effects in WDM transmission. In *Optical Fiber Telecommunications IVB*. I. P. Kaminow and T. Li, eds., San Diego: Academic, 611–641, 2002.

[78] F. J. MacWilliams and N. J. A. Sloane, Pseudo-random sequences and arrays, *Proc. of the IEEE*, 64:12, 1715–1729, 1976.

[79] N. S. Bergano and C. R. Davidson, Circulating loop transmission experiments for the study of long-haul transmission systems using erbium-doped fiber amplifiers, *J. Lightwave Technol.*, 13: 5, 879–888, 1995.

[80] Y. Sun, I. T. Lima, Jr., H. Jiao, J. Wen, H. Xu, H. Ereifej, G. M. Carter, and C. R. Menyuk, Study of system performance in a 107-km dispersion-managed recirculating loop due to polarization effects, *IEEE Photon. Technol. Lett.*, 13:9, 966–968, 2001.

[81] Y. Sun, B. S. Marks, I. T. Lima, Jr., K. Allen, G. M. Carter, and C. R. Menyuk, Polarization state evolution in recirculating loops. In *Proceedings of the Optical Fiber Communications Conference 2002*, Anaheim, CA, ThI4, 2002.

[82] G. M. Carter and Y. Sun, Making the Q distribution in a recirculating loop resemble a straight-line experiment. In *Proceedings of the Optical Fiber Communications Conference 2002*, Anaheim, CA, ThQ5, 2002.

[83] S. Lee, Q. Yu, L.-S. Yan, Y. Xie, O. H. Adamczyk, and A. E. Willner, A short recirculating fiber loop testbed with accurate reproduction of Maxwellian PMD statistics. In *Proceedings of the Optical Fiber Communications Conference 2001*, Anaheim, CA, WT2, 2001.

[84] H. S. Kim, S. H. Yun, H. K. Kim, N. Park, and B. Y. Kim, Actively gain-flattened erbium-doped fiber amplifier over 35 nm by using all-fiber acoustooptic tunable filters, *IEEE Photon. Technol. Lett.*, 10:790–792, 1998.

[85] C. R. Doerr, R. Pafchek, and L. W. Stulz, 16-band integrated dynamic gain equalization filter with less than 2.8-dB insertion loss, *IEEE Photon. Technol. Lett.*, 14:334–336, 2002.

[86] B. Eggleton, A. Ahuja, P. Westbrook, J. A. Rogers, P. Kuo, T. N. Nielsen, and B. Mikklesen, Integrated tunable fiber gratings for dispersion management in high bit-rate systems, *J. Lightwave Technol.*, 18:1418–1432, 2000.

[87] M. Shirasaki, Chromatic dispersion compensator using virtually imaged phased array, *IEEE Photon. Technol. Lett.*, 9:12, 1598–1600, 1997.

[88] C. R. Doerr, L. W. Stulz, S. Chandrasekhar, L. Buhl, and R. Pafchek, Multichannel integrated tunable dispersion compensation employing a thermooptic lens. In *Proceedings of the Optical Fiber Communications Conference 2002*, Anaheim, CA. FA6, 2002.

[89] Y. Su, G. Raybon, L. K. Wickham, R.-J. Essiambre, S. Chandrasekhar, S. Radic, L. E. Nelson, L. Gruner-Nielsen, and B. J. Eggleton, 40-Gb/s transmission over 2000 km of nonzero-dispersion fiber using 100-km amplifier spacing. In *Proceedings of the Optical Fiber Communications Conference 2002*, Anaheim, CA, ThFF3, 2002.

[90] S. C. Grubb, T. Strasser, W.Y. Cheung, W.A. Reed, V. Mizrachi, T. Erdogan, P.J. Lemaire, A.M. Vengsarkar, D. J. DiGiovanni, D. W. Peckham, and B. H. Rockney, High-power 1.48 μm cascaded Raman laser in germanosilicate fibers. In *Proceedings of Optical Amplifiers and their Applications 1995*, Davos, Switzerland, SaA4, 1995.

[91] L. Leng, Q. Le, L. Gruner-Nielsen, B. Zhu, S. Stulz, and L. E. Nelson, 1.6 Tb/s (40 × 40Gb/s) transmission over 500 km of non-zero-dispersion fiber with 100-km amplified spans compensated by extra-high-slope dispersion-compensating fiber. In *Proceedings of the Optical Fiber Communications Conference 2002*, Anaheim, CA, ThX2, 2002.

[92] T. Terahara, T. Hoshida, J. Kumasako, and H. Onaka, 128 × 10.66 Gb/s transmission over 840-km standard SMF with 140-km optical repeater spacing (30.4-dB loss) employing dual-band distributed Raman amplification. In *Proceedings of the Optical Fiber Communications Conference 2000*, Baltimore, MD, PD28, 2000.

[93] A. K. Srivastava, S. Radic, C. Wolf, J. C. Centanni, J. W. Sulhoff, K. Kantor, and Y. Sun, Ultra-dense terabit capacity WDM transmission in L-band. In *Proceedings of the Optical Fiber Communications Conference 2000*, Baltimore, MD, PD27, 2000.

[94] J. P. Blondel, F. Boubal, E. Brandon, L. Buet, L. Labrunie, P. Le Roux, and D. Toullier, Network application and system demonstration of WDM systems with very large spans (error-free 32 × 10 Gbit/s 750-km transmission over three amplified spans of 250 km). In *Proceedings of the Optical Fiber Communications Conference 2000*, Baltimore, MD, PD31, 2000.

[95] Y. Zhu, W. S. Lee, C. Scahill, C. Fludger, D. Watley, M. Jones, J. Homan, B. Shaw, and A. Hadjifotiou, 1.28 Tb/s (32 × 40 Gb/s) transmission over 1000 km with only 6 spans. In *Proceedings of the European Conference on Optical Communication 2000*, Munich, post-deadline paper 1.4, 2000.

[96] W. Idler, S. Bigo, Y. Frignac, B. Franz, and G. Veith, Vestigial side-band demultiplexing for ultra-high capacity (0.64 bit/s/Hz) transmission of 128 × 40 Gb/s channels. In *Proceedings of the Optical Fiber Communications Conference 2001*, Anaheim, CA, MM3, 2001.

[97] T. Tsuritani, A. Agata, I. Morita, K. Tanaka, and N. Edagawa, Performance comparison between DSB and VSB signals in 20 Gbit/s-based ultra-long haul WDM systems. In *Proceedings of the Optical Fiber Communications Conference 2001*, Anaheim, CA, MM5, 2001.

[98] S. Bigo, W. Idler, J.-C. Antona, G. Charlet, C. Simonneau, M. Gorlier, M. Molina, S. Borne, C. deBarros, P. Sillard, P. Tran, R. Dischler, W. Poehlmann, P. Nouchi, and Y. Frignac, Transmission of 125 WDM channels at 42.7 Gb/s (5 Tbit/s capacity) over 12 × 100 km of TeraLight Ultra fiber. In *Proceedings of the European Conference on Optical Communication 2001*, Amsterdam, PD.M.1.1, 2001.

[99] G. Charlet, W. Idler, R. Dischler, J.-C. Antona, P. Tran, and S. Bigo, 3.2 Tb/s (80 × 42.7 Gb/s) C-band transmission over 9 × 100km of TeraLight fiber with 50-GHz channel spacing. In *Proceedings of Optical Amplifiers and their Applications Topical Meeting 2002*, Vancouver, PD1, 2002.

[100] K. Rottwitt, A. Stentz, T. Nielsen, P. Hansen, K. Feder, and K. Walker, Transparent 80 km bi-directionally pumped Raman amplifier with second-order pumping. In *Proceedings of the European Conference on Optical Communication 1999*, Vol. II, Nice, France, 144–145, 1999.

[101] V. Dominic, A. Mathur, and M. Ziari, Second-order distributed Raman amplification with a high-power 1370 nm laser diode. In *Proceedings of the Optical Amplifiers and their Applications Topical Meeting 2001*, Stresa, Italy, OMC6-1, 2001.

[102] L. Labrunie, F. Boubal, E. Brandon, L. Buet, N. Darbois, D. Dufournet, V. Havard, P. Le Roux, M. Mesic, L. Piriou, A. Tran, and J.-P. Blondel, 1.6 terabit/s (160 × 10.66 Gbit/s) unrepeatered transmission over 321 km using second order pumping distributed Raman amplification. In *Proceedings of the Optical Amplifiers and their Applications Topical Meeting 2001*, Stresa, Italy, PD3-1,

[103] S. Bigo, Y. Frignac, G. Charlet, W. Idler, S. Brone, H. Gross, R. Dischler, W. Poehlmann, P. Tran, C. Simonneau, D. Bayart, G. Veith, A. Jourdan, and J.-P. Hamaide, 10.2 Tb/s (256 × 42.7 Gb/s PDM/WDM) transmission over 100km Teralight fiber with 1.28-bit/s/Hz spectral efficiency. In *Proceedings of the Optical Fiber Communications Conference 2001*, Anaheim, CA, PD25, 2001.

[104] Y. Frignac, G. Charlet, W. Idler, R. Dischler, P. Tran, S. Lanne, S. Borne, C. Martinelli, G. Veith, A. Jourdan, J.-P. Hamaide, and S. Bigo, Transmission of 256 wavelength-division and polarization multiplexed channels at 42.7 Gb/s (10.2-Tb/s capacity) over 3 × 100 km of Teralight fiber. In *Proceedings of the Optical Fiber Communications Conference 2002*, Anaheim, CA, postdeadline paper FC5, 2002.

[105] T. Ito, K. Fukuchi, K. Sekiya, D. Ogasahara, R. Ohhira, and T. Ono, 6.4 Tb/s (160 × 40 Gb/s) WDM transmission experiment with 0.8 bit/s/Hz spectral efficiency. In *Proceedings of the European Conference on Optical Communication 2000*, Munich, PD1.1, 2000.

[106] L. Nelson, Challenges of 40-Gb/s WDM Transmission. In *Proceedings of the Optical Fiber Communications Conference 2001*, Anaheim, CA, ThF1, 2001.

[107] L. Nelson, T. Nielsen, and H. Kogelnik, Observation of PMD-induced coherent crosstalk in polarization multiplexed transmission, *IEEE Photon. Technol. Lett.*, 13:7, 738–740, 2001.

[108] B. Zhu, P. B. Hansen, L. Leng, S. Stulz, T. N. Nielsen, C. Doerr, A. J. Stentz, Z. J. Chen, D. W. Peckham, E. F. Rice, L. Hsu, C. K. Kan, A. F. Judy, and L. Gruner-Nielsen, 800 Gb/s NRZ transmission over 3200 km of TrueWave fiber with 100-km amplified spans employing distributed Raman amplification. In *Proceedings of the European Conference on Optical Communication 2000*, Munich, Vol. 1, 73–74, 2000.

[109] S. Knudsen, Design and manufacture of dispersion compensating fibers and their performance in systems. In *Proceedings of the Optical Fiber Communications Conference 2002*, Anaheim, CA, WU3, 2002.

[110] A. Hasegawa, Numerical study of optical soliton transmission amplified periodically by the stimulated Raman process, *Appl. Optics*, 23:3302, 1984.

[111] L. F. Mollenauer, J. P. Gordon, and M. N. Islam, Soliton propagation in long fibers with periodically compensated loss, *IEEE J. Quantum Electron.*, QE-22:1, 157–173, 1986.

[112] M. Nissov, C. R. Davidson, K. Rottwitt, R. Menges, P. C. Corbett, D. Innis, and N. S. Bergano, 100Gb/s (10 × 10 Gb/s) WDM transmission over 7200 km using distributed Raman amplification. In *Proceedings of the European Conference on Optical Communication 1997*, Vol. V, Edinburgh, UK, 9–12, 1997.

[113] H. Suzuki, J. Kani, H. Masuda, N. Takachio, K. Iwatsuki, Y. Tada, and M. Sumida, 25-GHz-spaced, 1 Tb/s (100 × 10 Gb/s) super-dense WDM transmission in the C-band over dispersion-shifted fiber cable employing distributed Raman amplification. In *Proceedings of the European Conference on Optical Communication 1999*, Nice, France, PD2-4,1999.

[114] D. F. Grosz, A. Kung, D. N. Maywar, L. Altman, M. Movassaghi, H. C. Lin, D. A. Fishman, and T. H. Wood, Demonstration of all-Raman ultra-wide-band transmission of 1.28 Tb/s (128 × 10 Gb/s) over 4000 km of NZ-DSF with large BER margins. In *Proceedings of the European Conference on Optical Communication 2001*, Amsterdam, PD.B.1.3, 2001.

[115] D. F. Grosz, A. Agrawal, S. Banerjee, A. P. Kung, D. N. Maywar, A. Gurevich, T. H. Wood, C. R. Lima, B. Faer, J. Black, and C. Hwu, 5.12 Tb/s (128 × 42.7 Gb/s) transmission with 0.8 bit/s/Hz spectral efficiency over 1280 km of standard single-mode fiber using all-Raman amplification and strong filtering. In *Proceedings of the European Conference on Optical Communication 2002*, Copenhagen, PD4.3, 2002.

[116] I. Tomkos, M. Vasilyev, J.-K. Rhee, M. Mehendale, B. Hallock, B. Szalabofka, M. Williams, S. Tsuda, and M. Sharma, 80 × 10.7 Gb/s ultra-long-haul (+4200 km) DWDM network with reconfigurable "broadcast and select" OADMs. In *Proceedings of the Optical Fiber Communications Conference 2002*, Anaheim, CA, postdeadline paper FC1, 2002.

[117] N. Shimojoh, T. Naito, T. Tanaka, H. Nakamoto, T. Ueki, A. Sugiyama, K. Torii, and M. Suyama, 2.4-Tbit/s WDM transmission over 7400 km using all-Raman amplifier repeaters with 74-nm continuous single band. In *Proceedings of the European Conference on Optical Communication 2001*, Amsterdam, PD.M.1.4, 2001.

[118] J.-X. Cai, M. Nissov, C. R. Davidson, Y. Cai, A. N. Pilipetskii, H. Li, M. A. Mills, R.-M. Mu, U. Feiste, L. Xu, A. J. Lucerno, D. G. Foursa, and N. S. Bergano, Transmission of thirty-eight 40-Gb/s channels (> 1.5 Tb/s) over transoceanic distance. In *Proceedings of the Optical Fiber Communications Conference 2002*, Anaheim, CA, postdeadline paper FC4, 2002.

[119] C. Rasmussen, S. Dey, F. Liu, J. Bennike, B. Mikkelsen, P. Mamyshev, M. Kimmit, K. Springer, D. Gapontsev, and V. Ivshin, Transmission of 40 × 42.7 Gb/s over 5200 km UltraWave fiber with terrestrial spans using turn-key ETDM transmitter and receiver. In *Proceedings of the European Conference on Optical Communication 2002*, Copenhagen, PD4.4, 2002.

[120] K. Fukuchi, T. Kasamatsu, M. Morie, R. Ohhira, T. Ito, K. Sekiya, D. Ogasahara, T. Ono, 10.92 Tb/s (273 × 40-Gb/s) triple-band/ultra-dense WDM optical-repeatered transmission experiment. In *Proceedings of the Optical Fiber Communications Conference 2001*, Anaheim, CA, PD24, 2001.

[121] L. Nelson, Optical fibers for high-capacity WDM, long-haul systems, *IEICE Trans. Electron.*, E86–C:5, 693–698, 2003.

[122] C. Fludger, A. Maroney, N. Jolley, and R. Mears, An analysis of the improvement in OSNR from distributed Raman amplifiers using modern transmission fibers. In *Proceedings of the Optical Fiber Communication Conference 2000*, Baltimore, MD, FF2, 2000.

[123] J.-K. Rhee, D. Chowdhury, K. S. Cheng, and U. Gliese, DPSK 32 × 10 Gb/s transmission modeling on 5 × 90 km terrestrial system, *IEEE Photon. Technol. Lett.*, 12:12, 1627–1629, 2000.

Index

Springer Series in
OPTICAL SCIENCES

Springer Series in
OPTICAL SCIENCES